Panzerschrank
NG - 114
(NJ-

The economics of coastal management

The economics of coastal management
A manual of benefit assessment techniques

Edmund C. Penning-Rowsell
Colin H. Green
Paul M. Thompson
Annabel M. Coker
Sylvia M. Tunstall
Cathy Richards
Dennis J. Parker

Belhaven Press
London and Florida

Belhaven Press
(a division of Pinter Publishers)
25 Floral Street, Covent Garden, London WC2E 9DS, United Kingdom

First published in Great Britain in 1992
© Crown Copyright, reproduced with the permission of the Controller of HM Stationery Office

Apart from any fair dealing for the purposes of research or private study, or criticism or review, as permitted under the Copyright, Designs and Patents Act, 1988, this publication may not be reproduced, stored or transmitted, in any form or by any means, or process without the prior permission in writing of the copyright holders or their agents. Except for reproduction in accordance with the terms of licences issued by the Copyright Licensing Agency, photocopying of whole or part of this publication without the prior written permission of the copyright holders or their agents in single or multiple copies whether for gain or not is illegal and expressly forbidden. Please direct all enquiries concerning copyright to the Publishers at the address above.

British Library Cataloguing in Publication Data
A CIP catalogue record for this book is available from the British Library.

ISBN 1 85293 161 2

Distributed in North America by CRC Press Inc., 2000 Corporate Blvd., N.W., Boca Raton, Florida, 33431

Library of Congress Cataloging-in-Publication Data
A CIP catalog record for this book is available from the Libary of Congress.

Typeset by Communitype Limited, Kettering, Northants.
Printed and bound in Great Britain

Contents

List of figures viii

Foreword xii

Preface xiii

Acknowledgements xiv

Glossary of terms xvi

1   **Introduction and summary** 1

      1.1   The rationale, purpose and structure of this volume: the 'Yellow Manual' 1
      1.2   The aims and assumptions of benefit-cost analysis. 6
      1.3   Coast protection and sea defence: strategies, differences and relationships . . . 8
      1.4   The range of techniques and data 10
      1.5   The phasing and timing of each part of an assessment 11
      1.6   Resources for surveys and data processing 12
      1.7   Updating 13

2   **Theoretical perspectives** 14

      2.1   Summary of information. 14
      2.2   Introduction: why do project appraisals? 15
      2.3   The bases of economic analysis 17
      2.4   Stages in the application of economic analysis 22
      2.5   Basic methods of evaluation 25
      2.6   Application to sea defence and coast protection 32
      2.7   Risk and uncertainty 37
      2.8   Summary and 'health warnings' 39

3   **Coastal erosion: evaluating potential losses and benefits** 42

      3.1   Summary of information. 42
      3.2   Problem definition 43

vi THE ECONOMICS OF COASTAL MANAGEMENT

3.3	Erosion rates and erosion 'contours'	44
3.4	Delays in losses, and the extension of the lives of property and land	46
3.5	Infrastructure benefit assessment procedure	50
3.6	Valuing non built-up land	53
3.7	Summary, 'health warnings', data needs and procedures.	55

4 Coastal recreation: the impacts of coast protection and sea defence projects — 58

4.1	Summary of information.	58
4.2	The problem defined: coastal erosion, protection and enhancement	58
4.3	The methodology, execution and results of the research base.	68
4.4	The recommended approach to the assessment of recreational benefits: a two staged framework	79
4.5	The recommended methods and techniques	79
4.6	The recommended methods: analytic techniques and the interpretation of the results	86
4.7	Summary: assessment, limitations and guidance checklist	92

5 The benefits of flood alleviation: sea defence at the coast — 94

5.1	Summary of information.	94
5.2	The calculation of flood alleviation benefits: the basic inputs	94
5.3	Flood damages and losses in coastal situations	98
5.4	Flood impacts, damage and loss: data sources and other information	104
5.5	Operational methods and techniques for project-specific data collection	108
5.6	Analytical methods	112
5.7	Calculating the agricultural benefits of sea defence projects	114
5.8	Guidance checklist and caveats	116

6 The potential environmental gains and losses from coast protection and sea defence works — 118

6.1	Summary of information.	118
6.2	The problem defined	118
6.3	The state of the art of ecological and environmental evaluation	126
6.4	Archaeological, geological and landscape values	131
6.5	The recommended procedure for incorporating environmental values into decision-making	134
6.6	Methods and techniques.	138
6.7	Summary, assessment and guidance checklist	143

7 Integrated computer-based analysis — 146

7.1	Summary of information.	146
7.2	Multiple levels of analysis	146
7.3	Models for assessing the benefits of flood alleviation	148
7.4	Models for assessing the benefits of coast protection	151
7.5	Summary and checklist	154

CONTENTS vii

8   Case studies **156**

    8.1   Introduction 156
    8.2   Hastings. 157
    8.3   The case of Peacehaven 166
    8.4   Fairlight Cove, Sussex 171
    8.5   Herne Bay case study 175
    8.6   The case of Hengistbury Head 181

References **193**

Appendices **199**

    3.1   The derivation of the Extension of Life Factors (ELFs) and Formula 3.1 199
    3.2   Discounted Extension of Life Factors (ELFs) for property losses 201
    4.1   1989 Coastal recreation survey research questionnaire 210
    4.2a Standard site user questionnaire 239
    4.2b Standard residents questionnaire 253
    4.3   Show cards 266
    4.4   Visit count record sheet 276
    4.5   Checklist of recreational uses 278
    5.1   Revised land use classification to be used in the appraisal of flood alleviation projects . . 279
    5.2   Standard flood damage data for residential properties (salt water flooding) . . . 309
    5.3   Standard questionnaire for the assessment of site-specific flood damage
          potential: the business site interview schedule 352
    5.4   Land use coding form for the recording of data on the flood affected benefit area . . 361
    5.5   Standard household questionnaire for flood affected households 366

Index **375**

Figures

1.1	Summary of recommended methodologies and data sources for evaluating coast protection and sea defence projects	5
2.1	The time profile of projects	22
2.2	Three key stages in project appraisal	22
2.3	The externalities associated with coast protection	23
2.4	Uncertainty as to the area under the loss-probability curve and the upper and lower hydraulic bounds	38
3.1	Flowchart of assessment process	42
3.2	Hypothetical coastal erosion contours	45
3.3	Schematic diagram illustrating the benefit assessment framework	50
4.1	Procedure for the assessment of recreation benefits	59
4.2	Theoretical model of range of conditions at coastal sites	61
4.3	Location of user and residents surveys	67
4.4	Theoretical model underlying 1989 seafront survey	69
4.5	Drawings of the same section of beach under different conditions (Clacton, Essex)	72
4.6	Drawings of the same section of beach under different conditions (Morecambe, Lancashire)	72
5.1	Simplified flowchart of the stages involved in the assessment of the benefits of sea defence schemes	95
5.2	Comprehensive flowchart of the stages involved in the assessment of the benefits of sea defence (from Parker et al., 1987)	96
5.3	An example of a loss-probability relationship used to calculate annual average flood losses (from Parker et al., 1987)	97
5.4	Event tree – deaths from floods (from Green, Parker and Emery, 1983)	106
5.5	Causal model of relationships between impacts and judged overall severity of flood (from Green, 1988)	107
5.6	Example of event flood losses calculated from return period, land use and depth damage data	113
5.7	Example of standard table for calculating annual average and discounted benefits of flood alleviation	113
5.8	Drainage improvement project evaluation model (courtesy of Silsoe College)	115
5.9	Agricultural benefit assessment model (from Morris et al., 1984)	116

FIGURES

6.1	Environmental components divided into three functional categories, termed utilitarian, user and non-user components	121
6.2	The recommended procedure for consultation on the environmental effects of proposed coast protection and sea defence projects	134
6.3	Habitat-criteria matrix for evaluating ecological status of project affected area (with and without proposed projects)	139
7.1	Main components of ESTDAM and Micro-ESTDAM	149
7.2	Example of a loss-probability relationship (Swalecliffe, Kent)	150
7.3	Sensitivity analysis of benefits and costs (Swalecliffe, Kent)	151
7.4	Suggested spreadsheet frameworks for coast protection project appraisal	152
7.5	Derivation of discounted erosion delay benefit using the Middlesex spreadsheet format	153
8.1.1	Locations of case studies	158
8.2.1	Removal of beach material by longshore drift, Hastings	158
8.2.2	Overtopping of groyne due to longshore drift, Hastings	159
8.2.3	Deteriorating groynes and beach quality, Hastings	160
8.2.4	Hastings seafront showing phasing of coast protection works and sampling zones for user surveys	160
8.2.5	Seawall construction with 'bullnose' wave return profile, Hastings	161
8.2.6	Construction of seawall foundations, Hastings	162
8.2.7	Groyne reconstruction, Hastings	163
8.2.8	Drawings of the same section of Hastings beach under different conditions	164
8.3.1	Peacehaven showing existing coast protection and unprotected section of coast	166
8.3.2	Scenarios of cliff erosion at Peacehaven used in residents surveys	168
8.4.1	The location of the property at risk at Fairlight Cove, showing erosion contours for the 7 year erosion rate and the alternative road layout following erosion: working diagram	170
8.4.2	Residential property at risk from erosion at Fairlight	171
8.4.3	Part of the Middlesex spreadsheet used in the Fairlight benefit-cost appraisal	172
8.5.1	Herne Bay study area showing 100 year erosion line and 1000 year flood extent	175
8.5.2	Herne Bay drawing A: the beach in current condition	177
8.5.3	Herne Bay drawing B: the beach after erosion	177
8.5.4	Herne Bay drawing C scheme A version 1: the reef without watercraft	178
8.5.5	Herne Bay drawing D scheme A version 2: the reef with watercraft	178
8.5.6	Herne Bay drawing E scheme B: rock groynes and no reef	178
8.6.1	Hengistbury Head and Christchurch Harbour showing main features, existing coast protection, and works proposed in the scheme appraisal	181
8.6.2	Main ecological divisions of the Hengistbury Head-Christchurch Harbour complex in 1986	182
8.6.3	Main ecological changes predicted in the Hengistbury Head-Christchurch Harbour complex following a breach at Double Dykes	184
8.6.4	Ecological summary matrix for current conditions (1986) at Hengistbury Head-Christchurch Harbour	185
8.6.5	Ecological summary matrix for future conditions with coast protection at Hengistbury Head-Christchurch Harbour	185
8.6.6	Ecological summary matrix for conditions without coast protection at Hengistbury Head-Christchurch Harbour	186
8.6.7	Features of archaeological importance threatened by erosion at Hengistbury Head	188

Tables

1.1	The 'Yellow Manual': where to find what	4
1.2	Comparison of economic efficiency benefit-cost analysis and 'extended' benefit-cost analysis	9
2.1	The key differences between economic analysis and financial analysis	17
2.2	Judged relative importance of different factors in deciding which archaeological and heritage sites should be preserved	24
2.3	Judged relative importance of different factors in deciding which nature reserves should be preserved	25
2.4	Basic techniques for the economic evaluation of goods	25
2.5	A classification of loss of sites and/or site features	34
2.6	Willingness to pay via increased rates and taxes for coast protection (pence/year)	36
2.7	Chapter summary: key points	40
3.1	Residential property prices by region	56
3.2	UK residential property prices by dwelling type	56
4.1a	CVM surveys of coastal recreation 1987–1990: value of enjoyment of this/today's visit	68
4.1b	CVM surveys of coastal recreation 1987–1990: willingness to pay (WTP) increased rates and taxes to protect cliff-tops	69
4.2	Factors important in the choice of recreation site to visit (mean rating) in 1988 beach surveys	70
4.3	Standard data on losses and gains with erosion and protection of the seafront (1989 prices)	73
4.4	Comparison of values of enjoyment of visit obtained in different surveys (£ per adult visit)	75
4.5a	Local visitors: determinants of attitudes and behaviours regression equations with beta coefficients (= standardised regression coefficients)	76
4.5b	Day visitors: determinants of attitudes and behaviours regression equations with beta coefficients (= standardised regression coefficients)	77
4.5c	Staying visitors: determinants of attitudes and behaviours regression equations with beta coefficients (= standardised regression coefficients)	78
4.6	Guidance checklist for recreation benefit assessments	91
5.1	The additional effect of salt water damage on flood losses (all residential properties: flooding less than 12 hours)	99
5.2	The additional effect of salt water damage on flood losses (all residential property: flooding greater than 12 hours)	100

5.3	The effect of salt water damage to garden plants – loss per household (garden).	101
5.4	Event stress induced by sewerage flooding.	108
5.5	Damage reducing effects of flood warnings for residential property, April 1991 prices.	112
5.6	Checklist of points to guide an assessment of the benefits of flood alleviation at the coast.	117
6.1	Ecological site designations	124
6.2	Sand-dune systems	127
6.3	Guidance checklist of steps in assessing environmental values.	144
7.1	Multi-level modelling methods available for evaluating the benefits and costs of urban flood protection projects	147
7.2	Comparison of data source needs for BOCDAM and ESTDAM.	148
7.3	Average property values in 'Newton-on-cliff' (January 1987 values).	152
8.1.1	Summary of characteristics of case studies.	157
8.2.1	Reasons for choice of Hastings for visit	164
8.2.2	Perception of beach characteristics at Hastings [from strongly disagree (–5) to strongly agree (+5)].	164
8.2.3	Valuation of beach enjoyment, Hastings (percentages)	165
8.2.4	Valuation of beach enjoyment based on drawings of alternative beach scenarios, Hastings	165
8.2.5	Willingness to pay for coastal protection at Hastings through 'increased rates and taxes'.	165
8.3.1	Value of enjoyment of cliff top, Peacehaven	169
8.4.1	Summary of the benefit-cost analysis for the Fairlight Cove coast protection scheme	173
8.5.1	Summary of erosion and flood protection benefit estimates, Herne Bay	176
8.5.2	Mean value (£ in 1990) of enjoyment of a visit to Herne Bay seafront in different scenarios	178
8.5.3	Mean gain or loss (£ in 1990) in enjoyment per visit to Herne Bay seafront between present and future scenarios.	179
8.5.4	Annual incremental benefits from preventing erosion and additional recreational attributes of alternative schemes for Herne Bay	180
8.6.1	Aggregate travel cost results, Hengistbury Head (£ value per year).	190

Foreword

The Ministry of Agriculture, Fisheries and Food has been at the forefront over many years in supporting research directed at developing techniques for enhancing the 'value for money' obtained from the nation's investment in river and coastal management. This Manual is a further contribution to this effort, and represents current thinking as far as investment appraisal for coast protection and sea defence projects are concerned. It also makes an important new contribution to the valuation of recreational gains and losses at the sea front.

With increasing pressure on our coastline for recreational, industrial and urban development – as well as the need to safeguard nature conservation reserves at the coast – it is more important than ever that we invest wisely in the protection of the coast from erosion and from flooding from the sea. For Britain, as an island, the coastal zone is a vital national asset, and its management and wise development are a national priority.

With the prospect of sea level rise consequent upon climatic change – to add to the secular sea level rise that we have experienced already in the southern and eastern areas of England – this concern has even greater focus. We cannot afford to ignore the dangers inherent in these threats, but similarly as a nation we cannot afford to spend more than is necessary on the required protection works. We need a balanced approach whereby the resources that can be made available for coast protection and sea defence projects are spent as wisely as possible, commensurate with proven need, sustainable development, and in good time.

In this respect this Manual fills a significant vacuum, and builds on previous research at the Middlesex Polytechnic Flood Hazard Research Centre. It will not be the last word on gauging the economics of coastal management, but it will be an important guide to engineers and others in the pursuit of the cost-effective protection of our coasts, with their complex assemblages of assets, populations and valued environments.

The application of strict economic analysis will not be without its critics, and some of the material here will be controversial. But I am glad to see that much of this has been systematically discussed – and agreed as being useful – with a wide range of organisations and groups concerned with coastal management. Moreover, at least there is now a body of knowledge, techniques and data here to use as a sounder base than hitherto for such discussion, as well as for guiding the key processes of investment appraisal. I therefore commend this volume to all students and practitioners of the art and science of coast protection and sea defence.

Reg Purnell
Chief Engineer
Ministry of Agriculture, Fisheries and Food
February 1992

Preface

This Manual is the culmination of over four years of research and development work at the Middlesex Polytechnic Flood Hazard Research Centre.

The publication of this volume coincides with the transformation of Middlesex Polytechnic into Middlesex University, and we feel some sense of pleasure that our research at FHRC over many years has helped our institution to become recognised as a University and thereby privileged to stand alongside other centres of advanced learning in Britain.

This is the third 'coloured' Manual that we have produced which has been concerned with the benefits of environmental protection and enhancement. The first – the 'Blue Manual' – was concerned with the benefits of flood alleviation and was the culmination of research between 1973 and 1977, sponsored by the Natural Environment Research Council. That Manual has enjoyed a wide circulation, both in Britain and abroad, and remained in print for over a decade.

The second Manual was the 'Red Manual', and was the product of research sponsored by the Ministry of Agriculture, Fisheries and Food between 1981 and 1986 concerned primarily with the indirect effects of floods and flood alleviation projects. That Manual, too, continues to enjoy a growing reputation as having tackled systematically and successfully one of the more difficult areas of applied economics.

This Manual is intended to be known as the 'Yellow Manual', for obvious reasons, and we thus exhaust the trio of primary colours. This alone could give us some excuse for stopping here! The research on which this volume is based has been funded by the Ministry of Agriculture, Fisheries and Food between 1986 and 1991, following the transfer of responsibilities for coast protection from the Department of the Environment to the Ministry in 1985.

We thus seek to complement our work on flooding (which continues herein in Chapter 5) with material on the economics of protecting the coastline against erosion. That new thrust has brought added complexity to the analysis and, in all honesty, we feel that this volume is just the beginning of the process of obtaining a thorough understanding of the economics of coastal management. There is more research and development work that needs to be done to take this subject to the level that we have achieved in the flood alleviation field.

What is apparent, however, is that the coast is different from many other areas of the natural environment, and different from the rivers that we have studied before. There are many more interdependencies that exist, and many more delicate balances between the forces of nature. Changes in one area may have profound effects many kilometres or even hundreds of kilometres 'downdrift'. This means that any projects that are designed to affect the coastal zone need to be scrutinised very carefully to identify these 'external' effects. There are also many more trade-offs in the coastal zone than in the riverine environment, and this affects the impacts the projects can have and the feasibility of different project options. Herein lies the fascination of the coast, but also the complexity of its analysis.

This manual contains many thousands of data items. Although checking has been careful, the authors apologise for any errors that may have escaped us in this process. These are our responsibility alone.

Edmund C. Penning-Rowsell
Colin H. Green
Paul M. Thompson
Annabel M. Coker
Sylvia M. Tunstall
Cathy Richards
Dennis J. Parker

Acknowledgements

The research on which this Manual is based was funded by the Ministry of Agriculture, Fisheries and Food (MAFF) from 1986 to 1991, but the views expressed are those of the authors and must not be taken as reflecting the views of the Ministry. We are very grateful to all the staff of MAFF who have assisted us throughout this period: I. R. Whittle, B. R. Streeten, H. A. Fearn, T. Yates and G. Millar. The research project was overseen by a Steering Group which in addition to the above has comprised:

M. V. Hughes, S. Harding and M. Casale	H.M. Treasury
A. Roberts	Canterbury City Council
I. Townend, J. Gardiner and Dr C. Flemming	Sir William Halcrow & Partners
Professor R. K. Turner	University of East Anglia
M. West	National Rivers Authority (Southern Region)
G. Stephenson	CIRIA
M. G. Barret, J. Andrews and J. Brooke	Posford Duvivier
M. Bramley	CIRIA, now National Rivers Authority

We are also grateful to all the coastal local authorities who responded so willingly to our initial survey. In particular, we appreciated the help of the many members of staff of the local authorities and other organisations in the areas used for detailed surveys, including the following:

Scarborough	Mr Hall and Mr Jones	Scarborough Borough Council
The Naze	Mr R. Stoddard	Tendring District Council
Spurn Head	Mr A. Chilton	Holderness Borough Council
	Mr B. R. Spence	Yorkshire Wildlife Trust
	Mr M. Marshall	Heritage Coast Warden
	Mr S. Shuttleworth	Humberside County Council
Dunwich	Mr Tricker	Suffolk Coastal District Council
Hastings	Mr M. Newton and Mr R Griffiths	Tourism and Leisure Department, Hastings Borough Council
Bridlington	Mr Knapp	East Yorkshire Borough Council
Morecambe	Mr R. Eckersley	Morecambe Borough Council
Hunstanton	Mr J. Williams and Mr M. Child	National Rivers Authority, Anglian Region
	Mr J. Barrett	King's Lynn & West Norfolk Borough Council
Fairlight	Dr R. Kosmin	Fairlight Coastal Preservation Society
Peacehaven	Dr R. Benson	Lewes District Council
	Mr J. Scatchard	Posfod Duvivier

Hengistbury Head	Mr R. E. L. Lelliott and Mr M. Holloway	Bournemouth Borough Council
Herne Bay	Mr P. Brookes	Canterbury City Council

We are also grateful to many other people who gave advice, and in particular thank the following for the helpful information provided during the project:

Professor K. Clayton	University of East Anglia
Dr I. Joliffe	Bedford College, University of London
Mr C. T. H. Sharpe	Loss Adjuster
Mr T. Masterson	Chartered Surveyor

Chapter 6 also benefited from the contributions of Mr C. Wright and Dr A. McKirdy (English Nature), Mr B. Startin (English Heritage) and Dr P. Bigmore (Middlesex Polytechnic). In addition, all those who attended the workshop on environmental evaluation contributed their ideas to the final version of this chapter.

Our thanks are due to the many professional interviewers throughout the country who collected the data on which much of this research is based.

A number of staff from FHRC worked on the project at various stages including: J. Cuadra, I. Dos Santos, M-P. Fouquet, M. Hill, M. Horne, P. Lewin, A. Rogers, M. Smith and D. Wheeler.

We are especially grateful to A. N'Jai for his contribution to the research throughout much of the period; and to C. Ottmann, J. Difrancesco and K. Ingrey for seeing the book through the many stages to completion.

The staff of the Technical Unit, School of Geography and Planning, Middlesex Polytechnic have contributed in a variety of ways throughout the project. We would like to thank S. Chilton, N. Beesley and A. Ellis. We particularly appreciate the work of I. Slavin in the preparation of the illustrations for this publication.

Finally we would like to thank Belhaven Press for their patience during the completion of the Manual, and most importantly thank the thousands of coastal zone users and residents who took time to answer the many questions in our questionnaire surveys.

Notwithstanding these acknowledgements, the contents of this Manual are the responsibility of the authors alone.

Glossary of terms

Above design standard benefits – the benefits from reductions in flood losses from events which exceed the design standard of protection, expressed as an annual average benefit.
Benefits – the returns on the investment in the project; the gains, or the avoided losses, in consumption which it achieves.
Benefit-cost ratio – the ratio of the present value of benefits to the present value of the costs.
Capitalised value – the sum of the discounted income flow; its Present Value (q.v.).
Consumer surplus – the difference between the total amount an individual must pay for a given quantity of a good and the value the individual puts upon the availability of that given quantity of the good.
Contingent Valuation Method (CVM) – a method of evaluating goods using social survey methods (see Table 2.4).
Cost – the resources or alternative consumption which must be sacrificed for the end in view to be achieved.
Direct methods of valuation – methods of estimating the value of a good either by asking individuals or by observing their behaviour in relation to the consumption of that specific good.
Discounting – any pair of project options will have consequences which differ in terms of when the different streams of impacts occur. In order to compare the two options it is necessary to bring these two project streams to a common basis. In benefit-cost analysis, this is done by bringing all streams to a common base date by discounting. Discounting is the reverse of compound interest; in effect, the further into the future a benefit or cost occurs, the lower its equivalent value now. The appropriate rate of discount is called the **Test Discount Rate**.
Economic analysis – considers the changes in the flows of all goods and resources, whether or not they are priced, to all individuals and organisations to estimate the desirability of the net changes, as evaluated using economic values, which would occur as a result of an action.
Economic efficiency – occurs when there is an optimal allocation of all goods and resources, as defined by some objective function, and subject to any relevant constraints technical or other. In these terms, a Pareto optimum allocation of goods and resources occurs when any change would leave at least one person worse off without leaving anyone better off.
Economic value – the value to society of a good, expressed in money terms.
Environmental goods – resources which have some or all of the following characteristics: they are unique and constitute public goods; decisions concerning their availability have irreversible consequences; and they are valued less for their use value than for other reasons.
Environmentally Sensitive Area (ESA) – areas in which (UK) government subsidies are paid to encourage landowners to follow land uses sympathetic to an environmental or landscape objective.
Equity – the distribution of resources and goods is excluded from the concern of economic efficiency analysis. Distributional questions may therefore require to be taken into account along with the results of the economic analysis.
Evaluation – an assessment of the relative importance or significance of a site, or species, in scientific or other non-economic terms. As distinguished from 'valuation' wherein the assessment is made in economic, or monetary, terms.
Existence value – one term used in economics for 'non-use' values (q.v.). The term 'non-use' value is preferred because the reasons why individuals value goods other than for their use value are unknown.
Exploratory Data Analysis (EDA) – methods of looking, usually using graphical techniques, for patterns in

a data set prior to undertaking detailed analysis.

Financial analysis – undertaken by an individual or organisation to assess only the changes in flows of income or goods to that individual or organisation in consequence of a proposed course of action.

Good – general term used to cover all those things for which at least one individual has a preference for more (or less). A good may be tangible, like a cup of tea, or intangible, like an attractive landscape. Sometimes the term 'bad' is used to cover all those things for which at least one individual prefers less to more, and the term 'good' is restricted to those things of which more is preferred.

Hedonic Price Method (HPM) – a method of inferring values for public (environmental) goods (q.v.), based on prices of marketed goods, typically house prices which entail different levels of that environmental good (see Table 2.4).

Indirect loss – a flood loss which is consequent upon direct flood loss/damage.

Indirect methods of valuation – a method of inferring the value of one good by evaluating the consumption of another good or a bundle of goods with which the first good is necessarily linked.

Inherent value – the concept that a species or a member of a species has a right to existence irrespective of any human preferences for the existence of that species.

Intangibles – any consequence of a scheme option for which it has not been possible to estimate the economic value of that change.

Inter-generational equity – a concern for the rights of future generations when the quantity and nature of goods, particularly environmental goods, available to them will be determined by the actions of this generation.

Intrinsic value – a synonym for 'non-use value' (q.v.)

Log mean – mean of data transformed by taking logarithms, used where the variable in question has a very skewed distribution. The log mean is not expressed in the same units as the arithmetic mean.

Market price – that price for which a good is bought and sold in a market. If restrictive conditions are satisfied, this price may be used to estimate the economic value of the good. Or, the market price may need to be corrected, a 'shadow price' (q.v.) derived, in order for the economic value of the good to be estimated.

Monte Carlo techniques – techniques which take the average of a large number of simulations, where the values of variables used in each simulation are randomly drawn from a distribution of possible values.

Net Present Value (NPV) – the stream of all benefits net of all costs for each year of the project's life discounted back to the present date.

Non-monetary impacts – used to describe those impacts of flooding on households which do not have direct financial impacts; for example, stress and health damage. Consequently, they represent a sub-class of 'non-priced goods' (q.v.).

Non-priced goods – those goods which are not bought and sold in a market and for which, consequently, there is no market price from which to estimate their value.

Non-use values – the value given to a good over and above the use value the consumer/user attaches to that good.

Numeraire – a yardstick whereby the different impacts of a project can be compared to each other and to those of alternative projects. In economic analysis, money is used as the numeraire.

Option value – a premium an individual may be willing to pay in order to preserve the option of consuming or accessing a good which s/he does not consume or access at present.

Opportunity cost – in economics, the cost of a good is the value of whatever second good you have to give up to free the resources in order to have the first good.

Opportunity cost of capital – it is not desirable to invest in one project when alternative projects would yield a greater stream of future increases in consumption. The return in consumption from such alternative investments (in theory, after adjusting for externalities) is the opportunity cost of capital: this is the basis for discounting.

Palaeontological – related to the study of fossils.

Pecuniary externality – a change which simply has the effect of transferring goods or resources from one person, firm or sector to another without affecting the total supply of goods or resources available.

Perfectly competitive market – a model of a market based upon highly restrictive assumptions and for which it can be shown that it would result in prices which are economically efficient.

Present value – the value of a stream of benefits or costs when discounted back to the present time.

Private good – one which if consumed by one individual is not available to others, and which the present holder of that good can restrict access to the good by potential consumers. Consequently, the good can be bought and sold in a market.

Public good – one which cannot be marketed because the holder of the good cannot restrict the consumption of the good to particular individuals. Nor does its consumption by one individual diminish the availability of the good to others.

Quasi-option value – the value of the additional information which would be gained as to the consequences of an action by deferring taking that action.

Quasi-public good – one which has some of the characteristics of a public good whilst also having some of the features of a private good.

Rationality – in economics, rationality refers to the consistency or reliability of decisions rather than the process by which these decisions are made. The 'rational economic person' is axiomatically assumed to maximise their utility.

Return period (recurrence interval) – the average length of time separating flood events of a similar magnitude: a 100 year flood will occur on average once in every 100 years. Alternatively can be expressed as the **exceedance probability** – the probability of a flood of given or greater magnitude occurring in any one year.

Revealed preference methods – method inferring individuals' preferences, or the value they place upon a good, by observing their behaviour.

Sensitivity analysis – a procedure for testing how robust are the conclusions of an analysis to uncertainties.

Shadow project – a method of valuation applied to the loss of site, usually of ecological significance, as the cost of providing a site of equivalent significance elsewhere to replace that which would be lost (the Least Cost Alternative, see Table 2.4).

Shadow prices – if a market in a good is not perfectly competitive (q.v.), then market prices will not reflect the opportunity costs of that good. The price of the good, as corrected to equal its opportunity cost, is termed its shadow price.

Social time preference – a common assumption in economics is that the individual prefers consumption of some good now rather than at some later date. Since social choices should simply reflect individual preferences, social time preference is simply that of the individuals who comprise that society. This is one of the two reasons why benefits and costs occurring in the future are discounted.

Substitute good – a good which fulfils the same or nearly equivalent want or desire to that in question.

Sunk costs – a cost incurred in the past and which cannot be recovered whatever decision is taken now. Consequently, sunk costs are omitted in benefit-cost analyses.

Sustainability – the philosophy that economic development should take into account the rights of future generations and those of the less developed countries, particularly in the use of non-renewable natural resources.

Technological externality – occurs when a decision by one person or organisation affects the utility of another and those economic consequences are not borne by the decision maker.

Test discount rate (TDR) – the minimum rate of return which a project must achieve, equal to the opportunity cost of capital.

Time dependent preferences for consumption – the difference in the level of utility from the consumption, or use, of the same quantity of a good at different times by an individual. Individual time preferences, as embodied in 'social time preference' (q.v.), reflects one possible form of time dependent preference.

Transfer payments – see pecuniary externality.

Travel Cost Method (TCM) – a method of imputing values for the benefits derived from public goods (q.v.) based on the distance travelled and hence costs of visiting a particular site (see Table 2.4).

Uncertainty – the degree of ignorance about the consequences of some action either as a result of scarcity of data, possible inadequacies in the modelling techniques used, or simply because we cannot control the future.

Use value – the value an individual gains from consuming or accessing some quantity of a good.

Utility – the subjective gain an individual receives from the use or the existence of some good.

Value – the relative desirability of a good to some individual as measured against the yardstick of money.

Willingness to pay – the amount an individual is prepared to pay in order to obtain a given improvement in utility, expressed through the Contingent Valuation Method (q.v.).

1 Introduction and summary

1.1 The rationale, purpose and structure of this volume: the 'Yellow Manual'

1.1.1 Rationale

The rationale of this Manual is **to aid and improve decision-making** about investment at the coast in policies, plans or schemes to alleviate both coastal flooding and the erosion of the land by the sea. These decisions should be seen in the context of an integrated approach to coastal zone management.

This improved decision-making is needed because there is some evidence that coast protection and sea defence projects have been poorly appraised in the past, and that some investment has therefore been unwise. Undoubtedly this has been a function of the state of the art of project appraisal at the time: the range of methods and data that have been available until now for investment appraisal for these coast protection and sea defence projects has been very limited.

This rationale should not be misinterpreted. Although much of the focus of this Manual is aimed at 'projects' or 'schemes', by which it is usually meant engineering schemes, this is by no means the only outcome from using this Manual. There are many alternative strategies that should be reviewed in the coastal zone, including the **'do nothing'** option which would let flooding and erosion continue.

Moreover there is a tendency when evaluating projects to consider just the site in question, but it is clear that a **regional perspective** is necessary in the planning of coast protection and sea defence policies and works. This arises because there is a measure of hydrographic interdependency between locations along a coastline, such that erosion in one place can result in protective deposition in another. Flooding in coastal areas can be a function of erosion there resulting from the removal of sediments that previously protected natural sea defences. A regional perspective allows consideration of these factors, which may be missed in a purely local analysis.

In addition, the **land uses** at the coast are themselves interdependent, and this can affect their hazardousness and the potential for hazard alleviation. Thus the development of recreation areas is dependent on the availability of locations near to the sea for residential and holiday accommodation, which may be erosion-prone or flood-prone. In turn the recreational use of the coast is dependent on access and on the availability of suitable beaches for recreation, sheltered water for water sports, and cliff-tops and other open land for informal recreation. Each of these environments may be liable to flooding or erosion. Wildlife reserves are part of this complexity, with coastal locations being essential for the survival of many bird species, inshore fisheries and other flora and fauna. These are areas where each activity is subject to the forces of the sea. The coastal zone cannot be seen as un-interconnected individual parts, but should be seen as a whole, and investment appraised in this context.

Thus many alternative strategies need to be appraised, be they tighter land use control at the coast, tighter designation of environmental 'no go' areas such as Sites of Special Scientific Interest, or structural engineering schemes to prevent flooding and reduce or delay erosion. An **integrated approach** is needed, rather than one which looks only at local issues and uses a single discipline approach to analysis and policy making. This is not easy, and makes many of the decisions on investment very complex and difficult. But only with such an approach to evaluation will decision-making improve, unintended consequences be minimised, and the nation's stock of economic and environmental resources at the coast be enhanced.

1.1.2 The purposes of this Manual

This Manual has three main purposes, which interconnect and permeate the whole volume.

First, we aim to explain the limitations and complications of benefit-cost analysis to potential users. This is done so that they can use its techniques in a thoughtful and critical way to guide their decision making about investment in coastal works for alleviating **erosion of the land** by the sea (**coast protection** works) or alleviating **flooding (sea defence** works). Therefore the users of this Manual will be those who are analysing problems of erosion or flooding at the coast (and in estuaries) and are considering the possibility of proposing some plans or works to lessen these risks or reduce their impacts.

We hope, therefore, that this Manual's contents will be of use to local authority engineers and planners, staff within the National Rivers Authority and other Coast Protection Authorities, harbour and navigation authorities, agencies for the promotion of recreation, nature conservation agencies with interests at the coast, and individuals and landowners who have an interest in protecting their property and land from the effects of the sea.

It should be noted, however, that the approach to appraisal adopted throughout the Manual is the efficiency of investment for the *nation* – not the agency or individual – and users must take this important point into consideration in their use of its contents (see Section 2.3 for a discussion of financial versus economic appraisal). The focus is on benefit-cost analysis to aid the allocation of the nation's scarce economic resources, not the financial appraisal of investment for individual or corporate private gain.

This Manual adopts a critical perspective about investment appraisal: the data and techniques presented do not aim to *justify* predetermined designs, but to *analyse* the extent and character of a coast protection or sea defence problem along a coastline or at a particular site prior to making decisions about whether to intervene or whether to do nothing. The main vehicle for the message about the critical use of economic appraisal is Chapter 2 herein, but throughout the Manual we hope to stress sufficiently that there are pitfalls and complications throughout the benefit assessment part of benefit-cost analysis, and that an understanding of its principles is essential to its appropriate use.

Furthermore, benefit-cost analysis inherently involves some assumptions about the distribution of income and wealth in society (Mishan, 1971) – notably that the current distribution is optimal – and thus its use involves some moral decisions as well as the purely technical. Moreover, even at the technical level incorrect understanding, incorrect assumptions, and the wrong data can produce benefit-cost analysis results which are meaningless and indeed dangerous. We hope to provide some guidance so that these dangers are minimised.

Secondly, the Manual aims to describe the results of research at Middlesex Polytechnic Flood Hazard Research Centre between 1986 and 1991 which produced the data and techniques that are contained in this Manual. This research project was termed the COPRES project (Coastal Protection Research Project), and was aimed at developing methods, data and investment and planning appraisal systems.

Inevitably in any major research project there is much research done that does not find its way into the resulting publication, but we have attempted to include as much of the original research findings as possible here, and to explain their limitations, to give a context to the methods that we have produced and recommend. We also give references here to the many papers and reports that have been produced as part of this COPRES project, so that readers can pursue in more detail various aspects of the research, if they so wish, on the derivation of the data and techniques that are a product of this research.

The third aim, therefore, is to present the user with a range of techniques and data that can be used in a practical way to assess the benefits of projects to alleviate flooding and erosion at the coast. The Manual is oriented towards practising engineers or coastal zone managers who wish to see that their projects represent good value for the nation's money. In this way we hope to promote good design through emphasising the need to evaluate the impacts of the plans or works that are proposed, both in economic and environmental terms. Indeed, this Manual may well be used to determine that works to alleviate flooding or erosion are not appropriate, either because they are not cost-effective or because they would be so environmentally damaging that they should not proceed, or a mixture of both.

In this respect, to re-emphasise the point made above, the use of the benefit-cost analysis techniques and data given here should **not** be seen as a vehicle for justifying predetermined options, be they engineering works or doing nothing. Rather, benefit-cost analysis should be used more as a sieve to aid the filtering of possibilities, and to determine which are in the nation's best interest.

1.1.3 Benefit-cost analysis (BCA) and Environmental Assessment (EA)

There is a range of techniques that can aid decision

making, and benefit-cost analysis is just one technique amongst many. Similarly there is a range of techniques that can be used to quantify the environmental impacts of a number of policy options, and this is an important component of the data that the decision-maker needs.

Benefit-cost analysis can be seen as a part of the more general procedure termed Environmental Assessment (previously termed Environmental Impact Assessment or EIA). This is because BCA is concerned with a particular 'product' (the benefit-cost ratio or net present value), whereas EA is first and foremost more concerned with a process of incorporating information on all environmental attributes, values and changes into the decision-making sequences.

Benefit-cost analysis had the disadvantage that it is narrow in focus, but leads to a simple parameter on which choices can be based. It is also constrained by the fact that many environmental attributes cannot be quantified in money terms. Environmental Assessment, on the other hand, has the disadvantage that it tends to produce volumes of data that the decision-maker has difficulty in assimilating.

1.1.4 The structure of this Manual: how it should be used, and how to get started

This Manual is not intended to be read right through from start to finish. Rather, the component parts can be used in a 'stand-alone' fashion, and have been written with that in mind. However, the user should be wary of using parts of the Manual in a stand-alone fashion without considering the research basis of the methods and data that are presented, or the explanations that are contained in Chapter 2 and certain of the Appendices. The user needs to be able to understand and explain the basis of their results, rather than just present the numbers that are derived.

In summary, the user should therefore proceed as follows:

- Read Section 1.2 and then as much of Chapter 2 as possible;
- Analyse carefully the nature of the coast protection and sea defence problem in question by site visits and consulting those with most knowledge of these topics;
- Decide on the scope of the analysis by consulting Section 1.3.1; bear particularly in mind the need to take a regional perspective by evaluating the whole of the hydrographic unit in which the problem site is located;
- Evaluate Figure 1.1 and Table 1.1 to determine what areas of analysis are needed;
- Consult the introductory 'summary of information' sections of all the chapters that are relevant to the analysis to be undertaken;
- **Consider carefully the time required for any studies by reading Section 1.5, taking particular notice of, first, the need to commission baseline data collection surveys as much as 12 months ahead of the need for the results and, secondly, the need for staff training in the use of certain techniques; and**
- Consult the checklists provided at the end of each of these chapters, paying particular attention to any 'health warnings' contained therein.

Beyond this introductory chapter, **Chapter 2** presents a critical analysis of the basis of benefit-cost analysis, and a number of warnings about its use (see the Checklist, Table 2.7). **Therefore Chapter 2 should be read by all users of this Manual, even if they use just one small part of its other contents.** It is recognised that this chapter is more difficult than those presented in previous Manuals that we have produced (Penning-Rowsell and Chatterton, 1977; Parker *et al.*, 1987), but this is inevitable as we progress from the 'easy' areas of benefit assessment such as flood damages avoided, to the difficult and maybe impossible areas such as valuing recreational experiences and gauging nature conservation values.

Thereafter, Chapters 3 to 7 analyse the different aspects of appraising coast protection and sea defence projects, following the breakdown shown in Figure 1.1, and give the approaches and techniques that we recommend. Table 1.1 gives guidance as to where to find what and, together with Figure 1.1, should form a 'map' of the route towards the final appraisal results.

In **Chapter 3** we present our recommended methods for assessing the economic values of loss of property, land and other sites to the sea as a result of erosion. We analyse this phenomenon partly as a case of market failure, whereby the value of property at risk is not a true reflection of the fact that there is this risk. We have therefore devised a method which values property in terms of its use value, and gauges the benefit value to that property as a result of erosion control projects in terms of the increased life and potential use that this property will enjoy.

It is likely that most coast protection investment will rely heavily on property loss benefits such as are described in this chapter, although we believe that recreation benefits and 'environmental benefits' will be much more important in the future than they have been in the past.

The Fairlight Cove case study in **Chapter 8** (Section 8.4) presents an example of the application of these methods, and a discussion of some of the controversial points inherent in assessing the future value of residential property. In addition the Herne Bay case study

4 THE ECONOMICS OF COASTAL MANAGEMENT

Table 1.1
The 'Yellow Manual': where to find what

Topic	Theoretical perspectives	Description of method(s) and its application	Data requirements, etc
Coast protection			
Benefits of delay of erosion to property liable to erosion	2.4; 2.5.1; 2.5.2; 2.6; 2.7	3.2; 3.3; Residential: 3.4 Infrastructure: 3.5 Non-built-up land: 3.6	Erosion rates: 3.3; House prices: 3.4.4 Land use classification: Appendix 5.1
Benefits of delay of erosion to land etc.	Ditto	Non-built-up land: 3.6	Agricultural land 3.6.2
Recreation benefits from delay of erosion	Ditto, plus 4.2; 4.4	4.3; 4.4; 4.5	3.3; 4.5.1; 4.5.2; 4.5.3; For questionnaires: Appendix 4.2 a and b
Impacts of erosion, and coast protection schemes, on environmental values	2.4; 2.5; 6.2; 6.3; 6.4	6.5; 6.6	6.5; 6.6; 8.6 Hengistbury Head case study
Case studies			Hastings 8.2; Peacehaven 8.3; Fairlight 8.4; Herne Bay 8.5; Hengistbury Head 8.6
Alleviation of flood damage at the coast			
Assessing flood losses to property	2.6.2, 2.6.3, 5.2 plus other references cited in Chapter 5	5.2, 5.4, 5.5 and 5.6	5.4 plus Appendix 5.1, 5.2, 5.3, and 5.4; 8.5 Herne Bay case study
'Writing off' property loss	5.4.6	5.4.6	5.4.6; 8.5 Herne Bay case study
Agricultural benefits of sea defence	5.7.1	5.7.2	5.7.2
Environmental impacts of flooding and sea defence schemes	2.4; 2.5; 6.3; 6.4	6.5, 6.6	6.6; 8.6 Hengistbury Head case study
Case studies			Herne Bay 8.5; Hengistbury Head 8.6

INTRODUCTION AND SUMMARY 5

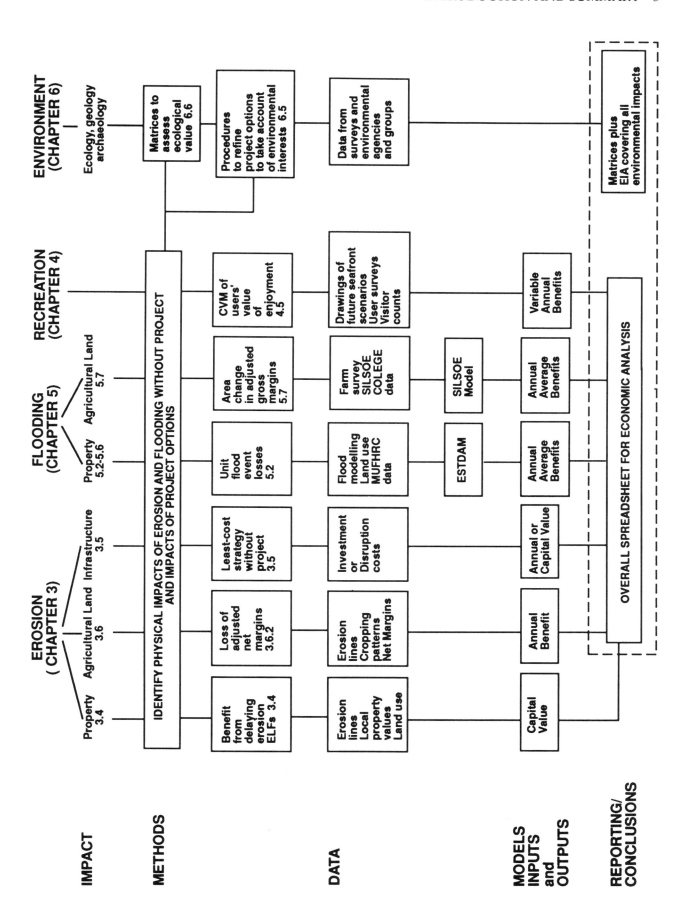

Figure 1.1 Summary of recommended methodologies and data sources for evaluating coast protection and sea defence projects

(Section 8.5) gives an account of a large project, and the interactions between coast protection and sea defence investment.

In **Chapter 4** we present our research results and recommended appraisal methods for assessing the benefits of recreation values at the coast. A great deal of effort within the COPRES project has been given to this subject, and we have undertaken a large number of large surveys to determine the best methods for use in the future. This is because we believe that the maintenance and enhancement of recreation on beaches and cliff-tops will be an important justification for coast protection projects in the future, and that the methods to be used are complex and difficult to apply.

We have illustrated the application of these methods in four case studies presented in **Chapter 8**. For beach recreation these include the studies at Hastings and Herne Bay; the examples at Hengistbury Head and Peacehaven illustrate analyses of, respectively, general open space recreation and specifically cliff-top recreation.

In **Chapter 5** we present the standard methodology for assessing the benefits of flood alleviation, and our results from the research on specific coastal data sets of depth/damage data. This chapter in effect summarises and builds on much work that has been presented elsewhere, but highlights specific difficulties of assessing sea defence benefits, including the question of salt water damage in flooded properties and the difficulties of predicting flood extents when these are based on breaches of the existing sea defences or some other cause for which the probability of occurrence is low. In this respect the Herne Bay case study (Section 8.5) shows how flood alleviation benefits can be assessed, based on an engineering analysis of the probability and severity of defences breached or progressively overtopped as erosion reduces their effectiveness.

Our contribution in **Chapter 6** is somewhat different. Extensive investigations and consultations have determined that the time is not yet right – and may never be so – for all environmental values to be assessed in monetary terms for inclusion within benefit-cost analyses. We have made good progress – as have others – with regard to the quantification of **use values** (the values that people give to environmental resources because they value their use, such as land or water for recreation). But the quantification in money terms of **non-use values** is more problematic.

These non-use values are the values that people give to resources that they may not use now but the wish to see passed to future generations (bequest values) or simply wish to see continue to exist (existence values) or those that they might use in the future even if they do not use them now (option values). With regard to these non-use values there are moral, methodological, technical and operational constraints to their valuation in money terms. Therefore we have, through widespread discussions and consultation (Coker and Richards, 1992), developed a *procedural* approach to incorporating a proper consideration of environmental values into decision-making about coastal investments.

We believe, in this way, that these values can be identified, and quantified to some extent, but that they cannot be valued in the same way as can other resources and goods. Nevertheless rigorous application of the procedure that we recommend will implicitly be an evaluation of these resources, although it cannot be explicitly in money values. A partial application of this kind of approach is illustrated in the Hengistbury Head case study (Section 8.6).

In **Chapter 7** we have separated out discussion of computer analysis of benefits, by describing our EST-DAM computer software for gauging the benefits of flood alleviation, and discussing how proprietary spreadsheets can be used to derive a system for assessing erosion control benefits. The results of these systems are used throughout the Manual and its case studies. In addition, the **Appendices** to the Manual contain data sets and survey instruments that will be needed by the user, and the text in the relevant chapters highlights the operational problems and potentials of their use.

As indicated above, **Chapter 8** includes case studies that illustrate the appraisal methods that we are recommending. Some of these case studies have been developed through analysis for actual decisions concerning investment; others have been developed as part of this research where we have investigated particularly interesting coast protection or sea defence issues. Inevitably, in the space available, the descriptions of the cases are somewhat abbreviated, but we have aimed at teasing out the key issues involved and drawing together lessons for use in future appraisal.

To give a measure of consistency throughout this Manual, the user is given, at the start of each chapter, a **summary of information** to be found in that chapter. In addition, at the end of each main chapter there is a section which includes a **checklist** of important points, and some **'health warnings'** about the problems of using the data and techniques that are recommended. **The user should not apply the techniques that are presented in the Manual without evaluating the relevance of those 'health warnings' to the case being considered.**

1.2 The aims and assumptions of benefit-cost analysis

This section is designed to be a brief introduction to Chapter 2. It is not intended to replace a close reading of that chapter, but to give the reader an initial guide to that field, prior to a more careful analysis of the contents of Chapter 2. The key points from Chapter 2 are summarised in Table 2.7.

The basis of Chapter 2 is that the principal purpose of project appraisal methods is to identify and clarify the issues involved in a particular decision; the purpose is not simply to calculate the benefit-cost ratio for a particular project. It is therefore argued that the development and adoption of particular project options should be appraisal-led rather than design-led. Moreover appraisal methods should always be used critically and, in a pluralist society where there are conflicts between the views of different individuals, one of the functions of appraisal methods is to aid communication between these individuals and different groups within society.

We would also like to emphasise that the choice of project options must be set in the context of the concept of sustainable development. This concept is therefore briefly outlined in Chapter 2, and some possible criteria for projects deriving from that concept are analysed.

Chapter 2 therefore begins by identifying the range of decisions involved in coastal zone management where the help of project appraisal methods may be sought. A number of criteria against which the performance of project appraisal methods can be compared are identified, and we argue that no single appraisal method is the most useful in all circumstances. All available methods have both strengths and weaknesses; the choice of the form of project appraisal to use in a particular case should therefore be determined by identifying which will be most useful in those particular circumstances. This choice will be governed by the relative importance of the different criteria in relation to the particular choice being made.

However, all project appraisal methods involve the comparison of at least two options or alternatives; one is the baseline option which is usually the 'do-nothing' option. This is also, in effect, a comparison of the 'with project' and 'without project' scenarios that are fundamental to the application of benefit-cost analysis. The impacts of different options usually differ in their nature, when they occur, and differ as to who is affected. Any method of comparison or evaluation must encompass and allow for this point, and be able to analyse these differences.

The methods which economic analysis uses to bring such differing impacts to a common base are also discussed in Chapter 2. In particular, the economic concepts of value and of discounting the value of impacts occurring at a future date back to some common base date of analysis are discussed, and reference is made to standard texts by which to pursue these subjects further. In addition, the differences between financial and economic analyses are outlined, and an understanding of these differences is crucial to the proper appraisal of coast protection and sea defence projects or schemes. In particular, also, the distinction between use and non-use value is drawn, where the latter does not involve use of the resources or access to a particular site. The non-use values are shown to be particularly important where sites are of environmental or heritage significance. It is additionally emphasised that with regard to non-use values, the economic concepts of value are particularly likely to be challenged since economic value is based upon individual preferences. Because this is an ethical and axiomatic assumption, it cannot be held that any alternative concept of value is invalid.

In a subsequent section (Section 2.6.5) we show that individuals appear to attach moral significance to the principle of retaining some classes of sites of environmental or heritage significance. For the reasons described in that section, we do not therefore believe that the economic valuation of non-use values on a site-specific basis is generally possible. Thus, Chapter 6 recommends a procedure which does not rely upon economic valuation to embody environmental factors and values into the appraisal process of the coast protection and sea defence options at a particular site.

In Section 2.6.5, and indeed throughout Chapter 2, our emphasis is upon outlining the basic principles and assumptions underlying economic analysis, rather than either providing a 'cookbook' approach or giving a detailed theoretical analysis of the underlying economic issues. The emphasis is upon giving the reader an understanding of the principles involved so that the usefulness of economic analysis to the analysis of a particular choice can be critically assessed.

Any project appraisal involves the processes of **identification, quantification and evaluation**. To apply economic analysis to compare a particular option to the baseline option, therefore, it is necessary, first, to identify all the changes that will occur, determine when these changes will occur, and distinguish who will be affected by the change. Section 2.4 discusses the principles and problems involved in identifying these three parameters and measuring them. In particular, the importance of identifying any benefits of allowing erosion – or coastal flooding – to continue is emphasised.

A number of different techniques can be applied to the valuation of the different changes resulting from a proposed project option, but not all can be applied to the valuation of every possible change. Section 2.5, therefore, describes the basis of the different possible valuation methods, identifies the types of change or impact to which they can be applied, and analyses their limitations. Once again, the emphasis is upon understanding the principles and assumptions upon which the methods are based so that the techniques are not used uncritically.

Section 2.5 concludes by recommending those particular valuation techniques which appear to be the most suitable for the valuation of particular classes of project impacts or changes. It is these techniques which are then developed and applied in subsequent chapters. Conversely, we conclude that reliance cannot be placed upon two of the possible evaluation techniques: the Travel Cost and Hedonic Price methods.

Section 2.6 introduces the principal conceptual issues involved in identifying the changes involved in coast protection and sea defence projects. Some rules of thumb are given there to identify whether the benefits from a particular coast protection project are likely to be large or small. The likely desirability of the changes caused by erosion, project construction and coast protection or sea defence works to different types of site and site features are then summarised in Table 2.5. This table also identifies the economic technique or range of techniques that we recommend should be used to value each of these changes.

The incorporation of risk and uncertainty into appraisal, and the management of uncertainty, are discussed in Section 2.7. We suggest that uncertainty is only important if its resolution would make a difference to which option is chosen: that is, whether the preferred option is 'robust' to remaining uncertainties.

Therefore, we recommend that appraisal should be an iterative process of progressively refining the estimates of the benefits and costs of the alternative options being considered. At each stage it is necessary to decide whether any reductions in the uncertainty concerning the estimates of the benefits of the option justify the costs of the work necessary to improve those estimates: project appraisal itself should be pursued in so far as its benefits justify its costs. A number of suggestions are given as to simple methods of testing the robustness of the preference between options to the uncertainties that then remain.

Finally, after the most cost-effective methods of reducing uncertainty have been exhausted, there will still remain uncertainties. This uncertainty must be managed. In addition, the compatibility of the options with the principles of sustainable development must be tested. We therefore suggest three criteria by which to test whether an option is both robust against remaining uncertainties and is compatible with the principles of sustainable development.

1.3 Coast protection and sea defence: strategies, differences and relationships

1.3.1 Strategies and the scope of the analysis

The two aspects of protection against the sea – coast protection and sea defence – can be difficult to separate. For example, sea walls may protect both against erosion and against the flooding of low-lying areas. Coupled with the need for consistent project appraisal, this means that even if a coastal management authority is primarily concerned with only one of these problems, it should be aware of the benefits and appraisal techniques relevant to both, since its projects may provide both types of benefit.

An important objective of evaluating works that aim to protect land from the effects of the sea is to choose the best policy, project and scheme from a national viewpoint. There is a wide range of choice available in project planning:

- Projects in different locations;
- Different projects which provide the same level of protection for an area, but at different costs (both economic and environmental);
- Projects at a given site which offer different levels of protection or have different risks of failure;
- Different standards or levels of upgrading or renovation to existing projects;
- Different dates when a project or scheme against erosion should be implemented;
- The option of taking no engineering or structural action, but improving information on erosion risks and providing flood warning systems, changing land use planning policy, and redistributing the burden of losses;
- Doing nothing.

Although it is clear that any option worth pursuing must be technically, socially and politically feasible, the fundamental presumption behind this Manual is that the choice of the most acceptable of these feasible options is best aided by an 'extended' benefit-cost analysis (see Table 1.2). The systematic framework provided by this technique is necessary when faced with the multifarious variety of possible options, problems and impacts which arise from projects to protect against the sea.

The benefits of such projects are determined by the losses and impacts which are averted by the works created by investment in projects and schemes. Hence this Manual is concerned with evaluating these impacts and losses. In general the appropriate comparisons are between projections of the situation **with** the alternative projects over their lifespans, compared with the situation **without** any of these projects over the same span of time.

However, it should also be recognised that a project or scheme may itself have adverse impacts (other than its immediate economic cost) and these also need to be evaluated. 'Extended' benefit-cost analysis involves evaluating, where possible, a wider range of impacts than those which can easily be identified as having eco-

Table 1.2
Comparison of economic efficiency benefit-cost analysis and 'extended' benefit-cost analysis

Method	Scope	Objective to be maximised	Consequences included in the analysis	Measure of value	Basis of inter-temporal comparison
Financial analysis	Single organisation	Usually profit	Any change which results in an inflow or outflow of money to the organisation; consequences to others are explicitly excluded unless these result in potential changes to organisation's own cash flow	Market prices after these are adjusted by subsidies, grants and tax allowances available to that organisation, and taking account of taxes payable by that organisation	Weighed average cost of capital to that organisation
Individual utility analysis	Individual	The individual's well being or utility	All those consequences of choosing an option which the individual considers will raise or lower his/her utility level	Subjectively assessed values and costs; these include market prices after taking into account subsidies etc allowable to that individual. But they also will include costs such as inconvenience, discomfort and any other subjectively assessed deterrents to choosing an option	Individual's preference for the distribution of consumption over time and the opportunity cost of capital to that individual
Economic efficiency benefit-cost analysis	Society as a collection of individuals, and usually as defined by national or regional boundaries	Social welfare based on present distribution of income and price structure	In principle, these are identical to those included in individual utility analysis and are limited to those so included. In practice, the consequences that are difficult to evaluate are often left as 'intangibles'	Opportunity costs of resources	Test Discount Rate (Social opportunity cost of capital)
Extended benefit-cost analysis	As above	As above plus distributional constraints	As above plus greater attention to the value to humans of the environment where this is unpriced; the use of 'minimum safe standards' and 'shadow projects' to take account of the interests of other species	Opportunity costs of resources	As above plus externally applied sustainability constraints to take account of the interests of future generations
'Ideal' project appraisal method	As defined by that society's ethics, or the different ethical stances within the society; may include interests of other species, as well as of humans, and also the interests of future generations	All consequences which are regarded as of ethical importance, or which affect the total availability of resources	Likely to apply multiple value frames and not be limited to individually derived values	Likely to apply multiple criteria.	Sustainability criteria will dominate

nomic – or monetary – consequences, for example the environmental and social impacts, and the irreversible consequences of decisions.

While this 'extended' form of analysis can be seen as widening the scope of project appraisal, the perspective of national economic efficiency might well be seen as narrowing the scope of the appraisal too far in the coast protection and sea defence context. The fundamental interest, then, is in the overall or net gains to the nation, which means that a purely local transfer, such as loss of holidaymakers from one resort to a neighbour, is not included as an economic loss.

However, entirely omitting such transfers is too simplistic. It is important that local effects are measured, and stated in the overall appraisal to guide decisions, but it is also important to ensure that only national economic losses are included within the benefit-cost figures presented (see the Herne Bay case study for an example of such local benefits: Section 8.5).

1.3.2 Differences and relationships between coast protection and sea defence

The fundamental difference between alleviating flooding and alleviating erosion is that the first can be permanent while erosion is an on-going process that can normally only be delayed. The exception is where erosion can be countered by reclamation from the sea, but this, in effect, is merely a variant on the delay of erosion (i.e. it puts back the process of erosion still further than simply protecting the current line of the land/sea interface).

Thus flood losses have the potential to be **recurrent** – there is a risk of flooding in a given low-lying area in any single year, and it is possible that floods of different magnitude may recur in any combination of years. Hence we generally assume in assessing the impacts of flooding that the same land uses will continue in that area and that they will repeatedly suffer flood damage.

However, where, as a result of coastal flooding problems, more land might become permanently inundated, or is flooded repeatedly with a frequency of, say, once every two years, then it is almost inevitable that the existing land use will change permanently. In these cases property will be abandoned and in benefit assessments it is 'written off', just as in the case of coastal erosion where property lost to the sea is deemed to be lost forever.

Losses from erosion, then, arise where land which is above normal high tide sea levels slips, or is undermined, and therefore falls into the sea, or becomes so unstable that its normal use is no longer viable. This type of loss is very different from flood losses: erosion is a progressive or **sequential** process that is largely irreversible (but see the point above about reclama-tion). Although the erosion process can be halted for a while, once land and property has been eroded it cannot usually be recovered, and thus the losses are one-off and permanent.

The situation is, however, not as simple as this. One-off erosion effects can trigger a sequence of flooding and damages that arise from this flooding (as in the breach situation at Hengistbury Head (Section 8.6)). Erosion can, however, lead to deposition downdrift, and this can reduce erosion there or reduce the risk of recurrent flooding that also might be occurring there. It is also possible that recurrent flooding, perhaps with sea level rise, could result in scour at the coastline which creates erosion that could result in one-off losses. Thus the two phenomena are different but the coastal processes that create one may affect the other, either to enhance the second effect or to diminish it.

In terms of this Manual, the impacts of these differences and relationships are many and varied. First, it has led to the different approach to assessing erosion losses than flood losses (cf. Chapter 3 and Chapter 5). Secondly, it means that **the user of this Manual must evaluate the whole of the coastal unit under examination, to see how the coastal processes there mean that projects at the point of immediate focus are interrelated with other areas both 'updrift' and 'downdrift'.**

1.4 The range of techniques and data

1.4.1 Pre-feasibility and feasibility analyses

The philosophy underlying this Manual, and previous works concerned just with flood alleviation (Penning-Rowsell and Chatterton, 1977; Parker *et al.*, 1987; Suleman *et al.*, 1988), is to provide a range of techniques that fits several different needs. The key point here is not to commit resources disproportionately to the likely value of the results by failing to match the detail with which the analysis is done to the magnitude of the likely benefits. A 'horses for courses' approach is the one that we advocate.

Thus, first, there is the need for a **pre-feasibility analysis** gauging of benefits that can indicate whether a more detailed assessment of benefits and costs is likely to be worthwhile. There is no need to waste resources undertaking benefit-cost analyses in great detail if the result is that the ratio of benefits to costs turns out to be, say, 1:10. That order of magnitude of the result should have been determined by a pre-feasibility study, using techniques and data that quickly give a result that indicates the order of magnitude of costs and benefits.

On the benefit side of this pre-feasibility analysis it will be necessary to use generalised data and methods in a desk-based exercise. This data includes the standard or average data on the value of beach recreation (Section 4.1), the sector average data for flood damages and losses (Section 5.3) and average house prices for losses through erosion (Section 3.7). If computer processing is to be used at this stage (and this might be advisable on a medium to large scale study), then the BOCDAM computer assessment model might be used in the flood alleviation field, coupled with average data on the likely possible increase in adjusted gross margins for the order of magnitude of agricultural benefits.

On the other hand, we also provide data and techniques for application in detailed feasibility studies. In general these feasibility studies will take at least three times the resources of a pre-feasibility study (and often much more), and will involve field data collection, whether it is site specific damage surveys using the questionnaire in Appendix 5.3, or beach recreational surveys to assess the benefits of enhancing beaches or maintaining the status quo. The main bodies of each of the relevant chapters in this Manual give details of these more accurate but necessarily more time-consuming methods.

1.4.2 Different techniques for different benefit categories

It is an obvious point, but we present here different techniques for different categories of coast protection and sea defence situations.

This situation is summarised in Figure 1.1. For each problem there is a method of benefit computation, using data to be found either in this Manual or elsewhere. In the flood alleviation field these results come together within our ESTDAM computer model, including the results from assessments of the benefits of protecting agricultural land from flooding.

In the case of assessing the benefits of controlling erosion, we have devised a spreadsheet format which allows the incorporation of standard data on house prices. Into this framework can be fitted the evaluation of the impacts of erosion on roads, agricultural land and the recreational impacts arising from that erosion. This type of spreadsheet can easily be created using any of the proprietary spreadsheets such as Lotus 1-2-3 or Supercalc 5.

1.5 The phasing and timing of each part of an assessment

Careful **timing** and **phasing** of the various parts of a comprehensive benefit-cost appraisal for project assessment at the coast is vital. Different parts of a comprehensive assessment need to be started at different time periods ahead of results being available, and there is a real danger that the different parts of the assessment will be out of synchrony if careful attention is not given to this subject.

1.5.1 Tasks needing at least a complete calendar year

To assess accurately the potential environmental impacts of a proposed coast protection or sea defence project the investigations will need to start at least a year before the production of the final feasibility study report. This will allow, for example, the existing status and potential impacts on flora to be investigated over a complete growing season, and wildfowl and wader importance to be assessed over a complete cycle of breeding, migration and wintering seasons. It is often the case that this length of time is not allowed for in the feasibility study period, and this means that data is incomplete and liable to be the subject of serious error and criticism.

In addition, any survey of beach, water space and cliff-top recreation should encompass a complete year (or at least the period between Easter and the end of October). It is important to gauge the use of the water space, cliff-top and beach (and its associated promenade, which may have extensive winter use) over the whole of this period, so that annual use figures can be obtained to which contingent valuation and other survey results can be grossed-up.

1.5.2 Tasks needing at least 3–6 months

The main task that needs time to organise within surveys to assess the benefits of flood alleviation are the site-specific surveys of damageable properties within the benefit area. This is because these properties usually cannot be identified until after the land use survey of the benefit area is completed, and for a study with, say, 2,000 properties liable to flooding this alone could take at least two months to organise and complete.

Thus if site specific surveys are considered likely, it will be necessary to start the survey of land use in the benefit area perhaps six months ahead of results being required. This will allow three months for the survey of land use, followed by preliminary analysis of the results based on those survey details. Thereafter it will be possible to mount the site specific surveys, and as a rule of thumb guide it will take two person/days per site specific survey to initiate, interview the appropriate personnel, analyse the results and complete each survey.

Within both flood alleviation and erosion studies one of the tasks that consumes most time is the survey of impacts on utility infrastructure within the benefit areas. This is because individual interviews with the organisations concerned will be needed once the likely direct impacts of erosion have been identified, and these will take time to arrange, complete, and for the analysis of the data collected to be undertaken. The same applies to an assessment of the agricultural benefits of coast protection and sea defence projects (if this is identified as being worthwhile in the pre-feasibility study). Farm interview surveys will be needed in all but the most generalised pre-feasibility study, and these can take several days per interview to set up, conduct the interviews, and analyse the data thus collected.

1.5.3 Tasks that can be accomplished more quickly

Obviously the length of time needed to complete any analysis depends on the scale of the problem and the available resources.

Nevertheless there are tasks that can be completed more quickly than the 3–6 months listed above. This is largely because the tasks are not dependent upon many 'outside' sources of data which can delay the necessary processes.

Thus in many simple flood alleviation projects it is possible to undertake a land use survey of the benefit area and a small number of site specific interviews and analyse the data to obtain results within a three month period, assuming that the necessary hydrographic data is available within this period (and no less than one month before the final results are required). Likewise a basic survey of the impacts of erosion on cliff-tops can be undertaken within this time frame, assuming that data on property prices can be obtained reasonably quickly from local estate agents and district valuers and the information on erosion rates is readily to hand.

However, the number of appraisals that can be **completed** within a three month period is likely to be very small indeed. This is because it is almost impossible that a coast protection project does not have significant environmental or recreational implications, or that a sea defence project protecting low-lying land from flooding does not have any environmental impacts or involve surveys of farmers or recreationalists. The clear implication of this is that **in nearly all cases the users of this Manual should begin the preparation of the studies for the benefit assessments at least a year in advance of the results being required.**

1.6 Resources for surveys and data processing

1.6.1 Resources for surveys and data collection

For many of the users of this Manual the types of surveys involved will be new. This means that **in almost all cases the relevant personnel will require staff training and supervision.**

The most skilled tasks in terms of questionnaire surveys are the site-specific flood damage surveys, for which trained and experienced interviewers will be needed. For the contingent valuation surveys of recreational benefits it will normally be sufficient to recruit interviewers with experience of market research surveys, who should be properly briefed and trained for the purpose. It will normally be necessary to recruit a small number of interviewers additional to those strictly required to meet the sampling needs, in case of illness, etc., since the 'window' of times that are important for such surveys are often restricted (e.g. bank holidays).

The conduct of farm interviews used to assess the agricultural benefits of flood alleviation requires a high level of experience, and experts should be used from academic organisations (i.e. Silsoe College) or agricultural consultants.

Land use surveys of areas liable to flooding can be carried out by relatively inexperienced technicians, who need to be briefed with reference to Chapter 5 of this Manual and also the relevant parts of previous manuals (Penning-Rowsell and Chatterton, 1977; Parker et al., 1987) and the FLAIR report (e.g. N'Jai et al., 1990) to which the reader's attention is drawn in the relevant parts of the chapters below.

1.6.2 Data processing

Computerised data processing is recommended for all but the simplest and quickest of pre-feasibility studies (Chapter 7). This is because the calculations generally become extremely complex and tedious in most benefit assessments, since it is inevitable that computations have to be re-run many times as data is refined and assumptions are changed.

For assessing the benefits of flood alleviation there are two software packages available from Middlesex University. The BOCDAM suite of programs is a 'high level' assessment system which requires limited data and produces approximate results, and is only suitable for pre-feasibility studies. The ESTDAM system, running on personal computers or workstations, is more powerful, and requires much more data, but allows assessments to be undertaken with all the detail contained in this Manual.

Programs are available from Silsoe College for assessing the benefits of agricultural drainage and flood alleviation for incorporation with ESTDAM results. In addition, a set of spreadsheets using proprietary software is available from Middlesex University for assessing the benefits of coast protection projects to delay erosion, if the user is not confident that they can adapt their own spreadsheets systems to undertake the necessary computations.

We recommend that the user decides at an early stage in the feasibility study what systems are to be used for data processing, and that the data collection and survey systems are then geared to that decision.

1.7 Updating

Most of the data contained in this Manual is given at 1989, 1990 or 1991 prices, and the date relevant is given with each data item.

Within periods of two to three years of this date, updating can be undertaken simply using the most appropriate index series. In most cases this is the relevant Retail Price Index available from the Department of Employment (1990). Exceptions are for user benefits (the standard data in Chapter 4 for use in pre-feasibility studies), for which the growth in GDP should be used (Central Statistical Office 1992); and for house prices (see Section 3.7.1), for which the Department of Environment BS4 price series should be used.

However, and thereafter, most of the data within this Manual will be updated on a regular basis through the publication of the FLAIR report (Flood Loss Assessment Information Report), for which the 1990 version was compiled by N'Jai *et al.*, (1990). This FLAIR report will be published regularly by Middlesex University, and will accumulate all the data produced by the Univesity's Flood Hazard Research Centre. The exception is data on agricultural benefits, for which the user is referred to Silsoe College.

2 Theoretical perspectives

2.1 Summary of information

This chapter sets out the theoretical background to the economic issues involved in assessing the benefits of sea defence and coast protection projects.

To this end the chapter discusses the rationale, methods and criteria of project appraisal (Section 2.2) together with issues such as sustainability. Section 2.3 examines the basis of project appraisal, as well as important issues such as discounting and the distribution of gains and losses from project investment. At a more practical level, Section 2.4 identifies the stages in a project appraisal, and emphasises the importance of the correct phasing of investigations. To complement that material, Section 2.5 assesses the different methods of evaluation in general, while Section 2.6 applies some of this analysis to coast protection and sea defence projects. Section 2.7 tackles the questions of risk and uncertainty, and finally Section 2.8 provides some important 'health warnings'.

Overall, the chapter sets out the basis for the economic valuation of the different impacts of such projects and for the comparison of alternative project options. The different economic techniques available for valuing the different impacts are reviewed and the most appropriate techniques are identified. Those project impacts that it is currently feasible and desirable to value in economic terms are defined, together with those impacts for which it is judged that the use of economic values is inappropriate.

In discussing the most appropriate economic techniques to use, emphasis is placed upon establishing the validity of any techniques adopted. The primary basis for determining the project impacts to which economic values should be applied is the usefulness of so doing in a pluralist society, where there is conflict between the views of different individuals. In particular, analysis will only be useful if it aids communication between different individuals and groups.

The material in this chapter is necessarily challenging, for three reasons. First, when dealing with things which have prices, because they are bought and sold, economic analysis is quite straightforward. However, for many of the consequences of coastal flooding or erosion, such as the loss of a beach, there are no market prices that can be used to put a value on each of the many consequences. Nevertheless, an economic value can in theory be estimated for all these consequences but to achieve a valid result requires knowledge of the economic principles involved.

Secondly, whilst economic theory argues that all changes can be valued (Pearce and Markandya, 1989; Pearce, Markandya and Barbier, 1989), several environmental groups have explicitly rejected economic approaches to valuation of some environmental changes (Bowers, 1990; Friends of the Earth, 1990; Hopkinson, Bowers and Nash, 1990). Equally, the basic economic theory has itself been subject to increasingly critical review (Daly and Cobb, 1990; Lutz and Lux, 1988; Sagoff, 1988; Sen, 1977; 1987). Since there is no universally accepted truth, it is necessary to understand all sides of the argument, not least in order to communicate effectively with those who reject some components of economic analysis. The ultimate test of such an analysis is often whether it stands up at a Public Inquiry, or, preferably, removes the need for such an Inquiry by producing an analysis and result on which all parties agree. Thirdly, choices are difficult because the issues themselves are complex.

A central message of this chapter, then, is that economic analysis should always be used critically as an aid to decision making, rather than as a substitute for thought. We have made recommendations as to those areas where economics can be useful. However, this is an issue about which the reader must finally make up his or her own mind.

2.2 Introduction: why do project appraisals?

The use of project appraisal methods is justified by arguments that suggest that they enable better decisions to be taken. To gauge whether decisions are better, the following three tests are usually applied:

1. The **scarcity** of resources requires that, as a society, we choose to make those investments of resources that give the highest **return**;
2. The requirement of **accountability**; the need to justify the choices and decisions that are made on behalf of the public to that public and their elected representatives; and
3. The requirement that these decisions should be taken as a result of a **rational comparison** of the available options and the different consequences of those options.

The development of formal methods of project appraisal is a reflection of the sheer difficulty of satisfying these tests consistently in the face of the number and complexity of decisions that must be made. By routinising the decision process, we seek as a society to ensure that there is consistency in decision making whilst reducing the effort that must be put into ensuring that this consistency is obtained.

All decisions are about choices, but the scope of choice varies. Broadly, it is possible to distinguish three levels of decision making (Coopers and Lybrand Associates, 1986). **Programme allocation** concerns the allocation of resources between programmes; for instance, between coast protection and land drainage, or between coast protection and new hospitals. **Project prioritisation** is the process of selecting which individual projects should be undertaken within the available budget, and the order in which those projects should be undertaken.

This could, for example, be the decision as to which parts of the East Coast of England should be defended and where erosion should be allowed to continue unhindered. **Project appraisal** is the assessment of which of the available options is the best alternative to deal with a specific local problem, for instance whether protection of a particular stretch of coastline is justified, to what standard, and when and for how long.

2.2.1 Methods and criteria

In response to the pressures to routinise decision making, a range of project appraisal methods has been developed to use as decision aids for both public and private investment. For public investment these methods include statistical decision theory (Schlaifer, 1959), Multi-Attribute Utility Analysis (Raiffa, 1968), Multi-Criteria Analysis (Goicoechea, Hansen and Duckstein, 1982), Environmental Impact Assessment (Department of the Environment, 1990) and Benefit-Cost Analysis (Mishan, 1971; Pearce, 1984; Sugden and Williams, 1978).

Methods that are more suitable for programme allocation include Planned Programming Budgeting Systems (Novick, 1967), whilst methods for project prioritisation are usually simplified methods of project appraisal, perhaps coupled with optimisation techniques drawn from operational research (Green, 1990).

To be useful, a project appraisal method must respond to the needs outlined above, and several criteria have been derived from these needs by which to judge project appraisal methods (Green, Tunstall and House, 1989; Lichfield, 1964; Nash, Pearce and Stanley, 1975). Thus, a project appraisal method, its analysis and results, should satisfy the following criteria:

1. **Elucidation of the issues.** Clearly, the decision maker and those to whom the decision maker is accountable should be left with a clearer understanding of the issues and trade-offs involved in the decision having done the analysis than before the analysis was undertaken.
2. **Simplification.** Decisions involve complex trade-offs between different objectives, and the processes involved may themselves be complex. As issues such as sustainability, environmental quality and public participation in environmental decision-making increase in importance, so too does the complexity of the decision itself. The project appraisal method must structure and simplify the presentation of data leading to that decision and supporting it, so that the volume of such data does not overwhelm the decision maker.
3. **Feasibility.** The resources required to carry out the appraisal must be appropriate to the importance of the decision, and the time available to take a decision. Equally, the method must not require more information than can be obtained.
4. **Completeness.** The method should be able to encompass all the significant differences in the impacts of the different available options. The impacts that should be included are those that are important, and not simply those that are

easiest to measure. For public investments, importance is determined by the goals of the whole of society, not of individuals or groups.
5. **Rigour.** The results of the analysis should not depend upon who undertakes the analysis.
6. **Value basis.** The values and the trade-offs in the analysis, both implicit and explicit, should be those of the public as a whole where public monies are involved.
7. **Reliability.** Any project appraisal method is a predictive method. It is reliable to the extent that what is predicted to occur, does occur. As a minimum, the predictions must be sufficiently detailed so that they can subsequently be tested against what occurs in practice.

No single project appraisal method fully satisfies all these criteria. Benefit-cost analysis (BCA) has developed by evaluating those impacts which are easiest to measure and leaving the impacts that are difficult to evaluate as 'intangibles'. Hence, benefit-cost analyses usually score poorly against the criterion of 'completeness'. Conversely, the ideal project appraisal method would begin by evaluating the most important impacts.

On the other hand, Environmental Impact Analysis scores highly on the criterion of completeness, but the values involved are usually those of the experts undertaking the analysis. Some techniques, such as the Leopold matrix (Leopold *et al.*, 1971) lack rigour (Bisset, 1978), and many analyses fail the weak test of reliability: the predictions are so lacking in precision as to render a post-project appraisal impossible (Wood and Jones, 1991).

In the absence of a project appraisal method that scores highly against all the above criteria, it is sometimes appropriate to use two different project appraisal methods to assess a single project. For many cases, including the assessments of coast protection and sea defence projects, BCA will be an appropriate and required method in this multiple method arrangement. This is particularly the case because it is rarely intuitively clear what decision to make, given that the basis of the benefit calculations usually includes some very low probabilities and some joint probabilities (e.g. the probability of high surge levels and large wave heights occurring together). Also, the choice between the available options usually involves complex trade-offs between the widely different impacts.

2.2.2 Sustainability

It is now necessary to apply a constraint to the results of the comparative analysis of different project options to ensure that the project satisfies the principle of sustainability. That any development should satisfy this principle was embodied in the Communiqué of the 1990 Dublin Summit of the Heads of Government of the European Community, and it is reflected in the 1990 British government White Paper on the Environment (Department of the Environment, 1990).

Exactly what is meant by 'sustainability' is more difficult to define in practice (Pearce, Barbier and Markandya, 1990). However, it includes the principle of managing resources so as to safeguard the rights not only of future generations but also of those in the less developed countries (World Commission on Environment and Development, 1987).

2.2.3 Generic issues in project appraisal

Any method of project appraisal involves the comparison of two or more options in order to select the most desirable option from those available. One of the options considered is usually the 'do-nothing' option, and this is commonly used as the baseline against which the changes that would result from adopting each of the other options identified are measured.

The different options will involve a variety of impacts, or changes, and these will vary in terms of **what** is the impact, **who** is affected and **when** in the future the impact is to be experienced. Choosing between the available options, therefore, involves complex trade-offs as to the desirability of the consequences of widely different impacts. Since this choice involves comparison, implicitly and explicitly these different impacts must be brought to a common base: some common yardstick must be used by which to compare them.

Clearly the first stage of any analysis, therefore, is to **identify** the different impacts of the 'do nothing' and the 'do something' scenarios; where possible the magnitude of each of these impacts or changes is then **measured**. Comparing the options whose impacts vary in their nature, and vary as to whom or what is affected, and when the change occurs, involves an implicit or explicit process of **evaluation**. Initially at least, this may not involve monetary valuation.

In economics, however, the problems of this process of comparison – of evaluation – are handled, first, by using money as a yardstick, or **numeraire**, with which to compare the different impacts. Secondly, in so far that these impacts differ in terms of who is affected, this is resolved by comparing net changes in resources and consumption. Thus, economics explores the efficiency with which resources are used and excludes the distributional – or equity – question of whom should have these resources. Thirdly, the differences in the distribution of impacts over **time** are resolved, at least at a technical level, by the procedure of discounting.

Superficially, because both report the results of their analyses in monetary terms, **economic analysis** and **financial analysis** appear to be similar, or even the same. In fact, the two differ significantly (Table 2.1) because of the differing standpoints that they take. Economic analysis takes the national viewpoint and addresses the question: What is the best option for the nation? A financial analysis examines which is the best option for the organisation contemplating the investment. The two questions are very different, and the results of the analyses can be very different.

Table 2.1
The key differences between economic analysis and financial analysis

Economic analysis	Financial analysis
Concerned with the total net change in resources and consumption across the nation in consequence of the decision or change.	Concerned only with those changes which affect the organisation for which the analysis is being undertaken.
Uses money as a yardstick to compare changes in the stocks and flows of goods, whether or not these goods are priced.	Only concerned with changes which have monetary consequences for the organisation for which the analysis is being undertaken.
Based upon the concept of opportunity costs; these may be reflected in market prices but usually are not. Changes in taxes and subsidies should be excluded where these are not associated with a change in the distribution of resources.	Based upon market prices. Taxes, subsidy payments and similar monetary transfers are included.

For example, the owner of a hotel could carry out a financial analysis before deciding whether or not it would be worthwhile to replenish the beach in front of the hotel. In doing so, the owner would need to predict the number of additional visitors who would be attracted to the hotel because of the improved beach, or the increase in charges that could be levied on existing visitors to the hotel.

The owner would then compare the predicted increase in profits, net of taxes, to the costs of the beach replenishment. From these replenishment costs any subsidies, tax allowances or other capital investment incentives would be deducted, and the discount rate used would be the interest rate at which the hotel owner could borrow money. Other than any legal obligations, the owner should ignore any increased erosion that the hotel's project might cause to beaches downdrift, or any damage to the environment unless that damage would affect the hotel's profits.

In an economic analysis, on the other hand, the effects on the national tourist trade as a whole would be the focus of the calculation and not simply the effects on the one hotel. If the result of the project were simply to attract visitors away from other hotels in other parts of the country, or even from the neighbouring hotel, then the national benefits of the project would be small and probably zero. Equally, downdrift effects would be included as a negative project benefit and so would any damage to the environment. The change in value added would be a measure of the project's benefits. However, any profit remitted overseas would be netted out; so too would any increase in imports drawn in to meet the increase in tourism.

These crucial differences between economic and financial analysis must be borne in mind throughout any appraisal of coast protection and sea defence projects.

2.3 The bases of economic analysis

2.3.1 Economic value (see glossary of terms)

Any process of evaluation, by definition, is based upon some concept of value and the different possible bases for value have been a concern of philosophers for centuries (Brown, 1984). It is clear, of course, that value is central to benefit-cost analysis, hence its detailed analysis here.

In neo-classical economics, by axiom, all values are subjective: the value of some good is given by the individual and reflects his or her subjective preference for that good. In this respect the shorthand term **'good'** is used to denote any commodity, resource or item which an individual prefers or desires (for example a coast protection project, a beach, a cliff-top house, a fish taken from the sea, or a recreational experience). It is further axiomatically assumed that each individual is the best judge of her or his best interests – the individual's **utility** – and the values assigned to a good reflect the relative contribution that this good makes to that individual's utility. What is desirable and how desirable is that good (be it a bottle of Scotch whisky or a walk on a beach) is entirely a matter for the individual to decide. Because the value basis of economics is determined by axiom in this way, the results of all neo-classical economic analyses are conditional upon the acceptance of the axiom of individually assigned subjective values.

Because the economic value of a good reflects the contribution that it makes to the individual's utility, economic values are relative: a good has a higher value if it contributes more to the individual's utility (or welfare) and vice versa. The value is also sacrificial: it reflects the degree to which the individual would be willing to give up an amount of that good in order to have more of another good. Values are therefore not absolute but reflect the basis upon which choices are made between the consumption of different goods.

Since axiomatically all values are given by the individual, it follows that the values that should govern societal decisions about the supply of goods, including those resulting from coast protection and sea defence projects, should simply be some aggregate of these individual values. To do this, some way of measuring the relative desirability of different goods is required: we need a yardstick or numeraire. Economics has adopted **money** as this numeraire, and as a numeraire it has a number of advantages and disadvantages.

The most obvious advantage is that money is also used as a medium of exchange. When we give up some of our leisure time to undertake work in order to obtain goods to consume, we use money to simplify this bartering process. Similarly, in choosing between goods – in deciding which has the greatest value – we make many of these choices in the marketplace. Thus, since we pay more for those things that make the greatest contribution to our utility, the amount we are willing to pay for any good directly reflects this value. Since preferences are psychological, they cannot be observed. However, we can observe how much an individual is willing to pay for a good.

The disadvantage of using money as the numeraire is that it is held that the marginal utility of income declines: the greater the existing income, the lower the utility of a given increase. Since income and wealth are not uniformly distributed across society, the consequence is that a given amount of money will represent different utility changes to different individuals. This is usually characterised as an extra pound being worth more to someone with very little money in comparison with a millionaire.

Under very restrictive conditions, known as a **perfectly competitive market**, the prices the individual has to pay for a good exactly equal its value to him or her. Moreover, this price also exactly equals the value of the resources used to produce that good. However, almost invariably there is some degree of **distortion in the market** for a good. This can be as a result of factors such as monopolistic supply, the presence of taxes or subsidies, or imperfect information on the part of the consumer as to the availability of goods.

In these cases of market distortion, the price of the good will be above or below the value of the resources used to produce it. The value of those resources – their **opportunity cost** – is the value of the goods that could otherwise have been produced with them. If the price is above that value, then some of those resources could have been used to produce additional goods, the consumption of which would have resulted in a higher level of utility to the individual. Conversely, if the price is too low then too many resources are being diverted to produce that good when their use for providing another good would have produced more utility for the individual.

The axiom of human rationality allowed economics to avoid the question of why an individual prefers one good over another while economic analysis was restricted to goods that were bought and sold in the market. For a market in a good to be possible, the good must have two characteristics: the consumption by one individual must reduce the supply of that good available to other individuals, and the holder of the good must be able to control access to that good by other individuals. Such goods are termed **private goods**. The derivation of economic analysis from the analysis of choices about private goods is shown by the use of **consumption** as a general term to describe the use, access to or knowledge of the existence of a good by an individual.

However, many – and perhaps most – goods in an economy governed by the principles of sustainability lack one or both of these two distinguishing features. The classic example of a **public good** is a lighthouse: if one shipowner decides to build a lighthouse in order to warn his or her ships of a dangerous rock, all other ships will also benefit from the warning given by that lighthouse. Neither has the original shipowner any way of preventing them from so gaining, nor is there any easy way of requiring such beneficiaries to pay for the service.

An important feature of public goods, therefore, is that if the enjoyment or satisfaction that one individual gains from a good in no way diminishes that remaining for other individuals, then creating a market in that good would cause economic inefficiency. Since the marginal cost of provision is zero as the number of users increases, there is no basis for setting a market price. Consequently, the price cannot be set so that marginal value equals marginal cost. Therefore, at any charge level, economic inefficiency will result because marginal value exceeds marginal costs.

The enjoyment that an individual gains from a visit to a beach is a more relevant example of a public good, at least until the point where congestion occurs. Similarly, the satisfaction that an individual may gain from the preservation of a Site of Special Scientific Interest does not diminish the satisfaction that others may also gain, nor is there any way of preserving such sites so that only some individuals have their wishes fulfilled.

There are also two additional variants of the values that individuals ascribe to resources that they use: **use values**. These are the **option value** and **quasi-option** value. If someone does not seek to use a good now but is willing to pay in order to preserve the option of its use in the future, then this payment (or willingness to pay) is described as the **option value** (Schmalensee, 1972). It usually applies to situations where a proposed change might irreversibly affect the availability of the good in question. For example, whilst there may currently be a surplus of agricultural production and therefore of land in agricultural use, an individual might attach an option value to the preservation of the Fens of Eastern England, with their high agricultural outputs, in case climatic change results in failure of worldwide agricultural output to meet demand at prices the English can afford. By ascribing this option value to that land we are saying that there is merit (i.e. value) in adopting a 'wait and see' policy.

A **quasi-option** value (Arrow and Fisher, 1974; Henry, 1974) measures the improvement in the predictions of the consequences of a change by delaying a decision. It is thus a measure of the value of information, as that concept is used in Bayesian analysis (Moore and Thomas, 1988). For example, archaeologists often prefer to delay excavation of sites because excavation destroys the information a site contains. By delaying their excavation, archaeologists hope that in the future better techniques will enable them to capture more of the information contained in a site.

Economists have accepted that individuals may value a public good for reasons other than the use they may make of that good, or the right of access to that good. These **non-use values** are sometimes termed 'intrinsic' or 'existence' values, and economists have speculated as to the nature of the motivations that may give rise to these non-use values (Arrow and Fisher, 1974; Brookshire, Eubanks, and Sorg, 1986; Krutilla, 1967; Krutilla and Fisher, 1975; Madariaga and McConnell, 1987). Unfortunately, there is very little empirical evidence to test whether these hypothesised motivations are either correct or sufficient to account fully for the reasons that lead people to value a public good such as a nature reserve.

2.3.2 Who gains and who loses?

Economics, by focusing on the changes in the total amounts of resources and consumption available, treats the question of the distribution of resources as an issue over and above its concerns. This concentration solely on the total availability of goods means that transfers of money that are not accompanied by transfers of resources are excluded from economic analyses, including from benefit-cost analyses. Such transfer payments are termed **pecuniary externalities**: taxes such as Value Added Tax and Excise Duties are examples of such payments.

We need to test, in an analysis, whether the total sum of goods has increased as a result of some change resulting from investment in a coast protection or sea defence project. The test to apply is the Hicks-Kaldor Compensation Principle, or that of a Potential Pareto Improvement (Mishan, 1971). A Potential Pareto Improvement is said to occur if the gainers from a change could fully compensate those who lose by that change and still retain some gains. It is this principle that is the basis of the use of the benefit/cost ratio and net present value criteria in benefit-cost analysis.

Quite clearly it is possible for a project to pass this test but, as a consequence, for the rich to get richer and the poor to get poorer. Economists have two answers on this point. First, any redistribution of wealth that society considers to be desirable would be more effectively achieved by taxation than through altering the appraisal process for individual projects. Secondly, distributional issues should be considered alongside and together with the results of an economic efficiency analysis in assessing and deciding on whether to proceed with a project. One way of considering the distributional consequences of project investment is by differentially weighting the impacts according to whom is affected: weights for the rich would be smaller than weights for the poor. Neo-classical economics cannot, however, provide a rationale by which to determine the appropriate set of weights.

2.3.3 When: discounting

In order to bring the stream of costs and benefits over time arising from a project to a common base, it is conventional to discount future costs and benefits back to a common base date. A cost or benefit that occurs in one year's time is thus treated as having a lower real value now than an identical benefit or cost incurred today. Since the annual amount that can be earned by investing a capital sum is termed the 'interest', and the yield from different forms of investment is the 'interest rate', the corresponding term used to calculate the present value of a flow of future benefits (and costs) is called the **discount rate**. The formula is as follows:

$$PV = \sum_{t=1}^{t=T} \frac{X_t}{(1+r)^t} \qquad \text{Formula 2.1}$$

where:
PV = the present value of the benefit or cost occurring in a future year;
r = the discount rate;
t = the number of years into the future the bene-

fit or cost occurs after the base date of analysis;
X_t = the benefit or cost in year t;
T = life of scheme.

The discount rate adopted in benefit-cost analysis in Britain – the **Test Discount Rate** – is usually set by the Treasury.

Two points here are important. First, discounting has nothing to do with inflation and is applied to values estimated in real prices: the values of goods after the effects of inflation have been removed. Secondly, what is discounted is *consumption*. Thus, if capital benefits or costs occur in the future, then the rational for discounting these rests upon the possibility of converting these capital changes into gains or losses in the annual amount of consumption. In recent years, the application of discounting has become more controversial, particularly when applied to environmental goods (Goodin, 1986; Nash, 1973).

Economics defines decisions to invest as being a choice between having goods now and having goods in the future. The resources to be invested (i.e. not consumed now) can only be generated by sacrificing the consumption of some goods now. The investment will result in an increased stream of goods in the future. The discount rate should then reflect two elements. First, it should reflect our preferences for the consumption of goods at different points in time. Secondly, it should reflect what future increase in goods we could get from investing resources elsewhere: the **social opportunity cost of capital**.

This opportunity cost should be estimated as the rate of return from other possible investments, after netting out transfer payments such as taxes, and any externalities, such as adverse effects 'downstream' from the project. If a sustainable economy is required, then some allowance needs also to be made for the exhaustion of non-renewable resources associated with those alternative projects. Thus, the commercial rate of return from investments may exceed the social opportunity cost of capital and the derivation of the social opportunity cost of capital from market rates is at present difficult.

However, the gains in consumption yielded by alternative projects may still be distributed over time in a different way to those of the project being appraised. Or, alternatively, the capital to be invested may be drawn not from other investment opportunities but from decreasing consumption now: for instance, by increasing taxation levels. Thus, it is also necessary to determine our preferences for consumption over time.

There are several reasons why, as individuals, we may prefer consumption of a good at a particular point in time to that at some later or earlier point in time.

First, the marginal utility of a good is likely to change over time. In so far as society may be richer in the future, and we can consequently consume more then, the marginal utility of a given increment in the availability of a good will be lower in the future than it is now. Conversely, if the supply of a good is currently very high, then the marginal utility of an increment in the availability of that good will be greater at some future time when the good is scarcer. In addition, since discounting applies to consumption, a very large increment in consumption at one point in time may be less desirable than smaller increments at different points in time.

Thus, offered a choice between consuming an additional 1 million of goods this year, and an additional £33,000 (in real terms) of consumption each year for the next thirty years, it would be logical to prefer the latter. Obviously, were we to be offered the choice between £1 million today to invest or consume, and £33,000 a year for 30 years, we would prefer the former on the grounds that we could invest the sum to yield more in real terms – with interest – than the £33,000 a year.

Secondly, the utility attached to some good relative to other goods may change over time: as adults we attach a lower utility to going to children's playgrounds – and perhaps beaches – than we did when we were younger. In particular, since the demand for some goods is income-elastic, in the future, if incomes are higher in real terms than now, then the marginal utility of increments of those goods will be higher then than it is now. Conversely, the demand for some other goods is income-inelastic, and for other goods income elasticity is negative: higher incomes resulting in lower demand.

Thirdly, the consumption of a good requires the joint supply of both the goods and time for their consumption. It can be easier to shift consumption over time to match the availability of time than to make time available for consumption. Thus, if offered a free holiday now or next summer, the individual may prefer to take the holiday next summer during the school holidays. The rapid penetration of the consumer market of video recorders indicates the perceived advantages of being able to reschedule the availability of a good – in this instance, television programmes or films – to fit in with the individual's time schedule.

No clear direction of preference between consumption now and in the future necessarily results from the above three reasons. But a further three reasons imply a preference for consumption now rather than in the future.

First, whilst economics assumes that income expenditure can be substituted for capital expenditure, for many people the ease with which they can borrow against future income – the elasticity of substitution of capital for income – is very low as a result of market imperfections in the loan market. In consequence of this induced inelasticity, such individuals appear to discount future expenditure at a very high rate. Since this inelasticity results from market imperfections, it should not be used as a basis for selecting the discount rate to be used for public investments. However, it should be considered when private sector investments are being assessed.

Secondly, there are risks attached to the receipt of future gains. An obvious risk is that in the long run we are all dead. So, consumption now may be preferred to consumption in the future on the 'bird in the hand is worth two in the bush' principle: this is risk aversion. However, it is argued (Arrow and Lind, 1970), given the wide spread of public investments, that this public investment should be made on a risk neutral basis.

It is also generally considered by economists to be a self-evident truth that consumption now is preferable to consumption at some later date: this is termed **individual time preference**. Overall, these motivations do not provide an unambiguous rationale for the sign and magnitude of the discount rate.

Since economics is based upon the presumption that social values are some aggregate of individual values, any individual preferences for the distribution of consumption over time should be reflected in decisions as to public investment: this is **social time preference**. In principle, therefore, the Test Discount Rate is derived from a consideration of the opportunity cost of capital and of social time preference (Lind, 1982; Markandya and Pearce, 1988). In practice, this is a difficult exercise. However, the setting of a Test Discount Rate creates its own truth: if a project is only viable at a lower discount rate, it is probable that there exist projects that are viable at a higher rate. The latter should be undertaken in preference to the project under assessment.

One objection to discounting future losses of environmental goods is that social time preference can only reflect the preferences of individuals alive today. Consequently, with projects that impose costs on, or benefit, future generations, there is no basis within the axiomatic base of economics to take into account the interests of those future generations: this is the problem of **inter-generational equity** (Goodin, 1986). It has been argued, therefore, that lower discount rates should be applied to changes in the availability of environmental goods, in particular to take account of the interests of future generations, or, more strictly, to reflect our willingness to make sacrifices now for the benefit of future generations.

Markandya and Pearce (1988), however, have pointed out that it is by no means clear that the application of a lower discount rate for changes in the availability of environmental goods would necessarily lead to decisions that were preferable from the putative viewpoint of future generations. Instead, they propose the use of sustainability constraints. One definition of such a constraint is that given in the Brundtland Commission Report: 'Sustainable development meets the needs of the present, without compromising the ability of future generations to meet their own needs.'

In addition, in a benefit-cost analysis, whilst real prices should be used, allowance should also be made for predicted changes in the relative prices and values of different goods over time (HM Treasury, 1991). Such changes will take place because of the increased scarcity of some goods and improvements in the technological efficiency of production of other goods. The real cost of computers, for example, has significantly declined relative to other goods over the last ten years. For the above reasons, it is argued that the relative real values of environmental goods are likely to increase over time.

Applying growth rates to different goods is adopted in the benefit-cost analysis of new roads (Department of Transport, 1981). However, the benefit-cost analyses for roads are very simple and include very few of the impacts of roads. Applying the approach to the wider forms of analysis described in this Manual would involve predicting the appropriate growth rates for a wide range of goods. Consequently, for analytical simplicity and feasibility, we recommend that growth rates are not routinely applied to any goods.

Instead, first, the Test Discount Rate as set by the Treasury should be used as the discount rate for all goods (i.e. all costs and benefits). Secondly, to assess the inter-generational equity implications of a project, a time profile of the net benefits of the project should be drawn up (Figure 2.1). If a project's net benefits fall and stay below zero in some future years (i.e. it imposes costs then that are greater than the benefits that accrue (figure 2.1b)), then it must be questioned whether that project satisfies the criterion of sustainability. It is likely to require some change so as to return the time stream of net benefits in future years to a level that does not fall below zero.

22 THE ECONOMICS OF COASTAL MANAGEMENT

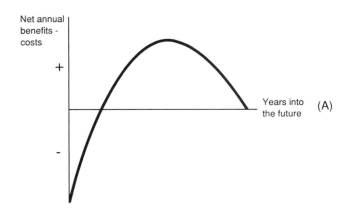

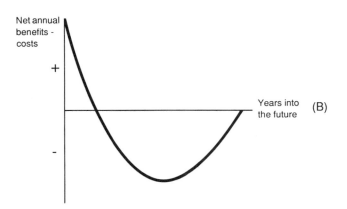

Figure 2.1 The time profile of projects

2.4 Stages in the application of economic analysis

As Figure 2.2 shows, there are three important stages in any project appraisal applying economic analysis: identifying project impacts, measuring those impacts, and evaluating the measured impacts. These are discussed below, in turn.

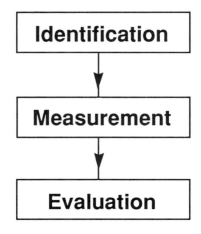

Figure 2.2 Three key stages in project appraisal

2.4.1 The identification of the change with or without the project

The criterion of completeness (Section 2.2.1) requires that all significant changes in the availability of any good, as a result of one or other project option, should be included in the analysis. Implicit in this statement, from the axiom of individual subjective value, is that all goods should be included for which at least one individual prefers more to less (or less to more).

Hence, the boundary of the analysis should be set by defining the points where the effects of the project become negligible. Since the basis of economic efficiency analysis is changes in the national availability of goods and resources, the theoretical boundary is now the United Kingdom (and possibly, after 1992, the boundaries of the European Community). But many of the effects of the type of project generally implemented at the coast will have been dissipated well before those boundaries are reached.

For both sea defence and coast protection projects the appropriate boundaries for impacts may well be far wider than the geographical boundaries of the project. If, for example, a proposed tidal barrier would increase the risk of flooding on coastal land to seaward of the barrier, then the project benefits should be reduced to take account of this increase in risk 'upstream' or indeed anywhere else. Similarly, coast protection projects typically interfere with existing patterns of sediment movement, and change the risk of erosion at locations downdrift. Any such effects of a project should be included in the benefit assessment for that project.

Figure 2.3 illustrates an area (A) that is currently experiencing erosion. The eroded material is carried by longshore drift along the coast to form the beaches at B. The material eroded from and drifting away from this beach is then carried out to sea, to be deposited as an underwater sand or shingle bank at C.

If it is proposed to construct a coast protection project at A, then there is likely to be some increased erosion downdrift at D – the 'last groyne' phenomenon. Erosion at B may increase, and less material will be deposited at C. The economic value of all these changes must be included in the benefit-cost analysis for the proposed project at A; they should either be counted as negative benefits, or as costs, of the project. Since the discount rate reflects the opportunity cost of capital, it is generally more appropriate as well as easier to count them as negative benefits.

In the case of the increased erosion at B and D, these negative benefits are the present value of the losses resulting from the increased rate of erosion there, or the cost of a coast protection strategy that would maintain erosion at its present rate, whichever of these two is the least. The potential economic consequences of any change to the rate and type of deposition at C need also to be examined. The changes might, for example, adversely affect its value as a fisheries resource. Or the deposit might be useful for aggregate dredging, or the change might require less dredging to maintain a shipping channel, etc.

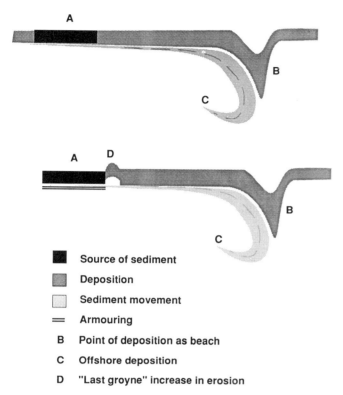

- ■ Source of sediment
- ▨ Deposition
- ▢ Sediment movement
- = Armouring
- B Point of deposition as beach
- C Offshore deposition
- D "Last groyne" increase in erosion

Figure 2.3 The externalities associated with coast protection

In consequence, the economic value of allowing an area of land to continue to erode will, sometimes, exceed the value of protecting that land, because of the effects and values 'downstream'.

In setting the effective boundaries of the analysis, then, the rule is to determine whether any consequent amplifying or compensating change occurs beyond a proposed boundary. Checking for compensating changes is particularly important since, if the market were perfect, all the changes locally would be balanced by compensating changes elsewhere. For example, if the market were perfect all visitors to a beach threatened by erosion would be able to transfer their visits to another beach, gaining precisely the same enjoyment at exactly the same cost. However, coastal recreation does not equate to a perfect market, and the extent of compensating changes will depend on the characteristics of each coastal site.

A second problem in setting the boundaries of the analysis occurs when, as is the case for both recreational and non-use values, the beneficiaries are a diffuse population. If the value of the change is estimated per visitor then to estimate the total economic benefits it is necessary to multiply this value by the number of people who benefit: to 'gross up' this value.

With non-use values, the problem – in principle – is even more complex in that an individual can accord a site a non-use value without visiting the site. There is, in consequence, no reason why a resident of John O'Groats should not have a non-use value for Land's End; indeed, it is very likely that they have! Were an estimate of the non-use value of preserving Land's End to be required, then *a priori* the sampling frame would at the minimum be the whole British population.

Whilst some studies of non-use value (Willis, 1990) have detected a distance-decay function – the further the individual is from the site, the lower the non-use value he or she places upon it – this implies the anomalous conclusion that an important site far from a large centre of population would have a very low value.

2.4.2 Measuring the change

A change will affect both the stock of a resource and the flow of goods from that resource. However, since the value of some stock is the discounted value of the future flow of goods from that resource, care must be taken not to include both changes in the analysis since this would be double counting. Thus, if some agricultural land would be lost through erosion, then either the shadow price of the lost agricultural production, or the shadow price of the land, is an appropriate value of that loss, not both.

For sea defence and coast protection projects, the measure of the loss that is most appropriate may change as erosion proceeds. Initially a property may be at risk of flooding; as erosion progresses the risk of flooding increases and at some point the property is lost through erosion to the sea. Following normal practice in flood alleviation benefit-cost analysis (Penning-Rowsell and Chatterton, 1977), the losses from flooding are initially counted by flow changes: the expected average annual flood losses.

When the property becomes uninhabitable through erosion, the stock change – the loss of both the property and the land upon which it is sited – is used to measure the loss. However, before the property is lost through erosion the discounted value of average annual losses to that property are likely to exceed the capital value of that property. At this point, the owner seeking to minimise his or her losses would choose to abandon that property, and to use annual average damages, rather than the capital value of the property, would overestimate the loss.

A change may be either marginal or result in a change of state: the losses consequent upon the latter are usually much greater than the former. Therefore, it is more important to identify and measure changes of state. A trivial example of a change of state is the case of a house affected by erosion: the whole value of the house and land is lost once erosion reaches the safety margin for that property. Conversely, the gradual erosion of a clifftop recreation area is more likely to result only in a marginal change in the recreational value of that site until it becomes entirely unusable. That is unless, for example, one fairway or green of a 9-hole golf course is lost: an 8-hole golf course has relatively little value.

Major changes of state are likely when the land behind a protection structure is low-lying, or involves a geological feature such as a shingle bank or sand dunes. Sites of ecological significance are particularly likely to suffer changes of state, rather than a marginal change, in consequence of even an apparently small physical change.

What, in principle, it is necessary to measure is any change for which at least one individual has a preference. If a change does not affect at least one individual's preference, then no economic gain or loss has occurred. There are a number of consequences. It is not always self-evident that anyone has a preference one way or another about a change, or whether they prefer the situation before the change to that after the change, or even whether they will notice that the change has occurred. Hence, in developing methods of evaluating the economic loss resulting from the erosion of beaches it was first necessary to determine what characteristics of beaches contribute to the individual's enjoyment of that beach (Section 4.3.1), and then proceed to their valuation.

A second problem area is with sites of ecological or heritage significance. Specialists in these areas have developed their own methods of assessing the scientific importance of a site (Section 6.3). However, unless the basis of these preferences reflects the opinions of lay people, then those sites that the public values and wishes to preserve and those that specialists value may well be very different. Furthermore, the economic value of those sites which only a few experts seek to preserve would be low because those experts comprise only a very small proportion of the population.

For example, it might be suggested that the public's apparent enthusiasm for the conservation of plants and animals is, in reality, limited to those which are attractive: there is a preference for furry mammals and birds, rather than beetles. Fortunately, this perception appears to be misplaced. In a survey of approximately 300 households living inland from the coast, respondents were asked to rate different characteristics of both nature reserves and archaeological sites in terms of their importance as reasons for conserving those sites (Green and Tunstall, 1991b).

For both types of site, the relative importance given to the different reasons (Tables 2.2 and 2.3) correlates quite well with the relative importance specialists seek to attach to the different features (Ratcliffe, 1977). It is particularly noticeable that the respondents attached least importance to those features that might be expected to add to the recreational value of those sites. This suggests that they give greater importance to the non-use values than to the use values of these sites.

Table 2.2
Judged relative importance of different factors in deciding which archaeological and heritage sites should be preserved

	Mean	Standard Deviation
Its educational value to school children	4.1	1.0
That there are very few similar sites like it in the country	4.0	1.0
That the site has not yet been fully investigated	3.7	1.2
How old the site is	3.4	1.2
Its association with important historical figure (e.g. William the Conqueror)	3.3	1.2
The state of preservation	3.2	1.3
The ease and cost of preserving the landscape	3.0	1.5
How much it adds to the landscape	2.9	1.4
The number of people who visit it	2.7	1.3

Scale:
    0 = least important
    5 = most important

n = 327   alpha = 0.7914

Note: Alpha is a measure of the agreement between respondents and can have a maximum value of 1.

THEORETICAL PERSPECTIVES 25

Table 2.3
Judged relative importance of different factors in deciding which nature reserves should be preserved

	Mean	Standard Deviation
It contains wildlife or plants that are disappearing in the UK	4.5	0.8
It contains a very rare species of wildlife or plant	4.3	1.0
It includes a natural landscape rather than a man made landscape	4.1	1.1
The wildlife or plants it contains have always been rare in Britain	4.1	1.0
The variety of wildlife and plants it contains	4.0	0.9
The wildlife and plants it contains are typical of the countryside as it used to be	3.8	1.1
The reserve contains a large proportion of the plants and animals of that kind in the UK	3.8	1.1
There are no other sites like it locally	3.8	1.2
It contains wildlife or plants that are attractive to look at	3.5	1.2
The amount there is to see when visiting	3.1	1.3
The number of visitors to the site	2.7	1.3

Scale:
0 = least important
5 = most important
n = 327   alpha = 0.8345

Note: Alpha is a measure of the agreement between respondents and can have a maximum value of 1.

2.4.3 *The valuation of the change*

Before discussing the available techniques for estimating the economic value of a change (Section 2.5), it should be re-emphasised that the validity of such values is dependent upon acceptance of the axiom that value is solely assigned by the individual. Those who assert that value is ascribable to a good for other reasons, such as an inherent right of a species by reason of existence (Brennan, 1988; Callicott, 1985; Fox, 1990; Regan, 1981; Rolston, 1985), will not accept economic values applied to those goods.

Economic values are also descriptive: the techniques described below estimate either what individuals have been willing to pay in the past for a good, or what they say they are willing to pay for that good today. However, classically 'is' does not prescribe 'ought', and hence it is a legitimate objection to an economic analysis to say that although the values may be accurately estimated, people ought to value the good more (or less) than they do.

Similarly, economic analysis presumes a consensus as to the direction of preferences for a change (Green and Penning-Rowsell, 1989). However, in some cases of coastal erosion, there may be no such consensus as to the desirability of one or more of the component changes involved. For example, when the Hengistbury Head case study was undertaken (Chapter 8), the geomorphological and ecological changes anticipated were regarded as undesirable. More recently (Coker and Richards, 1992), the view has been expressed that these changes might have been preferred, and that 'nature should have been allowed to take its course'.

2.5 Basic methods of evaluation

There are several basic techniques for estimating the economic value of a particular quantity of some good: these are listed in Table 2.4.

Table 2.4
Basic techniques for the economic evaluation of goods

Technique	Uses and limitations
Market prices	Only possible for private goods; depends upon the existence of a perfectly competitive market
Shadow prices	Standard method for the valuation of priced goods: prices are 'corrected' to those which would occur if the market were to be perfectly competitive
Travel Cost Method	Indirect method which can only be used to evaluate recreational benefits
Hedonic Price Method	Indirect method which can only be used to evaluate amenity benefits
Contingent Valuation Method	Direct method which can be used to evaluate all goods
Least Cost Alternative/ 'Shadow Project'	Cost of providing the same good by other means
Case specific approaches (e.g. Environmentally Sensitive Area payments)	Can often be used to set lower or upper bounds on the value of goods

2.5.1 Market prices

This is the simplest possibility. Under the very restrictive conditions of the perfectly competitive market, the resulting equilibrium prices measure the marginal value of the goods. In a perfectly competitive market, the equilibrium price reflects the cost of supplying the last unit provided, not simply the average cost of providing the total number of units. Equally, the price obtained reflects the marginal value of the last unit consumed (i.e. bought).

Unfortunately, perfectly competitive markets are very unusual. Consequently, market prices are usually only the first stage towards estimating 'shadow prices' – those prices that would exist if the price to the consumer did reflect both the marginal cost of supply and the marginal value.

2.5.2 Shadow prices

Markets can be imperfect for many reasons, including the effects of monopolies, taxes and subsidies, and imperfect information for consumers as to the availability of goods. In consequence, in almost all cases, market prices have to be corrected.

For example, the effects on prices of Excise Duties and Value Added Taxes, subsidies, and production or import restrictions should usually be removed. Similarly, when estimating the contents losses from flooding, it is usual (Penning-Rowsell and Chatterton, 1977) to use shadow prices, calculated as the average remaining value of the goods damaged, because the second hand market for household contents is believed to be distorted (Akerlof, 1970). Detailed discussion of the problems of deriving such shadow prices are given in the standard texts on project appraisal (Little and Mirrlees, 1974; Squire and van der Tak, 1975).

One other area where it is necessary to use shadow prices is for agricultural land values. Agricultural land values are inflated by the existence of the European Community's Common Agricultural Policy. Since land prices reflect the market prices of the production from that land, the subsidy element of the production is reflected in land prices.

Again, the approach is usually piecemeal. The shadow prices of goods affected are estimated; the quantity of goods affected is then multiplied by the unit shadow price; and the results are summed to give the total value. Since the approach can only be applied to goods for which there is a market price, it is partial rather than comprehensive. It should be used sparingly, that is only when there is good reason to believe that market prices are grossly distorted.

Both the market price and the shadow price techniques can only be used for priced goods. It is sometimes supposed that the creation of a market in a good will necessarily result in an increase in economic efficiency. This is not, however, true, either for public goods as a whole, where prices will not lead to economic efficiency, or sometimes for private goods.

2.5.3 The Travel Cost Method (TCM)

The Travel Cost Method (TCM) has been used extensively to evaluate the recreational benefits from different types of site, including beaches (Caulkins, Bishop and Bouwes, 1986). First proposed by Clawson (Clawson, 1959; Clawson and Knetsch, 1966), the procedure is limited to the evaluation of the recreational benefits of a site **as it is now**. The procedure is no more than a regression analysis of two variables: visitor rates to the site and visitor origin. Consequently, as will be discussed in more detail below, the method has two advantages to the analyst: a valuation is guaranteed and the analyst has the opportunity to choose the value within sometimes wide limits.

The basic assumption underlying the procedure is equivalent to the gravity model developed in nineteenth century geography (Haggett, 1965) and later applied in planning models (Wilson, 1968). This assumes that the number or frequency of visits to some destination from some origin depends upon the relative attractiveness of that destination compared to other destinations and the 'frictional' resistance of the distance between the origin and site: the opportunity costs of travelling to the site. In gravity models, this friction is measured by the travel distance whilst in the TCM, some surrogate measure of disutility of travel is employed, usually the cost of undertaking the journey. This cost of travel usually increases as some function of the distance of the origin from the site.

For a given site, it is assumed that the preferences for visiting that site follow some statistical distribution over the population. In addition, as with other goods, it is assumed that the marginal utility of a visit decreases: the n_{th} visit in a given period of time gives more utility than the n_{th+1} visit. It is also assumed that the equivalent of a perfect market equilibrium exists: individual visitors visit at frequencies such that individually and collectively, for the last visit undertaken, the costs incurred equal the marginal value – enjoyment – obtained from that visit. Given this assumption, it follows that the area under the curve of visit rate graphed against travel distance equals the consumer surplus. This area, therefore, can be estimated theoretically by regression analysis, with the visit rate being regressed upon travel distance.

In terms of a basic preferences-constraints model, preferences are assumed to be determined by some characteristics of the population, as are some constraints, such as income. The main constraint considered is then distance to the site; these factors, together with the availability of substitute sites, are considered to determine visit frequencies. As a method it is inductive: by manipulating data on the population in the different zones and their characteristics, along with travel distances and visit rates, an attempt is made to draw some conclusions as to the demand for a particular site. In principle, therefore, the objections to the inductive method made by Popper (1968) apply to the Travel Cost Method.

To apply the method, it is necessary to know as a minimum the origins of visitors to the site. Visitor origins are determined from a sample survey, either of visitors to the site or of residents of potential origins. The latter is preferable but much more expensive. In the former case, the normal problems of obtaining a sample representative of different visiting patterns occur (Tourism and Recreational Research Unit, 1983), as these vary over the week and year.

Conventionally, origins are aggregated as census zones in concentric circles or polygons centred on the destination. If, as is likely, there are differences between identifiable sub-groups of the population in the strength of their preferences for the site, and these sub-groups are differentially distributed between origins, then additional data is required. If, for example, different socio-economic groups have different preferences, then we require to know the breakdown of the population by socio-economic group for every origin zone. These additional parameters are then entered into the regression analysis.

In the absence of any empirical data on the determinants of preferences and constraints, and generally the absence of any hypothesised causal model, the selection of variables for inclusion is somewhat arbitrary. Where the predominant usage of a site is as a beach for informal recreation, or a local park, it may be hypothesised that the disabled are less likely to use the site and that families with children are somewhat more likely to use the site.

Whilst higher income should be expected to be associated with greater usage, partly through its association with a greater propensity towards car ownership and travel, income may be positively or negatively correlated with preference towards particular types of site. Socio-economic class, however, is not likely to be a useful explanatory variable because the classification system mixes both income and educational attainment, and the two are not necessarily correlated (Tunstall, Green and Lord, 1988).

Normally, the ratio of visits to population from the different zones is calculated simply as a combination of the proportion who visit the site at all and the number of visits made each year by those who visit the site. Since the analysis is usually based upon visitor sample data rather than population data, there is little choice in the matter. However, in principle, it is undesirable to assume that there is no real difference in the preferences of those who have never visited a site from those who visit only very infrequently: the assumption being made is that the population preference function is continuous and, thus, were the costs sufficiently low, everyone would visit a given site.

Applying the TCM requires the acceptance of several assumptions and confronts the analyst with a number of problems. These have been reviewed by several authors and are not repeated here in full (Common, 1975; Cheshire and Stabler, 1976; Duffield, 1984; Gibson, 1978; Harrison and Stabler, 1981).

The estimation of the opportunity costs of travel presents some problems. These opportunity costs are usually treated as having two distinct and separate components: the resource costs of travelling and that of the time taken up by travelling. The valuation of travel time has been a persistent problem in economics which is by no means resolved (Chevas, Stoll and Sellar, 1989). The estimation of the resource cost component of the opportunity cost of travel appears somewhat easier, but only in so far as the analysis is restricted to priced travel. What this component might be for those who walk or ride a bicycle to a site is problematic, and most local visits to the coast are made on foot.

Since travel costs generally rise according to distance from the site, to use it as a surrogate for value implies, first, that the value of visits undertaken from distant origins is greater than for origins nearer the site. Thus, it implies that some increasing functional relationship should exist between utility and distance or visit cost. This assumption has been shown to be untrue for some coastal sites (Green et al., 1990). Similarly, it implies that visit frequencies per household should be inversely related to distance to the site.

As described, several assumptions must be met if the results of an analysis are to be valid and all have been questioned and found to be invalid in some circumstances. First, all forms of consumption take a given quantum of time: the enjoyment of a visit does not result from simply reaching a site but upon spending some time at the site. Similarly, consumption depends upon the joint availability of the good and time for consumption.

The increase in real income and the supplies in goods suggest that consumption may be limited by time constraints as much as by income: on a two week holiday, one cannot undertake more than 14 day trips. If the number of recreational visits undertaken are constrained away from the point where marginal value equals marginal cost, then the estimate of the consumer surplus will theoretically be distorted. However, since the time constraint cannot be relaxed, this estimate of consumer surplus may be acceptable.

In addition, the method values the enjoyment gained from a visit on the assumption that in the event, equilibrium was achieved: perfect information is assumed as to the costs that will be incurred and the enjoyment that will be obtained. It is an ex-ante rather than an ex-post evaluation, and the results will be in error to the extent that visitors have imperfect information: if the costs of undertaking the visit were found to be higher than anticipated, or the enjoyment from the visit was less than hoped, then the results will be flawed. Information is likely to be better when visitors are making repeat visits, rather than first visits, and the method assumes that visitors set out with the intention of visiting a specific site. However, the number of first visits to a site may be relatively high (Tunstall *et al.*, 1990; Fouquet *et al.*, 1991) and, in the United Kingdom, a similarly high proportion of day trips are undertaken as recreational motoring, without a specific destination initially in mind.

Secondly, the travel distance or cost is assumed to involve disutility. This involves several problems and issues. Costs are generally assumed to rise as some linear function of distance and to have two components: resource costs and time costs. In evaluating these, we require to use perceived values of both, which require to be determined. For resource costs, it is likely that the public perceives the costs of travelling by car as simply the marginal costs of fuel and oil consumed, rather than including vehicle depreciation, etc.

It is usually assumed, moreover, that the spatial friction is homogeneous. However, congestion, and bridge, tunnel or road tolls are all likely to be included into the individual's cost function. Such considerations may also influence the individual's choice of which site to visit. Inhabitants of north London, for example, may have been deterred from visiting sites which require travelling through the Dartford tunnel not only because of the anticipated time delays but also because of the unpleasant driving conditions imposed by congestion.

Because it is necessary to use perceived resource and time values, the relationship of perceived cost to perceived distance is unlikely to be continuous. For long journeys, it is likely that visitors include the cost of overnight accommodation. Moreover, the position of food and drink consumed on the journey is questionable, given that the issue is the perceived cost of the journey before it is undertaken, rather than the economic cost. However, the problem of discontinuities in travel cost functions, where known, can be handled fairly easily by analysing the sub-populations such as day and staying visitors separately. Again, some of these issues could be resolved by using empirically derived data rather than relying upon untested assumptions.

Moreover, it has been pointed out that the journey may itself have utility (Cheshire and Stabler, 1976): the visitor may have chosen one site over another because the journey to it is scenically attractive, and consequently they gain utility from the journey. Or they may choose to take a longer route to a site because it is scenically attractive in spite of the higher costs involved. If travel time is assumed to have disutility, then the value of the site will be overestimated in those cases (Chevas, Stoll and Sellar, 1989) and a key assumption of the travel cost method will have been violated.

Theoretically, for the working population, the hourly value of leisure time at the margin equals the individual's wage or salary, net of deductions. This follows from the assumed equilibrium condition between work and leisure, where the number of hours the individual chooses to work in a week is such that marginal income received from the last hour worked just equals the marginal value of the hour of leisure given up. This assumption does not help at all in valuing the time of the non-working population.

Moreover, it is a surrogate measure for the disutility of the journey where the disutility is assumed to be a linear function of travel time. Quite clearly, the non-working population may also experience disutility from the journey. More seriously, there is little empirical evidence to show that the disutility of a journey is a linear function of travel time. In particular, given that any visit consists of a journey out and a return journey back, the (dis)utilities of both journeys need not be equal. It is further assumed that the utility of the journey does not change irrespective of the frequency of visiting.

In general, it is not known to what extent any of these conditions is true in the particular case being analysed and so a generalised value of time has to be assumed.

Theory does not, moreover, prescribe the functional relationship to be expected between visit rates and distance: consequently, any one of several different functional forms may be applied (Rosenthal, Loomis and Peterson, 1984), since any function may be fitted to any set of data more or less imperfectly (Aachen, 1982). Similarly, the confidence intervals about the curve are greatest at the extremes of the curve (Wetherill *et al.*, 1986) and it is precisely upon these points that the estimate of the area under the curve depends most crucially. The different functional forms give different estimates of the area under the curve – the consumer surplus – to the recreational value of the site. The most commonly adopted functional form is the log-linear. However, this is asymptotic to the distance axis and, consequently, it is usual to limit the curve to some distance when estimating the area under the curve (Clawson and Knetsch, 1966).

Harrison and Stabler (1981), draw the apt but perhaps overly dismissive conclusion by comparing the use in recreational studies of the TCM to a 'supermarket manager who would not expect to determine the elasticity of demand for one of his products by a study of the distances travelled by his customers'. Nevertheless, in the United Kingdom, where competing recreational facilities are located close together and many journeys are by foot, this conclusion seems largely fair.

However, in the very different context of the United States, it might yield useful results. Since, in its current form, even in the United States, the TCM is an inductive procedure, rather than a theoretically grounded and experimentally tested method, its validity remains to be determined. A pre-requisite of such use is the derivation of some general equation that can be used reliably to predict visits to sites other than those for which it was derived. At present, analyses are undertaken upon a site-by-site basis, rather than the opportunity being taken to test any basic hypothesis.

In addition, arbitrary assumptions as to the determinants of recreational behaviour must normally be made to apply the method in the absence of adequate empirical data. Since social surveys are needed to form a sound basis for the procedure, the Travel Cost Method may be best considered as embodying a partial hypothesis upon which to base Contingent Valuation Method studies. The advantages and disadvantages of the method are summarised in Table 2.4. As it can only be applied to the recreational benefits of a site, it is a partial method and must be applied in a piecemeal fashion along with other techniques to estimate the other components of value.

2.5.4 The Hedonic Price Method (HPM)

The Hedonic Price Method (HPM) is a technique that has been used to separate out some components of amenity gains or losses from other determinants of house prices (Rosen, 1974). It has been used in a number of topic areas. For example, it has been used to estimate the effects of aircraft noise on house prices (Nelson, 1979), and the premium that location close to a park or other recreation facility attracts to a house (Li and Brown, 1980). The method has also been used to gauge the effect of floods and other risks (Donnelly, 1989; Shabman and Damianos, 1976), and for assessing the effects of air or water pollution on house prices (Brookshire *et al.*, 1982; Wilson, 1984).

The basic assumption underlying the method is that any house is a bundle of attributes, and the price fetched by the house reflects the desirability of the particular combination of attributes offered by that house. The attributes may, for example, include proximity to a river corridor or a sea view. If the prices fetched by different houses, representing a wide range of combinations of attributes, are regressed upon the values of each attribute then it should be possible to separate out the relative value attached to each attribute.

However, what amenity gain or loss is involved in a particular case can be unclear. For a sea view, or location on an eroding cliff, what is measured by the method presumably includes the value of visits to the coast, views out over the sea, and perhaps the knowledge that no undesirable development will take place on that site. However, there is likely to be a high degree of multi-collinearity between these effects: they are all likely to be highly correlated. With such high multi-collinearity between variables, it is difficult to obtain reliable estimates of the contributions of any one individual variable (Wetherill *et al.*, 1986).

The lack of clarity as to what is being evaluated has two implications. There is, first, the risk of double counting, for example, the amenity benefits from living near the sea and the recreational benefits to the local population. Secondly, the regression equation itself may be mispecified by omission of important variables.

However, as a market, the housing market is the most complex to which to apply this procedure. Clearly, it should, for example, be possible to account for the differences in the prices of different models of cars by regressing the prices of the different models upon the different characteristics (e.g. reliability, speed, safety, economy). However, unlike the market for new cars, the demand for houses is not met by production but largely by the resale of houses from the existing stock. Since price is a function of supply and demand, it is necessary to consider those conditions under which people will be willing to sell their existing house, as well as those combinations of attributes that a potential purchaser seeks in a new house. Or we could assume that an individual sells one house simply because another offering a more desirable combination of attributes is offered for sale, since most purchasers must sell their existing house to buy another.

Consequently, the Hedonic Price Method would be most free from such distortions in the market for first homes. However, this is only a small proportion of the total market and, because of development constraints, the range of attributes offered, especially site characteristics, is likely to be distorted. Housing markets are also 'sticky': the transaction costs of buying a house are significant, and much greater when an existing property must be sold first.

House prices in the United Kingdom are also badly distorted by fiscal factors. Periodically, the housing market is subject to high rates of inflation. The Bank of England Housing Model (Dicks, 1989) suggests this is primarily due to losses of control over money supply to the housing market: the combination of mortgage tax relief, the income multiplier applied to mortgage limits and the use of interest rates as the sole economic regulator, all cause prices to inflate in some periods without leading to the corresponding deflation in others. 'Mortgage leakage' is a further complicating issue; this is the practice of people borrowing more than is required to pay the additional cost of the new home over the price reached for the home that has been sold, so as to use the low cost loan to buy other goods.

Since houses are seen as both sources of consumption and as investments, with the pure consumption element being a small fraction of the annual cost of a house (Department of Employment, 1990), this also adds a further complicating factor. As an investment, the rational individual will consider what he or she believes to be other peoples' preferences in choosing to buy a house. If preferences were homogeneous, this would not be important: that preferences are not apparently homogeneous (Munro and Lamont, 1985) is in reality a more serious problem with the use of the Hedonic Price Method.

In principle, the regression analysis should be informed by some empirically grounded theory as to the attributes of a home desired by individual occupiers. This bundle of attributes is likely to include characteristics of the neighbourhood as well as the home itself and the immediate environment. Since the Hedonic Price Method is another inductive method, and thus relying upon existing data, adequate surrogate measures for the perceived attributes must also be available from existing data sources. Omissions of variables will lead to mis-specification of the regression equation.

Whilst a great deal has been published upon the determinants of residential satisfaction and residential preferences (Hourthan, 1984), Hedonic Price Method analyses have not necessarily been informed by such studies. In an analysis of some fifteen Hedonic Price Method analyses, Freeman (1979) remarked that: 'the selection of explanatory variables seems to be almost haphazard. . . . Convenience and data availability appear to be the major determinants of this part of model specification'.

In preparing this Manual, many Hedonic Price Method studies in the United States were reviewed. We discovered very little agreement between studies either as to the explanatory variables to be included or in the functional form of the equations to be adopted. If, however, the basic hypothesis is correct, then within each homogeneous housing market there should be a single explanatory equation that can be used to evaluate all the components of the bundle of housing attributes. Lack of consistency between the models adopted must, therefore, cast doubt upon the validity of all, or imply that homogeneous markets are highly localised. There is some doubt whether adequate surrogate measures for some attributes can be obtained. It is possible to measure the quantity of public open space from planning data, but it is less clear how the quality of this open space might be assessed.

In summary, whilst the HPM has a face validity for use in a perfectly competitive market, market imperfections are likely to limit its use. Neither, as yet, has a reliable method been established that has been tested by replication. Therefore, to value the property loss resulting from erosion, the shadow price approach is to be preferred.

2.5.5 The Contingent Valuation Method (CVM)

The Contingent Valuation Method (CVM) can, in theory, be applied to evaluate the use and/or non-use values for any good. It essentially involves a social survey approach: that is, a sample of respondents is asked by way of an interview survey or postal questionnaire what value they place on a particular good. Usually they are asked to do so by stating how much they would be willing to pay, by a specific payment mechanism, for a clearly specified change in the availability of that good.

The method is widely used in the United States, where much of the basic research work was done under contracts from the US Environmental Protection Agency. The US Water Resources Council (1983) and the US Corps of Engineers (Moser and Dunning, 1986) have both published guidelines on its use. However, the Corps guidelines have been extensively criticised (Tunstall, Green and Lord, 1988; Mitchell and Carson, 1989). Major methodological and theoretical reviews have been published, largely based upon United States experience (Bateman *et al.*, 1991; Cummings *et al.*, 1986; Mitchell and Carson, 1989; Tunstall, Green and Lord, 1988). Recent theoretical and methodological developments are discussed in the papers edited by Petersen *et al.* (1988). Studies in the United Kingdom using the method are listed in Green *et al.* (1990).

Because the CVM is essentially a social survey methodology, the established social survey procedures must be adopted, including rigorous sampling design, fieldwork control and questionnaire design (Tunstall, Lord and Green, 1988). So, too, must the social survey practices of establishing a questionnaire's validity and reliability be adopted (American Psychological Association, 1954). Here testing for validity involves establishing that the questionnaire measures what it is intended to measure, and testing for reliability involves establishing the questionnaire's replicability and absence of measurement error.

From a scientific viewpoint, these characteristics – validity and reliability – are basic criteria to be applied to any measurement method. It is a virtue of the CVM, as opposed to the Travel Cost and Hedonic Price Methods, that such tests can be applied. Nevertheless, it is essential, particularly at this stage in the development of the CVM, that all applications embody basic tests of validity and reliability rather than make unsubstantiatable assumptions in these respects.

So far, whilst tests of the reliability of the CVM have proved satisfactory, tests of its validity have not yet been wholly so (Green and Tunstall, 1991a). Often, the proportion of the differences between respondents' stated willingness to pay that can be accounted for by differences in the theoretically expected causal variables has been below 20 per cent. A reasonable target, based upon experience in other areas of social survey work (Ryan and Bonfield, 1975), would be that 40 per cent explained variance is achievable and desirable in the longer term. Mitchell and Carson (1989) suggest that 25 per cent should be the current target for a well designed CVM study.

However, other tests of validity have proved more satisfactory. This applies, in particular, to tests of divergent validity. This is the test that, in this case, shows that the evaluation of different goods using the same willingness to pay questionnaire yields different values, rather than the same, and presumably arbitrary, value (Green *et al.*, 1990).

At this stage on the learning curve with the CVM (Mitchell and Carson, 1987), the indications are consequently that the method is fundamentally valid, but not yet very accurate. Indeed, one basic problem with the method is that, whilst there are many different ways of formulating the question of willingness to pay, there has not yet been a baseline methodology study to determine that form which has the greatest validity and reliability (Green and Tunstall, 1991b). Additionally, the use of the methodology has exposed several theoretical problems with the underlying neoclassical theory. In particular, there is the unpredicted discrepancy between the amount an individual is willing to pay for a given change (**'willingness to pay'**) versus the sum that the same individual would demand in compensation for an equal and opposite change (**'willingness to accept compensation'**).

In consequence, great care and time is required to undertake CVM studies. As a method, depending upon how the change is defined, it can be used in either a partial or a comprehensive way, either to derive a holistic value or a piecemeal valuation of changes in individual goods. In this respect some guidelines for the application of the CVM are given in Chapters 4 and 6. The minimum requirements for reporting CVM studies are the inclusion of the instrument used – the questionnaire – and reporting the results of the validity tests undertaken (Green and Tunstall, 1991b). A comprehensive analysis of the methodological issues inherent in the CVM concluded that the method is not yet sufficiently mature to be used as an 'off-the-shelf' technique, and that it should only be used in a research context (Bateman *et al.*, 1991).

2.5.6 *The Least Cost Alternative method*

The economic value of providing a given quantity of a good in one way cannot exceed the cost of providing exactly the same quantity and kind of good in the cheapest alternative way. This is a powerful concept for evaluating some goods in some instances, but clearly it is a principle rather than a specific technique. It is also usually used when other methods are not available, or not feasible within the resources available. It is usually applied by determining what would be the opportunity cost of using other inputs to provide the same quantity of goods. The procedures used in this Manual to value the loss of roads and infrastructure through erosion are an example of this method.

The limitation of this method is clearly that it has to be assumed that the marginal value of the goods exceeds the costs by the alternative method, otherwise those goods should not be supplied at that cost. Thus, for example, the economic value of land in a National Park is not its market price in its current permitted uses, but that which would pertain if there were no planning controls. The presumption is that the reason for planning controls is that maintenance of the park in its current use has a value at least as great as the value that could be obtained from any other use. So, similarly, the land value of a SSSI is not its market value as an SSSI but its value in the next highest alternative use to that proposed.

Again, if a historic building were to be lost as a result of some change such as coastal erosion then one bound on the value of protecting that building is the cost of disassembly, removal and reconstruction elsewhere.

A special sub-class of the least cost alternative approach is the concept of a 'Shadow Project' (Klassen and Botterweg, 1983). This method is usually holistic and comprehensive. It has been developed specifically to apply to the evaluation of changes, and particularly damage, to sites of ecological significance. The value of this change is then estimated as the cost of creating, or re-creating, exactly the same ecosystem in question elsewhere. The loss of an area of marshland would thus be valued as the cost of buying the same area and type of land elsewhere, and then establishing the same ecosystem.

There are some problems with this concept, notably the difficulty of establishing exactly the equivalent ecosystem. A second problem occurs where part of the value of the present site lies in its history, or its contiguity with a second site nearby, or because it has remained relatively undisturbed over a long period. Arguably the value of the Arctic Char in Lake Windermere would not be adequately represented by the costs of re-establishing them in another lake, simply because part of their significance lies in their uniqueness as a relic of past glaciations.

2.5.7 Case specific approaches

Turner and Brooke (1988), when evaluating the benefits of coast protection at Aldeburgh, used payments by government to farmers under the Environmentally Sensitive Area payment scheme to estimate the conservation value of agricultural land that might be lost to the sea.

This is a case where a geographically specific system of payment has been made for a combination of nature conservation and amenity reasons. These payments can be used as a surrogate for the value of those resources.

There are problems with this approach, however, concerned with determining exactly for what the payment is made. If the reasons are other than the nature conservation value of the resource, then the use of these payments as a surrogate for that value is questionable. Thus in the case in point, it is often unclear within the politically somewhat arbitrary agricultural support systems what is the basis of any one individual payment. Nevertheless the figures can perhaps give a lower bound for conservation value. At the very least such payments can be shown to have the support of government and therefore to be social/economic rather than private values.

2.6 Application to sea defence and coast protection

2.6.1 Sea defence

The basic principles involved in the evaluation of sea defence benefits are the same as those routinely used in the assessment of the benefits of flood alleviation projects.

The benefits of a sea defence project are the expected value of the losses that would otherwise be anticipated to result from flooding. This expected value is calculated by estimating the losses that would result from each of a series of events of differing return periods, or exceedance probabilities, usually up to the proposed design standard for the project. If the event losses are then plotted against the reciprocal of the return periods of the events, then the area under this curve is the expected value of the losses that will be avoided by the project during each year of its life: the **average annual benefit**. Discounting this stream of benefits over the anticipated life of the project, therefore, gives the present value of the benefits of protection.

Rather than limit the range of events considered to some pre-planned design standard, it is preferable, for two reasons, to calculate the event losses for a wider range of events. This applies to both the 'with project' and the 'without project' options. The annual average benefit is then the area between the two curves and is discounted in the usual way.

The first reason for preferring this approach is that some project options may reduce the extent or depth of flooding from events greater than the design standard. To the extent that such a reduction does occur, these **Above Design Standard Benefits** are legitimately countable as part of the total benefits. Conversely, when the project would fail under an event of some magnitude it is desirable to check whether the consequences of the resultant flooding will be more severe than would be the case in the absence of a project. Equally, any damages that are anticipated to be suffered by the project from events less severe than the design standard event should also be netted out of the average annual benefits (e.g. the costs of annual maintenance). The second reason for preferring this approach is that design standards should not be set arbitrarily but at that level which offers the greatest economic efficiency.

Compared to riverine flood alleviation benefit assessments, those for sea defence projects involve some additional complexities as discussed below (Sections 2.6.2 to 2.6.3).

2.6.2 Sea defence: estimating event losses

A flood can potentially cause damages to property and their contents through several mechanisms. For river flooding in the United Kingdom, the predominant damage-inducing mechanisms are depth and duration of flooding. Sea flooding is more likely to involve three additional damage inducing mechanisms, so that the damage from a given depth of flooding may exceed that from river flooding.

First, sea water contains at least one contaminant, namely salt. Inundation in saltwater has been shown to cause more damage than would inundation to a similar depth in freshwater (Chapter 5; Penning-Rowsell, 1978). Secondly, some properties may be exposed to the effects of spray, and more particularly the effects of debris borne by the waves. Thirdly, in some areas the depth and velocity of the floodwater may be sufficient to result in structural damage and perhaps failure. This is typically the case for caravans and chalets. However, given that the customary response by the public to a flood is to shelter in their home, project assessments should include an examination as to whether any properties are likely to suffer partial or complete structural failure.

In estimating the benefits of sea defence projects, consideration should also be given to the potential loss of life that might occur. In this Manual no attempt is made to place a 'value on life' (Dalvi, 1988), not least because there is no adequate basis upon which to estimate the number of lives that might be lost in a given event at a given locality. However, a check list of factors that are likely to contribute towards the risk to life is given (Section 5.4.5).

2.6.3 Sea defence: estimating annual average benefits

Compared to riverine flood alleviation benefit assessment, there are two added potential complications in assessing the benefits of sea defence arising from the frequent association between the flood risk and coastal erosion.

One effect of erosion is that the risk of flooding to areas further inland increases over time and thus annual average benefits vary over time (sea level rise has the same effect).

Secondly, if erosion is unchecked, land that is now at risk of flooding will first become unusable because of the frequency with which it is flooded, and then it will be lost through erosion. So, for the first part of the time horizon, the benefits of protecting a property equal the annual average benefits; then there occurs a one-off capital loss when the land becomes first abandoned and then eroded.

2.6.4 Coast protection

Coast protection is a misnomer: what these projects can hope to achieve is to slow down or prevent erosion during the life of the project, or until the project is destroyed by an extreme event. So the erosion is simply deferred and the 'protection' is for a finite period only, whereas the word implies that it is for ever. The benefits of any coast protection project are essentially the difference between the losses that will be experienced without the project compared to the same losses occurring at some time in the future with the project (Chapter 3; Thompson *et al.*, 1987b).

Whereas there are few, if any, benefits caused by flooding by the sea, the situation is different for coast protection. Continued erosion is desirable for some types of sites, particularly geological and palaeontological sites but also some types of ecological site. Hence there may be negative benefits, or disbenefits, to a project in consequence of a reduced rate of erosion, or as a result of the process of project construction.

Table 2.5 summarises the likely direction of the impacts of erosion, the construction process and the project upon different types of site. This table also summarises the nature of the use and non-use values associated with each category of site and the methodology recommended, if any, to estimate this value for a particular site.

In contrast to analysing the direction of the impacts, it is more difficult to generalise about the relative magnitude of the different benefits in so far as it is currently possible to value them. However, some general rules of thumb may be as follows:

1. A 'change of state' change will result in more benefits than any marginal change;
2. Agricultural benefits will rarely be of any significant magnitude (and probably only for high grade agricultural land about to be affected by a change of state), except potentially for their option value;
3. The recreational benefits of protecting beaches are generally large but depend critically upon the number of visitors using that beach and the rate of its deterioration;
4. Buildings are not usually worth protecting before they are within a few years of being lost, unless they are very valuable (e.g. large hotels); and
5. In general, when the loss is likely to occur is critical to the magnitude of the present value of the benefits.

The difficulty with the first rule of thumb is determining whether a marginal change or change of state will occur as a result of erosion. The change in question is to the stock of such sites or to the flow of consumption from such sites, rather than the change to the site itself. The two questions only become identical if it can be argued that the site is unique.

For example, Turner *et al.* (1992) evaluated the potential loss, by the Least Cost Alternative method, of a Martello Tower through erosion as the costs of acquiring and refurbishing another Martello Tower. In this case, the argument is that the loss of one of many Martello Towers constitutes only a marginal change. If there were only one remaining Martello Tower in the country, then it could be argued that the loss of that Tower would be a change of state. Then, one approach to evaluating that loss, by the Least Cost Alternative method, would be the cost of taking the Tower apart and transporting and reconstructing it at an alternative site.

There is sometimes scope for argument about whether a particular site is unique. Thus, many ecologists argue that all SSSIs are individually unique (Coker and Richards, 1992) so that the loss to be con-

Table 2.5
A classification of loss of sites and/or site features

Characteristics	Effects of erosion	Effects of construction	Effects of protection	Methods of economic valuation
Agriculture production functions	Destruction	Risk	Desirable	Use value – agricultural productivity or land values after subsidy elements removed (shadow price)
				Option value – upper bound set by present value of reclaiming equivalent value of land at some later date (least cost alternative)
Archaeological significance	Destructive	Risk	Desirable	Quasi-option value of unexcavated sites – not possible to value archaeological value of site – cost of rescue archaeology gives a lower bound (least cost alternative)
				Recreational value of sites (CVM)
				Non-use value – not possible to value
Coast protection function	Destructive if loss at this site exposes other structures to greater degree of attack	Unusual	See first column	Cost of improving those other defences to required standard or reduction in benefits of those defences net of the present value of maintenance costs, whichever is least. Counts as a benefit of the proposed project (least cost alternative)
	However, if loss at this site will increase sediment supply to other site, then desirable			Increase in net present value of protection at that site – counts as a negative benefit of the proposed project (least cost alternative)
Downdrift effects	Desirable	Negligible	Undesirable	Present value of expected erosion downdrift or cost of protection downdrift – whichever is less (least cost alternative)
Ecological significance	Varies – necessary to determine	Risk	Desirable – except if erosion desirable	Various ecological use values (least cost alternatives, shadow prices, and site specific methods depending on type of use value)
				Total value ('shadow project' method)
				Recreational value (CVM)
				Non-use value – not possible to value

Characteristics	Effects of erosion	Effects of construction	Effects of protection	Methods of economic valuation
Geological significance	Desirable	Risk – particularly for shingle features	Undesirable, but may be acceptable alternatives	Use value for educational trips (CVM) Cost of 'shadow project'
				Recreational value (CVM)
				Non-use value – not possible to value
Heritage significance	Destructive	Risk	Desirable	Recreational value (CVM)
				Historic value – removal and reconstruction of building; or cost of renovating/preserving an equivalent site (least cost alternative)
				Non-use value – not possible to value
Landscape significance	Normally destructive	Depends upon nature of feature	Depends upon nature of protection	Recreational value (CVM)
				Non-use value – not possible to value
Palaeontological significance	Desirable	Not usually	Undesirable	Scientific value – not possible to value
				Recreational value of sites (CVM)
				Non-use value – not possible
Moorings	Undesirable	Low	Desirable	Cost of moorings if no surplus of supply over demand
Offshore effects	Depends upon effect	None – unless dredging undertaken	Depends upon project	Any effects upon fish spawning grounds, offshore sites of archaeological or ecological significance, shipping channels or sailing and boating must be considered (shadow prices, least cost alternative)
Recreational facilities	Undesirable	Low	Desirable	Additional cost of supply elsewhere plus change in enjoyment (CVM)

Note: Some sites are likely to present all or some of these features in combination (e.g. an agricultural site may possess both a landscape value and an ecological significance, and, perhaps, some potential archaeological value). Care must therefore be taken to avoid double-counting. Equally, where separable, all features of significance require to be valued.

sidered is the loss of that site rather than the loss of one of many SSSIs: all changes are changes of state. This difficulty, and lack of consensus as to whether a marginal or non-marginal charge is involved, is one reason why we do not recommend the use of economic valuations for some potential consequences of coastal protection. In particular, we do not recommend the estimation of non-use values for specific site changes.

2.6.5 Coast-protection: non-use values of coastal sites

We have explored the nature of these non-use values through a CVM study of households at varying distances from the coast (Green and Tunstall, 1991a). Our study sought specifically to identify the following:

1. Who benefits: the definition of the population who would benefit through the preservation of particular types of site. We have researched, in particular, whether there is some distance-decay relationship for the values attached to different types of site;
2. What is the nature of the non-use values attached to those sites; and
3. What is the nature of the good that those sites offer and for which the public has a preference.

Respondents were specifically reminded of the 'greenhouse effect' and, overall, were willing to pay a substantial additional amount in 'rates and taxes' for coast protection (Table 2.6). A significant constraint upon whether the respondent was willing to make any additional contribution was the degree to which the respondent believed his or her household's expenditure was already totally committed to necessities, as opposed to being available for discretionary expenditure. The degree to which income was seen as available for discretionary expenditure also influenced the amount that an individual was willing to pay. Gross household income was less important than this discretionary element in influencing either willingness to pay or the amount that an individual was willing to pay.

Compared to visitors to beaches, who were also asked whether and how much they were willing to pay, but without being reminded either of the 'greenhouse effect' or of other types of site, respondents in this household survey were willing to pay a substantially greater sum. However, we found little or no evidence of any distance-decay effect, except for promenades, as to the relative importance respondents in the different locations attached to preserving the different types of site (Green and Tunstall, 1991a).

Exploring the nature of non-use values (Green and Tunstall, 1991c) reinforced earlier suspicions (Green *et al.*, 1990) that these values are associated with moral rather than utilitarian concerns. That is, we found evidence of a belief that sites of ecological value ought to be preserved because we owe a duty of care to other species, rather than because these other species might prove, in the long term, to be useful to the human race.

Table 2.6
Willingness to pay via increased rates and taxes for coast protection (pence/year)

	Starting point 50p/year	Starting point 20p/year
Percentage of respondents willing to pay an additional amount	......................69%......................	
logarithmic mean	0.25	0.37
logarithmic standard deviation	0.63	0.61
arithmetic mean	£4.27	£8.22
sample size	109	120

Notes: 1. The distributions are very skewed; taking the logarithm (to base 10) gives a normal distribution.
2. Willingness to pay was elicited using a bidding game: if a respondent stated that s/he was willing to pay an additional sum, the respondent was then randomly assigned to one of two bidding games. In each of these games, respondents were asked whether they were willing to pay an additional amount (the opening bids used were 20p/year and 50p/year): if they said yes, they were then asked about a higher amount; if they said no, they were asked about a lower amount.

Coupled with evidence that the most important factors that should be considered in deciding which nature reserves should be preserved were those associated with rarity, rather than those indicative of their attractiveness as places to visit, we concluded that non-use values were not likely to be site specific. Instead, it seems likely that respondents were expressing, upon moral grounds, a willingness to contribute towards the preservation of 'the coast' in general rather than specific coastal sites.

Whilst this conclusion is somewhat tentative, clearly the evidence is currently lacking to support the assumption that individuals attach specific non-use values to specific sites. In the absence of such evidence it would be unsound and unwise to assume that such values exist, and thus to seek to elicit them through a CVM study and to rely upon those results.

2.7 Risk and uncertainty

Risk and uncertainty are of concern to an analyst on three counts.

First, any analysis will be incomplete and therefore, by definition, the results will contain uncertainty. The analysis could always be further refined and made more precise thus making the estimates more certain. Analyses should therefore be undertaken upon an iterative basis of progressive refinement. The decision whether a further round of iteration is justified should itself be taken upon benefit-cost grounds: will the improvements in the estimates of benefits and costs affect the choice of the project option and can the costs of the extra analysis be justifiable?

Secondly, whether a project should be undertaken is necessarily uncertain since it depends upon the outcome of future events. The decision maker must decide whether the proposed project is likely to be justified by future events.

These first two issues are concerned with the progressive estimation and the reduction of the degree of uncertainty associated with the project, then determining what is the best way to improve the analysis, and determining when to stop.

However, the third and key question is as follows: What is the best way to manage uncertainty? Given the residual uncertainties inherent in coastal management, which project offers the best management response to this uncertainty?

2.7.1 The nature of risk and uncertainty

The terms 'uncertainty' and, to a lesser extent, 'risk' are themselves unclear and even uncertain (Rubinstein, 1981). That is, it is easy to say that we are not certain of an outcome or value, but much less easy to specify the reason why we are not sure. Conventionally, two sources of uncertainty are distinguished. There is, first, 'parametric uncertainty': these are statistical and other uncertainties as to the estimated value of a variable. Secondly, there is 'systemic uncertainty': this is the possibility that the model or procedure being used to estimate the variable, or forming the basis of the analysis, is itself incomplete or erroneous (Blockley, 1980). Parametric uncertainty is thus 'what you *know* that you don't know', whilst systemic uncertainty refers to 'what you *don't know* that you don't know'. 'Risk' is conventionally used as a synonym for probability but may be used to describe one or all combinations of the probability of an outcome and the desirability of that outcome.

Smithson (1985) has proposed a more complete categorisation that he defines as 'ignorance': this encompasses both uncertainty and risk. Parametric and systemic uncertainties are then sub-divisions of ignorance. Given the necessarily unknown nature of a future for which any project appraisal includes predictions for some 30–100 years, depending upon the engineering life of the project, ignorance is a more appropriate term than uncertainty in general systemic uncertainties in particular.

2.7.2 Refining analysis by reducing uncertainty

The design and development of any coastal investment project should be appraisal-led and both design and appraisal processes should be iterative: this will be a process of progressively reducing uncertainties. Each stage in this process should be governed by informal use of a benefit-cost approach, asking the question: Is the significance of the additional information to the outcome of the decision greater than the cost of collecting that information?

Any analysis is essentially a problem in sampling: the iterative process of refining the analysis is thus a question of developing a progressive sampling strategy for further investigation that gives the greatest increase in precision for the least additional cost.

For flooding, assessments of the event losses for some return period events are used to estimate the shape of the loss-probability curve and calculate the area under that curve. Increasing the number of events for which losses are estimated improves the precision with which the curve can be estimated, but at some additional cost in data collection and analysis. However, since the shape of the loss-probability curve is unknown before the analysis is complete, the choice of the additional events for which losses are calculated, and not simply the number of events analysed, determines the precision of estimation. The losses from an event of a given return period cannot be less than the losses from an event with a shorter return period. Therefore, the upper and lower bounds on the curve are known and defined by the enclosing rectangle between any two adjacent return periods (Figure 2.4). Consequently, additional return period events can be chosen for which the results of the analysis would be most likely to reduce uncertainty.

Similarly, in estimating the losses to an individual property, different levels of precision are possible (Parker, Green and Thompson, 1987). Again, deciding the appropriate level of precision to adopt for a particular property is a sampling question. The answer to this question depends upon whether greater precision would be likely to have any effect upon the choice of project option.

The iterative approach will start by using standard data to give first estimates of the likely project benefits, and the relative contributions from each class of benefits. Then, the second and any subsequent stages will refine the estimates for those classes of benefits which make the largest contribution to total benefits. Some rules of thumb to use in deciding the appropriate level of precision to adopt are given in Parker (1987).

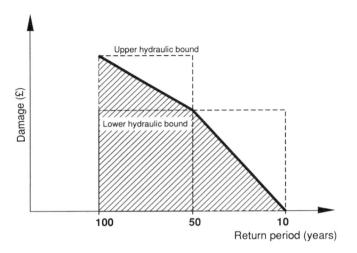

Notes: Whilst the loss-probability curve is undefined between any two adjacent event losses, it must lie within the rectangle defined by those two points

Figure 2.4 Uncertainty as to the area under the loss-probability curve and the upper and lower hydraulic bounds

2.7.3 Knowing when to stop: sensitivity analysis

The basic method for testing the sensitivity of the conclusions of the analysis to uncertainties is **sensitivity analysis**.

For parametric uncertainties, it would, theoretically, be possible to produce probability distributions for all the variables used. Thus, by making some assumptions as to whether the uncertainties in any one variable were independent of those of any other variable, it would be possible to calculate a probability density function for calculated benefits, or for the benefit-cost ratio (Goicoechea, Krouse and Antle, 1982). In practice, sensitivity analyses can easily degenerate into exercises of inserting different sets of values – often arbitrary – into the analysis and simply reporting the results.

The analytic method should be sensitive to variations in the values adopted for different variables, and, with experience, it is easy to predict to which variables the estimate of the benefits will be most sensitive. What is required instead is a test of the **robustness** of the conclusion of the analysis to errors in the data. A project option can be described as 'robust' if it still remains the preferred option even when relatively extreme values for those variables to which it is most sensitive are used. What it is necessary to test is *not* how sensitive the method is to errors *but* how great the errors have to be before the choice of the best option is changed. So, generally, the term 'robustness analysis' is preferred to 'sensitivity analysis'.

'Robustness analysis' can be described as evaluating how large a hammer you can hit the analysis with before it breaks. Knowing the differential sensitivity of the analytic method to the different variables involved gives an intuitive feel for the size of the different hammers.

For sea defence projects, the benefits will be very sensitive to the return period of the event at which flooding starts and the losses from short return period events (and relatively less sensitive to the magnitude of the damages from the more extreme events). Thus, changing the return periods of the events is a simple test of the robustness of the analysis and one that can be quickly undertaken using a spreadsheet program. If the return periods of key appraisal events have to be increased drastically before the project becomes economically inefficient, then the project is robust. It is usually necessary also to test for robustness against variations in the estimates of the probability of breaches or other modes of failure by any existing sea wall (Construction Industry Research and Information Association, 1991).

In some other cases, the discount rate can be used as a test of robustness. It is *not* a sensitivity test *per se* because the discount rate should be taken as a given, and testing for the robustness of the analysis in this respect should only be undertaken with great care. It should only be used where most of the costs occur in the first years of the period under consideration, and all the benefits flow in subsequent years. As a method it assumes that the uncertainty attached to all elements in the analysis increases at an equal compound rate; this assumption is not always plausible.

Finally, given the uncertainty that will always remain at the end of an analysis, it is unlikely to be appropriate to quote the results of a benefit assessment beyond perhaps two or three significant figures.

2.7.4 Managing uncertainty

Since systemic uncertainty cannot be removed, it must be managed. Whilst several somewhat unhelpful criteria have been suggested for choosing between alternatives that differ in the degree of uncertainty attached to the outcome of each (Moore and Thomas, 1988), a simpler approach is proposed here.

The first of these tests, derived from the principle of sustainability, is as follows: Which option forecloses least future options? On the sustainability principle, that option which leaves open the widest number of future options is to be preferred.

Secondly, the consequences of project failure should be assessed. Any option that results in a worse outcome if it fails than if the project had not been constructed at all is at least questionable. Such a failure might occur through overtopping or breaching. Thus, assessments of projects should not be restricted to the range of events at or below the design standard of protection, and the consequences of failure should be modelled and themselves be the subject of plans or projects.

Thirdly, the effects of altering the timing of the construction of the project should be explored. If a higher benefit-cost ratio or net present value would result from deferring the construction of the project for five years, then the project should not be built now. Indeed, the option of deferral should normally be considered for all coast protection projects. Similarly, for sea defence projects, the relative economic efficiency offered by different design standards of protection should be compared.

2.8 Summary and 'health warnings'

The key points in this chapter are summarised in Table 2.7. It is important in reading that table, and the remainder of this Manual, to appreciate that **economic analysis should always be used critically and only as an aid to decision making: it is not an end in itself**.

The two tests that should be applied to economic analysis are as follows:

1. Does it work?
2. Does it help to identify the best option?

As a generalisation, we recommend that the Shadow Price and Least Cost Alternative methods should be used where possible, and the Contingent Valuation Method should be used where the former are not possible. We do not consider, with the current state of knowledge, that either the Travel Cost Method or the Hedonic Price Method can be relied upon.

Where a large proportion of a site's value is believed to be in the form of non-use values, we recommend that an estimate is made of the total site value, via the Least Cost Alternative method. We consider it generally unwise, and probably unsound, to attempt to value the non-use values separately. Consequently, we recommend a group approach to implicit valuation, and for refining the possible project options, as the appropriate procedure to incorporate environmental values and issues into decision making (see the recommendations outlined in Chapter 4). It should be remembered that sustainability constraints should be applied to the choice of a project option. That option which maximises economic efficiency within the constraints of sustainability should be the one that is recommended.

However, finally, we recognise that there are theoretical deficiencies, technical difficulties and practical problems in all analysis. Nevertheless, and notwithstanding the present limits of economic analysis, the process of identifying and predicting the potential impacts of erosion or flooding, both during the construction of a project and upon the project's completion, are a necessary part not only of economic analysis but of any process of informed decision making and project design. Because it may not be possible to value some impact now is no reason for not fully pursuing its identification. Then there is at least some basis for making a decision. That decision making will be all the more informed after attempting to assess the magnitude of, and quantifying, that impact.

40 THE ECONOMICS OF COASTAL MANAGEMENT

Table 2.7
Chapter summary: key points

Project appraisal methods
- there is no best method of project appraisal for all circumstances: only the most useful method for a particular choice.
- key criteria for the selection of a project appraisal method are that it clarifies and simplifies the key issues involved in the choice.

Stages in project appraisal
- all significant impacts, or changes, as a result of a scheme option, must be identified, measured and evaluated.
- changes in state are more significant than marginal changes.
- care must be taken not to include a change which is counterbalanced by an opposing change which occurs as a direct result of the first change.
- care must also be taken to avoid counting both a change in stock and the change in flows which is a result of the stock change.
- identification of the population affected by a particular change is often more difficult than identification of the change and equally critical.

Values
- economic values are subjective, given only by the individual, sacrificial and relative. This assumption is ethical not scientific.
- it follows that those who believe in an alternative ethical basis to values will reject the results of economic analysis.
- consequently, we recommend that economic values only be applied in those circumstances where agreement exists as to the appropriateness of the ethical basis of economics.

Goods
- the different goods and values associated with different types of change as a result of coastal erosion are identified in Table 2.5.
- we advise that economic analysis ahould not normally be applied to non-use values.

Valuation methods
- the preferred valuation methods are: shadow prices, the least cost alternative, and the Contingent Valuation Method in that order.
- The Travel Cost and Hedonic Price Methods should not normally be used.

Discounting
- discounting has nothing to do with inflation.
- the appropriate rate of discount to be used is set by the Treasury.
- the foundations of discounting are somewhat shaky.
- in general, growth rates should not be used; where they are used they should be applied to all impacts.
- the time profile of benefits and costs should be examined: where, in the long run, the net effect is negative, then the option should be tested against the sustainability constraint.

Uncertainty
- uncertainty is another name for ignorance: ignorance over some areas can be reduced, but there are some areas where we do not even know that we are ignorant.
- the issues are: how most effectively to reduce ignorance as the consequences of each option; when remaining ignorance is unimportant to the choice; and how to best manage under ignorance.
- projects should be appraisal-led and the appraisal process should be iterative; at each stage, the value of further improving the estimates of the benefits and costs of each option should be compared to the costs of so doing.
- sensitivity analysis, better termed robustness analysis, should be used to determine whether the choice of options is robust to remaining ignorance.
- sensitivity analysis should include: in the case of sea defence schemes examination of the appropriate standard of protection; and for coast protection schemes, the date for construction.

- benefit figures should not generally be cited to more than 2 significant figures.
- since ignorance is irreducible, it must be managed: in comparing scheme options preference should therefore be given to that option which forecloses least future options.

Sustainability
- there is a presumption against any scheme option which has long run negative consequences.
- there is a presumption for scheme options which least reduce future options.

3 Coastal erosion: evaluating potential losses and benefits

3.1 Summary of information

This chapter discusses the procedures and techniques recommended for use in assessing the potential benefits of investment in coast protection. As indicated in Section 2.6.4, these benefits arise from delaying the process of erosion, and therefore delaying the permanent land and property losses, for the duration of the life of proposed protection works.

The main procedure (Section 3.3) presents factors ('extension of life factors' or ELFs) which are used to calculate, from current property or land values, the present value of the benefit from such a delay in the erosion process. The inputs required are estimates of the year in which property/land would be lost without protection, its current economic value, the anticipated duration of the life of the project, and the Test Discount Rate.

Agricultural land (Section 3.6) can be treated in a similar way, except that annual net economic margins for the period of extension land life should be used and these currently give very low benefits.

Infrastructure loss is more complex (Section 3.5), but only affects the other calculations if, in a 'do nothing' scenario, the least social cost would be to invest in infrastructure so that properties could continue to be used or other economic activities could continue.

Flow charts are presented in the chapter, summarising the overall process (Figure 3.1), and for the procedure for assessing losses to infrastructure (Figure 3.3). Some economic value data is presented, but the chapter concentrates on the methods and reasons for the approach, and on providing the extension of life factors.

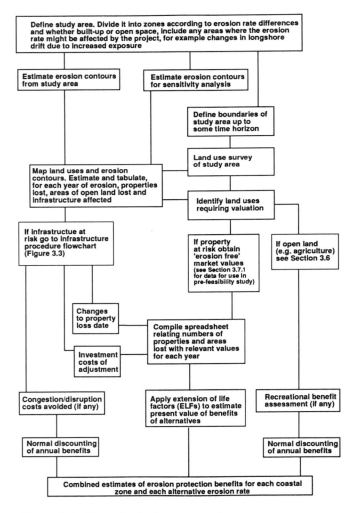

Figure 3.1 Flow chart of assessment process

3.2 Problem definition

The most obvious potential economic loss from erosion at the coast is loss of land and property. Buildings, including the land integral to the property, infrastructure, and non-built up land may all be lost to the sea, either directly through erosion or from land slips and falls following erosion of the coastline.

The important difference here from losses owing to flooding is that erosion is effectively **irreversible**: once land has been eroded, or the cliff-top is deemed too close to a property for that property to be used safely, then the land or property cannot be regained. Moreover, few British buildings can be moved away to a safer site, owing to their brick or stone construction, and land could only be recovered through reclamation from the sea (and then would have very different characteristics), which is inherently unlikely in areas with rapid rates of erosion.

This irreversibility means that future options for the use of that land and/or property are also lost. Hence there is a permanent loss at one point in time of a capital stock (i.e. land and buildings) which had offered a future stream of uses and therefore values. By comparison, flooding (Chapter 5) results in the loss through flood damage of an annual income flow (in agriculture and business, gross margins and value added respectively), and also the loss of a part of the capital stock which is repairable or replaceable (loss of contents of buildings or damages to structures, owing to the flooding).

In some situations, of course, floods may indeed destroy properties (such as caravans or wooden buildings) but even then the land is not lost, except by erosion accompanying the floods. In some coastal locations erosion may result in more frequent flooding of low lying areas (Section 5.4.6). If this leads to the abandonment of property the 'inundation' loss from frequent flooding is analogous in some ways to erosion loss.

However, even then the land might still have some residual uses which should be allowed for in assessing the net loss, although sound empirical evidence on the maximum acceptable frequency of sea flooding to different types of property is lacking, and thus information on residual uses is also lacking. This is not the case with property and land washed away into the sea: there is no future use!

Although the losses from coastal erosion are permanent and irreversible, coast protection works which are designed to arrest this process normally have finite lives before their effectiveness is so reduced that they too are lost through the same erosive forces; then new coast protection works are needed. Hence the benefit from a particular coast protection project should be seen as a **temporary**, but lengthy, extension to the useful life of the land and property. This is in contrast to the **permanent** loss of the property from coast erosion.

The effective life of coast protection works will, therefore, be important, in that this is the length of the extension of the life of the property and land protected. In this respect there will be uncertainty over the expected life of coast protection works but an estimate of their useful life is needed.

The most reasonable assumption thereafter is that the same erosion rate as before will start again, unless there is good reason to believe that the rate would be altered by the presence of the remnants of the old protection works or would be different owing to some other cause. Future decisions should not be prejudged by assuming that coast protection will continue indefinitely: fifty years from now a future generation may or may not wish to renew the works that now afford coast protection. Therefore any appraisal for a project being evaluated now should not assume that another project will be built when it comes to the end of its useful life.

While the number of properties now at risk from erosion in England and Wales is probably relatively small, compared to the number at risk from flooding, substantial parts of the coastline are eroding and/or have already some form of coast protection works, particularly the Holderness coast and parts of East Anglia and the South coast. Taking a long term perspective, there are also other areas where rates of erosion will result in significant losses over a 50-year time period. Consequently over the next 50–100 years a large number and a varied mixture of buildings, infrastructure, and open space land are all at risk from erosion by the sea.

The procedures and techniques given in this chapter offer appropriate ways of evaluating the benefits that should accrue from investment to delay this potential, permanent and sequential process of loss. The recommended approach is summarised in the flow chart in Figure 3.1. Within this approach are a number of procedures and techniques, together with a number of underlying methodological issues, which need some explanation in order to determine what is and what is not included within the recommended approach.

The key points embodied in the approach are as follows:

1. Estimates of **erosion** rates projected for 50 or 100 years into the future are needed, otherwise the other procedures and techniques cannot be applied and potential benefits cannot be estimated (Section 3.3).

   However, erosion events are often erratic and unpredictable in their timing (often depending on the pattern of severe storms), so such estimates may be difficult to make and be subject to uncertainty concerning geological conditions and future storm severity and frequency. **Techniques** for assessing erosion rate depend on geological surveys, analysis of historical records of erosion and old maps, and predictions of any changes in relevant coastal processes.

2. A **procedure** is provided for evaluating an **extension to the expected life and use of property and land** due to a delay in the erosion process resulting from investment in coast protection works (Section 3.4.2).
3. **Techniques** are provided for finding appropriate **values for properties** whose market prices are likely to be affected by perceived erosion risk (Section 3.4.3).
4. A **procedure** is provided for valuing potential **infrastructure losses** where these may be of several types (Section 3.5). First, the infrastructure may depend for its value on its proximity to the coast (for example promenades). Secondly, the infrastructure may serve erosion-prone property (e.g. in providing electricity supply). Thirdly, it may serve other areas which are not themselves erosion-prone but the route of the utility is liable to be eroded. Here the option of investing in alternative lines of service is an alternative both to coast protection and to accepting the disruption due to erosion.
5. **Techniques** are given for finding appropriate **economic values for agricultural land** threatened by erosion, based on its productivity for society (Section 3.6.2).
6. **Techniques** are provided for finding appropriate **economic values for other open spaces** threatened by erosion (Section 3.6.3, but see also Chapters 4 and 6).
7. The additional **methodological issue** that real values, such as property values, are predicted over a long period along with their attributes is discussed (Section 3.4.5).

The underlying assumption in all this analysis is that coast protection projects are compared with a 'do nothing' option which in fact is a least economic cost option having accepted that the erosion will occur. In other words, the public and private sectors are modelled as behaving rationally – to minimise their economic costs – if coast protection is not provided.

Most of our research undertaken on erosion losses has been methodological in nature and results in a set of procedures and techniques which simplify and standardise the appraisal process. Some data sets have been compiled, and other existing data sources are referred to, but in many cases site specific values for properties will be needed for the appraisal of particular coast protection projects. This is because the exact nature of the property being affected by erosion is crucial to the magnitude of the benefits of its protection, and to take averages for property values on a national or regional basis would not be appropriate for local circumstances in anything other than a preliminary pre-feasibility study.

3.3 Erosion rates and erosion 'contours'

The predicted rates of coastal erosion without further coast protection form the link between the physical processes along the coast and the economic benefits of coast protection. Without some detailed knowledge of erosion rates in the future it is not possible to calculate the benefits of delaying this process.

However, it is not the role of this Manual to specify in detail the available methods of modelling coastal erosion processes. The objective of that technical assessment of the erosion problem, so far as the benefit assessment is concerned, should be a set of **predicted erosion contours** for the coastline in question for at least the projected life of the proposed coast protection project.

Time horizons of 50 or 100 years of erosion are generally recommended for use in benefit assessments. The time horizon for erosion contour estimation need not coincide with the expected project life, since coast protection works delay all erosion up to the point in time when some naturally stable coastline is formed. However, distant delays of erosion, for example from year 150 from the present to year 200 from the present, yield benefits that are so heavily discounted using normal discount rates as to be insignificant. Hence using the project life, or the project life plus, say, 25 per cent, is an appropriate time horizon for most practical purposes. The only exceptions will occur where the value of property beyond the erosion contour representing project life plus 25 per cent is so massive that its value, when discounted back to present values, is still likely to be significant.

Obviously erosion predictions will not be certain and will have to average out the likely effects from storms of different magnitudes. Since the timing of erosive storms cannot be predicted the best that is likely to be achieved is predicted erosion rates based on past erosion rates, knowledge of the underlying geology, and any anticipated changes which might affect the erosion process, such as rising sea levels.

The appraisal should make clear the methods used and sources of information on which erosion predictions have been made. Alternative estimates of erosion rates should be made, so that a **sensitivity analysis** of potential benefits according to rate of erosion can be undertaken, since a major source of uncertainty in the benefit estimation is likely to be this erosion rate.

The results of this part of the analysis should be **erosion 'contours'** showing the predicted location of the edge of the cliff or sea after erosion at the average rates predicted (these rates may not be constant over time). In case studies the best compromise between detailed prediction and spurious accuracy has been found to be erosion contours at **five year intervals.**

Although erosion rates may be low, this level of detail is particularly useful where urban areas are at risk since losses are not uniform, according to the area of land lost, but patchy, according to the locations of the properties to be lost. Where agricultural land will erode slowly and at a constant rate, either wider erosion contour intervals or the area lost annually, can be used. For example, ten year erosion contour intervals can be used where the benefit assessment would not be affected by lesser accuracy.

For properties, the five year erosion contours/intervals are used to identify the band of years when a property would most likely become uninhabitable (i.e. either lost, or located so near to the edge of the cliff or sea as to be abandoned as unsafe). A property can then be assumed to be lost in the mid-year of that time period, but for property losses it is easier and probably more reliable to identify the erosion contour line which best represents the date when the property could not continue to be used. Over- and under-estimation due to grouping property losses at five year intervals should cancel out over a long appraisal period.

For properties at risk from erosion there will be some minimum acceptable **safety margin** ('sm' in Figure 3.2) between the unstable cliff top and the building, as indicated above. Once the cliff edge (and/or sea) moves closer, the risk to life, the problems of access, and the risk to home contents will make the property uninhabitable.

Some Coast Protection Authorities have defined minimum safety margins of a certain number of metres from the cliff edge. If this is the case this margin should be adopted in estimating the years when losses would occur. If there is no such declared safety margin then we recommend that a five metre safety zone is adopted for the benefit assessment.

Figure 3.2 illustrates a set of hypothetical erosion contours, and a safety margin. Inevitably some judgements will be necessary in deciding when properties would be 'written off' without coast protection, but the analysis should make the assumptions clear and the method should be consistent between properties and between years within the appraisal period.

This stage of the assessment is largely map based – relating erosion contours with land uses and property locations and types (the latter may need field surveys).

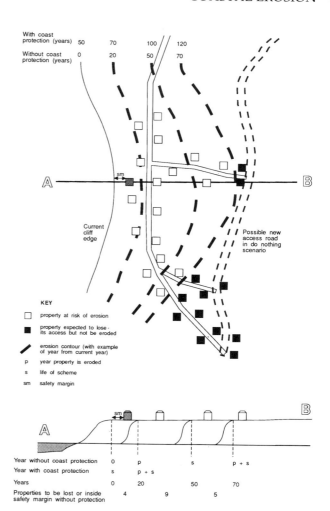

Figure 3.2 Hypothetical coastal erosion contours

From this, a list of the years in which properties and land would be lost should be derived which identifies the addresses and types of property, the types of land, and numbers of properties or floor areas involved.

For businesses, it may be more useful to estimate the floor areas of each property type involved (floor areas in square metres of shops, offices, and industry for example), since it may be easier to obtain property values on this basis. If individual property values are to be used then floor areas will not be necessary (they are only used when average property values per square metre of floor areas are used). For agriculture, the area of land of each grade and cropping pattern or land use lost in each year (or band of years) is the data required.

The only other complication to this part of the assessment is whether **public utilities and infrastructure** would be affected by the projected erosion, and particularly whether those infrastructure elements serving the properties concerned would be affected before the properties themselves would become unusable due to the direct risk of erosion. Section 3.5.1 discusses the possible types of infrastructure loss.

In this respect, where properties are at risk from erosion it will be necessary to establish the approximate locations of service lines to these properties. Often electricity cables and gas and water pipes are laid under the pavements outside houses so there may not be any difference in the date of loss due to properties being cut off from these vital services compared with the date when erosion reaches the safety margin for the property.

However, safety margins on access roads may be in force, or utility lines – or substations on which they depend – may be closer to the coast somewhere else other than at the location of the erosion-prone properties. This may mean that a property could be regarded as 'written off' before the property is itself touched by an erosion contour plus the safety margin. Taking these points on board, the procedure detailed in Section 3.5 ensures that losses associated with infrastructure erosion are estimated in the most appropriate way.

3.4 Delays in losses, and the extension of the lives of property and land

3.4.1 The concept of benefit as a delayed loss

The benefit of coast protection works is an extension to the life of, or the delay in the loss of, erosion-prone property and land for a period of time equal to the life of the protection works, assuming that erosion after the end of the works' life would proceed at the same rate as it would have done without the project. The key point, then, is that a property which is predicted to be lost from erosion in 20 years' time **without** protection would, **with** effective coast protection works having a 50-year life, then be expected to be lost in 70 years' time (Figure 3.2).

However, these are all estimates of the most likely timing; the problem is that of evaluating the benefit from a change in the risk, since the estimated erosion dates are based on estimated average erosion rates. Consequently, simply taking a current value for the property and land – for example the current market value – would give an estimate of the potential loss from erosion but not of the potential benefit from coast protection. The latter is based on an extension to the life of the property and the value of the benefits is profoundly affected by the timing of that extension.

The possible approaches to the problem of finding an appropriate economic value are as follows:

1. To find a figure for the **annual value** of the property or land, and from this calculate a present value for the additional time that the property can be used represented by the extended life of that land/property based on the erosion rate and intended life of the coast protection project.
2. To find **economic (market) values** for the property or land as if it were not at risk and to take the difference in discounted values of the loss with and without protection as the benefit of coast protection.
3. To obtain a **value directly** which reflects the present value of the loss due to an appropriate erosion risk – for example a value based on differences in house price according to differences in distance from the edge of a cliff.

The first two approaches are similar and have been adopted in the recommended method and procedure (Section 3.4.2) because they are more easily put into practice. The difficulties of the third approach are briefly discussed in Section 3.4.4.

3.4.2 The procedure for valuing property life extension

The formula on which this procedure is based is appropriate for any erosion loss which would be delayed by coast protection, although the text below is directed at assessing the benefits of protecting residential and other built-up property in cliff-top situations.

Ideally, we would like to find, in the marketplace, property prices which reveal the value of protecting property assuming that it would be eroded in 55 years' time rather than 5 years' time (i.e. a project life of 50 years). Such prices which properly reflect the two risks for a given property are, however, not observable because the market tends to ignore distant erosion risks or assumes that coast protection will be implemented, and only when erosion is imminent do property prices collapse (and in fact prices cannot often then be observed in such circumstances because there are no buyers).

The procedure recommended here to overcome this problem is relatively simple and depends on the following stages:

1. Finding a **market-based property** price which is not affected by erosion risk but otherwise reflects the attributes of the erosion-prone property concerned (Section 3.4.3); and
2. Using a **formula** which builds the property life extension into the assessment.

Formula 3.1 presents the extension of life factor (ELF) with which the present value of the benefits of delaying property losses can be derived simply from the 'erosion free property value', the Test Discount Rate, the erosion year without a project, and the expected life of the project.

$$PVB = MV \cdot [1/(1+r)^p - 1/(1+r)^{p+s}]$$

$$= MV \cdot ELF_{p+s} \qquad \text{Formula 3.1}$$

where:
- PVB = the present value of benefits;
- MV = the market value of erosion free property at base date prices;
- r = the Test Discount Rate;
- p = the expected life (in years) of the property with no coast protection project;
- s = the expected life (in years) of the proposed coast protection project;
- ELF_{p+s} = the extension of life factor for property p years from the current coastline or cliff-top and a protection project with s years' life, as defined in the formula.

This formula can be explained very simply as **the discounted value of the property loss without any proposed coast protection, less the discounted value of the same property loss with coast protection.** Obviously this will be a positive figure since losses further into the future are discounted more heavily. The ELFs are, however, smaller than the Test Discount Rate which effectively means that erosion delay benefits are discounted more heavily than flood alleviation benefits for example. This is because a capital loss is delayed and not prevented permanently (if coast protection works lasted for ever then the normal TDR could be used). By comparison flood alleviation prevents (and does not delay) losses in each year of the life of the project because the losses are recurrent - the capital stock at risk from flooding is assumed to be constant over time and damages are a relatively small fraction of the capital value and hence are repaired or replaced.

Since it is assumed that coast protection halts erosion until the end of the life of the project, and that thereafter erosion continues at the same rate as without a project, the ELF depends simply on the erosion contour for the property concerned (assuming no coast protection), the life of the proposed project, and the discount rate.

This formula is a generalised and reworked version of Formula 3 given by Thompson *et al.* (1987, 21), without including any particular discount rate. Formula 3.1 above is derived from a shadow valuation of property based on an annual property value derived from its erosion free market value, and assumes the equivalence of the discount rate implicit in that annual value with the Test Discount Rate (which is used in calculating the discounted benefit from the additional years of life given to a property by coast protection). The derivation of the formula is shown in Appendix 3.1.

A number of assumptions underlie this method. These are as follows:

1. The freehold **market value**, without erosion risk, reflects the current value to society of the property, including the land on which it is built. This value should reflect the discounted flow of services provided to society by the property and land. If there is reason to believe that market prices are distorted from economic or social prices then some adjustment should be made.

2. The method assumes that the future life prospects for the property are the same (**ignoring erosion risk**) whenever the view point. Thus the current (undiscounted) value is the same for a property 35 years from the cliff edge as it is for that property located at a point from the cliff edge representing 35 years plus a 50-year coast protection project. Hence no deterioration in the property is assumed, and the **property lifetime is assumed to be infinity.** This seems appropriate since the land would, without erosion, have an infinite life and similar uses could continue to be made of it.

3. Likewise the **social value of the property**, ignoring its future lifetime, is assumed not to change over time. Of course, as with land, there might be option price reasons for expecting this value to differ over time, or the implicit discount rate used in market prices could change. Thus as society becomes richer, more constrained, or poorer over time, the social value of the resources at risk could change. Since this is unpredictable, as with other relative prices, the assumption of **unchanging relative social values** appears justified.

Appendix 3.2 presents values for the extension of property life factor, ELF, for some likely project lives (20, 25, 30, 50, 75, 100, and 125 years) and for those years of current erosion risk likely to make any significant contribution to a discounted present value of benefits (years 1 to 100), and for both the Test Discount Rate (currently 6 per cent in 1992) and a series of alternative discount rates which may in the future be found to represent the real rate of return to investments. In general these ELFs are lower than the normal discount factor for the obvious reason that the discount factor for year p+s is subtracted from the discount factor for year p.

3.4.3 'Erosion free' property prices

The property (and land) prices required for the discounted erosion delay factor method presented in the previous section are **freehold market values, not adjusted for erosion risk**. That is, they are based on a normal expected life for that property type and an infinite life for the land on which the property is located.

The assumption is that the land and property combination would continue with no erosion risk indefinitely; even if the existing property were replaced at some date it would be replaceable with a land use of similar value. This appears realistic since, with the exception of any correlation between state of repair and age, property values do not appear to decline with property age.

This presents us with the difficulty that market prices for the actual properties at risk, where they are observable, may be affected by erosion risk (see Section 3.4.4 for the problems with risk-affected prices). Hence it may not always be appropriate to use the quoted price for a particular property: it may be lower than is intended for use in this method.

Highly aggregated data sets are presented in Section 3.7 for the main types of dwelling. However, in most cases greater reliability may be achieved by obtaining values locally for the specific types of property to be affected by the project since property prices vary considerably by location in Britain. The values used for residential property should reflect its location type – such as being near the sea – but it should be safe (based on properties which do not have an erosion risk).

Locally appropriate property prices can be obtained through the Coast Protection Authority's own valuation department, if it has one, or through local estate agents. It should be relatively easy for them to estimate typical or average values for this type of property which **ignore** the risk of the properties being lost through erosion without a project. On average, it is less time consuming and loses little in accuracy to take 'average' or 'typical' values for a property of a given type, or value per square metre in the case of businesses, rather than having individual valuations of properties.

3.4.4 House prices and risk

Since economic values are designed to reflect values to society, and economic analysis takes as its main axiom that people's preferences count (Chapter 2), an intuitively attractive way of assessing the benefit to society of reducing the risk and delaying the date of erosion would be to use variations in actual house prices relative to erosion risk. The difference in price for a particular house type according to erosion risk would then be a direct estimate of the economic benefit of delaying erosion by that amount.

A considerable body of research exists (Section 2.5.4) on the use of hedonic prices (Rosen, 1974), or property prices in this context, as a means of evaluating environmental resources and benefits (this is summarised, for example, in Freeman, 1979; and Pearce, 1978). Briefly, the idea is that environmental public goods, such as better air quality or reduced noise pollution, are not marketed but do vary spatially. People also make locational decisions: they consciously buy houses which are located in areas with better or worse air quality, for example.

Hence a statistical analysis of property prices could be used to establish a relationship between property prices and differences in the environmental attribute being evaluated, controlling for other factors, and from this **implicit prices** can be imputed for small changes in the environmental attribute. As Section 2.5.4 shows, however, there are a number of serious problems in the application of this empirical method, and the conditions necessary for it to give valid estimates of economic values are unlikely to hold in the British coastal housing market.

For the following reasons this method or procedure has not been adopted in assessing erosion protection benefits:

1. People are not free to make the choices implied in the Hedonic Price Method when they buy and sell houses. Transaction costs are not negligible, and there is not a continuous supply of similar houses with different erosion risks.
2. Environmental risk may not be linked in any simple functional form to the observable distance from the coast or cliff-top; also house owners and buyers may not perceive the statistical risks which are appropriate to public decision making.
3. There are many factors which determine the price of a house and, while not impossible, it seems unlikely that an empirical model would be effective in separating out the influence of erosion risk without a very large data set. Also there is no theoretical basis for deciding which factors to include in any such model.
4. There is a lack of data on current house prices in Britain, where sale values and property characteristics are not centrally recorded, and in any case turnover would need to be very high to give enough observable current prices. This is important where the study area might be small, such as for coastal erosion sites. If there are only 100 houses at risk in a given study, actual market values which could reflect the risks will be difficult to observe. The alternative is to model the problem in general at a national level and to allow for sufficient regional and local factors in the model specification. However, this method

could fall into the 'ecological fallacy' (Langbein and Lichman, 1978) of particularising an aggregate relationship to a case study.

5. Estate agents could be used to estimate market values, allowing for erosion risk, for the properties involved, and so avoiding direct household surveys and the lack of recorded market prices. However, these estimates will reflect the valuers' perception of risks and peoples' reactions, since the estate agent's brief would be to value allowing for the erosion risk or attribute in question. Moreover such a brief would draw attention to the issue being investigated so the results might be expected to be biased. It would be difficult to assess the validity of a measure which is a surrogate for a surrogate.

It seems likely from previous case studies that house prices are only affected by erosion risk in Britain when the risk becomes very obvious. Thus in the Fairlight Cove case (Section 8.4) properties were being sold and built at unaffected values despite a risk of being eroded in about 50 years' time, while properties close to the cliff edge could be unsaleable. Hence the much simpler procedure detailed in Section 3.4.2 has been adopted, so avoiding extensive surveys of house prices.

3.4.5 House price trends

Coast protection works – like most other projects – are appraised for an expected project life of perhaps 50 or even 100 years. Hence the benefit estimate is based on a prediction of the economic values of the losses which may be prevented over many years from the present. Whilst general inflation is ignored in social benefit-cost analysis, since real values are unaltered by it, potential changes in relative recent prices are relevant. For example, this is discussed in the appropriate appraisal guidelines (H. M. Treasury, 1991) which state that predictable relative price changes may be taken into account in benefit-cost analyses in Britain.

However, it is normal practice in flood alleviation benefit appraisals in England and Wales, for example, to assume that relative prices for damageable items and real gross margins for businesses will not change over periods of 50–100 years into the future (Parker *et al.*, 1987).

Despite this assumption some considerable changes have been shown over a ten year period (Suleman *et al.*, 1988) between potential damage values based on earlier inventories and retail price indices (Penning-Rowsell and Chatterton, 1977) compared with later inventories and relative prices. However, a predictive model of changes in prices would be needed in order to justify a different set of future relative prices.

The same arguments apply in the case of house prices, which are likely to form a large component of erosion mitigation benefits in the average coast protection project protecting urban areas. However, in this case a single value prediction forms the basis of much of the benefit estimate, rather than the many building fabric and contents components which are damageable in floods. Hence there might be a greater possibility of predicting relative house price movements into the future.

Kosmin (1988) analysed past data on national trends in house prices over various periods from 1946 to 1988 using a number of sources such as government statistics, financial institution and Building Society surveys. The cumulative growth rates in house prices in each case exceeded the retail price index increases over the same periods. A differential growth rate in house prices during these periods of approximately 2.5 per cent was found.

However, there are a number of problems with this result which mean that further analysis should be undertaken and that this approach to the problem is of limited use:

1. A disaggregated model of house price trends is more appropriate, allowing for house type and regional variations.
2. The method is based on the past only, without an explanation of why real house prices might increase in the future. Hence using such predictions depends on whether we believe past trends will continue.
3. It is clear that quite substantial fluctuations about any trend occur (as in 1989–92), so the time series method is prone to starting point and end point problems.

Ideally a more appropriate method would use a predictive model which is based on an explanatory function for house prices (that model might itself be calibrated with data from the past). For this method to be useful it should be possible to make predictions of the independent variables in the model with some degree of reliability.

A simple method would be to argue that houses are seen as personal investments and the supply lags behind the total demand for any given quality/set of attributes. Then, as real incomes grow, people are prepared to spend more on houses by buying better houses, hence the price of houses is pushed up in line with real per capita growth in Gross Domestic Product. If houses are a relatively good hedge against inflation, their prices must increase on average faster than general prices. Thus Dicks concluded that:

> Growth in the number of households and in real

incomes explains much of the long-run trend in house prices, with starts responding only very slowly to changes in the profitability of house-building, largely because of restrictions on the supply of land available for residential use ... (1989, p. 75).

It is probably possible to predict demographic changes reliably over the long periods needed in coast protection investment appraisal since the population of future house buyers for the next 20 years or more is already in existence. Predictions of long-run real economic growth rates will be more problematic; predictions based on a continuation of past trends will be inadequate, particularly given uncertainty over the implications of past economic growth for the environment and hence future economic development and welfare.

In addition the attributes of the goods being modelled are not constant. If the housing stock is improving over time then price changes may also reflect changes in the bundle of services provided by the average house of a given type. Thus the value of fitted carpets and kitchens and of extensions and conversions must be passed on to buyers since they generally cannot be removed when houses are sold.

Hence if a prediction can be made of changes in house attributes for a given house type for future years (that is, changes which do not involve demolition), then it might be possible to predict this component of real price increases. This might be possible if coastal areas have houses which have been less improved than other areas so that prices in other areas could be taken as a guide to real price changes over time as the coastal areas 'catch up'. However, there is no general evidence available to suggest that this is the case.

It seems likely that houses near cliff edges have received less investment because of the erosion risk and, therefore, they are of lower quality and will be lower priced anyway, irrespective of the direct risk factor.

The technique adopted in Section 3.4.3 avoids this problem by taking prices from properties of the same type without erosion risk, but in the same general area, accepting the typical qualities found in these non-erosion-prone conditions. This assumes that any differential in physical quality is due to erosion risk which depresses the overall value of properties.

The same argument is not appropriate to any price differential due to houses having a 'sea view'. Without coast protection pieces of land and properties will still have a sea view. They will just be different plots. There will be a redistribution of views: once the seaward properties have been lost, the landward properties will have the sea view that the others once enjoyed!

Hence no conclusive reason and no reliable method for making future predictions of long-term house price trends has been found, although it may be that the past relative price trend will continue. In the face of uncertainty the standard approach of assuming constant relative prices is recommended. But in a project which would mainly protect residential property the implications of an increasing real housing price are readily tested since they amount to a reduction in the discount rate (see the Fairlight Cove case study: Section 8.4)

3.5 Infrastructure benefit assessment procedure

3.5.1 Types of infrastructure loss

Different approaches to loss estimation – and thus benefit assessment – are appropriate to different forms of infrastructure loss. A procedure for assessing the benefits of protecting infrastructure from coastal erosion, where these would be additional to benefits calculated for other types of loss, is summarised in Figure 3.3.

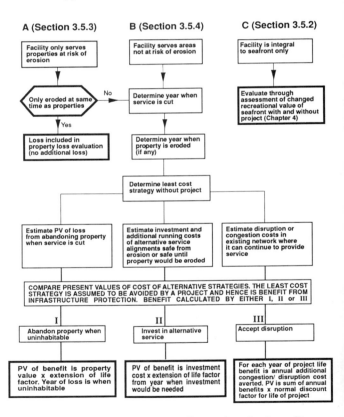

Figure 3.3 Schematic diagram illustrating the benefit assessment framework for infrastructure

The following types of infrastructure loss should be distinguished:

1. Infrastructure which is integral to the coast and

existing coast protection works (if any), in particular promenades and associated structures (Section 3.5.2);
2. Infrastructure which simply serves erosion-prone property (this includes access roads, and domestic power, water, and sewerage lines) and which is not itself erosion-prone, other than close by the erosion-prone property it serves (Section 3.5.3); and
3. Infrastructure which serves areas either not at risk from erosion or only at risk some considerable time after the service might be cut: typically through-route roads, and public utility main systems which, if cut, would affect larger areas (Section 3.5.4).

For each of these categories the following sections detail the problem posed for evaluation and the recommended approach. The interlinkages between these and the decisions needed in the appraisal process are given by Figure 3.3.

3.5.2 Infrastructure integral to the coast

Existing promenades and associated structures, and sea defence and coast protection works, are continually failing or falling into a state of poor repair. This is often the reason for an appraisal of the situation and the proposal for new coast protection work in the first place.

However, in a 'do nothing' scenario with erosion, similar works (i.e. in a state of disrepair) would not be built. The choice is either between a decaying coast or any proposed project, hence this type of infrastructure cannot be separated from the other attributes of the coastline in a 'do nothing' scenario, such as changing beaches and eroding cliff-top land. Therefore, it is more appropriate to treat the loss of such features as part of the loss of the coastal zone **recreation** facilities.

Since a major part of the potential benefits to coast protection can be the value to users of beaches and associated coastal features (including promenades), it would be double-counting to include both a figure for loss of part of the coastline (the infrastructure) and for the loss to users of use of the coastline. In other words the users can use, and therefore they value, the coastline because of the infrastructure it embodies.

Contingent valuation methods (Chapter 4) used for recreation loss or benefit estimation, for example, will include an analysis of the use of promenades along with the use of their associated beaches. Thus in the Herne Bay case study (Section 8.5) the basis for the recreational benefit appraisal was illustrations of the coastal zone under different scenarios which included deterioration of the promenade.

However, where a promenade is a combined access road or through-route (rather than spatially distinct from any such road) then additionally the implications to normal users of that road (rather than recreational users of the coastline) should be taken into account by using the methods in Sections 3.5.3 or 3.5.4 as appropriate for this potential additional loss.

Hence, infrastructure which is an integral part of the coastal zone (in terms of recreation experiences and coast protection works) should not be valued separately as a potential resource loss. Instead such losses should be incorporated in the assessment of recreational losses without coast protection, using the methods detailed in Chapter 4. This means that any illustrations or scenarios used in this method should incorporate the future condition of the infrastructure (promenade) in the 'do nothing' or 'without' scenario, and also should show the proposed condition of the seafront and promenade in the 'with project' scenarios.

In a less detailed appraisal (for example a pre-feasibility study) an assessment of recreational losses by the Contingent Valuation Method might not be used, and alternative approximate estimates of recreational losses may be uncertain. In such cases an estimate of the remaining value of the promenade (but **not** the existing coast protection works) can be used as an approximation (shadow value) for the loss.

That value should be in current prices for a promenade of the current specification in 'reasonable' condition (rather than any new design). The benefit of delaying this loss by providing coast protection can then be calculated by using the appropriate extension of life factor (ELF).

It is important to note that **the value of existing coast protection works 'protected' cannot be counted as a benefit of new works**. If the old works were fully effective there would be no need for a new project, whilst there is no direct benefit from saving defunct works from eroding since they have no value.

However, existing coast protection works may be incorporated physically into a proposed new project and this may have the advantage of reducing the costs of the new defences. Hence any residual value of old defences will be implicit in the cost calculation for a proposed new project, and would be double-counted if it were also included in the benefit calculation.

3.5.3 Infrastructure integral to properties at risk from erosion

Where the loss through erosion of power cables, telephone lines, water, sewerage and gas pipes, and access roads, would not affect properties other than those lost to erosion, and the impact would be approximately at the same time, there is **no additional economic loss**.

This is because the values of the properties assume that they have normal services provided and the past investment in this infrastructure was entirely to service these erosion-prone properties. Extending the lives of these services will not bring an additional benefit to society since they only benefit the properties which would by that time have been lost to erosion and which will have already been included and evaluated in the appraisal.

Although these infrastructure losses are already included in the assessment, the timing of property loss may need to be adjusted. It is important to identify the earliest time when the property would be deemed to be uninhabitable. This time may be determined by the proximity of the property itself to the cliff edge, or the proximity to the cliff edge of essential infrastructure serving that property. Where this is allowed for in the estimate of the year of erosion loss no other economic value is needed.

Hence the alignment of these service utilities should be mapped along with erosion-prone property and the erosion contours, and the earliest year (after an appropriate safety margin) taken when the property would be uninhabitable. Electricity, water, sewerage and road access are likely to be the most critical services – the lack of a telephone might reduce the property's value but not its usability.

3.5.4 Infrastructure serving areas not at risk from erosion at the same time

These other areas, which are not likely to erode when the infrastructure serving them would be lost, may either be erosion-prone themselves, or the demand served by the infrastructure at risk may be unaffected by erosion.

1. *Infrastructure serving distant erosion-prone properties* If the infrastructure is erosion-prone and serves properties which would themselves be eroded at some later time (all assuming no coast protection), it might be appropriate to regard these properties as lost when their services are lost.

The same technique as used in Section 3.4.2 would then apply – for example a property might be 'written off' five years before it would otherwise be (from direct erosion) because it would be without services. Thus a house without power supplies, water, sewerage and access would be uninhabitable, so its use would be lost to society. In this case the present value of losses from erosion would be greater than it would be if infrastructure were ignored, since an earlier loss is discounted less heavily.

However, where the service line is vulnerable to erosion a substantial time before the properties, it could imply 'writing off' properties 20 or more years before they would be predicted to erode. Hence it is necessary to seek the least cost strategy for society in the 'do nothing' scenario.

This method requires estimates of the feasibility and costs of providing alternative services or infrastructure for the disconnected properties so that they may be inhabited until a direct erosion threat makes them uninhabitable.

Hence, if it is feasible to provide replacement infrastructure to serve these properties when they are cut off, then estimates of the costs of doing this, in current prices, should be obtained. If the present value of these costs (using normal discount rates) is less than the difference between the present value of the property loss assessed for when the property's services would be lost and when the property itself would be lost, then the least social cost response to erosion without a coast protection project would be to invest in the new infrastructure.

Therefore the cost of additional infrastructure investment should be included in the project benefits as a social cost averted by the project, and the property life continued until it is directly uninhabitable due to erosion.

This is illustrated in Figure 3.2, where an alternative road alignment is possible to continue the use of some of the properties, examples of the calculations are given for roads in the Fairlight Cove case study (Section 8.4) and for sewerage provisions in the Herne Bay case study (Section 8.5). If there is no alternative route of supply with a lower cost than the advanced loss of the properties affected, then the loss which would occur when the services are lost may be the full value of the property or a large difference in the market value. To calculate the benefit from protecting such properties the appropriate extension of life factor should be applied for a loss without protection in the year of service loss.

2. *Infrastructure serving areas free of erosion risk* Where infrastructure would still be required to provide the services lost through erosion, to meet a continued demand in an area not at risk of erosion, the provision of least cost alternatives may be a suitable measure of the loss from their erosion.

How the loss is assessed and its magnitude will depend on the existing excess capacity in the relevant infrastructure system. Thus if a through road is threatened by erosion the 'do nothing' scenario will depend on what alternatives are feasible and on the strategy with the lowest social cost: either investment in a new road or road improvements, or accepting increased congestion costs on alternative routes. This is effectively a straightforward investment problem identical to normal choices over expanding infrastructure capacity.

The rationale for this approach is that simply including an annual economic value for the use of the infrastructure at risk may overestimate what the 'without project' economic loss would be, for example where there is spare capacity in other parts of the same infrastructure network. The losses might even be underestimated if alternative ways of coping with demand are not available and so congestion disrupts the service on non-eroded parts of the network.

If no spare capacity or existing alternatives exist then the economic costs of providing new infrastructure, if feasible, are the appropriate benefit (cost avoided) of coast protection. If people and utilities were to use alternative existing facilities following erosion, and the congestion costs are less than the costs of new construction, then it would not be rational to build new infrastructure.

In this case it is appropriate to take the annual additional costs (resources and time) of this reorganised service (including the effects of any congestion) as the benefit from coast protection which avoids this infrastructure service disruption. This could be assessed in the case of roads using the same methods, such as the COBA manual (Department of Transport, 1971, 1981), as for assessing the potential benefits of new roads/road improvements.

Hence the loss avoided by coast protection is the least cost option for society in a 'do nothing' scenario. This may be the cost of building new infrastructure, or the costs of increasing the capacity of existing facilities, or congestion costs from increased use of existing facilities. Unless the infrastructure impacts of erosion are likely to be very large it may be sufficient to discuss with local utility managers the implications of erosion and obtain an estimate either that the existing system could cope with the altered usage, or of the likely costs of maintaining services by installing new works.

While the assessment of the consequences of erosion risk to this type of infrastructure is more complex than that for property, it should not prove too difficult in practice. A flexible approach by both the agency appraising potential coast protection benefits and by those responsible for infrastructure is needed so that a fair estimate is made of what would be done if erosion were allowed to continue.

3.5.5 Infrastructure methods

The procedure given for infrastructure (Figure 3.3) presents firstly a decision: if the infrastructure is only part of the coastline or only part of erosion-prone properties then techniques and procedures given elsewhere in this Manual can be used and will include the implications of this loss (respectively, the recreation procedure of Chapter 4 and the extension of property life procedure of Section 3.4).

However, where the infrastructure loss could affect activities which would continue after the loss the least cost strategy for coping with this loss needs to be identified and evaluated by conventional means; the avoidance of this cost is then the potential benefit from coast protection. The application of this procedure will depend heavily on location-specific factors.

3.6 Valuing non built-up land

3.6.1 Scope of the techniques

This section is concerned with the techniques for valuing potential losses from erosion of, first, agricultural land and, secondly, other open spaces such as golf courses and recreational land. The latter type of loss may be more appropriately valued through the methods developed for assessing changes in recreational use of coastal areas (Chapter 4), or in the case of environmentally important land the procedural approach of Chapter 6 should be adopted.

In both spheres, most erosion problems are likely to result in marginal but continual losses of land. Hence the effects on individual farms or other open spaces should be marginal in terms of their enterprises and overall use of fixed assets and the overall value of the sites (Chapter 2).

In general the potential losses from erosion of open space land will be of much lower value per hectare than built-up areas (i.e. the losses to housing or businesses). Hence assessments should break the study area into zones/reaches/units based on land uses, not only to reflect physical processes but also to reflect the scale of losses. This will help in revealing the distribution of potential benefits.

Different techniques for obtaining values are necessary compared with those relating to properties since only marginal parts of a land unit may be lost in any one year rather than all of a property; also undistorted economic values are not readily obtained from market land prices.

3.6.2 Agricultural land

Estimating the national benefit of avoiding or delaying the loss of agricultural land by erosion is not a trivial task. There is no real consensus on a straightforward approach which captures all the value elements of land. In principle, the objective is to evaluate the change in value to society (i.e. the opportunity cost) which arises from erosion. A number of elements make up this land value (Chapter 2), including:

1. The expected value of its agricultural output (a

use value);
2. Its 'option value' (arising from the possibility that future market changes will affect the value of the land) (a future use value); and
3. Its non-use value (the value that society places on the continued existence of that land, aside from any present use it makes of that land).

In a competitive market we would expect the market price of land to capture its opportunity cost. However, agricultural land prices have for some time been regarded as distorted (by capital transfer taxes, Exchequer support, and land use planning controls (MacKinder, 1980)). They also reflect a number of private benefits (perhaps a 'quality of life' benefit for landowners, and possible speculative returns), which society as a whole might value differently. There is, as yet, no way of solving this valuation problem.

Hence, while not denying the importance of option and non-use values, the current recommendation (and MAFF guidance) is that the value to society of protecting agricultural land from erosion is the annual economic margin arising from agricultural production that would be lost without coast protection.

An estimate of the area of cultivable land lost in each erosion contour band by land type is the starting point for this type of appraisal. Estimates of areas which would be lost will be based on the predictions of erosion rates and the derivation of erosion contours as discussed in Section 3.3.

An average cropping pattern should then be compiled for that area, and predicted for future years for each category of land at risk. For outputs subject to quota restrictions (e.g. potatoes or sugar beet), the economic loss will not be to that crop but to the next best non-quota crop, since the quota allocation will be taken up by someone else: in most cases the next best crop will be food grain production.

Annual output quantities, producer prices and estimates of fixed and variable costs can be obtained from the latest edition of the *Farm Management Pocketbook* (Nix, 1991) or, for the eastern counties of England, from Murphy (1989). The MAFF agricultural land grades may be a useful first indication of likely productivity: in particular grades 1 to 3 should be distinguished from grades 4 and 5 since the two groupings have very different potential uses and values. If rotations include pasture or other inputs to livestock production the returns from these also have to be taken into account. Again, the relevant data can be obtained from Nix (1991) or Murphy (1989).

In order to use this data to estimate annual economic loss it is first necessary to adjust producer prices for the Exchequer support element: these adjustment factors are taken from MAFF (1985b). They are subject to periodic review and therefore it is advisable to contact MAFF staff from time to time to ascertain the current position. Adjusted producer prices are then multiplied by the output quantities to give adjusted revenue and adjusted net margin flows, estimated from:

Adjusted net margin = adjusted revenue *less* variable costs *less* fixed costs

In general the present value of the benefit is the discounted sum of the adjusted net margins between year p (when the land would erode in a 'do nothing' scenario) and year p+s (erosion year plus project life). The calculation needs to be carried out for each erosion band, and in effect this is applying the same ELF method as in Section 3.4.2.

In some cases changes in fixed costs will be small, and so agricultural benefits can be based on adjusted gross margins (i.e. adjusted revenue less variable costs). Generally, however, there will be changes in fixed costs as a result of erosion losses. It should be noted that with the economic prices prevalent in Britain at present (1992), adjusted net margins will be negative for a number of crops and land grades, indicating a negative economic return to society from their current use.

This implies that there would be an economic benefit from allowing erosion to take place. Since erosion losses are irreversible, this highlights the problem of non-use (existence and option) values, mentioned above. Currently small losses of land from erosion are probably of little importance to society, but cumulated and set in a context of rising sea levels and global climatic changes which could affect world agricultural production, the effects may become more significant and non-marginal. Hence there might be some benefit in maintaining the option of cultivating these areas in the future, although this would need to be a greater net economic benefit than from future intensification in other areas before the argument would be relevant.

However, land used for agriculture now need not continue in this use. If recreational and 'environmental' uses have higher social values, land with these uses could have a positive value to society. Hence it might be argued, in theory, that low output agricultural land should be valued, for example, as a low grade wildlife site (Section 3.6.3) which might or might not make it worth protecting and would leave open the option of improving it as an ecological site or converting it back to agriculture at some future date.

There is not currently an accepted reconciliation of this problem, but if agricultural option values or economic values for environmental losses become accepted then the social value of protecting agricultural land may become generally positive rather than negative. From the practical perspective of undertaking an appraisal it is recommended that negative net margins cannot be considered sound indicators of opportunity cost and should not be included in the estimate of economic loss.

3.6.3 Other open space land

This section briefly discusses the problems of assessing erosion losses for recreational land and 'environmental' land. There are much fuller discussions of the issues involved and methods recommended in Chapters 4 and 6 respectively. This section does not consider beach erosion which is covered in Chapter 4.

For informal recreational space, cliff-top erosion losses are likely to be marginal (see the Peacehaven case study: Section 8.3). If the procedures for estimating lost use value (Chapter 4) are followed it will be necessary to present in the interview surveys several scenarios reflecting different stages in the erosion process, as well as one with the proposed coast protection works, and to estimate visitor rates under each scenario. It should then be possible to estimate for erosion bands the loss to users inherent in the 'without protection' situation compared with the use value reported for the 'with protection' scenario.

However, recreational land will need to be treated separately by use. Formal recreational uses are likely to be marketed and so the costs of adjusting the business to cope with the loss or loss of net margin will be appropriate measures of marginal losses. For example, an 8-hole golf course would have little value but with some investment the 9th hole might be fitted into the reduced area following erosion.

However, at some point losses are likely to become non-marginal. Hence the alternative approach of estimating the costs of a shadow project – the opportunity costs of converting a similar but non erosion-prone area of land to this use, such as for a golf course – is an appropriate measure of the economic loss. In this approach, however, any residual value of the remaining erosion-prone open space should be deducted from the opportunity cost of creating the alternative site.

Where land of ecological value is at risk from erosion the methods of Chapter 6 should be followed so that a qualitative and semi-quantitative assessment of the ecological value of the site can be made, both with coast protection and for several stages of erosion. Economic evaluation of the sequential environmental losses from erosion is not currently possible. However, costing the creation of an alternative site into any proposed project (the shadow project technique) is one possible approach.

This approach – costing the creation of an alternative site – does not estimate the loss to society, but does indicate the cost of maintaining the same number of ecological sites of that type. Unfortunately this technique is limited since many coastal sites can be argued to be unique and hence irreplaceable to a greater or lesser extent. However, it should be remembered that the logic of this approach is reversed from its usual direction in the case of coast erosion.

If nothing is done then society must bear the loss of the ecological site or recreate another equivalent site. If coast protection is provided then this shadow project is avoided and the unique site is saved. The exception will be that in sites of geological interest coast protection is often seen as damaging and erosion is seen as valuable in itself (see Section 6.4.2).

Economic analysis in the cases of these open space land uses will clearly be complex and innovative. In some cases it will be virtually impossible, through lack of data or lack of clarity concerning the impact of erosion on the areas concerned and their values to society. However, the appraiser should not do nothing in these circumstances. At the very least a clear statement of the likely losses involved with and without a project should be catalogued, and their timing should be presented.

3.7 Summary, 'health warnings', data needs and procedures

This chapter gives the recommended method by which the benefits of protecting land and property from erosion can be calculated. The methodology recommended is based on assessing the value of the continued use of the land and property protected – temporarily – from erosion by coast protection works. The chapter presents, in Formula 3.1, a method whereby these benefits can be calculated.

It should be noted that the results of these analyses will be strongly influenced by the rates of erosion predicted for the future, and these must be the subject of careful attention. They are also dependent upon gaining data on the typical value of erosion-free property, and this again is an area where attention to detail will be important. In addition, finally, particular attention should be paid to the timing of erosion losses and also the timing of any proposed investment. Both factors will have a strong influence on the calculated benefits of coast protection projects.

3.7.1 Data requirements and availability

The data requirements for estimating erosion rates are not considered here: these are the domain of the engineer and geologist. The other main need for data is on the values of losses, and these requirements are briefly summarised below:

1. *Residential property values which exclude the risk of erosion.* For outline 'pre-feasibility' assessments where less reliability is required, the data in Tables 3.1 and 3.2 can be used, inflated to the appropriate base date (using one of the housing price indices; we recommend the index in the quarterly Department of Environment/Building Societies Association BS4 survey). However, regional and local variations in house prices are common, and the data in Table 3.1 combined with Table 3.2 may not match local price differences. Hence for greater accuracy and where many properties are involved either individual property values should be estimated or local values obtained for different categories of property.

Table 3.1
Residential property prices by region

	Housing land price (£/ha) 1989[1]	Average new dwelling price £ 1989[1]	Average (all) dwelling price £ 1989[2]	Average (all) dwelling price £ 1991 3rd quarter[2]
North	284,800	58,300	37,374	46,369
Yorkshire and Humberside	250,900	62,500	41,817	52,943
East Midlands	413,700	71,300	49,421	57,277
East Anglia	541,900	71,900	64,610	63,760
Greater London	3,090,900	85,200	82,383	86,356
Rest of South-East	760,000	96,500	81,635	79,769
South West	460,800	79,400	67,004	68,322
West Midlands	457,500	78,000	49,815	60,578
North West	382,000	70,100	42,126	55,049
England	451,600	78,300	NA	NA
Wales	180,000	64,900	42,981	51,414
Scotland	NA	55,700	35,394	48,910
Great Britain/UK	NA	75,700[3]	54,846[4]	63,926[4]

Notes: 1. Central Statistical Office (1991)
2. Government Statistical Service (1991)
3. Great Britain
4. United kingdom

Table 3.2
UK residential property prices by dwelling type

Dwelling type	Price (£) in 1988	% of average for all dwelling
Bungalow	56,297	104%
Detached House	81,152	150%
Semi-detached House	44,687	82%
Terrace House	39,218	72%
Flat or Maisonette	46,600	86%
All dwellings	54,280	100%

Note: Includes both old and new dwellings, for all UK.

Source: Government Statistical Service (1989)

Local valuation departments are likely to be the most able to give individual property price estimates (this may be useful for multi-occupancy units which may be of much higher value than adjacent similar but undivided units). Alternatively a sample of typical prices quoted by local estate agents may be used. In the latter case it is best to average the available prices from several estate agents where they differ. In both cases it is important to define the task as being to estimate a value for a property of the same quality as the one at risk from erosion but assuming that it is not at risk from erosion.

2. *Business property values.* This is a very diverse sector for which it is not possible to report average data, since there are no secondary sources which summarise average prices per m^2. For example, data collected from any one case study is likely to be strongly influenced by local factors and the types of business surveyed. Hence local estimates of these values are the only reliable option (although the per hectare values for residential development in Table 3.1 may give a first indication for use in a pre-feasibility study). Business property values may vary substantially according to the actual use and condition of the property. Therefore property-by-property values should be taken if, for example, a sea front with many business premises is likely to be eroded in the near future.

3. *Infrastructure losses.* These can often be included within the calculations for other losses; however, where investments in adjusting to erosion could be avoided by coast protection the relevant strategies and values will have to be estimated with the help of the local managers of the services involved. No standard or typical data is possible for these values.

4. *Agricultural land.* For rapid assessments the potential benefits from protecting agricultural land can be assumed to be zero. For more detailed assessments, where better grades of land are at risk, it may be worthwhile to assess the net margin loss. Cropping pattern and land use data can be obtained either from local farm management surveys (carried out by Agricultural Colleges, or University Agriculture Departments), or by interviewing farmers. For net margin calculations, and the adjustments to economic values, it will be necessary to use data in the Farm Management Pocketbook (e.g. Nix, 1991) and the latest MAFF note on agricultural output valuation adjustments (e.g. MAFF, 1985b).

5. *Other open spaces.* There are no standard data sets or sources for data. The methods referred to in Chapters 4 and 6 should be used if recreational or environmentally important land is at risk of erosion.

3.7.2 Operational methods for assessments

Relatively simple but repeated calculations are needed to compute discounted present benefits from coast protection. Computer methods are advised to ensure speed and accuracy.

Since annual benefits will vary according to the locations and types of property and land relative to the predicted erosion line, a spreadsheet is the most appropriate computer method when the study area is complex and where a number of alternative scenarios are to be assessed. This approach has the advantage that the relevant extension of life factors (ELFs) can be computed easily and then used in a spreadsheet.

For an area with very few properties at risk or with non built-up land and constant erosion rates then simple hand calculations using the data presented in this Chapter and Appendix 3.2 will be enough.

One assumption implicit throughout the method presented in this chapter is that the proposed coast protection works are effective for their expected life – that is to say, they halt erosion. There is nothing in the case of erosion analogous to residual 'with project' flood risks and losses (Chapter 5), although the extension of life method allows for variation in the design life of a project. Indeed this is one of the design parameters to be appraised.

In theory, however, the assessment should also include a factor for the risk of 'failure' of the coast protection works: a storm event resulting in prematurely renewed erosion. Incorporating this risk into the benefit calculation is complex since the risk may change over time. The simplest method of incorporating this risk is in the project costs: the cost estimates for maintenance should include the risk of extreme events in each band of years during the project life and an estimate of the costs of repairing the damage so that the project is again effective and erosion is prevented until the normal end of the project life.

4 Coastal recreation: the impacts of coast protection and sea defence projects

4.1 Summary of information

This chapter examines the issue of the recreation benefits from coastal sites, and the methods by which the recreation benefits of coast protection and sea defence projects may be evaluated.

Our research has indicated that these recreation benefits, particularly the benefits of preventing the erosion of beaches, are likely to be among the most important benefits of coast protection, in addition to the benefits of the protection of property and other major infrastructure works such as arterial roads and main sewers. In some locations these recreation benefits may be equal to or greater than the benefits of protecting property. The inference is that a significant number of coast protection and sea defence projects could be justified solely by their retention of recreation beaches.

In Section 4.2, the problem of recreation benefit assessment is defined and discussed in general terms. A programme of research was undertaken to develop and test methods of valuing coastal recreation benefits. This programme and its results are described in Section 4.3. The methodology, mode of application and some of the main results of the research are presented there. An example of one survey instrument used for beach surveys in the course of the research is included in Appendix 4.1. Further illustrations of the application of our recreation benefit assessment methods in particular circumstances are provided in the case studies in Chapter 8.

Section 4.4 outlines the recommended procedures for recreation benefit assessment and includes a flow diagram (Figure 4.1) indicating the recommended steps to be taken in carrying out a recreation benefit assessment as part of an overall project appraisal. In Section 4.5, the recommended methods for data collection at the pre-feasibility and feasibility stages of a project are discussed in detail. Recommended methods for carrying out Contingent Valuation Method (CVM) surveys and visit count methods are given. Examples of standard materials that may be used in conducting surveys and counts are provided: questionnaires (Appendix 4.2), show cards (Appendix 4.3) and a count sheet (Appendix 4.4).

Issues and recommended methods for the analysis and interpretation of the results are discussed in Section 4.6. Section 4.7 summarises the material presented in the chapter and lists some of the limitations of the methodology and data that are provided. The final section includes a guidance checklist of issues to be addressed in carrying out a recreation benefit assessment. This checklist is to be used in conjunction with the flow diagram discussed in Sections 4.4 and 4.5 when planning and conducting an assessment.

It should be noted that whilst much of the material in this chapter relates most explicitly to coast protection (i.e. erosion control), there will be recreational benefits from sea defence projects designed to alleviate coastal flooding. This will particularly be the case where beach recharge and other beach works are used for sea defence works, or where sea walls and associated promenades are proposed for the same reasons. The difference will usually be that sea defence leads to beach improvement, whereas coast protection is aimed at both halting beach deterioration and beach improvement.

4.2 The problem defined: coastal erosion, protection and enhancement

The recreational use of coasts is an important function

COASTAL RECREATION 59

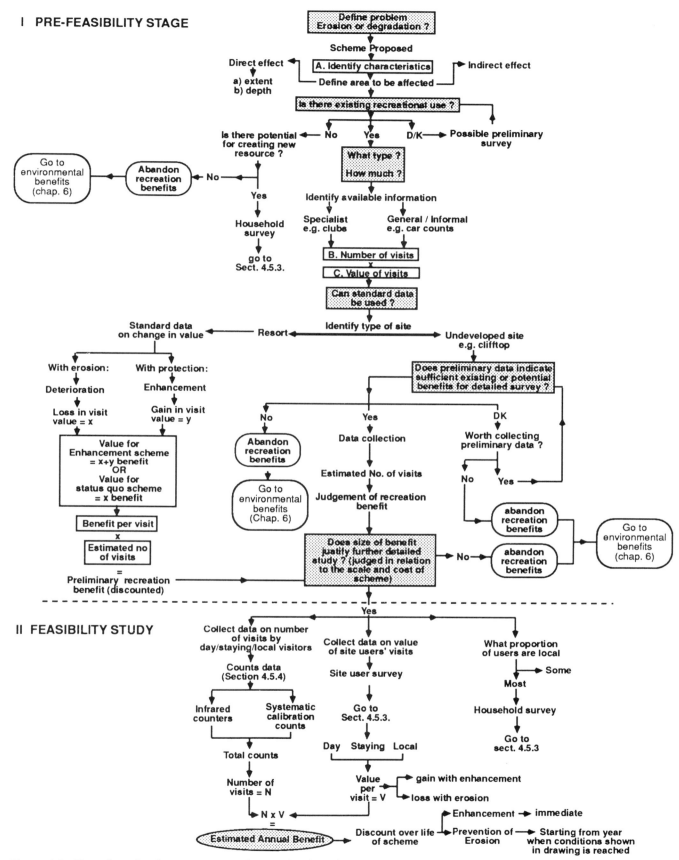

Figure 4.1 Procedure for the assessment of recreation benefits

which may be affected by coastal erosion as well as by coast protection and sea defence works. The functionality of beaches appears to be directly related to their physical characteristics and therefore changes brought about either deliberately by a project or occurring incidentally as a result of unchecked erosion could have significant effects on the recreational function of the beach and the enjoyment experienced by its users. A further factor concerning beaches is their physical complexity. The consequence of this is that both erosional processes and coast protection works also alter them in complex ways.

The first issue is how to identify the effects of current coastal erosion processes on the physical structure of the coastline. This requires the following:

1. An identification of the **extent and characteristics** of the area affected;
2. An estimation of the **nature and rate of erosion**; and
3. An estimation of the **type of physical changes** which are currently occurring.

The next issue is to identify the ways in which these physical changes are affecting different recreational users. The most significant changes from a recreational perspective are likely to be changes to cliffs, beaches and promenades. The third issue is to identify the possible physical effects of the proposed project or project options and variants. The fourth issue is to identify the effect of the project on the users.

The third and fourth require an outline of the nature of the proposed project from which its likely physical effects can be predicted, and the extent and characteristics of the area to be affected can be identified. A range of strategies exist for coast protection and these can have widely differing effects on the physical characteristics of the coastline and hence on the recreational benefits of any project.

The effects of both erosion and coast protection on beaches are particularly important because they represent the 'first line of defence' against the sea and are therefore likely to show the initial effects of erosion. In addition, the majority of recreation benefits have been found to come from beach users and hence the impact of changes to the physical structure of beaches could significantly affect the valuation of recreational enjoyment.

But coastal sites change naturally over time. The condition of coastal sites can be represented as a continuum, reflecting the degree of erosion of the site. The characteristics and the rate of change will vary from site to site and may be complex. A simplified representation of a site degrading as a result of erosion is given in Figure 4.2.

Any study of recreational benefits for a coast protection or sea defence project will normally be undertaken only when some deterioration has taken place at the site. However, a coastal site will not usually be allowed to become totally degraded before coast protection authorities intervene, or at least consider protection measures. Thus a feasibility study can be expected to be undertaken when the coastal site is not in a good condition – with regard to erosion (i.e. C in Figure 4.2) – but has deteriorated to some degree, a stage represented by the positions A – A_1 in Figure 4.2. A study will normally be initiated at a time when further deterioration can be expected to occur until the site is further eroded (i.e. B in Figure 4.2) or becomes very severely eroded (i.e. D in Figure 4.2).

Coast protection projects and their recreation benefits may have two components, as follows:

1. The prevention of further deterioration (i.e. from A – A_1, to B or D in Figure 4.2): this is 'do the minimum', or the maintenance of the status quo (A – A_1).

All projects will involve this first component but many projects will involve a second component:

2. A reinstatement of the condition of the site, from the current state to a better condition, such as that before deterioration took place (i.e. from A – A_1 to C in Figure 4.2), or to some other even better condition.

A project, therefore, may involve an improvement in the condition of the site rather than just prevent further erosion. Furthermore, a project may do more than just restore the site to its condition before the deterioration took place. Thus projects may involve an improvement beyond reinstatement or they may change the nature of the site in such a way as to offer an enhancement in recreational or amenity terms (i.e. a change from A – A_1 to C_1 in Figure 4.2).

The second component – an improvement or enhancement to a site as a result of a coast protection project – may only be eligible for central government grant aid if the improvement or enhancement is integral to an optimal protection project, that is it forms the best protection solution in technical and benefit-cost terms and is environmentally acceptable. For some sites, the optimal protection solution will involve changes to the site: beach nourishment or new structures such as fishtail groynes or an off-shore reef or a new sea wall which may enhance the amenity or recreational potential of the site. It is also recognised that protection projects may have to be given a design that is some way beyond the most basic treatment possible in order to be environmentally acceptable according to the standards of the day.

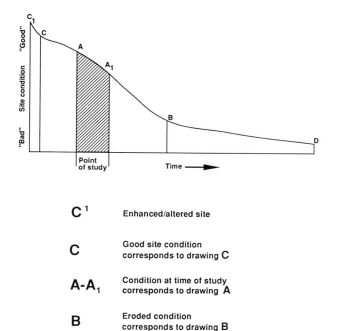

C¹	Enhanced/altered site
C	Good site condition corresponds to drawing C
A-A₁	Condition at time of study corresponds to drawing A
B	Eroded condition corresponds to drawing B
D	Severely eroded condition

Figure 4.2 Theoretical model of range of conditions at coastal sites (for drawings, see figures 4.5 and 4.6)

The fourth issue – and complexity – is how to value recreational enjoyment in order to assess the change in enjoyment with continued erosion or with a coast protection project. This is discussed further below.

4.2.1 The effects on recreation of changes to the physical environment caused by coastal erosion

The first consideration when attempting to gauge the effects on recreation of changes to the physical environment caused by coastal erosion is to define the extent of the area affected. Erosion may be site-specific or it may affect a more extensive area, as in the case of the Holderness coast. Where erosion results from the reduction in supply of sediment, this may be as a result of coast protection works further 'updrift' along the coast and the direct cause and effect relations are often difficult to identify.

The second consideration is to determine the characteristics of the area affected. The following classification is used (1–4).

1. *Beaches* The following characteristics of beaches may be altered by erosion:

- the composition of beach material;
- the beach profile;
- the beach's extent at low tide;
- the extent of beach above high tide levels; and
- the volume of beach material.

These are discussed below, in turn.

The type of beach material will change with erosion, as erosion removes finer material such as sand through wave action and leaves behind coarse material such as pebbles. This has an effect on enjoyment by recreational users, since the type of beach (whether it is sandy, pebbly, etc) is important in determining its suitability for a range of recreational activities.

The actual impact of changes in beach material on recreational use will vary depending on the characteristics of the beach users; for example, a sandy beach is more desirable for children playing. If a beach becomes too pebbly, its recreational value to older user groups for walking, sitting and lying will be similarly reduced. Local users are also more likely to be affected by changes in beach material than day visitors or staying visitors, as their choice of suitable locations is more restricted than that of other user groups.

The impact of erosion on the beach profile will be affected by the interaction of two factors: intensity of wave action, and beach composition in terms of particle size. In general, the larger the material, the steeper will be the beach profile, although beaches composed of the same material often show a wide variation in stable slope angle. Wave backwash drains down through coarse material with little effect on the profile whereas finer material is moved down-profile by wave action thus reducing the beach gradient. Hence wave characteristics have a greater influence on beach gradient than does beach material and in an eroding situation the outcome of changes to beach profile must be determined with reference to both beach material and the nature of wave action at that specific site.

The implications of changes in profile for recreational users are similarly influenced by the type of recreational activity being undertaken. As beaches steepen, their value for walking, sitting or lying will be reduced. However, the main impact on recreational users is more likely to result from changes in the amount of beach exposed at different states of the tide than on the profile *per se*.

Beaches which flatten as they erode will be at a lower level than adjacent structures such as promenades and groynes. Such beaches will be uncovered for less time at low tide, thus restricting the available recreational opportunities. As a beach profile flattens with erosion, less of the upper part of the beach will remain exposed at high tide and this will also restrict recreational opportunities.

A further aspect of the physical change of beaches due to erosion is that as the total volume of beach material decreases, the beach will become lower overall in relation to any fixed structures such as groynes, sea walls or promenades. This will reduce ease of access to the beach and may create dangers and inconvenience, such as where steps from the promenade to the beach become excessively long or steep. Additionally, where longshore drift is significant, different levels of beach material will build up on either side of a groyne thus creating further potential hazards.

2. *Promenades* As a beach is lowered by erosion, its role in protecting sea walls and the promenades they may support is reduced and such structures become undermined.

Initially, this creates hazards for beach users. If left unchecked, undermining will eventually cause surface cracking, slumping and collapse of the promenade resulting in loss of enjoyment for the recreational users of both the beach and the promenade itself. In extreme cases, access may have to be restricted altogether due to dangerous conditions, thus considerably restricting recreational opportunities. Even before this stage is reached, recreational enjoyment may be impaired by a deterioration in the visual appearance of the sea wall and in the condition of the promenade, for example when its surface and railings become in a poor or dangerous condition.

However, as promenades are almost always an 'optional extra' on an essential sea wall structure (except in so far as they provide essential access for maintenance of that structure), care must be taken when estimating coast protection benefits that there is no double-counting of benefits by including both the value of the promenade as well as the recreation benefits that result. That is to say, do not count as a benefit of coast protection the value of the promenade *per se* – even though the works will protect it – because this benefit is either captured as part of the benefits of the sea wall itself (in protecting property behind) or in the recreational value of the promenade (through the recreational survey). Further guidance on this point is given in Section 3.5.2.

3. *Cliffs* The effects of erosion on cliffs will obviously be influenced by the geological characteristics of the cliff material. Cliffs erode spasmodically and although average long-term rates can be estimated, actual rates of erosion are notoriously difficult to predict.

Where the beach in front of a cliff has eroded as a result of a reduction in material from the source of sediment, the erosion of the thereby unprotected cliffs is likely to proceed more rapidly in most locations. In the case of some types of cliffs, erosion may result in large pieces of material falling on to the beach below as the base of the cliff is progressively undermined by wave action. In other cases, there will be gradual slumping of the cliff face or a significant rotational slide of the whole of the seaward face.

Where the main cliff material is clay, the occurrence of these events may be determined more by the amount of rainfall than by the action of the sea. This makes it particularly difficult to predict the occurrence of dangerous erosional events, and remedial action, such as fencing off dangerous areas, may be correspondingly delayed. General recreational users of both cliff-tops and beaches may therefore experience a reduction in their recreational enjoyment as a result of cliff erosion. On the other hand, those visitors with a specific geological interest will prefer the erosion to continue in many locations in order to maintain exposures of strata which might otherwise become obscured by the accumulation of the products of weathering once the cliff toe has been protected.

For general users of cliff-top sites the increased danger resulting from instability at and near the cliff edge may reduce the enjoyment of their recreational visit. In some cases it may be necessary for access to be restricted and this will also detract from the recreational experience. For beach users, the potential danger from cliff falls on to the beach is similarly likely to detract from their recreational enjoyment of the beach. For walkers, and in particular long distance walkers, the loss of a coastal route in places and a deterioration in general amenity where a coast path cannot be moved back easily can seriously detract from their enjoyment of their recreational experience.

4. *Other coastal areas* The main loss of recreational enjoyment in other coastal areas affected by erosion is likely to result from any restrictions in access for recreational users, as in the case of eroding sand dune systems.

Other types of coastal area, such as saltmarshes, are little used for recreational purposes by the general public and what recreational use there is may be confined to minority interest groups such as bird watchers. However, it should be recognised that potentially significant losses in recreational benefits may exist from this category of user as the valuation of their recreational enjoyment may be quite high even though the numbers of such users are often low.

No detailed studies have been done on this aspect of user values and therefore no recommendations can be made except that in such cases a study of the valuation of recreational enjoyment by the specialist users should be carried out.

4.2.2 The effects on recreation of changes to the physical environment caused by coast protection or sea defence projects

There are two main options for coast protection and

sea defence: to do nothing or to carry out a protection project. There are three main options for these coast protection or sea defence projects, as follows:

a. Do the minimum;
b. Maintain the status quo (by preventing further deterioration due to erosion or returning to some safe condition); and
c. Improve the standard of protection.

There are several techniques for coast protection, as listed below:

- Beach nourishment;
- Conventional groynes;
- Rock armour protection of sea walls;
- Fishtail groynes;
- Off-shore breakwaters;
- Sea wall construction, reinforcement or upgrading; and
- Cliff re-grading.

Each of these may affect the physical characteristics of an area in a different way and therefore have a different impact on recreational users.

In general, the effects on recreational use range from extreme disruption in the short term (for example, through restriction of recreational access due to major engineering works), to minimal recreational disruption through smaller scale works outside the main tourist season. The former is likely to have higher long-term benefits (as in the case of Hastings: Section 8.2), whereas minimal disruption will result from projects such as those involving annual beach nourishment in winter by transferring existing beach material. However these types of projects may have lower long-term recreation benefits since they do not result in a significant improvement in the characteristics of the overall recreational resource.

1. *Beach nourishment* The effect of beach nourishment on beach users' enjoyment will depend on the nature and amount of the material used for the beach recharge project. Beach nourishment may be carried out to an extent that is more than is strictly necessary for reinstatement of previous beach conditions, resulting in enhancement of the beach and its use. This may not be allowable as a recreational benefit for central government grant aid purposes.

Where temporary beach nourishment takes place, the usual source of material is through recycling the material that has been transported along the coast by longshore drift. In this case it is likely that relatively coarse material will be used, since the finer particles will have been removed entirely by erosion. Where this occurs no upgrading of the beach quality will take place and there will be correspondingly little increase in recreational enjoyment derived from a change in the characteristics of the beach material.

Where nourishment is from outside sources, however, such as from offshore banks, it may be possible to exercise control over the mix of material used. In particular, the creation of a sandy beach is usually seen as providing an enhanced amenity for recreational users although its retention time will be less, and hence its maintenance more costly than with coarse material.

A further benefit from beach nourishment is due to the increased volume of beach material which will raise the overall height of the beach and thus increase the proportion remaining exposed and useable for recreation at high tide. The coarser the material, the better is likely to be the beach retention.

2. *Conventional groynes* Where groynes are enlarged and reinforced rather than being replaced simply to the same standard by reinstatement, they are more likely to retain fine beach material and therefore to maintain a sandy beach for a longer period. Then the recreational benefits of a sandy beach are likely to be retained for longer. This is particularly the case where longshore movement of material can be controlled through groyne design (as at Hastings: Section 8.2).

The presence of groynes may be seen as beneficial for informal recreational users by providing shelter from the wind and support when sitting. They also serve a useful function for families with small children by restricting the children from straying far and by delineating the 'home area' on the beach.

Groynes, however, may also detract from recreational enjoyment due to their aesthetic intrusiveness, in particular the loss of a beach view and a sense of openness, although our research has not been able to confirm this possibility. The categories of users most affected by a high level of groyning are those undertaking activities such as walking, strolling or walking the dog, who find the groynes restrict their free movement along the beach.

3. *Rock armour protection* Where this method is used to protect the sea wall or promenade, it will change the appearance of the beach in a way that some recreational users find intrusive. The rock armouring is perceived as being more dangerous and restrictive of access, although this need not be the case. In addition, a rock armour system may contribute to the trapping of litter.

4. *Fishtail groynes* Where these have been used (e.g. the Wirral Scheme (Davies 1989)) it is usually possible to remove and not replace the conventional groynes. This can have a considerable impact on the characteristics of the beach by giving it a more expansive and less enclosed feel.

Such structures also often offer the opportunity of creating new opportunities for access for water based activities such as angling and sailing, by providing access to the sea and also shelter.

To maximise these opportunities it is necessary to design the structures with recreational considerations as an integral part of the design brief. This may result, for example, in the construction of a walkway for anglers and strollers along the top of a rock groyne. The provision of ramps and even access to floating berths can also add to the recreational enhancement of such a project for water based users.

5. *Offshore breakwaters* Offshore breakwaters will provide areas of sheltered water for a range of water based activities. They may also increase access for beach users by the creation of a beach where there was none previously, as at Monk's Bay, on the Isle of Wight. In addition to the direct benefits to water based users, there may be additional benefits from the increased enjoyment created for general users of both the beach and promenade through watching such activities.

Some detrimental effects for recreational users may arise from the breakwater itself being a hazard to boating. There may also be indirect effects on recreational users over a wider area where such breakwaters reduce longshore drift and sediment supply and hence cause erosion of adjacent amenity beaches 'downdrift'. In addition, their existence has a substantial visual impact although the effect of this on user enjoyment has not been quantified.

6. *Sea walls* Sea walls seldom have any direct recreational benefits in themselves. The construction of a promenade on the sea wall will be included as a benefit in an overall recreation benefit assessment of beaches: double-counting must be avoided (Sections 3.5.2 and 4.2.1).

Almost invariably, any type of sea wall which will improve beach access will cause loss of beach as it enhances the drawdown erosive action of waves breaking against it. In particular, where sea walls exist without the presence of groynes they also detract from recreational enjoyment by increasing beach scour and hence reducing amenity beaches. They may also be perceived as scenically intrusive, although scenically attractive sea walls can be designed.

7. *Cliff re-grading* The main recreational benefit from coastal treatment by cliff re-grading arises from the potential for the creation of an amenity walkway which will enhance access for the same range of users as outlined in the previous sections. However, dangerous conditions or blocking of access may occur without adequate expenditure on maintenance: cliff re-grading will need to be a continuous process.

4.2.3 Assessment of recreational benefits: national economic benefits versus local benefits

The next issue to consider is how to place a value on coastal recreation (Section 2.3.1) in order to assess potential benefits and losses due to erosion, coast protection and sea defence works.

Recreation benefit assessment for use in benefit-cost analysis is concerned with national economic benefits and not with benefits to a particular local economy. If changes at a particular coastal site simply result in a transfer of recreation from one site to another without any overall gains or losses in the value of the enjoyment of the recreation once travel costs have been taken into account, then no national economic gain or loss will be involved in the change.

The availability of substitute sites for recreation must therefore be considered when the benefits of coast protection or sea defence are being assessed. If, when a site deteriorates, visitors can transfer their visits at no extra cost to a nearby site which they find equally attractive and enjoyable to visit, there will be no national economic benefit. Similarly, a national economic benefit will arise, when a visitor transfers visits to a reinstated coastal site from another site, only if the gain in the value of enjoyment in visiting is greater than any additional costs incurred in visiting the reinstated site.

Furthermore, recreational benefits must be distinguished from the benefits of tourism to the local economy. An increase in the number of day and staying visitors at a coastal site may generate through their spending additional business activity in the local economy. These benefits that tourism may bring to a local economy through multiplier effects should not be included in any assessment of project benefits. For example, the Herne Bay case study (Section 8.5) includes an assessment of the effects of a coast protection and sea defence project on the local economy, but these effects were excluded from the benefit-cost analysis and were only assessed for use in a financial appraisal to determine what extra enhancement of the seafront should be undertaken by Canterbury City Council as part of its town improvement project (i.e. not as part of the coast protection project).

4.2.4 Assessment of recreational benefits: the approach

In order to assess changes in recreational value consequent upon coast protection or sea defence works it is necessary to follow certain **key steps**. It is **first** necessary to put a value on the recreational experience of an individual and his/her experience, **then** gross this value up to reflect the total recreational use of a site by the population using that site. It is **then** necessary to ascertain how this value would be altered by a project and its options, and **finally** it is necessary to calculate the change in benefits for the whole area which would be affected. **None of these steps can be omitted.**

In this analysis, access to a site is an important determinant of its recreational use. It is necessary to identify the current availability of access and any changes which would result from a project, to determine whether the project would diminish or increase access opportunities, or leave them the same. The five categories of information required are discussed below.

The scale of the area to be affected by the proposed project It is necessary to identify the area which will be both directly and indirectly affected by any project in terms of its recreational use. This may include knock-on effects outside the immediate project area caused by changing patterns of recreational use resulting from the project itself. The extent of these impacts will depend on the nature and type of project proposed. Effects may be felt at the macro-scale (throughout a region) or at the micro-scale (at a site level). Both the lateral extent of the project influenced area and also its extent in terms of distance behind the seafront (project width) should be identified as far as possible at the outset.

The area affected by a project is referred to here as the 'seafront'. This comprises both the beach and promenade or sea wall if one is present. The beach is the most important area to protect in recreational terms. If the beach deteriorates, not only will recreational users suffer a deteriorating quality of enjoyment but the first line of coast protection will be undermined. The works for any coast protection and sea defence project will normally affect both these components of the seafront. It may also be necessary to determine the effect of the project on the water space in front of the beach and how far back from the seafront the project's impact will extend.

The types of recreational users The range of recreational activities currently being carried out at a site should be identified in order to categorise types of recreational user. This can be done using the checklist of recreational uses provided in Appendix 4.5.

The largest category of coastal user is generally the informal users of the beach and promenade, where one is present. Users of these two areas will be referred to below as 'seafront users'. Other categories of informal coastal user such as users of the cliff-tops and coastal footpaths are not usually present in sufficient numbers to contribute significantly to the recreational value of the seafront.

Although a range of recreational activities exists, two broad groups of users can be identified, as follows:

1. Specialist users, such as boardsailors and anglers;

and
2. Generalist users carrying out informal activities, such as strolling, sunbathing, picnicking and swimming.

Recreational use may be affected by a project in several ways. Existing uses will be maintained, or will be enhanced. Opportunities will be created for new uses, or areas will be restored where previous recreational use has declined or been lost altogether. As stressed above, a major factor contributing to the change in recreational use will be the extent to which the project affects access to the recreational resource.

Existing levels of recreational use The amount of recreational use which a site currently receives can be determined in terms of the number of recreational visits it receives. It is necessary to have an estimate of the way in which this use varies throughout the year in order to calculate total site use on an annual basis.

A pre-feasibility desk-top review should be carried out to define current levels of use. At this stage, an estimate of overall use levels in terms of the number of visits will be adequate for a preliminary quantification of benefits. The method employed to obtain this information will vary depending on the type of user and the nature of the site under consideration.

As far as users are concerned, where specialist activities are carried out the data on user numbers may be available from specialist clubs, although in many cases users may not be members or make use of facilities provided by such clubs (e.g. boardsailors may park in public car parks and launch from the beach). Therefore it is necessary to be aware that data from such private or semi-private sources is usually an underestimate of total use levels. Where sites are mainly used for general recreational activities, other sources of data will have to be employed and the availability of this data will largely depend on the nature of the site.

Sites can be roughly categorised into three types: commercial resorts; non-commercialised urbanised sites; and non-developed 'wilderness' sites. Empirical survey work has indicated that variations in the level of use of sites are so great that standard use data cannot be provided even according to broad site types. Therefore initial use estimates will need to be made, based on existing information. Data on levels of use may be available from previous site surveys or visitors counts from which current use levels can be extrapolated. Even such sources as student projects may provide basic counts of visitors which can be utilised for this purpose. If no such information is available, indirect measures of site use may have to be employed. Data which could be used for this purpose include such sources as car parking records and deckchair rentals.

If no secondary sources are available it is necessary

to determine at an early stage whether an intuitive estimate of user levels can be made, or whether some empirical data collection is required. If such data is needed its collection must be planned well in advance, especially where seasonal factors are likely to have a significant effect, which is usually the case at coastal sites.

Non-developed recreation sites are, by definition, those where the number of users is likely to be small and no sources of data generally exist on visitor numbers. These cases raise problems in determining at the outset whether or not it is worthwhile to measure recreational benefits. Once overall project benefits are known, if recreational benefits are then seen as potentially significant, there will usually be insufficient time to carry out field counts in order to obtain accurate use figures. This problem can only be overcome by exercising good judgement at the outset, based on the best estimates of site use available at that time which may require extrapolation from use levels at other similar sites.

Number of visits in terms of visitor types Existing levels of site use alone are inadequate for the calculation of the recreational value of a site, since different categories of visitor have been found to give different values for their recreational enjoyment (Table 4.3).

Three categories of visit type have been identified:

1. **Locals.** These are defined as those living within a three mile radius of a site. This distance has been employed because it represents a walk of one hour which is judged to be a realistic maximum for local users.
2. **Day visitors.** These are defined as anyone starting the trip from their normal place of residence. This category also includes visitors from the local vicinity but from outside the three mile radius who may consider themselves as 'local'.
3. **Staying visitors.** These are defined as anyone staying away from home for more than one night. In a resort, this would include holidaymakers staying in another location who might be deemed as making a 'day visit' to the site in question.

It is necessary to make some attempt to partition the total number of site visits into these visitor types. The only satisfactory way to obtain the necessary information to do this is to carry out sample counts, preferably at different times of the year, which identify users as belonging to one of these three categories of visitor.

The value of the enjoyment of an individual visit
Recreational use of coasts is generally regarded as a 'free good' even though some costs may be involved for a user such as car parking charges.

There are a number of methods available for attaching a value to such goods which have no market price (Section 2.5):

1. The Travel Cost Method (TCM); and
2. The Contingent Valuation Method (CVM), which includes the following versions:

    i) Willingness to pay
    ii) Willingness to accept, and
    iii) Enjoyment per visit

With the Travel Cost Method (Section 2.5.3) the value of recreational enjoyment of a site is equated with the costs incurred in travelling to the site. This approach has been widely used for valuing recreational activities in the United States but has not been found to be appropriate for coastal recreation in Britain. A number of reasons can be suggested for this. First, there are a large number of roughly comparable sites around the British coast and the distance-decay factors found to operate in other contexts are not applicable here. Also, decisions to visit the coast may be conditioned by decisions unrelated to distance, such as proximity to family, traffic delays and road conditions.

A second reason why the Travel Cost Method is inappropriate for coastal recreation is that in Britain visits to coastal sites may be combined with pleasure motoring or may be unplanned and result from a spontaneous stop at an appealing location. It is therefore not possible to separate out the different components of travel cost.

A further problem with the application of this method is that many people visit coastal sites in conjunction with other activities. These may be as varied as shopping, visits to inland sites and visits to relatives. It is therefore not possible to apportion that part of the travel costs which relates specifically to the recreational use of a coastal site. Travel cost methods are suitable only for sites where users incur travel costs in getting to the site, and will be inappropriate for sites which have a large number of local users who reach the site on foot.

The Contingent Valuation Method (CVM) (Section 2.5.5) is a survey method which has been found to be the most appropriate method of valuing enjoyment in the context of recreational use of coasts. It assumes that recreational users can express the value of their enjoyment in money terms. In the case of recreational enjoyment this is by reference to activities which have a known market value and give comparable levels of enjoyment (Section 4.3.1). The CVM is therefore an expressed preference approach in which people are asked how much they would be willing to pay for a

COASTAL RECREATION 67

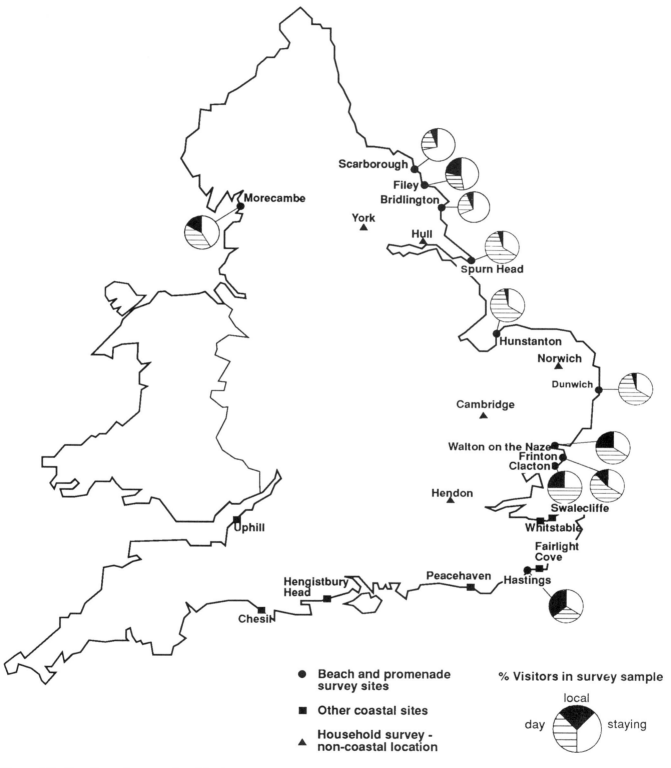

Figure 4.3 Location of user and residents surveys

given change in the conditions of the recreational site.

The method can be used in site surveys of current users or in general surveys such as household surveys to ascertain information about previous recreational visits. The problems with the application of the method include, first, the fact that individuals may be unwilling to express the value of recreational enjoyment in money terms: approximately 20 per cent of coastal site users were found to be unwilling to give a valuation for this reason in research for this Manual. Secondly, willingness to pay may be constrained by ability to pay and the method does not enable clarification of whether the value given is so constrained except in so far as values have been found to be significantly correlated with income.

Additionally, thirdly, the method assumes that the overall social value of the coastal resource is equal to the sum of all the individual expressed preferences. This is not necessarily the case, for example where a site has some scenic and environmental value which may be of more general value to society. A further problem with CVM is the nature of the good to be valued. In the case of coast protection and sea defence we are concerned with the change in the value of enjoyment both with and without a project. This requires valuation of hypothetical situations: of a deterioration in value without a project, and of a reinstatement or enhancement in value with a project. Clearly this makes for a complex series of questions and some potential for misunderstandings.

4.3 The methodology, execution and results of the research base

A programme of research was undertaken to develop methods and to test the practicalities of implementing the contingent valuation and travel cost methods of valuing coastal recreation, prior to developing simpler methods for continuing use by coast protection agencies in making their investment decisions. The theoretical and applications issues raised by these two valuation methods are also discussed in Sections 2.5.3 and 2.5.5. In order to test these methods, interview surveys were carried out with site users and households at the wide range of coastal and inland locations shown in Figure 4.3.

The research programme concentrated on beach and promenade recreation rather than cliff-top recreation for the following reasons. First, in general, seaside resorts and their beaches and promenades attract more visitors than do cliff-tops and therefore are likely to be much more important as a source of recreation benefits. According to the 1986 Countryside Recreation Survey (Countryside Commission, 1986), 65 per cent of the population of England and Wales aged 12 years and over had visited a seaside resort in the last twelve months; for cliff-tops and the seacoasts excluding resorts, the proportion was 36 per cent.

Secondly, beaches are often the frontline defence of the coast and their protection will often protect cliff sites and coastal landscapes as well as the property and other facilities behind them. Thirdly, beaches are often the first feature to be eroded and therefore will be more important in the discounted cash flow of benefits and costs. Cliff-top recreation areas may be quite wide and it therefore may be a number of years before cliff erosion can have a major impact on recreation by, for

Table 4.1a
CVM surveys of coastal recreation 1987–1990: value of enjoyment of this/today's visit

Year Sites	Size and type of sample[1]	Per cent able to value	Mean[2](£)	Log. Mean	Log. Std. Devn.
Beaches and promenades					
1988					
Scarborough	101 users	83	4.93	0.57	0.35
Clacton	170 users	90	9.96	0.89	0.33
Dunwich	101 users	61	6.87	0.59	0.48
Filey	88 users	88	3.64	0.47	0.29
Frinton	178 users	70	9.56	0.86	0.32
Hastings	247 users	66	7.72	0.70	0.42
Spurn Head	97 users	80	8.50	0.83	0.29
1989					
Bridlington	151 users	86	5.91	0.64	0.36
Clacton	146 users	67	10.52	0.85	0.40
Hunstanton	152 users	90	8.74	0.81	0.36
Morecambe	150 users	92	5.76	0.56	0.42
1990					
Herne Bay[3]	127 users	88	12.34	0.85	0.50
Herne Bay[3]	189 residents	83	3.59	0.42	0.49
Cliff tops					
1988					
Peacehaven[4]	214 residents	54	3.50	0.56	0.28

Notes: (1) Type of sample: users: UK resident adults aged 18+ interviewed on cliffs, beaches and promenades
residents: Adults aged 18+ in a sample of households in private residences.
(2) Mean value of enjoyment is calculated excluding those unable or unwilling to offer a valuation.
(3) Funded by Canterbury City Council.
(4) Included with the permission of Posford Duvivier.

Table 4.1b
CVM surveys of coastal recreation 1987-1990: willingness to pay (WTP) increased rates and taxes to protect cliff-tops

Year Sites	Size and type of Sample[1]	Per cent WTP	Mean(£) WTP[2]
Cliff-tops			
1987			
Walton on the Naze[3]	187 users	71	
£ per month (yearly equivalent)			17.11
£ per annum			6.73
1988			
Peacehaven[4]	214 residents	55	
£ per annum Starting point 50p			1.83
Starting point £1			2.83

Notes: (1) Type of sample: users: UK resident adults aged 18+ interviewed on cliffs, beaches and promenades
residents: Adults aged 18+ in a sample of households in private residences.
(2) Mean value of willingness to pay (WTP) is calculated excluding those unable or unwilling to offer a valuation.
(3) Pilot study.
(4) Included with the permission of Posford Duvivier.

example, causing the closure or diversion of a cliff-top path or the loss of a recreation area. More often cliff erosion will have only a marginal impact on recreation even in the medium term.

4.3.1 Beach recreation surveys

The research into beach and promenade recreation was carried out in two stages, with a series of beach recreation surveys carried out in the summer of 1988 and also in the following summer. In the course of the surveys, interviews were conducted with over 1,300 beach and promenade users at eleven coast sites in England. The locations where the interviews were carried out are listed in Table 4.1.

The Contingent Valuation Method is a particularly demanding example of the survey method and careful attention has to be given to all aspects of a CVM survey if reliable and valid results are to be achieved. This includes the design of CVM questionnaires, the sampling, and the supervision of fieldwork. For both the 1988 and 1989 surveys, a substantial structured questionnaire was developed, which took an average of 20 to 25 minutes to administer.

In each year, the questionnaire was based on a model of recreational behaviour and included questions designed so that tests of the validity of the valuations obtained could be undertaken in the course of the analysis (Figure 4.4). One aim of the research was to derive a shortened and simplified model questionnaire incorporating the key variables required for a CVM survey from the extended instrument developed and tested in the research.

Slightly different versions of the questionnaire were used for locals, day and staying visitors. The proportion of local, day and staying visitors interviewed at the different sites is indicated in Figure 4.3. As the interviewers were asked in the 1989 survey to attempt to undertake a fifth of their interviews with local residents, the proportions shown in Figure 4.3 do not represent those occurring in the population.

However, despite this constraint, marked differences are apparent in the proportion of day and staying visitors found at the different sites. It was hypothesised and found to be true that the different visitor types

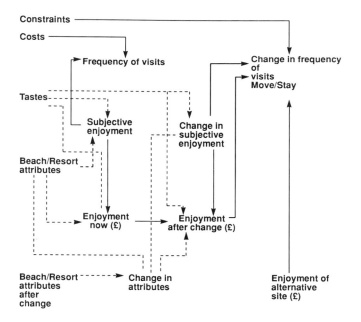

Figure 4.4 Theoretical model underlying 1989 seafront survey

would be attracted to the coast site by different factors, and that their visit characteristics – the frequency and duration of their visits and the activities engaged in – would be different. Local residents, for example, formed a higher proportion of those whose time was mainly (or only) spent on the promenade than of those using the promenade and beach equally, or of those spending most or all of their time at the seafront on the beach. Because of these differences, it was considered desirable to analyse the visitor types as separate subgroups in the user surveys.

Trained and experienced interviewers specially briefed on the project were employed to conduct the interviews. The fieldwork was closely supervised and follow-up telephone calls were made to respondents to check on the interviewers' work.

The survey sites were defined on maps and the interviewers were given detailed instructions on where and how to contact respondents on the site in order to ensure that respondents were randomly selected. In both years, interviewing took place over the summer period between July and late September. Interviews were conducted on 12 to 15 specified days, selected so that the beach was exposed during the interviewing day and also to provide an equal number of weekday and weekend days. One of the limitations of the research programme undertaken for the project is that it did not include any surveys conducted during the late autumn, winter or spring to test for any seasonal or weather effects in responses to valuation questions.

In both years, the survey population consisted of British resident adults aged 18 years and over using the site for recreation. Although it is known that children are prominent among beach users, and it is recommended that children should be included in outdoor recreation surveys because they are important users of open spaces (Tourism and Recreation Research Unit, 1983), contingent valuation questions were judged to be too difficult to ask of children and therefore, the survey population was confined to adults.

As there has been little previous research on the benefits of coastal recreation, the 1988 survey was designed to provide basic information on coastal recreation behaviour and characteristics of users. The survey aimed, first, to determine the relative importance of different features of coastal sites in attracting visitors, and, secondly, to determine the relative importance of the physical characteristics of a beach to visitors' enjoyment and use of that beach for different activities. Thirdly, we sought to obtain a valuation of the enjoyment of a visit to the seafront in its current condition. The ultimate aim of the research was to obtain valuations of a change in beach and promenade condition with erosion and with coast protection but the 1988 survey examined the potential effect of erosion upon users' enjoyment and frequency of visits in general terms only.

In 1988, three pairs of sites were selected which broadly represented three levels of development. These were large highly commercialised resort sites (Scarborough and Clacton), smaller less commercialised resorts (Frinton and Filey) and undeveloped

Table 4.2
Factors important in the choice of recreational site to visit (mean rating) in 1988 beach surveys

	Clacton	Scarborough	Frinton	Filey	Dunwich	Spurn Head
I wanted to visit the coast	8.5 (1)	8.1 (1)	6.3	8.2 (1)	8.5 (1)	7.0 (3)
The promenade/dunes/cliff is good for walking on	7.6 (2)	6.2	6.1	6.6	4.0	6.7 (5)
The beach is clean	7.3 (3)	7.8 (3)	6.7 (1)	8.0 (2)	7.0 (5)	5.5
The ease of access from home	7.2 (4)	5.7	5.6	5.9	6.0	4.1
The quietness	6.8 (5)	4.4	6.6 (2)	7.1	7.9 (3)	8.1 (1)
The type of beach (ie. sand, pebbles etc)	6.0	7.9 (2)	6.5 (3)	7.8 (3)	2.0	5.0
The beach is good for children	5.0	7.5 (4)	4.9	6.9	3.2	4.5
I have liked it when I have been before	6.5	7.5 (4)	6.4 (4)	6.0	6.2	5.2
The attractive, natural setting of the beach	6.6	7.0	6.4 (4)	7.3 (5)	8.3 (2)	7.2 (2)
The spaciousness of the beach	6.0	7.4	6.1	7.7 (4)	7.2 (4)	6.2
The beach is good for walks	0.5	6.3	5.3	7.1	5.0	6.8 (4)

Notes: (i) Scale is from 10 (very important to my choice) to 0 (not at all important to my choice).
(ii) Most important five factors shown for each site are indicated in brackets.

sites (Dunwich and Spurn Head). One site of each type was chosen on the north-east coast and one on the south-east coast of England. All the sites either had experienced erosion or had the potential to be affected by erosion.

Visitors were asked how important a number of factors were in their choice to come to the site. Factors were rated on a scale from 0 (not at all important) to 10 (very important). Overall, the desire for a specifically coastal recreational experience, as indicated by the statement 'I wanted to visit the coast', was the most important factor. But specific beach characteristics – the attractive natural setting of the beach and the cleanliness of the beach – were important attractors for visitors as a whole.

However, there were marked differences in the factors which attracted visitors to particular sites. At Scarborough, for example, the fact that 'they had liked it when they were there before', and that 'the beach is good for children', and 'the sandy beach' were important attractors. In contrast, at the undeveloped sites of Dunwich and Spurn Head, visitors were drawn by the 'attractive natural setting of the beach'. High levels of agreement with 'the quietness; not too many people' confirm that these were sites that people visited in search of a quiet natural experience (Table 4.2).

The 1988 survey also showed that the physical characteristics of the beach are important in determining the kind of activities undertaken by visitors and their rating of the beach for particular activities. At Dunwich, for example, few visitors went on long walks along the pebbly beach. We also examined the importance of groynes to the rating of the beach as good for walking on, for sitting and lying, and for children. A strong relationship was found between the presence of groynes and beach function: groynes were found to detract from a beach's rating for walking on, but were found to enhance a beach's rating for sitting and lying, and for children to play on.

In order to examine the potential effect of erosive changes on the beach, respondents were asked the following questions: 'With coastal erosion beaches usually become narrower. If this beach became narrower and uncovered for less time at low tide, would you get more or less enjoyment from your visits to it or would it make no difference?', and 'Would you visit more often, less often or as often as now, as a result of this change?'.

Overall, 55 per cent reported that their enjoyment would be reduced and 32 per cent that they would visit less often if the stated erosive changes were to occur. There was considerable variation in the effects of the erosive change reported for different sites. The impact of the erosive change appeared most marked for Scarborough and appeared to have least effect at the undeveloped sites of Spurn Head and Dunwich (Penning-Rowsell et al., 1989).

The 1988 survey established that the physical characteristics of the beach are important factors in attracting visitors to coastal sites and that these characteristics are also important in determining the kind of activities undertaken by visitors at the site and their rating of the beach for particular activities. It also showed that erosive changes were likely to have a deleterious effect on the enjoyment of coastal recreation and frequency of visiting and thence on valuations of beach recreation.

Two different questioning approaches for eliciting contingent valuations have been tested in this research: a willingness to pay approach, and the approach seeking the value of the enjoyment per adult visit. The latter approach was employed in both the 1988 and 1989 beach user surveys to obtain a valuation of current beach recreation. Respondents in the surveys were asked the following simple open question: 'What value do you put on your individual enjoyment of this visit to the beach today?'

It is recognised that valuing an unpriced public good such as beach recreation is a difficult and unfamiliar task for respondents. To help them to arrive at a valuation, they were asked to think of a visit or activity which had given them the same amount of enjoyment as their visit to the beach today. A list of possible visits and activities was shown to them on card D reproduced in Appendix 4.3. It was suggested to respondents that the costs of the alternative visit could be used as a guide to the value of their beach recreation.

After following this procedure, over 80 per cent of respondents were able to offer a valuation (80 per cent in 1988, and 84 per cent in 1989). As can be seen in Table 4.1, there was some variation in the proportions offering a valuation at different sites.

The findings of the 1988 and 1989 beach recreation surveys and the surveys conducted with seafront users at Hastings (Section 8.2) and Herne Bay (Section 8.5) indicate that visitors to seafronts value highly their recreational experiences there. The mean values given for enjoyment of 'today's visit' at the individual sites are shown in Table 4.1. They ranged from £3.64 at Filey (1989) to £12.34 at Herne Bay (1990). The overall mean values for all sites are £7.75 in the 1988 beach survey and £7.55 in the 1989 beach survey. With the exception of Spurn Head, the valuations offered at the northern sites are lower than those given at the southern sites. These differences may reflect lower incomes among visitors to northern sites, or lower costs of alternative visits and activities used as a guide to the value of beach recreation, or a mixture of both factors.

In a survey at Hastings in 1988, methods for quantifying losses and gains that would result from specific erosive changes and a specific coast protection project were developed and tested. The methods developed for the Hastings survey were subjected to further test-

72 THE ECONOMICS OF COASTAL MANAGEMENT

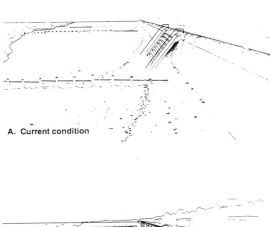

A. Current condition

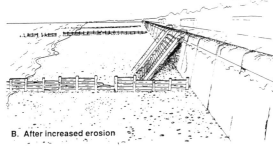

B. After increased erosion

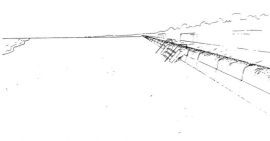

C. After coastal protection work

Figure 4.5   Drawings of the same section of beach under different conditions (Clacton, Essex)

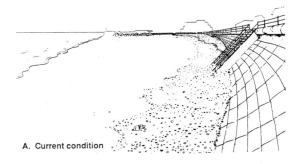

A. Current condition

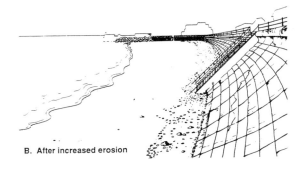

B. After increased erosion

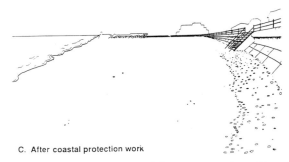

C. After coastal protection work

Figure 4.6   Drawings of the same section of beach under different conditions (Morecambe, Lancashire)

Text accompanying Drawing A. Beach in current condition

The drawing shows that:

i) There is little beach above high tide and the high tide mark is shown in the drawing
ii) The beach is reached by sets of steps down from the promenade
iii) The beach is mainly sandy with small patches of shingle
iv) There are few, low groynes which are in a reasonable condition
v) The sea wall is in a good condition

Text accompanying Drawing B. After increased erosion

Compared to Drawing A:

i) There is no beach above high tide and the high tide mark is shown in the drawing
ii) The drop from the promenade down to the beach is longer so the flights of steps are longer
iii) The beach is a mixture of sand and shingle
iv) The groynes are more exposed and in a poor condition
v) The sea wall is not in such a good condition and there are cracks in places

Text accompanying Drawing C. After coastal protection works

Compared to Drawing A:

i) There is a wide beach at high tide and the high tide mark is shown in the drawing
ii) The drop from the promenade down to the beach is shorter so the flights of steps are shorter
iii) The beach is mainly sandy with small patches of shingle
iv) There are no groynes visible
v) The sea wall is in a good condition

Text accompanying Drawing A. Beach in current condition

The drawing shows that:

i) There is little beach above high tide and the high tide mark is shown in the drawing
ii) The beach is reached by sets of steps down from the promenade
iii) The upper part of the beach is mainly shingle at low tide, a narrow band of sand is exposed and below this, towards the sea, there is a wide section of mud
iv) There are no groynes
v) The sea wall is in a reasonably good condition

Text accompanying Drawing B. After increased erosion

Compared to Drawing A:

i) There is no beach above high tide
ii) The drop from the promenade down to the beach is longer so the flights of steps are longer
iii) At low tide there is no sand and the beach is mostly mud with a narrow band of shingle at the top
iv) The are no groynes along this section of beach
v) The sea wall is in a poor condition, with cracks and holes and the sea is washing in underneath

Text accompanying Drawing C. After coastal protection works

Compared to drawing A:

i) There is a wide beach at high tide and the high tide mark is shown in the drawing
ii) The drop from the promenade down to the beach is shorter so the flights of steps are shorter
iii) At low tide there is a wide band of sand exposed with a band of shingle at the top of the beach
iv) There are no groynes visible
v) The sea wall is in a good condition

ing in the 1989 beach surveys. These were undertaken at four sites, two in the north (Morecambe and Bridlington) and two in the south (Clacton and Hunstanton).

In each part of the country, one site was chosen as being in a reasonably good condition, either because it had not as yet been greatly affected by erosion or because it had been recently improved. The second site chosen was in the process of eroding but in a state in which it which could deteriorate still further. The eroding sites were selected from sites where coast protection or improvement works were planned or were under consideration. It was important that the likely impacts of erosion and the possible protection works at the sites should be known, so that these changes could be represented to respondents in CVM scenarios.

Thus in the Hastings survey and the 1989 beach surveys, drawings of the beach in varying conditions and with accompanying explanatory texts which were read out by the interviewers were used to describe the changes. Examples of drawings used in the 1989 beach surveys are shown in Figures 4.5 and 4.6. The drawings represent the beach in current condition (A), after increased erosion (B), and after possible coast protection works (C). These drawings can be considered in relation to Figure 4.2, which shows the simplified representation of beach erosion over time. Drawing (A) of the beach in current condition represents position A – A₁ in Figure 4.2. Drawing (B) of the eroded beach represents position B in Figure 4.2 and drawing (C) represents position C.

For each of the drawings, respondents were asked to put a value on their enjoyment of a visit to the beach shown in the drawing.

After erosive changes in a beach resulting in a reduction in enjoyment, individual visitors have two options: they can continue to visit the site but with reduced enjoyment or they can transfer their visiting to an alternative site. In the first case, the economic loss is given by the loss in enjoyment. In the second case, an economic loss will not occur if there is a perfect substitute site available which the individual can visit at no extra cost and with as much enjoyment as the current site. Whether or not an economic loss is involved when visits are transferred to an alternative site will depend upon two factors: the value of enjoyment visitors get from a visit to the alternative site, and the value of any additional travel costs required in visiting the alternative site.

Therefore in the 1989 beach surveys, visitors were asked whether they would visit the eroded site 'much less often', 'much more often', or 'the same amount as now'. If they said they would visit less often, they were asked where they would visit instead. The value of the enjoyment of a visit to the alternative site and any additional costs in visiting there were then elicited.

Table 4.3
Standard data on losses and gains with erosion and protection of the seafront (1989 prices)

	Local visitors	Day visitors	Staying visitors	All visitors[1]
Mean loss in £ per adult visit (with erosion)[2]	1.58	2.37	5.55	3.59
Mean gain in £ per adult visit (with coast protection)[3]	1.04	1.80	1.31	1.49

Notes: (1) Based on data from seafront user survey in the summer of 1989 at Bridlington, Morecambe, Clacton and Hunstanton.
(2) Changes to the seafront with erosion as represented by drawings and text in Figures 4.5 and 4.6.
(3) Changes with coastal protection as represented in drawings and text in Figures 4.5 and 4.6.

From this data it was possible to estimate the economic losses per visit associated with coastal erosion at the beach sites, taking into account both changes in enjoyment of recreation at the current site and any loss associated with transferred visits.

Any economic gains from an improvement in the beach following coast protection consist of the difference between the value put on a visit to the beach in its current state and after improvement. The gains and losses per adult visit arising from the changes in beach condition presented to respondents in the 1989 beach surveys in the drawings are shown in Table 4.3. Where a coast protection project not only protects against further erosion but also results in a reinstatement of the site to its former good condition or produces an enhancement, the total benefit per adult visit will be the sum of the loss per adult visit prevented plus the gain per adult visit.

A seafront user survey and a residents survey were also carried out in 1990 at Herne Bay as part of a coast protection project benefit assessment for Canterbury City Council. These surveys (Section 8.5) provided a further opportunity to apply the methods developed in the 1989 beach surveys to a specific erosion problem and to a distinctive proposed coast protection project – an off-shore rock reef. This case study also provides an illustration of the methods employed to 'gross-up' from the total annual recreation benefit for the sample of respondents in order to arrive at a total project recreation benefit for the population from which that sample was drawn.

A second method of recreation benefit assessment, the Travel Cost Method, was also examined as part of the 1988 and 1989 beach recreation surveys. The theoretical and applications issues raised by this method are discussed in Section 2.5.3 and briefly in Section 4.2.3. A crucial assumption of the method is that the value of the enjoyment of a recreational site must be higher for those who travel further to reach the site and who, therefore, incur higher travel costs in visiting. This assumption was tested in the 1988 and 1989 beach surveys. The value that visitors put upon their enjoyment of 'today's visit' using the Contingent Valuation Method described above were compared to the costs incurred in making the visit. These travel costs were estimated by using the AUTOROUTE computer program to calculate the quickest, shortest and cheapest route between visitors' places of origin for the trip and the coast site (for day trippers only).

In the 1988 survey, the value of enjoyment was positively related to the distance travelled or the time taken to reach the site at only two out of the six sites: these were both the undeveloped sites of Spurn Head and Dunwich (Green et al., 1990). In the 1989 survey, a significant, positive correlation between the value of enjoyment of the visit derived by using the CVM and distance travelled was not found for any of the four resort sites included in the survey. It is concluded therefore that the use of the Travel Cost Method to value coastal recreation benefits cannot generally be advised, and that the method may well give misleading results.

4.3.2 Cliff-top recreation surveys

Two case studies were carried out to test methods of recreation benefit assessment for cliff-top sites.

In an early pilot case study, at **The Naze** in Essex in 1987, the sample consisted of visitors to the cliff site, since it was assumed that the site would attract day trippers and holiday-makers in large numbers as well as local residents (Table 4.1). The problem of the erosion of the cliffs was presented to respondents in a statement – the cliff edge was said to be 'retreating at a rate of about, on average, 1 metre per year so that in ten years time, an area of about an acre would be lost'.

The costs and effects of protection upon the geological character and attractiveness of the Naze were presented through statements and a photograph of a protection structure similar to that proposed for the Naze. Through a series of questions, respondents were asked to give their views on the erosion and protection proposed and their likely effects on respondents' recreational use of the cliff-tops. This provided respondents with the opportunity to consider the issues involved prior to the posing of the valuation question.

A willingness to pay questioning approach was used in order to elicit a valuation. Respondents were asked 'Would you be willing to pay increased rates and taxes to protect the Naze from erosion by the sea?'. Those who responded positively were then asked how much they would be willing to pay, in the form of a direct open question. Although asked of users, the willingness to pay question does not elicit a user value alone, since users may be willing to pay to protect the site for reasons other than their current or future recreational use of the cliff-tops. Average willingness to pay figures were calculated assuming that the undecided were unwilling to pay.

Visitors – local people, day visitors and holiday-makers – were estimated to make a total of 45,000 visits to the Naze site per annum. The different categories of visitor (local, day and staying visitors) were found to differ significantly in their average willingness to pay. Therefore, the mean annual willingness to pay estimates were grossed up separately for each visitor type with the number of visits per visitor type assumed to be in the same proportion as the visitors found in the survey. Gross annual benefit figures of £85,000 and £108,000 were calculated for reported monthly and annual willingness to pay, respectively. When these annual benefits were discounted over a 50-year project life, a total benefit of some £2 million was calculated (Penning-Rowsell et al., 1989).

The cliffs at the Naze are considered to be of international scientific importance. The protection works proposed could have the effect of damaging the geological interest of the cliffs. The potential disbenefits of the protection project to the specialist users – geologists who use the cliffs for teaching and research – were not evaluated in the exploratory case study although it would have been possible to do so using the Contingent Valuation Method. Therefore the results of the study must be regarded as exploratory and incomplete. Furthermore the methods used to assess the recreational benefits were not those recommended in this manual.

A second study of cliff-top recreation benefits that might arise from a specific erosion problem and proposed protection project was undertaken at **Peacehaven**, Sussex. This survey and its results are described in the case study (Section 8.3); only the key features of the research are summarised here.

At Peacehaven it was decided to conduct a residents survey because it was considered that the cliffs in Peacehaven, although part of the long-distance coastal path, would predominantly be used for recreation by local residents. Three out of four proposed phases of a coast protection project had already been completed at Peacehaven at the time the research was undertaken, leaving a short 300m section of cliff unprotected between the two stretches of cliff-base works.

The erosive changes likely to affect this unprotected section were described to respondents: the unprotected coastline was stated to be eroding so that approximately 13 feet (4 metres) of cliff were being lost every ten years and the possibility of a major cliff fall leading to the loss of up to 8 metres of cliff-top land was mentioned. A number of techniques were used to present the erosive changes to respondents in the survey. A map was used, showing where the cliff line would probably be in 50 years' time if no action were taken. Three drawings were shown, one showing the present cliff face, the second the cliff face after 50 years of erosion and the third the cliff face after protection. Statements were also used outlining the effects of such erosion and protection on cliff-top recreation.

The respondents were asked to value their enjoyment of a single visit to the cliff-tops for recreation, and to say whether they would be willing to pay 'increased rates and taxes' so that more money could be spent on preventing erosion of recreational land in the unprotected section. A bidding procedure was used to elicit a valuation of coast protection (Randall, Ives and Eastman, 1974; Randall, Hoen and Brookshire, 1983; Cummings, Brookshire and Schulze, 1986), rather than a direct open question. This procedure required respondents to answer a repeated question as to whether they would be willing to pay more or less than a given sum, until the maximum sum they were willing to pay was reached. Two starting point sums were used for this – £1.00 and £0.50 – as a check on distortion in the valuation due to starting point value bias.

The results showed that the respondents valued their cliff-top recreation quite highly, and a majority were willing to pay to protect the coast at Peacehaven. The average individual willingness to pay figures obtained using the two starting points for bidding were £1.83 and £2.83 per annum. Gross annual benefit estimates of £10,625 and £16,430 were calculated using the total number of electors in Peacehaven as a proxy for the total adult population. The latter benefit figure based on the higher starting point was thought to be the more accurate because the valuations elicited in response to the lower starting point were constrained by the low range of bids, particularly by the low level of the top bid offered.

Although it was indicated to respondents that the access to the cliff-top recreation area would be closed as a result of erosion after 50 years, there was considerable uncertainty regarding the future rate of erosion, and hence the time at which access would be lost and the benefits would come on stream. Therefore, it was not possible to arrive at a reliable total benefit figure for protecting the coast. But given that the number of people benefiting was not large and some of the benefits were likely to be realised only after a considerable number of years, total discounted benefits were unlikely to be substantial. That result led to the decision not to proceed with the protection of the hitherto unpro-

Table 4.4
Comparison of values of enjoyment of visit obtained in different surveys (£ per adult visit)

Date of survey	1989	1988	1988	1987	1987
Sample	Beach users – 4 sites	Beach users – 6 sites	Beach users – Hastings	Maidenhead residents	River corridor users – 12 sites
Goods valued	Beach now	Beach now	Beach now	River corridor improvements: Maidenhead Ditch	Water quality improvements
Mean	7.55	7.75	7.72	0.94	0.51
Log mean	0.70	0.74	0.70	−0.25	−0.23
Log standard deviation	0.40	0.37	0.42	0.39	0.38
Number of cases	603	737	198	242	837

tected stretch of cliff at Peacehaven.

It was concluded on the basis of this research into cliff-top recreation that although individuals attached quite high values to cliff-top recreation and were, for the most part, willing to pay something for coast protection, the benefits of protecting recreational land on cliff-tops from erosion are unlikely to be substantial unless a combination of the following apply:

1. The number of visitors using a site for informal recreation is very large;
2. There are special recreational features such as a golf course or fun fair on the cliff-top site likely to generate extra recreational benefits; and
3. The rate of erosion is very high so that short term changes in the recreational area would be more than marginal and the benefits of protecting the site would come on stream at an early stage.

Table 4.5a
Local visitors: determinants of attitudes and behaviours
Regression equations with beta coefficients (= standardised regression coefficients)

Explanatory variables:	Dependent variable:					Explanatory variable: form of question
	1. Frequency of visits per year ($\log_{10}$)	2. Overall rating of resort	3. Overall rating of beach	4. Logenj	5. Logenjb	
Main reason	−0.47	−0.18				main reason for outing to make visits elsewhere not to seafront
Time	1.34					time taken to get to beach
Logtime	−1.55					$\log_{10}$ time to beach
Town		0.37				town as advantage to living here
Play	−0.26					activity at seafront: playing with sand, stones etc
Long walk				0.30		activity on beach: long walks
Promenade			0.10			importance of quality of promenade in choice of site to visit
Quality/walk			0.31			quality of beach to walk on
Quality/sit			0.17			quality of beach to sit or lie on
Quality/play			0.49			quality of beach for children to play on
Beach/overall		0.43		0.34		overall rating of beach
Sex	0.19					male respondent
Logenj					0.77	$\log_{10}$ (£ value of enjoyment today)
Reduce					0.37	proportional reduction in enjoyment after erosion
R^2 (adjusted)	0.37	0.45	0.79	0.16	0.73	
F	10.45	22.66	81.35	6.72	69.72	
df	5,76	3,78	4,79	2,60	2,50	
p	<0.001	<0.001	<0.001	<0.001	<0.001	

Note: (applies to Tables 4.5a – 4.5c)

Where form of question is a statement this is a dummy variable such that:
agreement = yes = 1, not applicable/no = 0

Logenj = $\log_{10}$ of £ value of enjoyment after erosion

Table 4.5b
Day visitors: determinants of attitudes and behaviours
Regression equations with beta coefficients (= standardised regression coefficients)

Explanatory variables:	Dependent variable:					Explanatory variable: form of question
	1. Frequency of visits per year ($\log^{10}$)	2. Overall rating of resort	3. Overall rating of beach	4. Logenj	5. Logenjb	
First	0.19					first visit
Logtime	−0.45					$\log_{10}$ time to go to beach
North				−0.33		northern site
Main reason	−0.14		−0.21			main reason for outing to make visits elsewhere not to seafront
Promenade				−0.12		importance of quality of the promenade in choice of site to visit
Town/facilities		0.20				importance of the town and its facilities in choice of site to visit
Scenery		0.11			−0.09	importance of scenery and places in area in choice of site to visit
Convenient journey		0.15				importance of the convenience of the journey in choice of site to visit
Cost					−0.08	importance of cost of trip in choice of site to visit
Decision	−0.10					I/others made decision to visit
Quality/walk		0.38	0.34			quality of beach to walk on
Quality/play			0.57			quality of beach for children to play on
Sex				0.12		male respondent
Logenj					0.81	$\log_{10}$ (£ value of enjoyment today)
Reduce					0.28	proportional reduction in enjoyment after erosion
Age		0.18				age of visitor in years
R^2 (adjusted)	0.23	0.39	0.75	0.18	0.80	
F	24.33	27.86	252.67	13.42	182.51	
df	3,237	6,248	3,249	4,220	4,173	
p	<0.001	<0.001	<0.001	<0.001	<0.001	

4.3.3 Validity and reliability of the Contingent Valuation Method

The contingent valuation method is well established as a recreation benefit assessment method in the United States (Cummings, Brookshire and Schulze, 1986; Mitchell and Carson, 1989). In Britain, although there is now a considerable body of experience in its use, it is necessary to be cautious in its application (Green et al., 1990). In particular it is essential that checks on the reliability and validity of the valuations obtained should be incorporated into the design and analysis of CVM surveys.

In the beach surveys undertaken as part of the research leading to this Manual, a test of the reliability of the method was provided when a beach survey at Clacton in 1988 was repeated in the following year during the same summer period. There was no significant difference in the mean valuations found in the two years (£9.90 in 1988 and £10.50 in 1989). Ideally, a test

Table 4.5c

Staying visitors: determinants of attitudes and behaviours
Regression equations with beta coefficients (= standardised regression coefficients)

Explanatory variables:	Dependent variable:					Explanatory variable: form of question
	1. Frequency of visits per year ($\log^{10}$)	2. Overall rating of resort	3. Overall rating of beach	4. Logenj	5. Logenjb	
Local group				0.16		number in group who live locally
Group composition				0.30		group composition
Main reason				−0.18		main reason for outing to make visits elsewhere not to seafront
Time spent				−0.21		time spent only/mainly on the beach/promenade
Swim/paddle	−0.24					importance of the suitability of the sea for paddling and swimming in choice of site to visit
Walk on prom	−0.18					activity at seafront: walking/strolling on the promenade
Town		0.11				importance of the town in choice of site to visit
Promenade		0.33				importance of quality of the promenade in choice of site to visit
Scenery	0.26			−0.13		importance of scenery and places in area in choice of site to visit
Quality/walk	−0.14		0.22			quality of beach to walk on
Quality/play		−0.24	0.55			quality of beach for children to play on
Quality/sit			0.20			quality of beach to sit and lie on
North				−0.20		northern site
More/less sandy	−0.21					beach is more/less sandy
Disposable income	0.13					disposable income
Logenj					0.78	$\log^{10}$ (£ value of enjoyment today)
Reduce					0.27	proportional reduction in enjoyment after erosion
R^2 (adjusted)	0.15	0.38	0.81	0.20	0.69	
F	6.14	49.93	313.39	7.81	192.43	
df	6,172	3,228	3,224	7,181	2,171	
p	<0.001	<0.001	<0.001	<0.001	<0.001	

of reliability would include samples taken at different seasons as well as repeat observations on the same sample after the passage of time but this is impossible in the case of beach users.

The same contingent valuation question form has been used to obtain valuations of markedly different goods in quite different settings: the value of the enjoyment of visits with improved river water quality, and following enhancements to an existing flood alleviation channel, as well as the value of enjoyment of coastal visits. As Table 4.4 shows, quite different results have been obtained using the same measurement method to value different goods in different settings, indicating that the method has discriminant validity and that the values produced are not simply a function of the measurement method. Significant differences have also been found in the valuations of the enjoyment of the visits evaluated at different coastal sites.

If the results obtained in contingent valuations are not simply random numbers, then it should be possible to explain the differences in the values offered by individuals in terms of a causal model of recreational behaviour and value. The model on which the 1989 beach survey was based is shown in Figure 4.4. Tests have been conducted to establish whether the valuations obtained are consistent with the hypothesised relationships in the model.

The results of these regression analyses are shown in Table 4.5, presented separately for local, day and staying visitors because it is assumed that the constraints on their behaviour and their preferences will be different. For all three groups, overall rating of the beach, as predicted, was mainly dependent upon ratings of the quality of the beach or promenade for certain activities and the proportion of the variance explained was high for this variable. Results were less satisfactory for the value of the enjoyment of the beach in its current condition. For all three groups, the value of the enjoyment of the visit after erosion was clearly anchored to the valuation of the visit to the beach in its current condition and was also explained by the extent to which respondents considered that their enjoyment of the visit would be reduced with erosion. Again the proportion of the variance explained was high for this variable.

The results of these tests of validity are good, but not entirely satisfactory. The proportion of the differences in individual valuations of enjoyment of the current visit that can be accounted for by the explanatory variables determined by the model is not very high. Mitchell and Carson (1989) suggest that 20–50 per cent of variance explained should be regarded as an adequate target for a CVM study at present, and while we do not find this totally satisfactory it is useful that our results are within or near this range.

4.4 The recommended approach to the assessment of recreational benefits: a two staged framework

The research that we have done on beach recreation, in the context of the benefit-cost analysis of coast protection or sea defence projects, leads us to recommend a two staged approach to benefit calculation. The two stages that should be adopted are as follows:

1. *Pre-feasibility stage* The first step at this stage is parameter definition. This consists of defining the project characteristics and those of the area which will be affected, including its current recreational use.

The second step of this first stage is to carry out a desk study to collect existing information on current levels of recreational use in order to judge whether it is worthwhile conducting a detailed study to quantify the recreational effects of the proposed project.

At the pre-feasibility stage it may also prove necessary to collect some empirical data by means of user counts where there is inadequate secondary information on current recreational use.

2. *Feasibility study stage* This stage requires a detailed site survey of current users or a survey of local residents in order to determine the existing value of their recreational enjoyment and to quantify the change in value with the proposed project.

A figure for the recreational benefit of the project can then be calculated by estimating the annual benefits from the total number of visits multiplied by enjoyment per visit values, and discounting over the life of the project.

Figure 4.1 shows the procedures to be followed at each of these stages.

4.5 The recommended methods and techniques

Different methods are suggested for application at the two different stages in the project appraisal process. The first set of methods is to be used in the pre-feasibility stage. The second set is to be used at a later stage, after it has been established in the pre-feasibility stage that the proposed project has the potential to provide sufficient recreation benefits to justify the use of further resources in data collection. The methods recommended for the two stages are discussed in the sections below. These sections are designed to be read in conjunction with the flow diagram – Figure 4.1 – which sets out the procedures to be followed.

4.5.1 Recommended methods for the pre-feasibility stage

At the pre-feasibility stage, a desk top review should be undertaken using secondary sources and standard data where available and appropriate. The information required to assess the potential losses with erosion and gains with coast protection or sea defence, and the procedures and methods to be followed at the pre-feasibility stage to obtain this information, are given below and in Figure 4.1.

Step 1

The **nature and rate of erosion** and the **characteristics** of the proposed project should be outlined in order to determine the nature of their effects and the extent and characteristics of the area affected. Both direct and indirect effects should be identified.

Step 2

The **existing recreational use** should be determined and this may indicate that:

a) current recreational use exists;
b) there is no current recreational use; or
c) the extent of current recreational use is unknown.

In the case of 2(a) above, it is necessary to identify the type and amount of usage. In the case of 2(b) above, it is necessary to consider whether the proposed project has the potential for creating a new or significantly enhanced recreational resource, and if so, whether the recreational benefits which would result can be included in the benefit assessment as acceptable for central government grant-aid purposes (Section 4.2).

It is also essential at this stage to consider whether a new or enhanced recreational resource is likely to generate new recreational use and thence national economic benefits or merely to attract visitors from other comparable facilities elsewhere and therefore create only local benefits (Section 4.2.3). In the case of 2(c) above it may be necessary to carry out a preliminary data collection exercise in order to identify whether existing levels of use are sufficiently large to justify more detailed consideration in terms of recreational benefits.

Step 3

Where recreational use exists, the **type and amount** must be determined next. The amounts of generalist and specialist recreational users should be estimated. An estimate of the annual number of adult visits is required at this stage. Detailed data about seasonal differences in the pattern of visiting are unlikely to be available and an estimate based on the best available data may have to be made. It is recognised that children form a significant proportion of the visitors to coastal sites and children are often the reason for adults making such visits. But because of the difficulty of obtaining valid valuations from or for children, the recommended methods of recreation benefit assessment are confined to adult visits.

It is not possible to provide standard data on the number of adult visits to sites of different types even for use at the pre-feasibility stage. The number of visits to a site is not determined by easily identifiable characteristics of a site alone and sites have been found to vary greatly in the number and type of visits they attract.

It is therefore recommended that the best available local or national records should be used to derive an estimate of the number of adult visits to a site per annum. Data on levels of use are likely to be more easily available from secondary sources for specialist users (e.g. sailing, boating, sailboarding, fishing enthusiasts, geologists or birdwatchers) than for people using the coast for informal recreation (e.g. sitting, strolling, paddling and swimming or sunbathing), or for both.

Local or national clubs and organisations should be approached to provide estimates of the number of specialist users and of the availability of alternative sites for the activity. Estimates of the informal use of the coast might be based on the following secondary data sources for visit numbers:

a) Visitor survey or count data: Planning or Recreation and Tourism Departments of local authorities may have carried out surveys in the past from which current use levels can be extrapolated;
b) Car and coach park records multiplied by an average adult car or coach occupancy rate;
c) Records of entry to coast facilities: visitor centres, pier, funfairs, etc. Data may be available from the area Tourist Board, or organisations such as English Heritage or the National Trust;
d) Estimates derived from surveys of coastal businesses: records of available bedspaces, and hotel and guest house occupancy rates.

If no direct survey or count data are available, the indirect measures of site use listed above may have to be employed. All these data sources are likely to have severe limitations as guides to the number of visits to a specific coastal site as opposed to direct counts, but they will at least give a first-order indication suitable at this pre-feasibility study stage.

If no secondary data are available, it will be necessary to determine whether a judgemental estimate of user levels can be made or whether data collection through a survey or counts is required even at the pre-feasibility stage. If such data collection is needed, then timing is important: the data collection should be planned well in advance especially where seasonal factors are likely to have a significant effect, as is usually the case at coastal sites. It will be crucial to have count data for the summer holiday period when the level of use is highest.

At sites where no data on the level of use exists but the number of visitors is thought to be small, as may be the case at undeveloped sites and cliff-top sites, the issue arises as to whether it is worth attempting to collect data on the number of visits and to assess the potential recreation benefits and losses at the pre-feasibility stage, before the overall project benefits and costs are known, given that the number of visits and hence the recreation benefits and losses are likely to be small. However, if it is later found that recreation benefits and losses might be significant in the overall benefit assessment, there might then be insufficient time to carry out counts in order to obtain accurate estimates.

Step 4

The next step is to **value the changes in enjoyment per adult visit with changing coastal conditions**. These changing conditions may be with continued erosion or with protection. The value of a visit in terms of user enjoyment must be determined either using 'standard data' (if the requirements for using such data are met) or if they are not met then the only way to obtain such data is to carry out an interview survey of recreational users or local residents.

Table 4.3 provides standard data (at 1989 prices) for the values of changes in enjoyment per adult visit with changing coastal conditions for typical sea fronts. This data should be updated to the project appraisal base date using the rate of real growth in the economy – change in Gross Domestic Product (GDP) – which can be found in the serial 'Economic Trends' published by the Central Statistical Office, HMSO. There are two components to this standard data, as follows:

a) a value per adult visit of loss in enjoyment with erosion; and
b) a value per adult visit of gains in enjoyment with protection.

If the effect of the project proposed for the site being assessed is limited to reducing or preventing further erosion and maintaining the status quo, then only the first of these components will apply. If the project is expected to reinstate the site to its former good condition or otherwise enhance the site in amenity and recreational terms, then the second component (b) will apply as well, and the value of the project per adult visit will consist of both component (a) and component (b).

Our research has shown that the recreational enjoyment, and hence the value attached to a visit to the coast under varying conditions depends upon a wide range of factors. These include the characteristics of the users, the site, and the alternative sites available. Therefore, the use of the standard data even at the pre-feasibility stage is appropriate only if the following conditions are met:

1. The site is a developed resort type site, but *not* a cliff-top site or undeveloped site such as Spurn Head; no clear cut definition of a resort site can be given as coastal sites are on a continuum from large, highly commercialised sites through less developed mainly residential sites which attract few day and staying visitors to completely undeveloped cliff-top and beach sites. The sites to which the standard data are applied should be broadly comparable in level of commercial development to the sites at which the data were collected (Bridlington, Morecambe, Clacton and Hunstanton).
2. Where the anticipated changes in the coast with continued erosion approximate to the changes presented in the drawings and texts (Figures 4.5 and 4.6: Drawing B) used in the 1989 seafront surveys from which the standard data for erosion losses (component (a) above) are derived.
3. Where the anticipated changes with coast protection approximate to those presented in the drawings and texts (Figures 4.5 and Figure 4.6: Drawing C) used in the seafront surveys from which the standard data for coastal protection (component (b) above) were derived.

If the site involves only cliff-top informal recreational areas then the standard data in Table 4.3 will not be applicable. However, in the absence of any better data, figures from the Peacehaven case study (Section 8.3) may be used at the pre-feasibility stage.

However, if the proposed project for a given site involves special structures such as fishtail groynes or an offshore reef, then the standard data for changes in recreational value with coast protection will not necessarily be valid and a site specific survey will be required. In these circumstances the losses with erosion may still be evaluated using the standard data (component (a)). A preliminary estimate of the losses with erosion should be made in this way in order to determine whether a site specific survey to evaluate the benefits is worthwhile.

The drawings and text presented at two of the sites in the 1989 survey are given in Figures 4.5 and 4.6. These should be consulted to determine whether the expected erosive changes and the protection project at the site which is the subject of the benefit assessment approximate to the changes and projects presented in the surveys.

Standard data are given in Table 4.3 for different visitor types and for all visitors. The surveys undertaken to collect this data indicated that the recreational value of visits with changes in the coast varied according to the type of visitor. The visitor types on which the standard data are based are defined and discussed in greater detail in Section 4.2.3, and can be summarised as:

Local users – defined as those living within a three mile radius of the site;
Day visitors – defined as anyone living more than three miles away and starting the trip from their normal place of residence; and
Staying visitors – defined as anyone staying away from home for one night or more.

Staying visitors as defined in the standard data are not necessarily staying in the resort or place where the coastal site is located but may be on tour or staying elsewhere in the region. Therefore data on bedspaces and occupancy rates for the resort or town will provide only an approximate guide to the number of staying visitors as defined in the standard data.

Where estimates of the number of visits by different visitor types per annum are available, the standard data for different visitor types should be used in preference to the overall data since this will enhance the precision of the assessment. However the overall figures may be used at the pre-feasibility stage if estimates of the number of visits per annum cannot be obtained for the different visitor types.

Step 5

Estimation of **annual benefits** can be made from the change in value per visit multiplied by the total number of visits per annum. The annual benefit may have one or two components:

a) Annual benefit of preventing a loss with erosion = (value per adult visit of loss with erosion) x (number of adult visits per annum);
b) Annual benefit of the gain with coast protection project = (value per adult visit of gain with coast protection project) x (number of adult visits per annum).

If the proposed project has as its objective the prevention of further erosion only, then only component (a) of the annual benefits will be estimated. If the proposed protection project not only prevents further deterioration in the coast but also reinstates the site to its former good condition or results in its enhancement in recreational or amenity terms, then the annual benefits will comprise both component (a) and component (b).

The total recreational benefits of a proposed project are estimated by discounting the estimated annual benefits over the life of the project according to the Test Discount Rate. A different approach is required here to derive estimates for the two components of the annual benefit, where both apply.

In the case of (a), above – the annual benefits of preventing a loss with erosion – the annual benefit would not arise immediately upon completion of the project and continue throughout the life of the project. This is because in the absence of a project, a number of years would have to pass before the site would have deteriorated to the state depicted in the drawings of the eroded beach used for the standard data. The annual benefits of preventing this erosion might only arise after the passage of, say, ten or fifteen years. The annual benefits would then accrue over the remainder of the life of the proposed project, discounted by the Test Discount Rate to give the recreation benefit from the project preventing this further erosion.

In the case of (b), above – the benefits of a gain with protection – then the benefits from the project will be immediate upon completion of the proposed project and should be discounted over the total life of the project by the Test Discount Rate to give the recreation benefit from the project of a gain with protection. For projects which not only prevent further erosion but also offer an enhancement to the site, the total recreation benefit of the project comprises the two sums added together.

Step 6

If the pre-feasibility stage investigations using standard data indicate that the size of the potential recreation benefits is sufficient in relation to the scale and possible cost of a proposed project, a decision to proceed with the feasibility study stage recreation benefit assessment methods should then be taken.

For undeveloped and cliff-top sites and for other sites or projects to which standard data for gains and losses cannot be applied, a decision will have to be taken at the pre-feasibility stage as to whether the recreation benefits of a project are likely to be large enough to justify further detailed study. This judgement will have to be based on the following information:

1. The number of visits to the site; if the number of visits to the site is small it is unlikely that the recreation benefits will be substantial even if the individual visitor places a high value on a visit.
2. The presence or absence of specialist users and facilities such as golf courses, sailing facilities or funfairs likely to generate high values for recreation;
3. The rate of erosion: unless cliff-top sites are eroding rapidly so that erosion will have a more than marginal impact within the short to medium term (5-10 years) such sites are unlikely to

yield high recreation benefits over the life of a project;
4. The availability of alternative sites: undeveloped sites may be unique or have characteristics that cannot be found at other sites in the region. Such sites may be highly valued for recreation for these reasons.

4.5.2 Recommended methods for recreation benefit assessment at the feasibility study stage

The categories of data required and the procedures to be followed at this stage remain the same as for the pre-feasibility study, but site specific data collection methods are recommended for the feasibility stage, as follows:

1. Details will be needed on the extent and rate of erosion, the area affected and detailed information on the coast protection or sea defence projects proposed, so that these data can be used to develop and present CVM survey scenarios;
2. Data on the number of adult visits per annum – count data collected using infra-red counters (calibrated by manual count data) or count data collected manually on site;
3. Data on the value of changes in enjoyment per adult visit with changing coastal conditions (with erosion and protection) – a CVM survey will be needed of seafront or cliff-top users, specialist users or of local residents.

The procedures to be followed during the feasibility study are outlined in Figure 4.1 and the site specific data collection methods are described in detail in Sections 4.5.3 and 4.5.4.

4.5.3 Contingent Valuation Method (CVM) surveys

Methods of economic evaluation including the Contingent Valuation Method (CVM) have been considered in theoretical terms in Section 2.5.5. The results of developing and testing the CVM and the Travel Cost Method in coastal surveys are described in Section 4.3. The decision to recommend the CVM as the main method for recreation benefit assessment, and the detailed methods recommended below, are based on the experience gained in our research and their application to specific sites is illustrated in the case studies (Sections 8.2, 8.3, and 8.5).

The following issues need to be addressed in order to carry out the CVM surveys:

1. *The conduct of surveys: pilot survey and main survey* A decision has to be taken as to whether to carry out the survey(s) in-house or to contract out the survey work to a consultancy or market research firm or to academic researchers with experience in this field. In-house surveys can be cheaper but it is essential to ensure that survey research expertise is available for the design of samples and questionnaires (although standard questionnaires are provided in Appendix 4.2) and that staff are available to supervise fieldworkers, and to process and analyse the data. Where an outside organisation is employed, it is desirable that it has previous experience in undertaking CVM surveys.

Experience shows that CVM surveys are not easy to carry out, and therefore it is important to ensure that trained, experienced and briefed interviewers are used to administer the questionnaires. Research shows that such interviewers achieve better results than interviewers without professional training and skills. It is important, too, to decide at the outset what will be the computer system to be used in the processing and analysis of the data so that the questionnaires can be designed to facilitate these procedures.

Planning and Recreation departments within local authorities may have experience and expertise in carrying out surveys and may have an interest in co-operating in undertaking surveys into coastal recreation.

It is recommended that preliminary exploratory research such as discussions with local resident groups, interviews with users and other informants and a pilot survey to test the functioning of the questionnaire, the other survey materials such as drawings, texts and maps showing schemes and the sampling and other survey procedures should be carried out before the main survey is conducted. Pilot surveys should comprise at least 30–50 interviews.

2. *The area affected* The area affected by the erosion and by the proposed protection project will need to be defined in detail. It will need also to be described on a simple map and in text which is clear and straightforward enough to be presented to respondents in the survey.

3. *The population* A decision will need to be taken as to the relevant population of those likely to benefit from the recreational site to survey: the site users or residents living in the catchment area for the coastal site. This will depend on the level and type of site use identified at the pre-feasibility stage.

Where the site attracts large numbers of day and staying visitors, a site user survey is to be preferred. Where the site has mainly specialist use – such as a golf course or for fishing – surveys of these users may be appropriate. Where the site mainly provides an amenity and recreational resource to local residents, a residents survey using the electoral register as a sampling

84 THE ECONOMICS OF COASTAL MANAGEMENT

frame for addresses can be used. This approach will need a selection procedure to select a random sample of adults within each household, or instructions to interview 'the householder or housewife'. For further information on sampling methods and survey methods in general we recommend that Moser and Kalton (1971), Hoinville and Jowell (1982), Kish (1965) should be consulted.

Another circumstance in which a residents survey might be desirable is where there is no current use of the site for recreation or where a considerable improvement in the recreational facilities, likely to attract new users, is proposed in the project. A survey of local residents would at least provide a valuation of the improved site for local residents, although it would not be possible to generalise from this to the value of visits from day and staying visitors that might be generated by the improvements.

A very extensive and hence expensive survey would be required to evaluate the new recreational use that might be generated outside the local area by new or radically improved coastal facilities such as a marina. Where financial resources permit, both a site user and a residents survey could be mounted, since a residents survey will provide a more accurate estimate of local residents' valuations of the change in the site because the number of local residents included in a site survey – particularly one conducted in the summer – is likely to be small. Furthermore, a survey of local residents may be judged to be desirable as part of the process of consulting local people about possible changes affecting the local seafront. The Herne Bay case study (Section 8.5) illustrates the use of both a site users and residents survey.

The recommended survey population has been confined to adults because it was not considered to be possible to elicit valid valuations from children. Nor is it judged to be desirable to ask adults to offer a valuation for children in their party or household, or to make an arbitrary assumption about the value of children's enjoyment of beach visits by judging children's visits at, for example, half the value of adult visits.

4. *Sample size* The size of the sample for the survey will need to be determined with regard to the following factors:

1. The financial resources available for the survey;
2. The number of sub-groups to be examined. It is recommended that valuations should be calculated separately for local, day and staying visitors; it is generally recommended also that sub-groups should not be less than 50–100 in size (Hoinville and Jowell, 1982); and
3. The level of precision required of the estimates. Where the recreation benefit may just make the difference as to whether the overall benefits are sufficient to justify a protection project, it will be important that the survey estimates are as accurate as possible.

As is apparent from Table 4.3, valuations obtained in CVM surveys have large variances, and therefore large samples are required if good estimates are to be obtained. Given the kind of standard deviations shown in Table 4.1, **samples of at least 500–600 in size are required** to provide estimates of the current valuation of a visit to the coast which could be estimated with 95 per cent confidence to be within + or – £1.00 to £1.50 of the mean. For smaller samples, the confidence intervals would be wider. Therefore, samples of less than 200–300 are not recommended. At sites with a very light level of use, the absolute number of visitors to the site may restrict the sample size that can be achieved. The pilot survey can be used to estimate the variances on the valuation questions at a particular site and this information can be used to determine the sample size required to achieve estimates to a given level of precision at that site.

5. *The timing of surveys* Since recreational use of the coast is known to vary with the seasons, ideally surveys should be carried out at different times of the year to reflect possible differences in the valuation of coastal recreation at these different times and therefore in different weather conditions. Our research and other CVM research has not examined the possibility of seasonal effects on valuations. If only one survey time can be chosen, the summer holiday period is recommended because that is the time when the largest number of visits occur. For site surveys, it is recommended that surveys be carried out on a minimum of eight survey days, divided equally between weekdays and weekend days (Tourism and Recreation Research Unit, 1983).

For further guidance on the conduct of site surveys, reference should be made to the Recreation Site Survey Manual (Tourism and Recreation Research Unit, 1983). Time needs to be allowed for exploratory research, a pilot survey to test the methods at the particular site and the main survey. An absolute minimum of four months should be allowed to carry out the surveys from inception to completion of a report. The overall project appraisal programme needs to take into account the summer timing and duration of the survey work in the overall work planning.

6. *The definition and representation of the coastal change in the CVM surveys* A major advantage of the CVM is that it is possible using this method to obtain valuations of the good – a coast site – in current conditions but also to value future changes in the site: the degradation or reinstatement or enhancement with coast protection or

sea defence, and with alternative projects. This is, of course, not possible with the Travel Cost Method.

Thus CVM surveys should evaluate both the deterioration that might follow if the 'do nothing' option were adopted, and any improvements that may result from a coast protection or sea defence project or alternative projects, unless the effect of the project(s) would simply be to maintain the coast in the current state. The likely changes need to be clearly defined so that they can be presented to respondents in the survey in CVM scenarios.

Our research tested the use of carefully controlled drawings and explanatory texts to represent changes in seafronts and cliffs with erosion and protection project(s). Examples of the drawings and the texts used in the 1989 beach surveys are given in Figures 4.5 and 4.6 and further examples of drawings are in the Hastings, Peacehaven and Herne Bay case studies (Sections 8.2, 8.3, 8.5).

It is recommended that simple texts worded in non-technical, neutral language, with drawings and possibly photographs of protection structures comparable to those proposed for a site, be used to describe the anticipated changes at the coastal site. A balance has to be maintained between providing very full information to the respondent and avoiding information overload. Furthermore, it is known that the kind of information provided can affect the responses (Mitchell and Carson, 1989).

The materials used to represent changes to respondents in the surveys need to be carefully tested through techniques such as group discussions or interviews with a range of appropriate respondents and in the pilot survey to ensure that the materials, pictures and texts, are comprehensible and are understood as intended.

It is important to note that the valuations obtained in the surveys will be valid **only** for the changes as represented in the scenarios. If the predictions concerning the erosive effects and the proposed project were to be revised radically after the CVM surveys were undertaken the recreation benefit estimates would not be valid for the revised situation. Therefore it is vital that the CVM scenarios accurately represent the likely changes.

7. Asking the valuation questions: the CVM questionnaire
Two broad approaches to asking valuation questions were developed and tested in our coastal recreation research: these were willingness to pay questions, and the value of the enjoyment of a visit questions. It was concluded that the latter questioning approach should be adopted as the recommended method of obtaining valuations, partly because willingness to pay questions do not provide a valuation of user recreation benefit alone. Other motivations such as the desire to protect local businesses or housing from the threat of flooding or erosion may enter into peoples' willingness to pay, which introduces the possibility of an element of double counting in the benefit assessment.

It is recommended that respondents in the survey should be asked the sequence of questions shown in the standard residents and site user questionnaires provided in Appendix 4.2(a) and (b). These questions are required in order to provide the following information needed to estimate the economic losses and gains resulting from changes at a coastal site with erosion or protection:

1. The value of enjoyment of today's visit (user survey);
2. The value of enjoyment of a visit in current conditions as depicted in a drawing (residents survey);
3. The value of enjoyment per visit after erosion as depicted in a drawing;
4. Where a coast protection or sea defence project(s) will result in reinstatement or improvement: the value(s) of enjoyment per visit with coast protection project(s) as depicted in drawing(s);
5. For those who would transfer their visits to an alternative site after erosion:
    (i) The value of the enjoyment of a visit at the alternative site;
    (ii) The difference in costs associated with a visit to the alternative site compared with the current site; and
6. For those who would transfer their visits to an alternative site after protection:
    (i) The value of the enjoyment of a visit at the alternative site;
    (ii) The difference in costs associated with a visit to the alternative site compared with the current site.

Thus it is recommended that respondents should be asked initially to put a monetary value on their 'individual enjoyment of this visit to the seafront today' (for site surveys) and a visit to the seafront in current condition (for household surveys). To assist them in this task, respondents should be prompted to consider a visit or activity which had given them the same amount of enjoyment as their visit to the seafront and they should be shown a list of possible alternative recreational activities. Card D showing a possible list of alternative activities is reproduced in Appendix 4.3. Subsequent questions seek valuations of a visit to the seafront in varying conditions as presented in the drawings and texts.

The standard questionnaires in Appendix 4.2 (a) and (b) incorporate questions on visit frequency, duration and the activities undertaken as part of coastal visits. The residents questionnaire includes questions on site visit frequency in the Spring/Summer months and Autumn/Winter months so that visit rates can be calculated for the population for the survey taking broad seasonal variation in visiting into account. Both questionnaires are based on the model developed during this research (Figure 4.4) and include ratings of the seafront and the proposed project(s) and questions to classify respondents according to their visitor type, age, education and income. These questions are needed so that the validity of the valuations obtained can be tested in the analysis. The questions are given in the form and sequence that has been tested and found to perform well in our research and therefore it is recommended that this question wording and the structure of the questionnaire should, as far as possible, be retained. The questionnaire wording and structure may be amended to fit local circumstances and points at which such changes might be incorporated are indicated in the questionnaire. There is flexibility in the standard questionnaires to allow site specific questions to be added and for the scheme valuation section to be repeated for a number of scheme options. The questionnaire should be tested in a pilot survey to ensure that the wording of questions is appropriate to local circumstances.

4.5.4 Data on the number of visits

Two data collection methods are recommended for obtaining accurate estimates of adult visit numbers at the coastal site at the feasibility study stage.

1. *Infra-red counters* It is recommended that infra-red or similar counters should be installed at the site. The counters should be located at key access points such as steps down on to the beach, or main points on the promenade, or on gates or stiles on cliff-top paths.

Ideally the counters should be left in place over a twelve month period so that the seasonal pattern in the number of visits can be established. The counters require regular reading and some maintenance as they can easily be damaged. Further details on these counters can be found in the Recreation Site Survey Manual (Tourism and Recreation Research Unit, 1983). Clearly the use of counters needs to be planned at an early stage in the project appraisal process if a full year's recording is to be achieved.

2. *Manual counts* The records of the infra-red counters will have to be calibrated with manual counts conducted periodically in order to translate the data from the counters, which merely record passages past a point, into information on the number and type of adult visits to the seafront or cliff-top.

Manual count information at each of the locations with a counter should be collected on the number of adult visits, the number of multiple passages, and the type of visitor involved. Ideally, these manual counts should be undertaken over a period of several days during each of the seasons so that any seasonal variations in the relationship between counter records and manual count data can be identified. A model visit count record sheet for use for manual counts is included in Appendix 4.4.

Manual counts may be used as an alternative method to infra-red counters to obtain estimates of the number of adult visits to a site. The Recreation Site Survey Manual (Tourism and Recreation Research Unit, 1983) also provides information on conducting manual counts at recreation sites. Specific site characteristics will determine the appropriate location for the counters and the detail of procedures for conducting the manual counts.

4.6 The recommended methods: analytic techniques and the interpretation of the results

4.6.1 Checking, coding and processing the CVM survey results

The questionnaires and all the responses should be checked for consistency and completeness on the day or day after they are completed so that any errors made by the interviewers can be corrected as the survey proceeds. Corrections and additional coding may be carried out at this stage.

It may prove necessary to discard some questionnaires, for example interviews wrongly conducted with non-British residents or those under eighteen years old, or partially completed interviews. Checked and coded data should then be entered on to the computer database directly from the questionnaires, using the data entry program chosen at the survey planning stage. Further computer checks should then be run to check for coding or data entry errors, and to ensure that the data is consistent and that codes are within the specified ranges.

4.6.2 Exploratory data analysis: outliers and non-response

Initially, the frequency distributions and summary statistics of all the variables in the survey should be examined using Exploratory Data Analysis techniques (Tukey, 1977). Two issues will have to be addressed in examining the CVM valuation variables: how to treat item non-response to these questions, and how to deal with 'outliers' or extreme values.

The responses to the questionnaire will provide several valuation variables to be used in calculating losses with erosion and gains with protection. Respondents may have failed to provide an answer to one or more of these questions and a decision will have to be taken as to how to treat these question non-responses. In the 1989 beach recreation survey, for example, about 20 per cent of respondents were unable or unwilling to provide a valuation for 'today's visit'. Some of the respondents objected in principle to the CVM procedure and considered that it was not possible to put a monetary value on an unpriced public good. Some of these respondents rated their beach recreation experience very highly. Others found the thought processes involved in arriving at a valuation too difficult or too time-consuming, and were unable or unwilling to make the effort to arrive at a monetary value.

It is desirable to carry out some analyses to identify whether those not offering valuations are distinctive in their behaviour or characteristics: for example they may be predominantly elderly, or less educated people, or infrequent users of the seafront. As there was little evidence from our research that those who did not respond to the valuation questions did not value their beach recreation, it was considered inappropriate to include the non-responses in the frequency distributions as zero responses. Therefore, it is recommended that all the non-responses to the valuation questions should be excluded from the analysis and from calculation of mean values of gains and losses. This, in effect, is to assume that the non-respondents have placed an average rather than a zero value on their recreational experience. If the non-responses were included as zero values the means of gains and losses would be considerably reduced. It is desirable to make the calculations both ways, as a kind of sensitivity analysis.

A second issue concerns the treatment of 'outliers' or extreme values in the valuation responses. The mean values, which are the most commonly used summary statistic in CVM surveys, are sensitive to outliers: a few very high valuations – for example the respondent who values his or her day's enjoyment of the beach at £100.00 – can have a considerable effect on the mean valuation. But such high valuations may be legitimate: the respondent may view his or her day at the seaside as equivalent to an expensive day at the races or a night at the opera. Alternatively, these high valuations may be a form of 'protest vote' against the CVM process.

Three possible approaches to this issue are discussed in the CVM methodology literature: the deletion of outliers on the basis of the researcher's judgement of each case; the use of robust statistical estimators such as the alpha-trimmed mean (Mitchell and Carson, 1989) to exclude extreme values from the calculation of the mean; and the use of regression diagnostic techniques to identify whether or not the 'outliers' should be regarded as part of the distribution or be excluded. Any of these approaches can be recommended, with the proviso that the methods adopted to deal with extreme values should be carefully reported. It is again possible to derive estimates including and excluding extreme values as a form of sensitivity analysis.

Zero and negative valuations, however, are to be included in the calculation of the mean.

4.6.3 The assessment of the losses with erosion and gains with coast protection

As outlined in Section 4.5.1, the recreation benefit arising from coast protection may have two components: (a) the loss avoided through preventing erosion; and (b) the gain achieved through improvement in the facilities for recreation provided by the coast protection project.

Where a coast protection project can be expected to do no more in preventing erosion than to maintain the status quo – with the seafront or cliff-top remaining in its current condition with no enhancement of the recreational facilities – then only the former benefit may be calculated. However most coast protection projects would be expected to result in some enhancement to the coast and its recreational potential rather than in simply maintaining the current state (not least because coast protection studies are usually a response to a deteriorating situation). Therefore an additional positive recreational gain (that is, (b) above) would be included in the calculation of the recreational benefit. The overall recreational benefit of a project would consist of the loss avoided plus the recreational gain from the project.

1. *Losses with erosion* In calculating from the survey data the economic loss resulting from changes at a coastal site, it is necessary to take into account not only the loss of enjoyment that may follow from the change but also the possibility that, given a change in the coastal site, users will decide to transfer their visits to an alternative site. In the first instance, the economic loss is measured by the loss in enjoyment; in the second, it is measured by the difference between enjoyment at the site in its current condition and enjoyment at the alternative site plus any increase in cost involved in visiting the alternative site.

The loss or benefit in the first instance is:

$$B_1 = E_0 - E_1 \qquad \text{Formula 4.1}$$

In the second instance the benefit is:

$$B_2 = (E_O - E_A) + (C_A - C_O) \quad \text{Formula 4.2}$$

where:

- B is the benefit per person (in cases 1 and 2);
- E_O is the value of enjoyment of 'today's visit'/'a visit in current conditions';
- E_1 is the value of enjoyment per visit after erosion;
- E_A is the value of enjoyment per visit at the alternative site (visited after erosion);
- C_O is the cost incurred in visiting the present site; and
- C_A is the cost incurred in visiting the alternative site (visited after erosion).

The total economic loss is given by Formula 4.1 for those who continue to visit the site and by Formula 4.2 for those who would transfer their visits elsewhere.

The axiom of rationality (Section 2.3.1) implies that the individual will choose to adopt whichever behaviour results in the least reduction in utility; that is, to continue to visit the first site if the value from Formula 4.1 is less than the value from Formula 4.2, and visit elsewhere if the reverse is the case and the value from Formula 4.1 is greater than that from Formula 4.2.

The axiom of rationality also provides a way of checking the consistency of respondents' replies as to the change in enjoyment and costs involved in visiting alternative sites. The loss to those who choose to visit another site after erosion, instead of continuing to visit the eroded site, cannot be less than zero. If the individual would gain more enjoyment from another site, net of the increase in costs incurred through visiting that site rather than the present site, then the 'economic rational' individual should already be visiting that site.

Typically, only a small proportion of respondents in user surveys give 'irrational' responses: they should not be visiting the site but going elsewhere. Whilst these respondents are labelled as 'irrationals', this is simply a label, and not a definition, which indicates that their behaviour is inconsistent with economic theory.

If responses appear to be 'irrational', then the responses need to be investigated further. Where the proportion of such respondents is low, then an obvious explanation is measurement error. A second obvious explanation of their behaviour, that they were imperfectly informed at the start of their journey about the enjoyment they would gain from a visit, can be ruled out if respondents are local residents. A third hypothesis is that the measure used in the survey instrument of the marginal costs of visiting the preferred alternative site is inadequate. The question is directed towards the monetary costs involved. To the extent that the respondents' existing behaviour is partly determined by considerations of convenience and the opportunity cost of time taken to travel to the other site, then the equation used to estimate the losses to movers is in error. Where convenience and the opportunity cost of time are significant, the losses estimated for movers will be underestimated.

If a small proportion of the responses given in the survey do not conform to the axiom of rationality, it is recommended that the cases should be excluded from the analysis. But if a substantial minority of the responses are 'irrational', as was the case in the Herne Bay residents survey, it is recommended that the irrational respondents should be included in the calculation of the mean losses with the value of their loss set equal to zero. The issue of the treatment of 'irrational' respondents is discussed further in the Herne Bay case study (Sections 8.5.4 and 8.5.6).

2. *Gains with protection* The gains with protection will in most cases be simply the difference between the value of enjoyment of 'today's visit'/'a visit in current conditions' and the value(s) of enjoyment per visit with the coast protection project(s). If a number of alternative projects have been presented in the survey, there will be a corresponding number of alternative values for the gain with protection.

When users continue to use the site, the gain after protection with project n is:

$$B_1 = E_{Xn} - E_O \quad \text{Formula 4.3}$$

In some cases, particularly where a scheme radically alters the recreational facilities that a site offers rather than reinstates the site to its former condition, some individuals may dislike the new scheme site and experience a loss of enjoyment with the scheme. In such cases, it will again be necessary to take into account whether or not the users in this situation anticipate that they would transfer their visits elsewhere if the scheme were installed.

When users transfer to an alternative site, the gain after protection with project n is:

$$B_2 = (E_O - E_{An}) + (C_{An} - C_O) \quad \text{Formula 4.4}$$

where:

- B is the benefit per person (in cases 1 and 2);
- E_O is the value of enjoyment of 'today's visit'/'a visit in current conditions';
- E_{Xn} is the value of enjoyment per visit after coast protection with project n;
- E_{An} is the value of enjoyment per visit at the alternative site with project n;

C_{An} is the cost incurred in visiting the alternative site with project n; and

C_O is the cost incurred in visiting the present site.

The comments above concerning the rationality of the respondents replies also apply to responses on gains with a protection scheme.

Gains and losses should be calculated for each individual in the survey and then the mean value should be derived.

4.6.4 Checks on the reliability and validity of the survey results

As has been pointed out in Section 4.5 the questionnaire and the analysis should be based on an underlying model of recreational behaviour and value which defines the expected relationships between the variables included in the survey.

A basic form of validation is to test whether the results from the survey conform to the underlying theoretical model (Figure 4.4). This can be done using correlation and regression analysis. In order to carry out the regression analysis, it may be necessary to transform the variables to normality using a logarithmic or other form of transformation. It is recommended that the theoretical, or construct, validity of the valuations and gains and losses obtained in the survey which are the dependent variables which the model seeks to explain, should be tested using regression analysis. The valuations and gains and losses can be regressed on the group of independent or explanatory variables which are predicted by the model to be determinants.

The size and sign of the beta-coefficients can be examined to test whether the results are consistent with the theoretically expected relationships. The proportion of the variance in the individual responses for the dependent variables explained by the independent or explanatory variables will indicate the extent to which the results conform to the underlying model. However, an analysis of the residuals should also be undertaken in order to check this further. Explained variances or R^2s of the order of 0.20 to 0.50 can be taken as indicating a reasonably good level of fit or explanation, suggesting that the valuations offered are consistent with the preferences, behaviour or characteristics that are theoretically expected (Mitchell and Carson, 1989). An example of the kind of analysis recommended is given in Table 4.5.

The convergent validity of the valuations obtained in a survey should also be tested by including in the survey alternative measures of the same variable. However, as already outlined, willingness to pay questions do not measure exactly the same variables as enjoyment per adult visit questions, nor does the Travel Cost Method compare directly with Contingent Valuation Methods. Therefore, although, in principle, it would appear desirable to include both TCM and CVM measures in a survey, the experience of coastal recreation research outlined earlier (Section 4.3) indicates that the TCM may well give misleading results and consequently its use as a test of convergent validity cannot be recommended.

A further test of validity is to compare the results obtained using the same methods in different settings. The data on valuations presented in Section 4.3 and the case studies in Chapter 8 provide some examples of valuations obtained using comparable methods to those recommended in Section 4.5 to value different goods and to value the same type of good – beach recreation – at different locations. These results can be used for comparison.

It is desirable but rarely possible to include tests of reliability in the design of a survey which require repeat measurements to be obtained from the same population. Ideally any available opportunity should be used to include such tests, even on a sub-sample of the total survey sample.

4.6.5 Grossing-up to estimate total annual recreational benefits

The aim of the CVM survey is to provide data from a sample survey from which to generalise about the total population from which the sample is drawn: all the households in an area or all the visits to a site. It is important, therefore, to take whatever steps are possible to check that the sample obtained is representative of the population from which it is drawn. It is recommended that household survey data should be compared to data on the number of electors in different parts of the survey area and to Census data for the survey area. It may be possible, by using Census data, to compare the social characteristics of respondents in the survey – their gender, age, and home ownership, for example – with the population in the survey area.

With site surveys of users, data on the survey population of visits is likely to be lacking and therefore checks on the representativeness of the survey respondents are likely to be problematic. It is possible that the data collected for the purpose of manually calibrating the counts obtained with infra-red counters will provide some data on the survey population which can be used to check the survey proportions, for example of men and women, or the proportion of locals, day and staying visitors.

It is recommended that data should be collected during the survey on survey non-respondents. This data will necessarily provide only limited details on non-respondents such as whether they are local, day or staying visitors for site survey non-respondents, and gender for visitors, and gender and location for household survey non-respondents. This data should be analysed to see whether there is any possibility of non-response bias.

Non-response will only be of concern if it differentially affects a sub-group who differ markedly from other respondents in the key variables of interest – the valuations and gains and losses derived from them. This might be the case, for example, if non-respondents were predominantly local or elderly people or from one particular area of the town or the seafront. If a marked non-response bias were to be identified, it would be possible to weight the survey data in order to ensure that the results were more representative of the population than was the raw survey data.

Once it has been ascertained that the survey sample is broadly representative of the survey population, it is necessary to gross-up from the survey data in order to obtain a gross annual benefit for the population. If the survey is of householders or residents, the electoral register or Census data may be used as a guide to the total population. Data on frequency of visits to the coastal site collected as part of the household survey can also be used to provide an estimate of the mean number of visits per annum per adult resident or householder in the survey area. The total number of visits per annum to the coastal site generated by the population in the survey area can then be calculated by multiplying the total population in the area by the estimated figure for the mean adult visits per annum. This figure for the total number of visits per annum can then be used to gross-up to obtain the annual recreation benefit.

If the survey is a site survey, the survey data should be grossed-up according to the estimated number of adult visits per annum. Methods of obtaining site specific estimates of this figure by the use of infra-red counters calibrated by the use of manual counts have been described in Section 4.5.4.

As has been detailed in Section 4.5.1, the annual estimated recreation benefit arising from a coast protection project may have one or two components. The benefit arising from the two components should be calculated as separate figures. Several alternative annual benefit figures for gains with coast protection may be calculated if separate valuations have been obtained for alternative projects. The Herne Bay case study (Chapter 8) provides an illustration of the calculation of a number of annual benefit figures for alternative coast protection projects offering differences in recreational facilities.

The estimated annual benefits should be calculated as at the pre-feasibility stage, but using site specific survey and count data:

a) Annual benefit of preventing a loss with erosion = (value per adult visit of loss with erosion) x (number of adult visits per annum);
b) Annual benefit of the gain with coast protection project = (value per adult visit of gain with coast protection project) x (number of adult visits per annum).

For a site survey, estimated mean losses and gains, and estimated total annual recreational benefits, should be calculated separately for local, day and staying visitors in order to enhance the accuracy of the estimates. This can be seen in Table 8.5.3 which shows the data for the Herne Bay study.

Confidence intervals may be calculated for the estimated mean losses and gains and thence for the annual recreation benefits. It can be argued that this is an inappropriate statistical procedure since the calculation of such intervals is based on the assumption of a normal distribution of the values for gains and losses and these usually have a skewed distribution. However, 95 per cent upper and lower confidence limits do provide an indication of the range within which the annual benefits are likely to lie and the lower limit can be employed if it is thought desirable to be conservative in making benefit estimates. These methods are discussed in the Herne Bay case study (Section 8.5.4).

4.6.6 Calculating the total project recreation benefits

Similar methods and procedures to those outlined in Section 4.5.1 for the pre-feasibility stage are followed in calculating the total project benefits at the feasibility stage.

Two factors have to be taken into account: the rate of erosion or degradation, and the assumed life of the project. More precise information on these factors is likely to be available at the feasibility stage than could be employed at the pre-feasibility stage, and this should be exploited. The total project loss with erosion and the total project gain with protection again have to be calculated as separate components.

When calculating the total project loss with erosion, a decision has to be taken as to how soon the coast would deteriorate to the condition described in the CVM erosion or degradation scenario and shown in the drawings and text used in the survey. This will depend upon the rate of change in the coast. It may be concluded that the scenario conditions would not arise until, say, a further ten or fifteen years had passed without any protective action being taken. Then, the losses would be calculated from year 10 or 15 of the project and would be assumed to be constant over the remaining life of the project.

The gains, however, would be assumed to start on

Table 4.6
Guidance checklist for recreation benefit assessments

A. **Pre-feasibility stage**

1. Define erosion or degradation problem.
2. Outline protection project(s) proposed.
3. Define the extent of the area affected by erosion/degradation and the project(s) – the site.
4. Determine whether there is existing or potential recreational use of the site.
5. Identify the current type of recreation use of the site: specialist or informal; use by local residents, day visitors or staying visitors.
6. Locate secondary data on level of current recreational use of the site: the number of adult visits per annum to the site, preferably for local, day and staying visitors separately where different user types are involved.
7. Identify the type of site: commercially developed resort, mainly residential site or undeveloped site.
8. For resort sites subject to changes comparable to those described in the 1989 beach survey, use the standard data on the value of gains and losses per adult visit.
9. For these resort sites, calculate a preliminary annual recreation benefit = (standard data for gains and losses per adult visit) x (number of adult visits per annum). This is computed separately for local, day and staying visitors, where possible, and discounted over the appropriate period of the life of the project to give an estimate of the preliminary total recreation benefit of the project.
10. For resort sites decide whether the size of the preliminary total recreation benefit (calculated according to 9, above) justifies proceeding to the feasibility study stage (judged in relation to the scale and likely cost of the protection project).
11. For non-resort sites such as cliff-top or undeveloped sites, judge on the basis of the preliminary data on the rate of erosion, the number of users and if there are specialist users, whether the size of the likely recreation benefit justifies proceeding to the feasibility study (judged in relation to the scale and cost of the protection project).

B. **Feasibility study stage**

12. Define the erosion or degradation problem in detail so that descriptive texts and drawings or photographs can be prepared for scenarios for use in CVM surveys.
13. Determine the details of the protection project(s) so that descriptive texts and drawings or photographs can be prepared for scenarios for use in CVM surveys.
14. Define the area affected by the erosion/degradation and project(s) and prepare descriptions and maps for use in CVM surveys and in count exercises. Select locations for infra-red or other counters and interviewers if a site survey is required.
15. Define the survey population: those who may benefit. Informal or specialist users will require a site users survey; local residents or new recreational facilities will require a survey of householders or adult residents selected from a sample of addresses from the electoral register. Conduct both resident and users surveys where resources permit.
16. Define sub-groups within that population requiring separate analysis – local, day or staying visitors for site surveys or different residential areas for household surveys.
17. Install infra-red or other comparable counting devices at appropriate locations on the coastal site if a site survey is to be conducted, or plan to undertake manual counts. Plan, if possible, to record data using counters over a twelve month period so that seasonality in visiting can be taken into account.
18. Decide whether to carry out the CVM surveys and/or manual counts in-house or whether to contract out the work. The two data collection exercises – manual counts and interview surveys – can be integrated by using the same interviewers.
19. Arrange for manual counts to be conducted to calibrate the results of the infra-red or other counting devices where used. Manual counts should be carried out over a period of several days, preferably in each of the seasons of the year.
20. Use samples as large as financial resources permit. Contingent valuations have large variances; therefore large samples are required to increase the precision of the estimates, and if many sub-groups are to be examined separately.
21. Plan to conduct site surveys on a minimum of eight survey days over the summer period if the survey can only be carried out at one time of year and plan for the exploratory, pilot and main surveys to take a minimum of four months to complete.
22. Take measures to ensure that a representative sample of the survey population is selected by specifying a random element in respondent selection procedures and by regulating the times, places and days on which interviews are to be conducted for site surveys.
23. Use trained, experienced and well briefed fieldworkers.
24. Supervise fieldwork closely and ensure that checks with a sample of the respondents are made to check on the interviewers' work.
25. Ensure that detailed records on survey non-response are kept.
26. Use the model questionnaires provided in this Manual. These may be adapted as necessary to local circumstances but should retain the question forms and sequences as far as possible. The questionnaires are based on an underlying model of coastal recreational behaviour and value, and include questions on behaviour and preferences for use in validity checks.
27. In the questionnaires, the main valuation question employed is the 'value per adult visit' approach. Willingness to pay questions and the Travel Cost Method can be included for comparison although they measure different things.
28. Use clear drawings or photographs and simple non-technical texts in CVM scenarios to represent changes in the coast to the survey respondents. Care must be taken to ensure that drawings, etc., accurately represent the likely erosive changes and projects.

92   THE ECONOMICS OF COASTAL MANAGEMENT

29. Check that the sample is representative of the population by comparing it with population statistics and by examining data on survey non-response.
30. Check the data carefully and conduct a full exploratory data analysis. Decide on how extreme values and non-response to the valuation questions should be treated.
31. Derive mean values for enjoyment per adult visit to a coastal site in varying conditions, and thence gains and losses with change in these conditions for local, day and staying visitors in site surveys or for other significant sub-groups.
32. Base the analysis on an underlying model: check that the results from any analysis conform to the underlying theory and that the statistical fit is adequate. Use regression analysis or other comparable techniques for this purpose.
33. Compare the results from the survey with results from other surveys of coastal recreation and other goods valued using comparable methods.
34. Calculate the estimated annual benefit of avoiding erosion and of improvements with the coast protection project(s) separately for local, day and staying visitors.
35. Derive the estimated total project recreation benefit for avoiding erosion and for improvements with coast protection project(s). Do this separately for sub-groups. Discount the annual benefits over the appropriate number of years of the project life.
36. Confidence limits may be calculated so that the upper and lower limits for the estimated total project recreation benefit can be indicated.
37. Where there is uncertainty over the estimated number of adult visits per annum to the site, different adult visit estimates can be used to derive estimates of the total project recreation benefit within sensitivity analyses.

---

completion of the project and continue through the life of the project. Both gains and losses should be discounted over the appropriate periods using the Test Discount Rate. The Herne Bay case study (Chapter 8) again provides an example of the calculation of the total project benefits using these methods.

4.7 Summary: assessment, limitations and guidance checklist

The estimated annual benefits from recreation use of a site should be calculated as comprehensively and accurately as possible. They are likely to be an important part of the justification of many coast protection and sea defence projects. A checklist is provided in Table 4.6 to guide these assessments.

A high value for recreation benefits may arise from the presence of a large number of site users, even when the value of individual visits is small. A high benefit value may also arise with a smaller number of users where individuals give a high value to the enjoyment of their visit. This is more likely to be the case where specialist rather than generalist recreational activities are being carried out.

The loss of recreation value with continued erosion in the absence of a coast protection project can be seen as one of the benefits of a project. The benefits of protecting a coast by different types of project can be compared where the survey has utilised different drawings portraying various project alternatives.

A major limitation of this approach is that in most feasibility studies the respondents to such a survey will be restricted to current site users or local residents who are non-users. In some circumstances existing users may prefer a project which will maintain the 'status quo' to one which will enhance the facilities, if such enhancement is perceived as changing the type of recreational user, the nature of recreational activity, or increasing the level of recreational use in the locality. This may be because current users anticipate increased crowding which would detract from their enjoyment.

However, this response will be influenced by the relative proportions of local, staying and day visitors in the current user population, since significant differences have been found between the responses of these three groups. Whereas staying visitors appear to prefer projects that will not attract increased usage, locals may favour such an increase if they anticipate it will provide enhanced tourist income for the locality.

A survey approach, based on a valuation by current users, clearly has limitations in that it cannot take into account potential new users who would benefit from a project providing enhanced facilities. Methodologically, there is no opportunity for incorporating the valuations of this group of potential users into the recreation benefit calculation without carrying out extensive (and hence costly) household surveys in the catchment area of the new facility.

A further aspect of the recommended approach to which attention needs to be drawn is that the valuations of recreation enjoyment, and hence the estimates of losses and gains from erosion and coast protection, are only valid for the scenarios represented in the drawings and descriptions used in the surveys. Therefore, if project proposals or detailing were to vary subsequently, the valuation results might not be valid.

Another aspect of the approach which should be borne in mind is that although the valuations given by recreational users have been shown to be consistent between user groups, and across sites and from year to year in the surveys so far undertaken, the theoretical models which underlie these are relatively new and are still being validated through wider application.

A further methodological limitation is the inability to compare benefit estimates derived from the 'willingness to pay' method with those obtained from valuations of 'visit enjoyment'. Because the willingness to pay method may incorporate both use and non-use values and hence benefits, and it is not possible to apportion these elements, there is no direct comparison with the user based valuation of enjoyment of visit. Although the non-use benefits are unlikely to be a significant factor at resort sites, they are likely to increase in importance in the case of less developed sites.

The estimation of total annual visit numbers at the pre-feasibility stage and the calculation of these numbers in the feasibility study will considerably influence annual benefit figures. Since these figures can only be as accurate as the available information, it is of great importance that the need for such data is anticipated at the **earliest possible opportunity** so that counts may be initiated where necessary in order to provide the most accurate basis for benefit calculations.

Even in the case of large resorts where detailed data is available for the summer months there are usually few figures for usage in other seasons. In addition, secondary source information such as car parking numbers or data for specialist club use are often of doubtful accuracy or even very inaccurate. A major issue raised by this is that, depending on the relative importance of the recreation benefits to the overall project benefits, so the need for accuracy of visit and valuation figures becomes important. However, by the time the stage is reached at which this becomes evident, it may be too late to collect accurate data on visit levels. Clearly, careful planning of the whole appraisal process is essential.

The mean valuation obtained and hence the estimated mean gains and losses per visit can be considerably influenced by 'outliers' or extreme values in the distribution. A decision to exclude extreme values may be taken if, after detailed examination of the responses, this appears justified. The treatment of non-response to the valuation questions can also affect the mean valuation.

Survey non-response may also bias the estimated valuations obtained from a survey. Estimates derived from relatively small scale sample surveys are subject to fairly large sampling errors. It is important, therefore, that sample sizes are as large as resources permit for CVM surveys. Use of confidence intervals gives an indication of the range within which the estimates might be expected to be found with a given level of confidence. Obviously, however, the confidence level selected, and the possible selection of the lower or upper limit of the confidence interval will have a considerable impact on the overall benefit estimates.

Finally, it is important to note that any benefit estimates may not cover all the potential recreational losses and gains that may arise from coastal erosion and protection. Losses may result from less frequent visiting or visitors simply staying at home and these are difficult to predict and therefore to value.

Equally difficult to predict and value are potential gains from increased frequency of visiting, transferred visits and new recreational uses that may be generated by a coast protection project. These could be important, particularly with project proposals that include the potential for new water-based recreational developments (Section 8.5). The disbenefits which might arise for residents and current users with such a project might well be outweighed by the benefits to new and transferred visitors. Thus, the annual benefit estimates obtained by the methods outlined in this Chapter may only represent a part of the total potential recreational benefit of coast protection and sea defence projects.

5 The benefits of flood alleviation: sea defence at the coast

5.1 Summary of information

This chapter contains a summary of the latest information and techniques concerned with gauging the benefits of flood alleviation at the coast.

In this respect, Section 5.2 outlines the stages involved in the calculation of annual average flood damages, and the nature of the loss-probability relationship that is central to this calculation. Section 5.3 gives the results of research into the extra effects on flood damages that are caused by sea water flooding as opposed to fresh water fluvial flooding.

Sections 5.4 and 5.5 give details of the available data and techniques for this assessment benefit modelling and calculation (and therefore complement Chapter 7). Section 5.7 gives the recommended approaches for the assessment of the agricultural benefits of flood alleviation at the coast. Finally, Section 5.8 provides a checklist of matters to be addressed in this field, and a number of 'health warnings' with regard to pitfalls that may prevent the user from obtaining reliable results.

The material in this chapter is inevitably abbreviated in comparison with other Manuals produced on the benefits of flood alleviation. Thus the user is necessarily pointed to a number of other sources, including the original 'Blue Manual' (Penning-Rowsell and Chatterton, 1977), which remains useful as an overview, and the later 'Red Manual' (Parker *et al.*, 1987) in which data and techniques are presented for all sectors except flooding to residential property.

In addition, therefore, readers will need to evaluate the later 'Update report' (Suleman *et al.*, 1988) and use the regularly published 'FLAIR reports' (Flood Loss Assessment Information Report) (N'Jai *et al.*, 1990). This latter publication contains data and information on all aspects of flood alleviation benefits, updated as closely as possible to the year previous to the date of publication. Finally, any evaluation of agricultural benefits will need to consult the numerous publications from Silsoe College on this subject (e.g. Morris and Hess, 1984; Hess and Morris, 1985).

Since these cross-references will be necessary for most projects of any size, the material here is designed to give, first, results from research specific to coastal situations, and, secondly, an overall picture of the stages and data needed for this type of benefit assessment. In addition, thirdly, there is guidance as to the lessons that have been learnt in the past about the importance of different steps within this process and their significance for the accuracy of the results.

5.2 The calculation of flood alleviation benefits: the basic inputs

The methodology for assessing the benefits of flood alleviation is well-known and relatively uncontroversial (Penning-Rowsell and Chatterton, 1977; Parker *et al.*, 1987). When applied to sea defence situations, as opposed to the riverine environment, there are a number of complications that require extra data and additional techniques, and these are described in this chapter, but the fundamentals remain the same.

The main area of difficulty is in estimating the frequency of flooding, particularly if this flooding results from the failure of existing defences, for which the return period of the failure event needs to be calculated or estimated and the flooding scenarios thereafter assessed. There may well be no historical information on flooding on which to base the estimate of frequency – or the return period of the failure event – yet this will profoundly affect the flood damage that will occur and the benefits of preventing this damage.

In this respect it is useful to remember that flood damage can be categorised as tangible or intangible, depending on whether monetary values can be assigned to that damage. Research has determined that the latter are very important and they should not be neglected (Green and Penning-Rowsell, 1989).

Secondly, flood damages can be also classified as to whether they are direct or indirect, depending on whether the damage is the result of direct contact with the flood waters (direct damage) or whether the losses result from disruption of economic activity consequent upon the flooding (indirect losses). These indirect losses can be the result of flooded premises being disrupted, or premises some way away from the flood being disrupted because communication is affected by the flooding. Indirect losses should not be confused with intangible losses.

5.2.1 The basic objective: deriving the loss probability relationship

Figure 5.1 gives a simplified flowchart of the stages that need to be followed in order to calculate the benefits of flood alleviation projects (or, put another way, the stages for calculating the capitalised value of flood losses that will occur in the future if a 'do nothing' option is adopted).

Figure 5.1, and the more detailed Figure 5.2, show that a number of different data inputs are required, including information on flood levels at different return periods within the benefit area, and data on the damageable items within that area, in the form of a land use survey. In addition, information is needed on the effects of the project or projects being appraised, in terms of the reduction in flooding frequency and/or severity, or the delaying of that flooding if the project is designed to replace a sea defence project that is nearing the end of its life.

The fundamental objective in these calculations is the calculation of the annual average flood loss to be suffered in the future if the 'do nothing' approach is adopted and flooding at the coast is to be allowed to occur without policies, plans or projects to prevent it from happening. This value for annual average losses is then discounted over the lifetime of the project or projects being appraised, to produce the figure for the capital sum it is just worth spending to yield a benefit-cost ratio of 1.0.

This annual average flood damage is the area under the graph of flood losses plotted against exceedance probability (the reciprocal of the return period in years). This graph is termed the 'loss probability relationship' and should be derived from an analysis of several future floods with a range of severities, or the

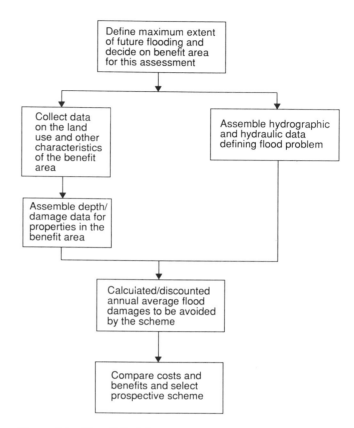

Figure 5.1 Simplified flowchart of the stages involved in the assessment of the benefits of sea defence schemes

results will be unreliable (at least three and preferably at least six flood events should be used). It is important, in particular, to determine the return period of that flood at which damage begins, assuming that some floods cause no measurable damage (Figure 5.3), and also the timing and return period of the failure of existing defences if that is the mechanism of flooding to be alleviated.

The benefits of flood alleviation must always be calculated with the 'with and without' criterion in mind: benefit-cost analysis looks at the difference in the nation's resources with and without the investment that is being appraised (Chapter 2). Thus if a project does not prevent all the flooding that might occur in the future, but residual overtopping of sea defences is likely or a failure of the project is possible, then the damage from such continuing or possible additional future flooding in these situations should be appraised.

For this to be done, it is, in effect, necessary to calculate two loss probability relationships for each option being evaluated: the relationship without the project (usually the 'do nothing' situation) and the relationship with the project in place and with the residual flooding that is expected to occur (Figure 5.3). The difference

96 THE ECONOMICS OF COASTAL MANAGEMENT

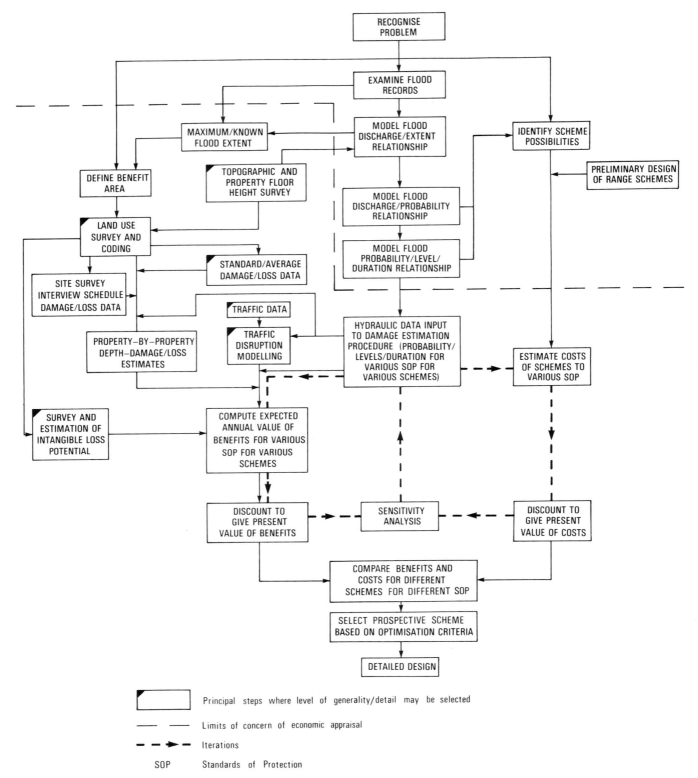

Figure 5.2 Comprehensive flowchart of the stages involved in the assessment of the benefits of sea defence (from Parker *et al.*, 1987)

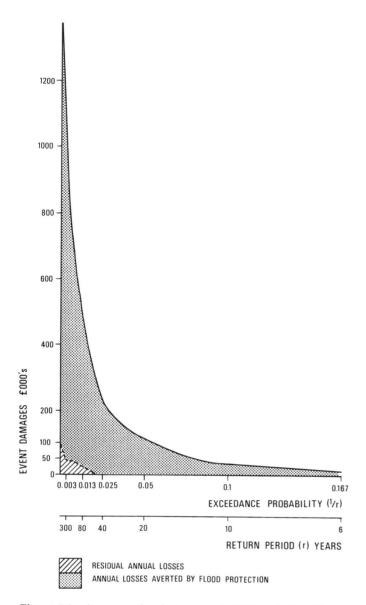

Figure 5.3 An example of a loss-probability relationship used to calculate annual average flood losses (from Parker et al., 1987)

between the areas under the graphs of these two relationships is the net benefit of that investment, on an annual average basis. That is the benefit sum – once discounted over the lifetime of the project – that should be compared with discounted project costs.

5.2.2 Basic inputs: the hydrographic analysis

A prediction of the extent and depths of flooding that will occur in the future is a basic input to the assessment of the benefits of flood alleviation. To derive this prediction will not be easy but it is necessary for project design, and will therefore need to be undertaken if any sea defence works are to be implemented.

There are perhaps two somewhat different basic scenarios that are commonplace in terms of sea defence situations. The first, and most dramatic, concerns the possibility of a **breach** either in the existing sea defence structures (owing to failure of design or the works coming to the end of their lives) or a breach in a natural or semi-natural coastal feature such as Chesil Beach (Penning-Rowsell and Parker, 1980), Hengistbury Head (Parker and Thompson, 1988), the Orfordness spit at Aldeburgh (Turner et al., 1992), or Whitstable (Parker and Penning-Rowsell, 1981). In only the Whitstable situation was the cause of future flooding a possible failure of existing artificial sea defences, but this is becoming a common cause of the need for new investment against flooding at the coast.

The second situation concerns **progressive overtopping** of existing sea defences or a situation where the sea defences are **non-existent or inadequate**. This now appears to be less common than the first situation, but is nevertheless an important cause of flooding at the coast and possibly an increasing problem if global climatic warming leads to sea level rise whereby existing sea defence projects are in danger of being overtopped to an increasing degree.

A related type of flooding is where riverine flooding and sea flooding occur simultaneously, for example where a river is 'backed up' into its estuary owing to high tide and/or surge levels preventing its flood water flowing into the sea.

The methods of predicting flooding in these situations are complex and are described and summarised elsewhere (Construction Industry Research and Information Association, 1991). Clearly it will be necessary to calibrate the predictions made against known flood histories, for which historical records will be necessary. The data that is required for this analysis is the extent of flooding in the area liable to be affected, so as to determine the area in which the benefits of flooding will be assessed (the benefit area).

Within this benefit area, data on the different depths (or altitudes) of the flood waters will be needed for each return period being appraised, allowing for any wave effects and flows within the area and between the area and the sea. The calculation of the return period of these events needs to take into consideration any joint probability that may be important (e.g. the probability of a certain still water level, and the combined probability of that still water level and waves of a certain height).

5.2.3 Basic inputs: land use survey

The calculation of annual average flood losses is highly place-specific: it is based on the land use in the benefit area at the time of the analysis. It is recognised, of course, that this land use will change during the lifetime of the project, which is one of the reasons for the 'standard' or average approach to depth/damage data (see Section 5.2.4 below).

Undertaking this land use survey is discussed in Section 5.3, below. What is needed is a record of all property that is liable to flood in the benefit area, classified by its type and function (see Appendix 5.1 for the land use classification currently used). This land use classification is hierarchical, and therefore the survey can be undertaken at different levels of accuracy or generality. This means that a decision can be taken to undertake, first, a **pre-feasibility survey** using the highest level of the classification, followed by a **more detailed survey** using all the sub-categories.

The land use survey includes recording the floor height (altitude) of all properties within the benefit area. This information needs to be accurate, since the accuracy of property floor heights has been found to be one of the major factors in determining the true shape of the loss-probability relationship. In addition, it will be necessary to record the heights of roads passing through the benefit area, so as to determine the depths of their flooding, and the locations of key links within other communication systems (e.g. telephone and electricity sub-stations, etc).

5.2.4 Basic inputs: depth/damage data applicable to coastal situations

The loss-probability relationship is a function of flood frequency (Section 5.2.2), the characteristics of the benefit area (Section 5.2.3), and the damageability of properties within that area. This last item is important, and damage to property from flooding differs in coastal locations from riverine environments.

The research for this Manual has investigated the most important of these differences (Section 5.3.1) to provide data on the susceptibility to flooding of a large number of the different property types shown in Appendix 5.1. Others will require a site-specific survey (Section 5.4.3) if the standard data in Appendix 5.2 is not adequate or not adequately detailed.

Clearly, the methodology being advocated here needs to match each item in the land use survey with depth/damage data specific to that land use item. This will not always be possible and to undertake the necessary detailed surveys of all properties for which standard data is not available may be excessively time-consuming. Judgement needs to be used here, to match the data required in the appraisal to the likely cost of the project and its likely cost-effectiveness.

5.2.5 Basic outputs: the annual average flood damages avoided by sea defence projects

The product of the inputs detailed above is the difference between the flood damages to be incurred in the future with a project to alleviate flooding and those future flood damages without that investment (which may or may not be worse than the status quo).

This difference can be calculated as an average for all years in the future, or for specific years if the flooding situation is known to be changing (perhaps as a result of a breach in defences occurring, natural erosion at the coast, sea level rise or fall, or for other reasons). The capitalisation of this difference between 'with' and 'without' flood damages is the sum that could be justified to be spent on the flood alleviation project to yield that stream of future benefits.

More details of the overall methodology and theoretical perspective on this assessment can be found in Penning-Rowsell and Chatterton, 1977 (Chapter 1), and Parker *et al.*, 1987 (Chapter 1).

5.3 Flood damages and losses in coastal situations

Research for this Manual has assessed in more detail than was hitherto available the extra damage costs that are likely to be incurred by sea water flooding, over and above that damage caused by fresh water flooding. This research complements and updates research undertaken some years ago (Penning-Rowsell, 1978), and uses survey information from a number of project appraisals undertaken in the last ten years (e.g. Parker and Penning-Rowsell, 1981).

5.3.1 Sea water flood damage to residential properties

Additional damage will be caused to the building fabric and inventory (contents) of residential property by the presence of salt in the water, over and above that caused by fresh water. The presence of sand, silt and sewage in the water also increases damage or losses. These damages do not include structural damage caused by the forces resulting from tides and waves or by the movement of flotsam and jetsam.

Salt water damage factors are presented in Table 5.1 (for flood durations of less than 12 hours) and Table 5.2 (for flood durations of more than 12 hours). These factors, showing the **extra** damage caused by salt water flooding, were derived by a team of professional surveyors/valuers, building on many years of experience in assessing values for property construction and rehabilitation. The application of these salt damage factors

THE BENEFITS OF FLOOD ALLEVIATION 99

Table 5.1
The additional effect of salt water damage on flood losses
(all residential properties: flooding less than 12 hours)

Components of damage	Depth above upper surface of ground floor (metres)																	
	-0.3	0.0	0.05	0.1	0.2	0.3	0.6	0.9	1.2	1.5	1.8	2.1	2.4	2.7	3.0			
Paths and paved areas	Nil	Nil	Nil	Nil	Nil	Nil	Nil	Nil	Nil	Nil	10	10	10	10	10			
Gardens/fences/sheds	Nil	Nil	Nil	Nil	Nil	Nil	Nil	Nil	Nil	Nil	Nil	Nil	Nil	Nil	Nil			
External main building	2	5	5	10	15	20	25	30	35	40	50	50	55	60	60			
Plasterwork	Nil	Nil	Nil	10	10	15	15	15	15	15	20	20	20	25	25			
Floors	Nil	Nil	10	10	15	15	15	15	15	15	15	15	20	20	25			
Joinery	Nil	Nil	Nil	10	15	15	15	15	15	15	15	20	20	25	25			
Internal decorations	Nil	10	10	10	10	10	10	15	15	15	15	15	15	20	20			
Plumbing and electrical	Nil	5	5	10	10	20	20	20	20	20	20	20	20	20	20			
Domestic appliances	Nil	Nil	20	30	50	50	60	75	75	80	100	100	100	100	100			
Heating equipment	Nil	Nil	5	5	5	5	5	5	5	10	10	10	10	10	10			
Audio/video	Nil	Nil	Nil	Nil	Nil	Nil	Nil	Nil	Nil	Nil	Nil	Nil	Nil	Nil	Nil			
Furniture	Nil	Nil	10	10	15	15	15	15	20	20	20	20	20	20	20			
Personal effects	Nil	Nil	20	20	20	20	50	50	50	50	50	50	50	50	50			
Floor cover/curtains	Nil	Nil	25	25	10	10	10	10	10	10	10	10	10	10	10			
Garden/DIY/leisure	Nil	Nil	10	15	25	30	30	30	30	35	40	40	50	50	50			
Domestic clean-up	Nil	Nil	20	25	30	35	40	45	50	60	70	80	90	100	100			

Note: Salt water damage factors assessed to nearest 5 per cent. Application conditional on damage not exceeding average remaining value (ARV).

Table 5.2
The additional effect of salt water damage on flood losses
(all residential properties: flooding greater than 12 hours)

Components of damage	Depth above upper surface of ground floor (metres)																
	-0.3	0.0	0.05	0.1	0.2	0.3	0.6	0.9	1.2	1.5	1.8	2.1	2.4	2.7	3.0
Paths and paved areas	Nil	10	10	10	10	10	10	15	15	15	15	15	15	15	15
Gardens/fences/sheds	Nil	Nil	Nil	Nil	Nil	Nil	Nil	Nil	Nil	Nil	Nil	Nil	Nil	Nil	Nil
External main building	2	5	10	15	20	30	40	40	45	50	50	55	60	65	70
Plasterwork	Nil	10	20	20	25	25	25	25	25	25	25	25	25	25	25
Floors	Nil	5	10	15	20	25	25	25	25	25	25	25	30	30	30
Joinery	Nil	Nil	10	15	20	20	25	25	25	25	25	25	30	30	30
Internal decorations	Nil	5	10	15	20	20	20	25	25	25	30	30	30	30	30
Plumbing and electrical	5	10	20	25	25	25	25	25	25	25	30	30	30	30	30
Domestic appliances	Nil	Nil	20	40	50	80	90	100	100	100	100	100	100	100	100
Heating equipment	Nil	Nil	10	10	10	10	10	10	10	10	10	15	15	15	15
Audio/video	Nil	Nil	Nil	Nil	Nil	Nil	Nil	Nil	Nil	Nil	Nil	Nil	Nil	Nil	Nil
Furniture	Nil	Nil	15	15	15	15	15	20	20	20	20	20	20	20	20
Personal effects	Nil	Nil	20	50	50	80	80	80	80	80	80	80	80	80	80
Floor cover/curtains	Nil	Nil	50	50	50	50	80	80	80	80	100	100	100	100	100
Garden/DIY/leisure	Nil	Nil	15	20	30	30	40	40	50	50	60	60	70	70	90
Domestic clean-up	Nil	Nil	25	30	35	40	45	50	60	70	90	100	100	100	100

Note: Salt water damage factors assessed to nearest 5 per cent. Application conditional on damage not exceeding average remaining value (ARV).

is conditional on their not resulting in calculated values for damage exceeding the average remaining value (ARV) for inventory items: damage cannot be greater than the value of the goods in question at the time of the flood. For the rationale of the use of average remaining values see Penning-Rowsell and Chatterton (1977), Chapter 2, and Section 2.5.2 herein.

1. *Building fabric* **Paths and paved areas**: Short duration sea water floods are unlikely to cause extra damage over and above fresh water floods at low depths, but penetration of water can occur, at greater depths, because of the pressure of water. For longer duration floods there may be some extra damage to reinforced concrete in the areas concerned. The soaking of the concrete areas with salt water could increase carbonation, as the steel reinforcement bars rust more extensively after contact with salt.

Gardens/fences/sheds: Damage to fences and sheds is likely to be no greater with salt water than fresh water flooding. Garden plants will be severely affected by salt water and plants will almost certainly be killed by both long and short duration floods, even at low depths when root damage can occur. Data for plants is not included in this category, but a separate assessment of damages can be made, based on a household survey. These are derived from short return period and generally shallow flood events, with sewage contamination, and are presented in Table 5.3.

Table 5.3
The effect of salt water damage to garden plants –
loss per household (garden)

Depth of flooding metres	Fresh water damages £	Salt water damages	
		% increase	£
–0.3	20	20	nil
0.0	60	80	33
0.3	113	226	100
0.6	164	328	100
0.9	198	396	100
1.2	209	418	100

(At 1988 prices)

External main building: Up to 55 litres of water can be held by a square metre of brickwork (i.e. the external component of a cavity wall). More water can be retained if the wall is thicker or is constructed with lightweight composite blocks. Once salt has impregnated, the possible damage is greater than with fresh water flooding. The face of the brickwork can be affected, and spalling processes (chipping and splintering) will be accelerated.

Mortars of different ages are constructed from different materials, with different reactions to salt. Old mortars are basically lime and sand and in seriously exposed conditions these will powder and erode more quickly after salt penetration. Any metallic finishes to buildings in the form of windows, doors, cladding, etc, would also be damaged to a greater extent after salt contamination due to electrochemical action. Embedded steels will also be affected. The metal corrodes and expands, causing brickwork to crack and move.

External decorations can be affected to a greater extent by salt contamination than with fresh water. Any salt which becomes trapped between layers of paint will cause the paint to blister at a greater rate than is induced by fresh water, which can pass and repass through paint film.

Plasterwork: There is a tendency for plaster to be impregnated to a greater extent by salt water than by fresh water. Surfaces of cement sand or cement lime and sand are not usually affected by flood water, but older types of lime and calcium sulphate plasters may soften when saturated. These defects are often followed by expansion and rippling of the plaster face. Plasterboard, insulation board and fibreboard become soft when wet and tend to warp and sag. Salts will also affect the surface of these materials, producing pitting. Cleaning can also damage these materials to a greater extent after salt impregnation and drying out. Hygroscopy (the absorption and retention of moisture from the atmosphere) is increased, extending the damage and the drying out periods.

Flooring and joinery: For flooring, the impregnation of salt will prevent the efficient drying out process due to the increased attraction of dampness from the atmosphere. Storm surges causing coastal flooding are more likely to occur during periods of high humidity, which pervades prior to the flood occurrence and afterwards, thus keeping timbers wet. Timber can take up to nine months to dry out completely. If the period when saturation exceeds 30 per cent is extended, then rot (which starts at 20 per cent), can be more extensive. Adhesives to woodblock flooring may also be slightly more damaged by salt water. The same assumptions have been made for joinery as for flooring above.

Internal decorations: Paint on timber and metal surfaces can blister if salt contaminates between layers. Metallic surfaces that are painted will degrade more quickly after salt contamination. Salts on the surface will lead to increased areas of damaged decoration as drying proceeds. Increased salt water absorption into plaster and brickwork will also lead to increased areas of damage. Owing to larger quantities of sand, silt, mud, and flotsam and jetsam, the discolouration of internal decoration will be greater with salt water than with fresh water flooding. The action of cleaning off the silt and mud, etc, will be more likely to damage internal decorations than the more simple washing down that can be possible after a fresh water flood.

Plumbing and electrical fittings: Electrochemical action in the presence of salt water is much accelerated and the salt deposits left after drying out will also be a problem with electrical wiring. The absorption of water by such salt deposits could continue to leave wiring in a dangerous condition, so that more extensive cleaning and inspections will be necessary. Steel piping and conduits will corrode more quickly with salt water than with fresh water. Plaster pipework and conduits will not be further affected by salt contamination.

2. *Household inventory items* **Domestic appliances:** The presence of salt in contact with metal and electrical wiring will increase the risk of corrosion by electrochemical action. Unless the affected metal can be washed down immediately with fresh water, corrosion will begin and replacement will be advisable. The presence of silt, mud or sand will also create problems with regard to the efficient cleaning and re-use of the equipment.

Heating equipment: Increased corrosion is likely, due to the presence of salt water on most metal surfaces. Oxidation of metals will occur, and weaknesses in tubing and jointing will be vulnerable.

Audio/video equipment: As television and radio sets and other similar equipment should be regarded as a total loss after any immersion, there will be no extra damages with salt water compared with fresh water.

Furniture: Due to staining from salts and the increased mud, sand and silt from salt water flooding, the damage is more severe than with fresh water. Damage to metal parts, such as springs, will also be worse owing to oxidation.

Personal effects: Damage to clothes and shoes will be increased due to staining by salts which may not wash out. Increased silt, sand, etc, will also increase staining.

Floor coverings/curtains: Again there will be increased damage due to salt staining and increased quantities of mud, sand, silt, etc. In floods of short duration at low depths, these effects will produce high extra damages to carpets, but at greater depths the higher fresh water damages will not show a proportional increase for salt water flooding. For longer duration floods, textiles and carpets may not be able to survive the extra cleaning required as a result of the retention of salt and silt and may be a write-off.

Garden/Do-It-Yourself/leisure equipment: There will be greater damage to metallic surfaces with the increased likelihood of corrosion to all items. More damage will also be caused by the presence of increased quantities of silt, sand, mud, etc.

3. *Domestic clean-up* The effects of salt water contamination will involve more comprehensive cleaning than for fresh water flooding. Cleaning up can involve the removal of quantities of sediment, in some cases shingle and cobbles, and other debris. Contamination by sewage can also add to clean-up costs. All metal parts and equipment will require hosing down with fresh water and checking before re-use. Textiles, such as soft furnishings, will require more thorough cleaning due to the retention of salt and silt.

Where salts have penetrated into timber, plaster and brickwork, even with extensive washing down with fresh water, the effects may not be immediate but can be apparent after some months, perhaps after redecoration. Deterioration of the fabric and of items of inventory can result from the effects of salt penetration and damage by the presence of extra silt, sand, mud, etc.

5.3.2 Salt water flood damage to non-residential properties

The progressive direct effects of flooding, including salt water flooding, on manufacturing plant and machinery have been assessed in the 'Red Manual' (Parker *et al.*, 1987). The amount of direct damage caused by flooding also affects the size of the indirect losses, with increased loss of production while repair or replacement of machinery is carried out.

The salt water damage factors used for residential properties (Table 5.1 and Table 5.2) can also be used for the additional damages caused by salt water to structure for retail businesses, such as cafés, shops, pubs, etc., in order to modify the data sets in the FLAIR reports (e.g. N'Jai *et al.*, 1990).

5.3.3 Storm and spray damage

Wave and spray damage Residential and non-residential properties can be vulnerable to wave and spray damage at high tides and in storm conditions, usually where they are situated behind existing coast protection or sea defence works which are inadequate or not designed to prevent spray overtopping.

The impact of waves and spray can damage property without actual flooding taking place. Sand and shingle can be thrown considerable distances by the force of the waves. Buildings with large areas of glass are particularly vulnerable, such as cafés and shops. Exterior decorations, roofing, exposed balconies and staircases can also be damaged in the zone at risk. Timber or part-timber properties can be considered to be at greater risk than brick-built buildings.

Structural storm damage of this kind is site-specific and dependent on the distance of the property from the point where waves break. The information required at each site will include the following:

1. Details of properties/area at risk at present;
2. Information on whether the project will prevent wave/spray damage after its completion;
3. An assessment of the value of the exterior property characteristics, based on repairing to pre-flood conditions.

It is unlikely that total property damage will occur, unless a breach or serious overtopping causes additional flood damage. The debris produced by the spray/wave damages can produce additional damage to other properties, in flood conditions. This additional damage would, however, be difficult to predict.

Velocity damage Structural damage due to the action of high velocity floodwater can occur in severe sea flooding events in which existing sea defences either breach or are seriously overtopped.

Structural damage may not be restricted to the seafront zone or the immediate vicinity of the breach or overtopping. The most severe structural damage can be expected to be close to steep land gradients and will be particularly severe where the land is located below sea defences. Barriers such as buildings can constrict and direct flow, with structural damage being aggravated by the flood water's debris load. The sediment load carried by the floodwater is also increased by high velocity, which will further add to the damages.

High velocity floods can be categorised into the following damage categories:

1. At velocities of under 2.0 metres per second and under 3.0 m^2/s (depth x velocity), minor or inundation damages only will occur;
2. At velocities over 2.0 m/s and at depth x velocity of between 3.0 m^2/s and 7.0 m^2/s partial damage will occur;
3. At velocities over 2.0 m/s and at depth x velocity over 7.0 m^2/s total damage will occur.

Where total destruction of property is likely in the zone at risk from a breach, the cost of rebuilding and refurbishing should be used as a measure of the economic loss. The market value of the property at risk includes the value of the land, so that full market value is inappropriate as a measure of loss. An assessment of potential property damage in the total destruction zones should be site specific – the number of properties at risk multiplied by their rebuilding/replacement value.

For partial damages, the characteristics of the property at risk should be assessed. External structural features, as with spray and wave damage above, will be affected. Most damage is likely to occur to caravans, mobile homes, chalets and lightly constructed homes without solid foundations. Scour around buildings leads to instability, which is increased by saturated soil. Prolonged flooding will lead to delay in the initiation of remedial action and will result in subsequent deterioration even where initial velocity damage is less severe. Other buildings that are largely constructed of timber will also be vulnerable due to natural buoyancy, and will add to damage from debris. Cars and dinghies can also be washed away.

Debris damage Additional damages are likely to be caused by flotsam and jetsam in floodwater. Damage is dependent on depth and velocity. The deeper the water, the greater its capacity to carry greater quantities of debris at high velocities. The destructive force of the water can move beach material, loosen soil and sediment and carry debris from already damaged buildings to cause further damage elsewhere.

It is possible to assess areas where the quantity of debris is likely to be high by the characteristics of the soil and vegetation and the existing buildings. However, as data on damage caused by high velocity floodwater will inevitably include damage caused by flotsam and jetsam, a separate assessment of debris damage can lead to double counting and is therefore not recommended.

Waves and tides Additional damages are likely to be caused by the movement of tidal water and waves in and around flooded properties. These damages will be higher for long duration floods, with the household contents that have not already been salvaged being liable to further damages. Further structural damage can also be caused by scour from the movement of floodwater around buildings which may already be weakened by the original surge. The assessment of these further damages over time is problematical as long duration floods are insufficiently frequent for comparative data to have been collected.

5.3.4 Emergency services

Depending on the severity of a flood event, several emergency services may be involved in both emergency works and clean up operations, during and after the flood event (for more details see Parker *et al.*, 1987, Chapter 10). Extra staff and resources may be required and additional administrative costs may be involved. Authorities and bodies providing emergency services include the following:

Local authorities
Police
Fire services
Ambulance services
National Rivers Authority

Voluntary services
Armed forces

Local District or Borough and/or County Councils' departments involved can include social services, emergency planning and highways. Evacuation and alternative accommodation may need to be arranged, social workers mobilised and voluntary and other bodies alerted. Public utilities need to be notified of the possible need to shut down supplies to the flood-prone area.

The emergency services have fixed sets of resources allocated to deal with the many types of emergency with which they are faced. It is the additional, or marginal, costs which may be incurred, in addition to the costs normally allocated, which should be assessed for including with the other benefits of sea defence.

Standard cost data for these emergency services has been compiled in the 'Red Manual' (Parker *et al.*, 1987). A scenario survey methodology was employed, for District Councils, the police and others, to determine the likely additional costs for these agencies of four floods of varying magnitudes, including coastal flooding.

The standard cost data are for use in the following two situations:

1. Where predominantly residential areas, including some commercial property such as shops, offices and factories, are affected; and
2. Where commercial properties dominate, in which case fire service, military, ambulance and voluntary service costs are excluded, as they are largely inapplicable.

In addition to the standard data, a standard checklist of questions is provided for situations in which a recent flood event provides an opportunity for investigating the actual marginal costs of such an emergency. All or parts of the standard data can be modified by the derived case-specific cost data.

5.4 Flood impacts, damage and loss: data sources and other information

Given the results of the research described above (Section 5.3) there now exists a number of sources of data on flood damages, both of a secondary source nature (i.e. the data in this volume) and guidance as to the primary data that needs to be collected from the field. These are discussed below.

5.4.1 Standard flood damage data

Appendix 5.2 gives sets of damage data for flooding to residential properties for different flood depths. These data sets have been assembled using the factors given in Tables 5.1 and 5.2 to modify the standard flood damage data sets available in the most recent 'FLAIR report' update of the Middlesex University flood damage data files (e.g. N'Jai *et al.*, 1990). However, Appendix 5.2, for simplicity, only gives flood damage information for:

1. The main types of property (detached; semi-detached; terraced; bungalows; and flats);
2. A sector average for the average of 'all houses'; and
3. The main categories of property by age (pre-1918; 1918–1938; 1939–1965; 1965 onwards).

There remains a more detailed set of data, using the different social classes of the occupants of the property concerned. This data is available from Middlesex University and is included in the data sets that the EST-DAM program uses automatically in its analysis of flood alleviation benefits (Chapter 7).

5.4.2 'Average' flood damage data from previous surveys

The term 'standard' depth/damage data is reserved for data assembled from secondary sources (such as that described in Section 5.4.1 above). The term 'average' data is used to denote data derived from previous site surveys, averaged to give a generalised indication of flood damages for the relevant property types.

Data is available in the most recent 'FLAIR report' on damage to non-residential property. This is not adjusted for salt water damage, as this refinement is not possible to quantify for the huge variety of non-residential property shown in Appendix 5.1. However, it would be possible to revise the data in the FLAIR report using the indices in Table 5.1 and Table 5.2 to allow for salt water damage, or use that data unadjusted knowing that it will give minimum flood damage values. In any event, note should be taken of the standard deviations given in the FLAIR report tables, as indicators of the likely error in applying this data to individual land use items in areas being appraised with this Manual.

5.4.3 Site specific surveys of flood damage potential

For properties liable to flooding for which standard or average data does not exist, or is in some ways inappropriate, there will be no option but to undertake a site-specific survey using the questionnaire given in Appendix 5.3 (or a variant thereof tailored to local circumstances), the application of which is discussed in Section 5.5.5.

An essential part of this approach to collecting potential flood damage data is an amalgam of the experience of the plant manager, who will know about the value of stocks, equipment and buildings, and the experience of the interviewer/surveyor who will know of the effect that floods will have on these items. Thus the completion of this questionnaire is a joint activity to produce the best estimate (and the fact that it is an estimate is to be stressed) of the economic impact of floods of different depths in the premises concerned.

In addition, through the completion of the parts of the questionnaire concerned with turnover and the impact of an interruption to the work of the premises through direct flooding or the indirect impacts of flooding elsewhere, an estimate is arrived at of the national economic indirect flood damage to be suffered by those premises per unit of time (i.e. per day) (Parker et al., 1987).

5.4.4 Sources of other indirect flood damage data

The 'Red Manual' (Parker et al., 1987) has shown how indirect flood damage can be a large proportion of total damage potential (a figure in the range 10 per cent to 25 per cent might be typical). In special circumstances these indirect losses can be much higher, in particular when breach events cut communication lines, such as was the case with the Chesil sea defence project where the indirect flood losses amounted to as much as 90 per cent of total estimated losses (Penning-Rowsell and Parker, 1987).

However, that case was exceptional. In most circumstances most of these indirect losses are from the flooding of industrial premises (Section 5.4.3), but other indirect damages arise from traffic disruption, the loss of utility outputs such as gas and electricity, and the cost of emergency services and provisions.

Techniques to assess these potential flood impacts are not likely to differ between floods caused by sea water and those from rivers, except in so far as floods from the sea may be caused by a breach in existing defences, and thus the structural failure of buildings may be very significant, or may have a tidal factor which may mean that floods have a diurnal rhythm which, for example, causes roads to be flooded twice every 24 hours while remaining passable at low tide.

Consequently, with these exceptions, the user of this Manual is referred to the results and data in Parker et al. (1987) for guidance in this respect.

5.4.5 The non-monetary impacts of flooding on households

We know from our research that the 'intangible' impacts of flooding from the sea can be very substantial indeed. The residents of Uphill, Avon and Swalecliffe, Kent, suffered severely from the flooding that they experienced. This research has yielded some data which may be useful if flooding from the sea is experienced in similar circumstances. It does not quantify the money value of those impacts directly, but does indicate the order of magnitude of the impacts as felt by the people involved (Green et al., 1985, Parker et al., 1983).

Thus, in addition to the damage done to the house and its replaceable contents, floods have other impacts upon households, which include the following:

The risk to life
Health damage
Stress
Disruption
Loss of memorabilia
The impact of evacuation

Whilst it is not currently possible to value these impacts in economic terms, these represent real economic losses, and to those affected these impacts can be more severe than the direct, financial impacts of the flood (Green and Penning-Rowsell, 1989).

1. *Loss of life* There has not been a reported loss of life directly as a consequence of coastal flooding since the 1953 floods, although residents of flooded areas routinely believe that the stress of the flood caused premature deaths (Green et al., 1985). Moreover, the storm conditions that usually coincide with major flooding at the coast often induce accidents during which people are drowned. The dividing line between these deaths caused in the generally chaotic storm conditions and people dying directly owing to the flooding is a fine and often arbitrary one. In addition, the absence of deaths solely attributable to flooding over nearly forty years should not induce complacency, and the potential for loss of life should be assessed in any area at risk.

Lives may be lost through a number of mechanisms (Figure 5.4). Of these, the most important are likely to be the following:

a) People being swept way whilst attempting to walk through flooded streets (including the risk of falling down a manhole from which the cover has been blown off by water pressure); and
b) People being trapped and drowned, or dying through exposure, within the flooded dwelling.

106 THE ECONOMICS OF COASTAL MANAGEMENT

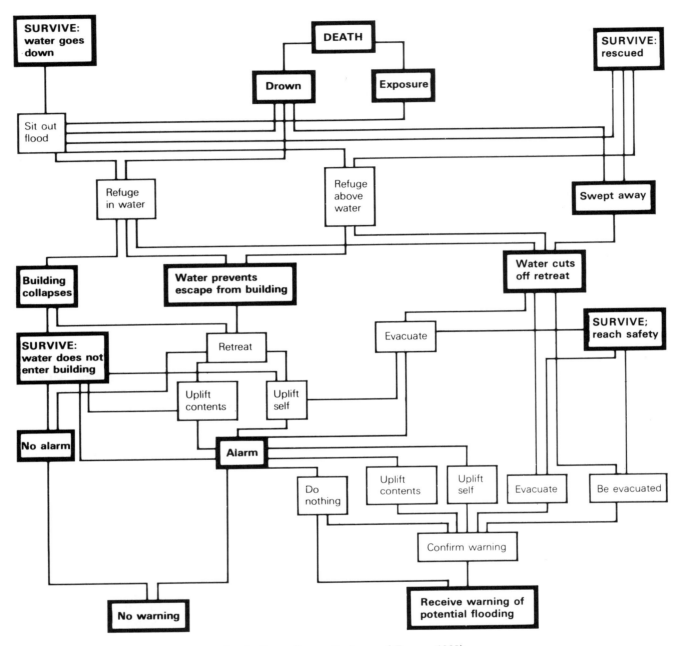

Figure 5.4   Event tree – deaths from floods (from Green, Parker and Emery, 1983)

Other than for cases where flood defences are breached, and in mobile homes and similar light structures, the risk of death as a result of structural failure of the property in which the individual has taken refuge is likely to be low in the United Kingdom. Based upon limited empirical evidence, Clausen (1989) suggests that a limit of the product of depth (metres) x velocity (metres/sec) of 10.0 be taken as the point at which complete collapse of a building will occur.

Data from the United States (Abt *et al.*, 1989) indicates that fit males will generally lose their balance at products of depth and velocity (metres times metres/sec) greater than 1.5, whilst a safer limit for the same group would be a product of 1.0 (for example, a 0.6 metre depth of water flowing at a velocity of 1.7 metres/sec). The risk of being trapped and drowned is clearly greatest in single storey dwellings. However, the force with which water can break into a property is such that the risk cannot be excluded for two-storey properties.

No estimates of the risk of death conditional upon the occurrence of coastal flooding are available and the variations in risk factors between areas would be likely to make any such generalisations of limited value. Nevertheless the correct economic basis for valuing reductions in the risk to life or limb have been set out by Jones-Lee (1976), based upon earlier discussions by Dreze (1962) and others (Shelling, 1968). Dalvi (1988) summarises the studies which have been undertaken to derive economic values for reductions in the risk from different hazards: commonly termed 'the value of a statistical life'.

However, as Dalvi notes, both economic theory and empirical evidence suggest that this 'value of the statistical life' will vary not only according to the initial probability of death, the reduction of risk involved and also between hazards. Consequently, we do not recommend the adoption of the Department of Transport's 'value of a statistical life', nor any other, not least because there is as yet no reliable basis upon which to predict the probability of death.

Consequently, we recommend that a general assessment is made of the relative risk in a given area, not least in order to focus attention upon how warning and other effective disaster planning would be undertaken. Intuitively, those at greatest risk will share some of the following characteristics:

a) Those living or working where overtopping or failure of the sea wall will lead to rapid and deep flooding;
b) Where the individual is infirm or disabled; and/or
c) Where the individual lives in a mobile home or single storey dwelling.

In addition, the life-threatening potential of the second order effects of flooding need to be considered. In the Health and Safety Executive's (1978) Quantitative Risk Analysis of the petrochemical installations on Canvey Island, flooding was found to pose one of the most severe risks, both in terms of the probability and the consequences of a chemical release. A similar potential for the release of toxic or flammable chemicals as a result of flooding may exist elsewhere.

In general, a plan for the effective warning and aid with evacuation of high risk populations should be prepared. Some recommendations as to the principles which such plans should embody have been given elsewhere (Foster, 1980). Care needs to be taken not to underestimate the time required for warnings to be disseminated and evacuation to be carried out (Sorensen, 1988; Sorensen and Rogers, 1989). Equally, the empirical evidence is that the probability of a warning actually being received by those at risk is typically low (Neal and Parker, 1988), and lowest in those low probability events, such as the failure of sea defences, when it is most needed (Smith, 1986).

2. *Other non-monetary impacts* The subjective severities of the different impacts of flooding have been shown to be partially interdependent (Figure 5.5): the degree of disruption experienced is partly dependent upon the damage done to the contents of the dwelling (Green and Penning-Rowsell, 1989). Conceptually, the total impact of a flood upon a household may be described as the difference between the challenge presented by the flood in comparison to the financial, personal and social resources that household can mobilise to meet that challenge.

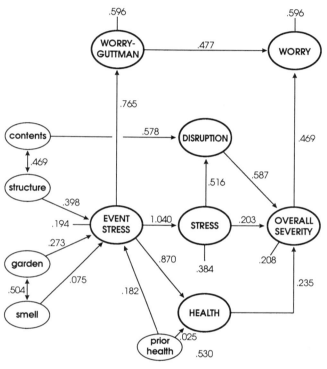

key:

Figure 5.5 Causal model of relationships between impacts and judged overall severity of flood (from Green, 1988)

As Figure 5.5 shows, a critical determinant of the household's assessment of the overall severity of the effect of the flood is degree of stress experienced. This directly affects the degree of health damage a household experiences as well as the worry or anxiety experienced as to the risk of flooding in the future. A scale to measure the stress resulting from the onset of flooding has been developed (Table 5.4). There is some evidence that warnings reduce the stress experienced during a flood (Doizy, 1991). Conversely, stress appears to be greatest when the onset of flooding is most rapid.

Up to 80 per cent of victims of sea floods have been found to report health damage as a result of the flood they experienced (Parker *et al.*, 1987). It has also been shown that elderly flood victims experience a lower health status than matched control samples (Green *et al.*, 1985). These effects may last for many years (Parker *et al.*, 1984) and are not adequately measured by the costs of medical treatment (Green and Penning-Rowsell, 1989). At greater depths of flooding than those for which Figure 5.5 was derived, households may be forced to leave home, and memorabilia such as photographs will be lost. Both these impacts have severe consequences, with the impacts of evacuation being particularly severe when the household is forced to split up (Milne, 1977; Western and Milne, 1979).

Table 5.4
Event stress induced by sewerage flooding

	% respondents agreeing with statement
I just wanted to run away	24
I didn't know what to do	33
I cried	25
I felt numb	26
I imagined the worst was going to happen	43
It was a nightmare	47
I was angry	68
I swore	53
I thought I was going to have a heart attack	10
I felt sick	33
My pulse raced	34
I was horrified	47
I thought that this was the worst possible time for it to happen	31
I knew what to do	29
I felt hot and cold flushes	18
I was afraid I was going to die	8
I was terrified	34

number of cases = 134
mean score = 2.48
std deviation = 3.49
standardised item alpha = 0.849
Source: Green, 1988

The longer duration, and frequently greater depths, of sea flooding suggest that the challenge posed will be greater than from river flooding. Those households with least resources (e.g. the elderly, the disabled, those on low incomes) can be expected to be particularly severely affected.

5.4.6 The 'writing-off' of frequently flooded property

An important decision in any benefit assessment involving coastal flooding is at what stage to 'write off' properties that are projected to flood very frequently.

This need arises where sea defences are breached and the 'do nothing' situation would result in property previously behind those defences being continually exposed to flooding, sometimes with every high tide. This was the situation at Whitstable (Parker and Penning-Rowsell, 1982), and Herne Bay (Section 8.5). In other cases natural sea defences may be altered with erosion, and property becomes exposed to flooding, as at Hengistbury Head (Parker and Thompson, 1988).

In these circumstances our recommendation is that where property is projected to flood more than or equal to once in every three years, it should be 'written off' and the capital value of that property used as the benefit of its flood protection. This is less extreme than might appear, because with repeated flooding at intervals of up to once every three years, the capital sum generated by the repeated flood damage will alone approach (and can exceed) the property's capital value (see the Herne Bay case study, Section 8.5.3).

There may be circumstances where properties currently suffer minor and periodic flooding, perhaps by overtopping of defences and flood waters accumulating from spray, and then, as defences deteriorate, flooding would become much more frequent, perhaps every one or two years or even more frequently. In these circumstances the 'do nothing' situation requires the calculation of the current flood damages to be undertaken, followed by 'writing off' the capital sum at the time that the sea defences deteriorate. This may result in a total loss of more than the capital sum value of the property, but this is not illogical since the damage is suffered (and goods and building fabric are repaired and replaced), and then the value of the renovated property is then lost.

The full capital value of the property is counted because it is assumed that the land on which it stands is not useable if flooded every two years. However, this assumption may need to be modified if the land could be used for other purposes (for example, a boat yard or some similar use where periodic flooding is not too problematic).

5.5 Operational methods and techniques for project-specific data collection

Figure 5.2 gives a detailed flow chart with the different stages in the assessment of the benefits of flood alleviation, and uses as one of the sets of inputs the data on flood damages to be found in the publications listed above. When applying this flowchart to specific proposed projects for flood alleviation, the following steps should be followed (Sections 5.5.1 to 5.5.7).

5.5.1 Phasing of the assessment operations

As in all aspects of assessing the benefits of coast protection and sea defence projects the phasing of the survey and analysis tasks is important for efficient execution of the benefit-cost analysis. A number of discrete stages can be recognised within the flood alleviation part of this task, and careful timing of these to fit in with other operations will be necessary.

1. *Defining the problem* It is important that the exact nature of the sea defence (i.e. flooding) problem is defined. This means that its extent needs to be modelled (Section 5.5.2) and the periodicity of flooding determined. This may involving modelling the probability of the breaching and other failures of existing sea defences (using techniques and data which are outside the scope of this Manual) and determining the likely flooding scenarios for post-failure situations. It is also important at this stage to derive a very approximate estimate of the likely benefits, so as to know in what detail to undertake subsequent parts of the benefit assessment (there is no point spending a great deal of time on the assessment if the project is likely to be hopelessly cost-ineffective). In this respect all phases are part of an iterative approach.

2. *Collecting basic data* There is certain data that is central to the benefit assessment including the land use data and the data on damage caused to the relevant properties. The latter data needs to be collected after the initial land use survey, because it depends on that survey to determine whether primary data is needed and to determine a sampling framework.

3. *Undertaking a preliminary benefit analysis* With the basic data, and some (or all) of the site specific surveys, a preliminary analysis can be undertaken of the likely benefits of flood alleviation. This should be undertaken before any detailed design work is started, since otherwise that design work may be wasted if the project has to be changed or abandoned if it is shown subsequently not to be worthwhile economically.

4. *Determining the post-project residual flooding* The benefits of flood alleviation investment must exclude the future flood losses that will continue after the project or projects are installed (for example, controlled overtopping, and spray damage unprevented). Thus at this stage it will be necessary for the hydrographic analysis to determine the post-project situation (effectively, the extent of any flood problem after the investment has been made). This may involve an element of engineering judgement.

5. *Refining the benefit analysis* This will be undertaken by tailoring the benefits to what exactly the project will achieve, and at what cost, and the analysis of project options (different design types; different design standards; different mixes of flood alleviation strategies, etc). Also at this stage it will be necessary for the flood alleviation benefit results to link in with other economic computations (in particular recreation surveys, and erosion impacts) for which the timing must be such that they produce results at this stage: a most important point. This is followed by sensitivity analysis, and the comparison of costs and benefits, and the selection of the prospective project.

These and other aspects of the survey work are explored in more detail below, and the phasing above should be borne in mind accordingly.

5.5.2 Defining the extent of the benefit area and the frequency of flooding

Defining the extent of the benefit area will be done using the hydrographic analysis of the maximum likely flood in the area affected. This is possible using historical records (for example the extent on the 1953 floods on the East Coast of England), or by using hydrographic and hydraulic models to estimate the extent of flooding caused by a range of floods of low exceedance probabilities.

Given that standards of flood protection at the coast tend to be high (a return period of 500 to 1000 years is not uncommon as a design standard) defining the extent of the flood with a long return period is to be recommended. The definition of the benefit area must include the areas upstream in rivers contributing to flooding at the coast, and in that respect the benefit area may be a composite of a 'ponded' lower part, affected mainly by sea levels, and a rising profile of flood waters along these contributing rivers. Care should be taken to ensure, however, that this upstream flooding will be alleviated by the sea defence works or associated river flood alleviation investment, or that it is assessed as residual flooding with perhaps lesser severity or frequency.

A particular problem arises in the case of the failure of existing defences, which is a common situation in which benefit-cost analyses are undertaken (in order to justify the replacement or enhancement of those degraded defences). There is, here, the need to determine, first, the probability of a breach or failure occurring in any one year in the future, and, secondly, the extent of flooding caused by the breach/failure event. This analysis is outside the scope of this Manual, but guidance can be found in Construction Industry Research and Information Association (1991), and there is no doubt that this is a complex exercise. A detailed case study is illustrated in Thompson *et al.* (1987) for the failure of cliffs at Hengistbury Head leading to increased flooding in Christchurch Harbour.

After the probability of failure has been determined, the flooding situation following that failure needs to be determined. This, then, defines the 'without project' situation: no flooding until failure occurs, followed perhaps by catastrophic flooding at the time and event of the failure, and then (with nothing at all done to prevent flooding) a continuation of periodic flooding by both frequent and infrequent events.

The 'with project' situation, in contrast, may involve small scale periodic flooding from overtopping of the existing defences, if they are inadequate to prevent this flooding, followed by virtually no flooding if the enhanced project is implemented almost completely to alleviate the flooding in the area.

5.5.3 Land use and topographic height survey and coding

The purpose of this survey is to identify and code, using the coding form in Appendix 5.4, all the property in the benefit area. In addition it is wise to extend the benefit area to cover a 'safety margin' of, say, 0.5 metres in altitude above the limit of likely maximum flooding, and extend the land use survey to take in these areas. This will allow for errors in the initial definition of the benefit area, and also perhaps for changing sea/land relationships caused by eustatic change or sea level rise.

An essential part of this land use coding is determining the floor height of all properties at risk. This can be done from detailed maps, or in the field, but it is advisable to level the floor heights of any properties for which the damage is likely to be crucial to the total benefit figure. This is a matter of judgement at the initial land use coding stage, but a rule of thumb might be that any industrial/commercial/retail premises which have floor areas greater than 10,000 square metres should have their floor heights levelled for a benefit area including up to 500 properties. For larger benefit areas it will generally be advisable to wait to do the selection of properties for this levelling until the first calculations of potential flood damages have been made, and then level the floor heights of all properties that contribute more than 5 per cent to total flood damages or flood alleviation benefits. In practice it is normal to level the floor heights of a further sample of properties throughout the flood affected area to confirm and adjust estimates from maps and by eye.

As indicated above (Section 5.2.3) the land use coding can be undertaken at several levels of generality, reflecting the type, category and sub-category of property in the land use classification (Appendix 5.1). In practical terms this means that if a pre-feasibility study is being undertaken rapidly, the basic sector or category land use information can be coded (often from maps, with minimal fieldwork). If, on the other hand, a pre-feasibility survey has shown sufficient benefits to go to the next stage, then the full feasibility survey should be undertaken with all land use coded in the field, using the full land use classification.

5.5.4 Data acquisition for traffic disruption surveys

The assessment of the economic impacts of traffic disruption due to flooding, which can be substantial, requires data on the roads likely to be affected by flooding, the depth of floodwater on those roads for the different return periods of the events being appraised, and the traffic flows on those roads for standard 24-hour periods. If there is a strong seasonal element in traffic flows (such as on promenades at the coast) this seasonality will have to be assessed.

The methodology for assessing the impacts of road traffic disruption caused by flooding is the same as that caused by erosion (Section 3.5) except that flooding is likely to be short-lived rather than permanent or semi-permanent in the case of erosion. If traffic is disrupted by flooding, alternative routes will be needed for diversions. These routes will also require data on traffic flows, to gauge the impact of increased flows on these diversions, and the congestion and reduction in flow speeds that this will cause.

The value of the disruption caused by the flooding is then calculated as the difference between the time costs plus the resource costs of traffic moving along the original route, and those costs for the diversion route(s). The formulae for these assessments are provided by Parker *et al.* (1987, Chapter 6) and the data used in these assessments – in the formulae and as the coefficients – are provided by the Department of Transport in the latest copy of their Highways Notes (Department of Transport, 1986).

5.5.5 Questionnaire surveys: site specific damage surveys

These surveys will be necessary when no appropriate data is available in the FLAIR reports or no results are available from previous research for the properties found in the benefit area which will contribute significantly to total benefits.

The relevant questionnaire will be found in Appendix 5.3. Properties at which to conduct interviews should be selected following the land use survey, and the criteria outlined above for selecting properties for levelling floor heights can apply equally well in this selection process.

In practical terms it has been found important to warn potential interviewees that they will be approached, by letter and in advance. The appropriate personnel to interview are the site manager, the finance manager, or the managing director of small concerns. The interviewee should be familiar with the plant and equipment and its value, and also should be knowledgeable about the turnover of the business and the presence or absence of competitors.

These site specific surveys are complex operations, and interviewer training and experience are essential. Further details of the problems inherent in this process of site specific surveys will be found in Parker et al. (1987, Chapters 4 and 5).

Parker et al. (1987) also give procedures for calculating potential direct and indirect damages for the properties that have been the subject of interview surveys. This analysis is designed to derive potential direct flood damage data from the questionnaire survey data on stock, equipment and buildings. The analysis will also yield the national economic indirect losses from flooding, and regional or financial losses, if appropriate. A number of steps are recommended if the data from the questionnaire survey is incomplete.

5.5.6 Questionnaire surveys: post-flood household surveys

Some insight into the 'intangible' effects of floods, and therefore the intangible benefits of flood alleviation, can be gained by undertaking surveys of those affected by flooding. Indeed such post-disaster surveys can serve several different purposes and before undertaking any such survey it is necessary to determine carefully the objectives of the study. A survey to identify the immediate needs of the flood victims would be both designed differently, and take place at a different time, to one designed to assess the long term effects of the flooding.

Great care should be taken in both the design and administration of any type of survey of flood victims. No survey should leave flood victims in a worse psychological state than they were before the interview. For this reason, interviewers must be carefully selected, trained and briefed.

After the Bradford fire disaster, post-disaster counselling has been introduced into the United Kingdom, in part as a reflection of previous experience in the United States (Lystad, 1985; National Institute of Mental Health, 1979). In addition, there has been upsurge of interest in disaster research in the UK (Duckworth, 1986; Parker and Handmer, 1992). This raises the risk that victims will be subjected to a number of surveys, where one of the stresses of the post-disaster period can be the different information demands of the different organisations in their post-disaster recovery operations.

Thus a survey should only be undertaken when there will be a clear benefit to either the victims of the flood under study or to the potential victims of future floods. Any survey should consequently be co-ordinated with the local counselling team and with any community group that may have developed in the aftermath of the disaster.

In many of our studies our interviewers were described as 'the first people who had "bothered" to ask the flood victims about their experiences', and the victims were both pleased and anxious to discuss their experiences. Flood victims were pleased that someone considered that their experiences were sufficiently important to be worth learning about, and United States experience is that describing the experience has some therapeutic value for flood victims (Lystad, 1985; National Institute of Mental Health, 1979).

Interviews should not be undertaken until at least six months after the flood, in part to allow for partial recovery to take place. Interviewers should be briefed that their role is that of a quasi-social worker rather than a market researcher, and the preferred previous experience of such interviewers should be medical or sociological surveys, or with groups like the Citizen's Advice Bureau or the Samaritans. Given that floods are not seen as natural events but as occurring as a result of failure by organisations (Green, Tunstall and Fordham, 1990), flood victims will have sought to identify those organisations who are 'to blame' for the flood which they suffered. If it is one of those organisations which is commissioning the survey, then it is advisable that the survey is undertaken through an independent agency.

We recommend that the household interview schedule given in Appendix 5.5 be used as a basis for such a survey. This schedule has been developed and validated through some 2,000 interviews with flooded households. Use of this survey instrument will enable comparison of the severity of the flood with those of a range of other floods (Parker et al., 1987).

5.5.7 The damage-reducing effects of flood warnings

The benefits of a flood alleviation or sea defence project are only legitimately compared with the costs of that project if the project really generates those benefits by preventing the predicted flood damage. If existing policies and plans – such as an efficient flood warning system – have an effect in reducing the likely flood losses that will occur in the future, then the benefits of additional flood alleviation projects or measures will be correspondingly lower.

Much research has been undertaken on the damage-reducing effects of flood warnings (Cole and Penning-Rowsell, 1981; Parker and Neal, 1988). Although much of this work is related to fluvial flooding, some is ger-

Table 5.5
Damage-reducing effects of flood warnings for residential property, April 1991 prices

Depth of flooding (m)	Total potential damage[1]	Total potential inventory damage	Up to 2 hrs	% of total damage	2–4 hrs	% of total damage	6 hrs	% of total damage	8 hrs	% of total damage
1.2	7,376	3,965	1,866	25.3	2,633	35.7	2,855	38.7	3,002	40.7
0.9	6,482	3,687	1,711	26.4	2,437	37.6	2,632	40.6	2,761	42.6
0.6	5,369	3,119	1,369	25.6	1,997	37.2	2,158	40.2	2,266	42.2
0.3	3,999	2,283	1,200	30.0	1,684	42.1	1,804	45.1	1,884	47.1
0.1	1,526	866	374	24.5	501	32.8	546	35.8	577	37.8

1 Building fabric and inventory damage.
Source: Parker, 1991

mane to the coastal situation. With warnings, householders will be able to undertake the same type of flood proofing works to prevent sea water entering their properties, as will the owners of retail and industrial premises. Valuable goods, stock and equipment can be moved to prevent their damage, just as in fluvial flooding.

However, the damage saving rates from warnings of 2–4 hours are generally found not to be dramatically large, and there is no evidence that much longer warning lead times significantly increase damage saving. Damage savings in the range 5 per cent to 20 per cent are typical, and the average tends towards the lower end of this range. These saving rates are lower than those originally suggested by Penning-Rowsell and Chatterton (1977) as more research has shown that flood victims are not able to react sufficiently to secure greater savings. More detailed information is given in Table 5.5 and by Parker (1991), and allowance should be made for this damage saving where it is known that there is a flood warning system that can deliver useful warnings at least 2 to 4 hours before the onset of flooding.

Analysis such as this naturally leads to questions as to how future flood damages should best be reduced, including the use of a range of policies simultaneously (i.e. a mixture of structural flood alleviation works and non-structural alternatives such as flood warning systems). Generally, the wisest use of scarce resources will be with an optimised mix of flood alleviation strategies. This mix can be designed, for example, such that warnings alleviate intangible effects including the threat of loss of life, and structural projects alleviate flooding occurring in the areas otherwise most frequently or most severely affected. This is just the kind of option review and optimisation process to which benefit-cost analysis should lead.

5.6 Analytical methods

The computation of flood alleviation benefits can be carried out in a number of ways. Purpose-designed software, running on personal computers, is available for these operations, and its use is desirable for all but very small projects. These programs are described in Chapter 7.

However, it is possible to undertake these calculations in a simplified manner using proprietary spreadsheets or by 'hand' calculation. However, if there are a number of return periods of flooding being analysed, and a changing flood scenario with time (perhaps as progressive erosion at the coast leads to more and more serious flooding) this hand calculation can become almost impossibly complex.

5.6.1 Calculation stages

The first stage in the calculation of the benefits of flood alleviation is identifying the flood levels and profiles for each return period event being appraised (at least three events and preferably six). Then, the properties affected by floods at each return period are identified, and the flood damages and indirect losses calculated from data such as in Appendix 5.2. This gives the event damages and indirect losses for each return period, which is displayed best in the form of the Table in Figure 5.6, which comprises the loss-probability relationship in tabular form.

The next stage is to assess the annual average benefits derived from that data, for which the Table in Figure 5.7 is the standard form of analysis. This identifies the exceedance probability, the probability of the flood in the interval between return periods, and the

THE BENEFITS OF FLOOD ALLEVIATION

```
SECTOR ANALYSIS OF EVENT DAMAGES - SWALECLIFFE        BENEFIT ASSESSMENT

YEARS           1.00      2.00      8.00     17.00     32.00    114.00    250.00    500.00
-------------------------------------------------------------------------------------
DIRECT DAMAGES
SECTOR 1           0         1        53       131       143       193       217       230    RESIDENTIAL
                   0       147     18663    123997    163121    323061    463964    541673    DAMAGE (£)
ROW %            0.0       0.0       3.4      22.9      30.1      59.6      85.7     100.0
COL %            0.0       0.1       4.7      18.2      21.4      31.1      35.7      36.1
SECTOR 5           0         5         7         8         9         9         9         9    RETAIL & RELATED
                   0      3713      5637      8641     16794     19580     23251     27648    DAMAGE (£)
ROW %            0.0      13.4      20.4      31.3      60.7      70.8      84.1     100.0
COL %            0.0       3.6       1.4       1.3       2.2       1.9       1.8       1.8
SECTOR 6           0         1         1         1         1         1         1         1    PROFESSIONAL
                   0      2529      6689      8342      8342      8342      8397      8485    DAMAGE (£)
ROW %            0.0      29.8      78.8      98.3      98.3      98.3      99.0     100.0
COL %            0.0       2.4       1.7       1.2       1.1       0.8       0.6       0.6
SECTOR 7           0       143       408       514       530       568       584       588    SITE SURVEY DATA
                   0     97253    366068    538904    574321    665507    716108    736657    DAMAGE (£)
ROW %            0.0      13.2      49.7      73.2      78.0      90.3      97.2     100.0
COL %            0.0      93.8      92.2      79.3      75.3      64.1      55.0      49.1
SECTOR 8           0         0         1         3         3         5         6         8    INDUSTRIAL
                   0         0         0         0         0     21363     89209    186540    DAMAGE (£)
ROW %            0.0       0.0       0.0       0.0       0.0      11.5      47.8     100.0
COL %            0.0       0.0       0.0       0.0       0.0       2.1       6.9      12.4
SECTOR 9           0         0         0         0         0         0         0         0    PUBLIC UTILITIES
                   0         0         0         0         0         0         0         0    DAMAGE (£)
ROW %            0.0       0.0       0.0       0.0       0.0       0.0       0.0       0.0
COL %            0.0       0.0       0.0       0.0       0.0       0.0       0.0
TOTAL              0       150       470       657       686       776       817       836    TOTAL
                   0    103642    397057    679884    762578   1037853   1300929   1501003
ROW %            0.0       6.9      26.5      45.3      50.8      69.1      86.7     100.0

CHALETS            0      5691     29081     45907     62323     99176    142394    154402

INDUSTRIAL         0         0         0         0         0     13694     17127    137861
INDIRECTS

INTANGIBLE         0         0     48377    290262    331728    691100    905341    988273
HOUSE COSTS
```

Figure 5.6 Example of event flood losses calculated from return period, land use and depth damage data

```
               CAPITAL SUMS (DISCOUNTED ANNUAL BENEFITS)
RETURN         EXCEEDANCE    BENEFIT    \PROBABILITY  AVERAGE \  INTERVAL BENEFIT     \ CUMULATIVE BENEFIT
PERIOD         PROBABILITY              \ OF FLOOD    BENEFIT \                       \
                                        \ IN INTERVAL.         \(ANNUAL)(DISCOUNTED)\(ANNUAL)(DISCOUNTED)
YEARS                          (£)      \              (£)    \   (£)      (£)      \  (£)      (£)
-----------------------------------------------------------------------------------------------------
 1.00          1.00000           0      \
                                         \ 0.50000    54666   \ 27333     498989    \ 27333    498989
 2.00          0.50000      109333       ---------------------------------------------------------------
                                         \ 0.37500   267735   \100400    1832894    \127733   2331883
 8.00          0.12500      426138       ---------------------------------------------------------------
                                         \ 0.06618   575964   \ 38115     695824    \165848   3027707
17.00          0.05882      725791       ---------------------------------------------------------------
                                         \ 0.02757   775346   \ 21379     390293    \187227   3418000
32.00          0.03125      824901       ---------------------------------------------------------------
                                         \ 0.02248   987812   \ 22204     405354    \209431   3823354
114.00         0.00877     1150723       ---------------------------------------------------------------
                                         \ 0.00477  1305586   \  6230     113734    \215661   3937088
250.00         0.00400     1460450       ---------------------------------------------------------------
                                         \ 0.00200  1626858   \  3253      59386    \218914   3996474
500.00         0.00200     1793266       ---------------------------------------------------------------
                                         \                    \                     \
RECTANGLE
               0.00200     1793266                              3586      65465      222500   4061939

TRIANGLE
               0.00200      198244                               396       7229      222896   4069168
```

Figure 5.7 Example of standard table for calculating annual average and discounted benefits of flood alleviation
(NB: 5% discount rate)

114 THE ECONOMICS OF COASTAL MANAGEMENT

average benefit for that interval. The annual average benefit is the sum of the annual benefit for the interval over all the return periods being analysed.

In this respect the type of analysis shown in Figure 5.7 should be done separately for each of the future flood scenarios, if these change, to yield the annual average flood damages for the individual flooding scenarios. Thus, for example, there may be three such situations. First, there is minor overtopping of an existing sea defence project, followed, secondly, by breaching of that defence and thereafter, thirdly, much more serious and continuing flooding without the previous sea defence protection (assuming the 'do nothing' situation, and therefore no sealing of the breach or other remedial measures).

The annual average flood damage needs to be calculated for the first and third of those scenarios, and the event damage in the year of the failure also needs to be calculated (assuming it is more serious than the floods that would occur in the third situation above). Then the appropriate damages can be applied to each of the years in the future to which they apply, and summed over the lifetime of the proposed project.

5.6.2 Interpretation of the results

Figures 5.6 and 5.7 give typical output from the ESTDAM computer model described in more detail in Chapter 7. Any calculation method used should derive similar results. The interpretation of these results is important to obtaining a correct understanding of the benefits of flood alleviation and sea defence.

Figure 5.6 shows the event damages calculated for each return period being analysed, showing that in this example tangible flood damages are dominated by flooding to the residential sector. It is important to ensure that the flooding is being correctly modelled by the ESTDAM program, such that the numbers of properties and the event damages match known events in the past or likely events in the future, particularly for the more minor but more frequent future floods (which contribute most to the annual average damages).

What Figure 5.7 shows is an important table in the interpretation of the results, and is often misunderstood. The benefits of flood alleviation for a given design standard are just those benefits accumulating to that design standard. The first flood considered must yield zero damages, or the loss-probability curve is not being fully quantified.

The 'triangle' and 'rectangle' results in Figure 5.7 denote those results for that part of the loss-probability curve above the flood with the highest return period being considered (see also Figure 7.2). These 'triangle' and 'rectangle' figures can only be counted if the loss-probability relationship is also calculated for the post-project residual flooding, when the difference between the two full benefit figures ('without-' minus 'with-project', both with the triangle and rectangle parts included) is the benefit of that project or projects (i.e. the difference between the areas under the two loss-probability curves). Further discussion of these points can be found in Parker *et al.* (1983).

5.7 Calculating the agricultural benefits of sea defence projects

The flooding of agricultural land, or its poor drainage, can inhibit crop growth, cause crop damage, restrict timely field access by machines or grazing cattle, and limit crop and livestock activities (Morris *et al.*, 1984). The salt contamination of land affected by sea flooding can also affect the crops grown there, at least for the following season, although gypsum treatment can speed up the process of recovery.

5.7.1 Benefits and controversy

The agricultural benefits of flood alleviation and drainage arise through a number of processes of agricultural enhancement and change. There is reclamation, where increased land can be put to agriculture, the improved production from existing enterprises, the change of enterprise (both change of crops and change in livestock or rotation), the reduction in costs (e.g. a reduced need for bought-in feedstuffs following pasture improvement), a reduction in crop damage caused by flooding, and a change in the economy of the whole farm from improvement to part of its area.

Thus the alleviation of this flooding, with adequate drainage, can yield both financial benefits to the farmer in terms of increased yields and returns, and economic efficiency benefits to the nation. Much land now defended from the seas is Grade 1 agricultural land and thus of considerable importance to the nation in terms of efficiency and/or other social benefits connected to the volume of food produced.

However, assessing the benefits of agricultural improvement through flood alleviation and drainage has been intensely controversial in the past (e.g. Black and Bowers, 1984). Agricultural drainage can have serious impacts on environmental values and these impacts have been inadequately counted in benefit-cost analyses. Critics have stressed that national benefits have been exaggerated, and that this has led to decisions promoting wetland drainage to the detriment of nature conservation and landscape values.

This controversy has focused on the need to differentiate between actual and potential benefits (and not exaggerate the former), on the rate of up-take of benefits by farmers improving their land with their own field drainage investment, and on the major question of financial versus economic valuing of the benefits (particularly the question of support policy effects on farm prices).

With the shift of support from the Ministry of Agriculture, Fisheries and Food away from agricultural drainage works in the 1980s, at least partly as a result of this critique, this controversy has subsided. At the same time, and in response to the controversy, the Ministry has derived techniques and data for moderating the farm gate prices previously used in benefit assessments, to allow for the subsidy effects of policies including the European Community's Common Agricultural Policy (Ministry of Agriculture, Fisheries and Food, 1985b).

As a result, sea defence and flood alleviation works for agricultural enhancement have declined. Nevertheless there may be cases where sea defence projects can be justified by agricultural benefits alone, not least where existing protection projects come towards the end of their design lives. We recommend that the methods applied by Morris *et al.* (1984) at Silsoe College, Bedfordshire, should be used to assess these benefits. These methods are summarised below and in Figures 5.8 and 5.9.

5.7.2 Assessing agricultural benefits and costs

Figure 5.8 gives the agricultural investment appraisal framework and Figure 5.9 illustrates the basic Silsoe model for the assessment of agricultural benefits. The basis of the evaluation is a farm survey of land within the benefit area to determine land use, soil type, farm survey data and the flood experience of the land in question (Figure 5.8). The degree of improvement consequent upon a flood alleviation or sea defence project will vary. This could be, for example, from low intensity grazing to arable cultivation, or from high intensity grazing to arable, or low value arable to high valuable arable cultivation.

In the benefit assessment model (Figure 5.9), the potential for arable land use change with flood alleviation is gauged, depending on the type of land and the market for the relevant products. This is undertaken with data from both the farm survey and that derived from a regional assessment of agricultural characteristics.

From this appraisal an assessment is made of the change in the value of the output, using producer prices adjusted by the factors published by the Ministry of Agriculture, Fisheries and Food (1985b). These changed values are compared with the value of the existing outputs being obtained, making any allowance necessary for existing and projected crop

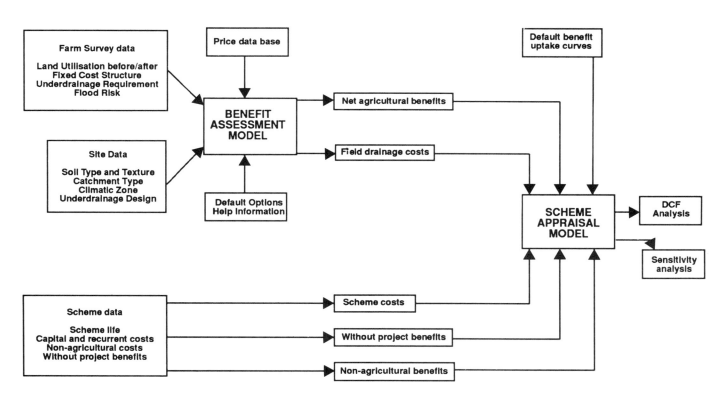

Figure 5.8 Drainage improvement project evaluation model (courtesy of Silsoe College)

damage from flooding, to derive a benefit figure calculated as the difference between these two output values: the 'with' and the 'without' situations.

Grassland production is treated separately in the model (Figure 5.9). Here the Silsoe methodology has developed a grassland energy model which evaluates the energy yield from grassland as converted into livestock weight. The net change in energy production from improved grassland with flood alleviation and drainage is calculated, and converted into economic values using output values and, again, the Ministry of Agriculture, Fisheries and Food adjustment factors.

The model is completed with a comparison of pre-project and post-project benefits, and the derivation of net return figures, allowing for the farm costs needed to achieve the up-take of benefits that are gauged from the farm surveys. This results in a net agricultural cash flow over the years following investment in arterial and sea defence structures, and the discounted value of that cash flow is the capital sum just worth investing to obtain those benefits.

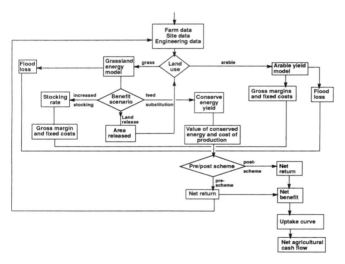

Figure 5.9 Agricultural benefit assessment model (from Morris et al., 1984)

5.7.3 Assessment

In any comprehensive survey of sea defence needs it is likely that there will be some agricultural land to be evaluated. The benefits – or otherwise – from protecting this land from the sea should be treated with care, particularly to gauge any trade-off between agricultural enhancement and landscape and nature conservation losses that might appear economically efficient but not publicly supported.

First, it should be determined whether the flooding will have an adverse effect on agricultural output (the incidence of flooding with a low exceedance probability is not likely to yield significant agricultural benefits).

Secondly, a pre-feasibility study should determine the possible advantageous agricultural and economic effects of flood alleviation, as a desk exercise, and a comparison made with the approximate costs of any engineering works.

This exercise will determine whether a full feasibility study should be undertaken, since these are liable to be time-consuming for large areas of agricultural land (and only large areas are likely to justify sea defence works, and thereby be worth evaluating fully). Finally, any evaluation of agricultural benefits should use experts in the field, follow standard methodology, and use accepted data sets unless it can be shown clearly that these are not appropriate to the circumstances in question.

5.8 Guidance checklist and caveats

It is likely that flood alleviation benefits will be an important justification for most coastal works, other than those which involve just erosion threatening cliff-top properties. In many assessments the benefits of flood alleviation will dominate the benefit-cost comparison, or be a major element of the total benefits (see the Herne Bay case study: Section 8.5).

Table 5.6 gives a checklist of points that should be considered in each of the benefit assessment areas considered in this Chapter. In addition, it will also be important to gauge the relative importance of the different types of benefit, and therefore the relative degree of attention to be given to each.

The simplified caveats in Table 5.6 need to be taken into consideration by the user of these techniques and data sets. These caveats are important. Thus without adequate height information on the floor levels of properties, benefit assessments gauging flood losses that would occur in the future can be very inaccurate. Also, the calculation of indirect losses will be badly flawed if financial impacts are assessed (i.e. the losses to the firm, rather than to the nation). Agricultural benefits must be assessed using the Ministry of Agriculture, Fisheries and Food adjustment factors applied to producer prices.

Finally, it is important to ensure that the future flood scenarios are as accurate as possible. Benefit figures are highly sensitive to, and will be transformed by, quite small changes in assumptions about sea level rise or other land/sea level shifts. Since these influences also affect the benefits of protecting the coast from erosion, it is vital that all aspects of a comprehensive benefit assessment of coastal works are based on the same assumptions.

Table 5.6
Checklist of points to guide an assessment of the benefits of flood alleviation at the coast

A. Direct flood damages

1. Ensure that the benefit area is correctly defined, in relation to the maximum flood likely to be taken as the design standard or to affect the benefit total.
2. Ensure that the floor heights of properties are accurately assessed (including the levelling of floors of key properties).
3. Use trained interviewers/surveyors for site specific damage assessment surveys.
4. Ensure that there are at least three floods of different return periods contributing data to the loss-probability relationship, and preferably six or more.
5. Ensure that the return period of the flood that just causes damage (the threshold of flood damage) is accurately assessed.
6. Give particular attention to assessing accurately the damage effects of the minor and intermediate floods: they contribute most to annual average flood damages, and thus to the benefits of flood alleviation.
7. Ensure that post-project residual flooding is assessed, and that the annual average value of these residual floods is deducted from the 'without project' annual average flood damages, to yield total benefits.
8. Ensure that damage caused by spray and storm effects are separated from flood damages and, if they are to be counted as project benefits, ensure that the project will alleviate these impacts.

B. Indirect flood losses

1. Ensure that the interviewers used to undertake site specific damage assessment surveys are familiar with the concepts and calculations necessary for assessing industrial/commercial indirect flood losses.
2. Ensure that all indirect flood losses included in benefit figures are national economic benefits (not financial benefits).
3. Ensure that any indirect benefits in the form of emergency services and utility costs are calculated taking account of basic services provided (i.e. use overtime and similar wage rates to yield marginal extra costs).
4. Obtain local advice and information of the likely diversion routes to be used when roads are affected by flooding, to ensure that indirect benefits assessed in this way are realistic.

C. Non-monetary impacts

1. Do not ignore these; they could be the most important benefits from the project.
2. Undertake any survey work at the right time (not immediately after a flood) and with the right staff who should be fully briefed.
3. Ensure that any surveys and assessment of these impacts ties in with other local initiatives concerned with flood victims.

D. Agricultural benefits

1. Use those expert in the field for assessing these benefits.
2. Ensure that the data collected is in such a form as to allow a measure of public scrutiny, commensurate with the needs of the confidentiality of farm surveys.
3. Use producer prices adjusted as advised by the Ministry of Agriculture, Fisheries and Food.
4. Ensure that the project will have the effects envisaged in terms of take-up of drainage improvements or other on-farm costs.
5. Calculate any likely crop damages in the same way as other flood losses are calculated, creating a loss-probability curve for this damage category.

6 The potential environmental gains and losses from coast protection and sea defence works

6.1 Summary of information

This chapter assesses the current methodology for the evaluation of the environmental significance of coastal sites. This is a more complex process than that of determining recreational user benefits because the concept of environmental significance and value contains elements of both use and non-use value.

Environmental values cannot, therefore, be measured adequately using Contingent Valuation Methods. Nevertheless, because of the increasing importance attached to the protection of the environment at the coast, the assessment of the environmental gains and losses from coast protection and sea defence works needs to be carried out in a comprehensive and systematic way. The approach adopted should reflect the current best practice of those actively working in the field.

Aspects of environmental value are therefore discussed in this chapter, and the potential gains and losses from coast protection and sea defence projects are examined. A *procedural* – rather than a strictly *economic* – solution is recommended for the incorporation of environmental values into project design. In essence, the recommended procedure represents a valuation process, even though the values that are derived are implicit since at this point in time there is no appropriate methodology for measuring environmental values in money terms. This procedure has been discussed and agreed with representatives of the main statutory and non-governmental environmental organisations at a Workshop held in 1990 (Coker and Richards, 1992).

Section 6.2 defines the problem. Aspects of environmental value which can and cannot be valued in monetary terms, at the present time, are discussed. Alternatives to monetary evaluations are evaluated in Section 6.2.3 and the priorities of the different environmental interests are discussed in Section 6.2.4. Section 6.2.5 summarises the pragmatic reasons for taking the environmental effects of coast protection and sea defence projects into consideration in project design.

Section 6.3 reviews the methodology by which ecological and environmental evaluation is carried out, usually for site designation and nature conservation purposes. Sites of known ecological value, recognised by site designation, are listed. The difficulties of using absolute numerical values or a single index or criterion of value, which could be used as a basis for monetary evaluation, are discussed. Two contrasting methods of incorporating environmental values, from the United States of America and The Netherlands, are summarised (Section 6.3.6).

In Section 6.4 archaeological, geological and landscape values are discussed. Section 6.5 summarises the recommended procedures for use with coast protection and sea defence projects. Potential consultees for the different aspects of environmental value are listed. Sections 6.6 and 6.7 describe the matrix which represents aspects of ecological value at a site and an assessment of their relative value (Section 6.6.1), together with the methods and techniques which are used for the prioritisation of archaeological sites (Section 6.6.2), geological sites (Section 6.6.3) and landscape significance (Section 6.6.4).

6.2 The problem defined

6.2.1 Introduction

Coast protection and sea defence projects often affect a variety of environmental components at a coastal site, such as its ecological significance, and its landscape and archaeological value. That there is an effect on these components is recognised, but is not always easy to quantify.

To be economically worthwhile, coast protection and sea defence projects need, in general, to show that the potential benefits of a project should exceed the costs. As engineering projects become increasingly more expensive, frequently extending to several millions of pounds, the search for potential benefits is being widened. Engineers and planners feel the need to include environmental benefits, as projects may involve the protection of ecological and other environmental sites. To do this, it would be more convenient if a common numeraire – monetary value – could be placed on the environmental effects of a project. Environmental benefits – and costs – could then be placed on the same scale as other benefits and costs, rather than being considered as part of an 'extended' benefit-cost analysis, as at present (Chapter 2).

6.2.2 The value of components of the environment

At the coast, coast protection and sea defence works are more likely to have an impact on the non-built environment than engineering works in many other situations. When evaluating the costs and benefits of coast protection projects it is most important to take into account the effects that any proposed project may have on the coastal environment right from the outset. In this section the term 'evaluation' will be used to refer to assessing the relative importance of environmental changes and the term 'valuation' will be used only when referring to attempts to value environmental significance in money terms.

Identification of components of the coastal environment likely to be affected by coast protection and sea defence projects
It is essential at the outset of a project proposal to obtain basic information on the environmental significance of a site in order to identify which components of the environment may be affected by the proposed project. All sites should be assumed to have at least four potential components of environmental significance, as follows:

1. Ecological significance;
2. Archaeological significance;
3. Geological significance; and
4. Landscape significance.

It should not be assumed that a statutory designation in any one of these topic areas reflects the total environmental significance of a site. It is always necessary to check whether other environmental components also have a value which is not reflected in that designation (e.g. a geological Site of Special Scientific Interest may also be of ecological significance). Similarly, the absence of any designation may be the result of a number of factors such as administrative limitations, resource constraints, ownership issues and problems of achieving a consensus, rather than reflecting the lack of importance of the site. Therefore designation should only be used as an initial indicator of site importance and the appropriate local expertise, whether from statutory agency or voluntary body (see Section 6.5.2), should always be asked to advise on each of the above site characteristics.

A further point which should be borne in mind is that sites may be regarded as of considerable importance from a local perspective even though they do not conform to any international or national criteria of site importance for designation. This raises the more problematic issue of whose definition of environmental significance is taken.

On the one hand we have clearly recognised national and international systems of designation which represent a general scientific consensus, and yet their implementation may be lacking. On the other hand, people local to a site may value that site for many valid reasons which are not recognised or applicable in a wider context. Given that every site is, in some senses, unique and that the importance given to the 'commonplace' by locals may be perfectly valid in a local context, this gives the proponent of a project a problem from the outset in defining what constitutes 'environmental significance' and who defines it. In practice, designations have to be used as the best available indicators of significance but at the same time the existence of differing perspectives should be recognised.

Assessment of the extent to which proposed works will affect the environmental significance It is necessary to identify the characteristics of the proposed project in order to assess the ways in which it may have an effect on each of the components of environmental significance. In particular, the area affected by the project should be outlined so that the scale of both direct and indirect effects can be identified. Indirect effects may arise in two ways. First, there may be indirect impacts within the area directly affected by the project. An example of this would be where changes to the habitat affect an organism upon which a rare species of bird feeds. Secondly, there may also be indirect impacts on adjacent areas which are not directly affected by a project itself, for example where the organism's habitat is affected by changes in the sedimentation pattern caused by 'downdrift' transfer of the effects of the project.

The nature of the project should be identified in as much detail as possible to enable its impact on the different environmental interests to be accurately predicted. This presents its own range of problems, since the processes influencing the environment are usually complex ones and therefore interactions and indirect outcomes are often extremely difficult, or even impossible, to predict. Both the scale and nature of the project should therefore be specified in sufficient detail at the preliminary stage to facilitate this process.

Definition of the environmental gains and losses Once the changes likely to be caused by the project have been outlined, it is necessary to determine whether these should be regarded as environmental gains or losses. Coast protection and sea defence projects can be categorised as generally beneficial or detrimental in terms of their environmental effects. In many cases the process of coastal erosion itself will be damaging to the environmental significance of a site as, for example, where a significant habitat, or archaeological or geological site is being removed by erosion. In such cases – and other things being equal – a coast protection project which reduces erosion will provide mainly environmental gains.

In other cases, the process of coastal erosion itself will be regarded as beneficial by at least some groups of environmental interests, as in the case of a significant cliff exposure being maintained through erosion. In these cases a coast protection project which reduces erosion will create at least some environmental losses. In yet other cases, some particular aspects of the ecological significance may be enhanced by erosion which will allow natural processes to take their course whilst other aspects deteriorate.

Although general agreement can usually be reached over whether a project will result in environmental losses or gains overall, there may have to be some trade-offs between different environmental interests to achieve this. This is particularly the case with the ecological component of environmental significance where there may be differences of opinion between those groups of ecologists who want to protect the status quo against erosion and those who want to allow natural processes to take their course. Hence defining overall environmental benefits may not be as straightforward as it initially appears.

Economic valuation of environmental value It is clear that not all of the environmental components of coastal sites will be held to be equally important by different groups in society. In order to make comparisons between these components, it would be desirable if they could each be measured in comparable terms, and monetary value potentially provides a common numeraire for this purpose.

But each of the environmental components has several different aspects to its value, some of which are easier to value in money terms than others. In order to consider how these various components might be valued in economic terms, they have been divided into three functional categories, termed utilitarian, user and non-user components (Figure 6.1). In practice, it is not so straightforward to allocate all components of environmental value to categories according to the function they perform since categories may overlap or be inadequate. For example, a different approach might make a distinction between 'existence for its own sake' independent of human values, and the other components of the 'intrinsic' category (Tunstall, Green and Lord, 1988). Neither is it equally easy to measure the value of each of the categories in economic terms.

For functions in the first category, the utilitarian component, we can generally use direct or indirect market mechanisms to give them a monetary value. In the case of the second category, the user component, we now have considerable experience in the application of Contingent Valuation Methods to give values in money terms to components of use that had previously been considered as 'environmental intangibles'. However, we are only just beginning to look at how to approach the valuation of functions in the third category, the 'intrinsic' components of value.

The most easily measured of the utilitarian functions are those associated with commercial production, the value of which can be measured directly in money terms, for example, the market value of fish harvested. Other functions may be at least partially valued in indirect terms such as the cost of seeking alternative fishing grounds when spawning grounds have been destroyed. However, due to the complexities of biological systems it is unlikely that it will ever be possible to measure direct causal relationships in such cases. There will always be an element of 'knock-on effect', sometimes of considerable economic importance, where actions in one area destroy spawning grounds for fish caught in another area or even by another country.

Therefore there will clearly be an 'unknown' element in any attempt at the valuation of such functions. The same may also be true of the value of the environment in disposing of waste products. It may be possible to calculate the cost of disposal by alternative means, in the event of biological systems having become so polluted that they can no longer absorb wastes, but an element of unknown 'cost' in environmental terms should still be acknowledged even though it cannot be measured. Sites may perform a protective function by directly sheltering other habitats or human developments, as in the case of a sand dune system which acts as a first line of defence against erosion or flooding. This protective function may be valued indirectly by conventional means used to measure the costs of flooding or erosion which would result from the absence of such protection. A similar approach could be adopted to indirect protective functions, such as in the case where cliffs provide a sediment source which contributes to protection at another location 'downdrift'.

Measurement of functions in the user category depends largely on the application of Contingent

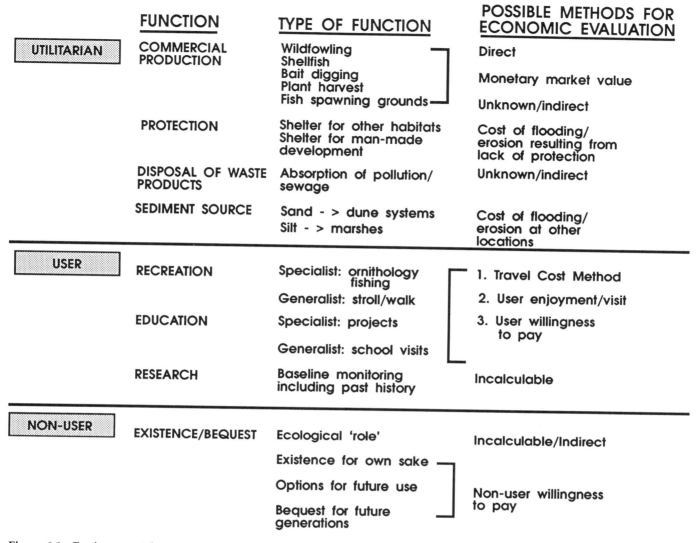

Figure 6.1 Environmental components divided into three functional categories, termed utilitarian, user and non-user components

Valuation Methods (CVM) for allocating monetary values. Functions such as recreation and educational use of a site may be valued by the Travel Cost Method (TCM), which uses journey costs as substitute values for measuring willingness to pay for the recreational or educational experience (Stabler and Ash, 1978). However, there are several limitations apparent when attempting to apply this approach to the valuation of recreational benefit at the coast in Britain (Section 4.3).

An alternative approach is the use of questionnaires to assess enjoyment per visit by comparison with recreational activities with a known cost attached which will provide comparable enjoyment value (Section 4.3.1). This approach could also be applied to valuing educational use, although its function as a surrogate valuation technique should be stressed since it would in no way seek to measure the 'total educational value' gained from the visit.

Willingness to pay in money terms is an alternative CVM approach which may be employed to measure user values and this is particularly appropriate in the case of specialist user activities carried out at a specific site, for example bird watching at a nature reserve. The direct willingness to pay for entry to or use of the facility or site may be taken as a measure of its value to the user, although this measure does not attempt to incorporate any component of non-use value and so provides only a partial estimate of environmental value.

Indirect willingness to pay (e.g. through the taxation system) for the upkeep of the environment in general, such as the creation and maintenance of Heritage Coasts, may also be measured by the use of questionnaires (Green, Tunstall and House, 1989). However, it must be stressed that this measure does not attempt to distinguish between user and non-user benefits, and it is therefore unclear whether any component of non-use value has been included. This means that the method again provides only a partial estimate of environmental value.

Putting a value on the research function of the environment is a far more problematic area to tackle and it may be seen as occupying a grey area between the 'user' and 'intrinsic' categories. Research is clearly an important component of environmental value which may be of direct use to society, for example as a baseline against which change can be monitored and as a knowledge base on which predictions can be made. It may also be considered as becoming part of the past history of a site itself and, other things being equal, to lose a site with a recorded history could be considered a greater loss than to lose one without. This component of value clearly overlaps with the concepts of both educational and historical value.

The components considered so far constitute all the ways in which we, both as a society and as individuals within that society, make use of the environment in a direct or an indirect manner.

In contrast, the third category contains other things about the environment, apart from its use, which we know are valued by some people (Brennan, 1988). Although the term 'intrinsic' has been adopted for this non-use category, we have to recognise that these components also reflect the values which we as humans put on the environment. One way of putting a money value on these components is through assessments of non-user willingness to pay, obtained in the same way as for user willingness to pay. However there is an obvious methodological difficulty here in that the sample population is the whole of society and not a specifically defined user group.

When we turn to the possibility of valuing the ecological 'role' or 'systemic value' of the environment, then the technique of assessing willingness to pay is a very inadequate vehicle for allocating monetary value. Since the concept of ecological role relates ultimately to maintaining the environmental integrity of the global ecosystem, we may consider this of incalculable value – it is beyond price.

Aside from considerations of ecological role, several attempts have been made to find indirect measures of how both individuals and society as a whole value the other components of intrinsic value. Willis and Benson (1988) determined how much individuals would be willing to pay to view the meadows of Teesdale, among other locations, both by direct willingness to pay and through the Travel Cost Method.

But arguably such measures relate more to the general amenity value or aesthetic value of the meadows (e.g. as the setting for a picnic) than to any component of 'intrinsic worth'. This view is supported by the fact that only the user population was questioned – we have no values from the non-users. We should also question whether or not these meadows would still have a value to society even if they did not have attractive flowers and were not suitable locations for walks or picnics. Even if they appear to have no value at present, this does not allow for the concept of bequest value for future generations.

Chapter 2 has discussed the problems of distinguishing between society's values and the aggregate values of individuals within that society. An alternative approach to the estimation of the value which society puts on some component of the environment is therefore given by taking as surrogate values the cost to the nation of compensating farmers and land owners for profits foregone (Section 2.5.7), through the environmentally sensitive management of designated areas under management agreements (for example in Environmentally Sensitive Areas or Sites of Special Scientific Interest). However, we need to recognise that such a value is wholly an artifact of the agricultural economy at a particular point in time. A second approach to social valuation is to take the value which statutory or voluntary agencies have paid to acquire a site in order to protect it. But this value is also an artifact of market conditions at one point in time and, as such, may grossly distort our attempts to gain a realistic measure of intrinsic environmental value.

In view of the issues outlined here, it is not surprising that in general we tend to put values or surrogate values on those aspects of the environment which are most easily measured and not necessarily those which are most important. This highlights the very real limitations on economic valuation of the environment which currently exist.

From an economic viewpoint, as indicated above, it may be desirable to place a monetary value on the various significances of the ecological, archaeological and other environmental components of sites. These then would result in a common basis for comparison with other values. Environmentalists, on the other hand, would generally consider that the low priority frequently given to such sites by developers and planners would mean that an inadequate valuation would be placed upon them. Not only would the placing of monetary value on a site be of little practical use in assessing changes or effects, but it may also be unrealistic to value in monetary terms the facets of environmental value which are unique and therefore considered by some to be 'above value'.

6.2.3 Alternatives to monetary evaluations: designations

If monetary evaluation is not possible, it may at least be possible to evaluate environmental components in relative terms.

In Britain there already exist implicit systems of relative valuation for sites of environmental value. These are incorporated into the system of site and area designations for all types of environmental value. At the coast, sites of ecological significance may be designated in terms of international, national and local value; archaeological sites may be scheduled as Ancient Monuments; geological sites may be designated as Sites of Special Scientific Interest; areas of high landscape significance may be designated as Areas of Outstanding Natural Beauty (AONBs), or Heritage Coasts.

However, using site designations as a guide to site value can result in the omission of important environmental information. **The fact that a site is not designated does not mean that it is of no value.** There are several explanations as to why this situation may have arisen:

1. The designation of sites frequently occurs for pragmatic reasons. The opportunity for purchase or management may have arisen in the past and this opportunity was taken. This does not mean that sites of equal value do not exist elsewhere which are not designated. A good example of this process is in the designation of Ramsar sites – internationally important wetlands. In the United Kingdom these have only been designated in areas which already have the protection of designation as SSSIs or National Nature Reserves (NNRs). Other sites which fulfil the criteria for selection, but are in areas without designated status, have not yet been recognised as of comparable importance if this is measured by designated status alone.

2. There may be boundary problems involved in site designation. For example, in Scotland SSSIs extend to the spring tide low water line, whereas in England and Wales these sites extend only to mean low water. If measured by designation alone, this would suggest that the part of the shoreline between low water spring tide and mean low water levels is of less value in England and Wales than in Scotland. Clearly this is not necessarily so.

3. The value of a site may not have been recognised. An unexcavated archaeological site, for example, may be far more valuable than supposed; rare species may not be discovered until a detailed ecological survey is carried out. The recognition of sites of marine importance, in particular, has been slow, in part due to the lack of resources for extensive research compared with terrestrial sites. In addition, knowledge of sites of marine ecological and archaeological importance is often dependent on proximity to areas which are popular for diving, rather than based on comprehensive surveys of the inshore area.

4. Some aspects of environmental value are under-represented by site designations. This is due to the strengths of interest groups and other bodies representing different environmental interests, which attract varying levels of public support. Unpopular specialisms or under-researched aspects of environmental value 'lose out'. The marine habitat, for example, is under-represented by site designations. Marine habitats have only recently been covered by protective legislation, and SSSI status does not extend below mean low water in England and Wales. Because of the complexity of the process of designation, only two Marine Nature Reserves have so far been designated in England and Wales.

5. The environment as a whole has a value outside protected areas. The isolation of ecological sites could lead to their being surrounded by development, to become 'islands' within which the maintenance of viable breeding populations of species becomes increasingly threatened. It is therefore important that the species diversity of the wider countryside is maintained.

6. Total value may be more than the sum of component parts, yet the designation system may only recognise one aspect of value. For example, the designation of an archaeological site usually involves the protection of a particular monument. However, an archaeological landscape may contain several separate features of comparatively low historic value if considered individually, but which, when taken together, represent an important example embracing a spectrum of the past.

Table 6.1 illustrates the range of statutory and non-statutory ecological designations in England and Wales. A site designation indicates that a site is of known value, but the environmental value of all sites should be taken into account. That a site is of known value does not necessarily mean that the site requires protection from erosion or flooding. Further information is required in order to assess the direct and indirect impacts of a coast protection or sea defence project on any site.

6.2.4 Different environmental priorities

From an environmentalist's point of view, the problems of coast protection and sea defence works may be more complex than those perceived by the coastal engineer and planner. Environmental value consists of several components of value – the priorities of the various

124   THE ECONOMICS OF COASTAL MANAGEMENT

Table 6.1
Ecological site designations[1]

A. **Statutory designation**

1. *International designations*

Ramsar sites	Wetlands of international importance designated under the Ramsar Convention. These designations were originally for bird significance, but have now been enlarged to include other wetland values, e.g., invertebrates.
Special Protection Areas	Designated under European Community legislation, these sites are of bird significance.

2. *National and local designations*

National Nature Reserves	Most of these sites have the additional protection of being in national ownership, and represent the highest nature conservation value. English Nature designated, (formerly Conservancy Council designated).
Marine Nature Reserves	This designation represents the highest value for the marine environment. However, the process of designation has remained too slow for the sites designated so far in any way to represent the most valuable areas. English Nature designated, (formerly Conservancy Council designated).
Sites of Special Scientific Interest (SSSIs)	The redesignation of these sites represents the minimum area required for a spectrum of wildlife value to be safeguarded across the country. English Nature designated.
Local Nature Reserves	The local county or district council, usually with the Countryside Commission, are involved at these sites, and their particular local value means that access to the features of significance is of most importance.

B. **Non-statutory sites**

Reserves of County Trusts affiliated to the Royal Society for the Promotion of Nature Conservation.
Reserves of the Royal Society for the Protection of Birds.
Wildfowl Trust Reserves.

Other local reserves	Any local interest group or landowner may have designated an area of land as being of particular ecological significance, and this may receive a measure of protection (e.g. in local by-laws).
Voluntary marine conservation areas	These areas are not covered by protective legislation or by-laws, but in view of the paucity of Marine Nature Reserves, should perhaps be considered as such.

Note: The ownership of these sites may be public or private bodies or individuals. Site management may be the responsibility of the owner or designating body of the site, or of another body or interest group. Possible groups for consultation are listed in Section 6.5.

interests involved at a site will not necessarily coincide. Ecological, archaeological, geological and landscape values may not be best served by the same policies. At any one site, however, the views of any or all of these environmental groupings need to be taken into account. The aims and priorities of these groups can conflict, as can the views of representatives of different specialisms within these groups.

From the point of view of a geologist, for example, the best interests of a site may be served by continued erosion, which will maintain the fresh exposure of geological strata. Coast protection would result in the obscuring of features of geological significance by vegetation growth as erosion ceased and new strata of significance would not be further exposed. But by slowing the rate of erosion, rather than its complete cessation, geological interest can be maintained in many cases.

Geomorphological significance would usually be maintained by the continuation of the dynamic processes which have resulted in the varied landforms around our coastline. By comparing the present situation with that shown in historic maps, it can be seen that the coastline is not in a fixed position but is constantly changing, especially along the 'soft' coasts of the south and east of England. The processes of change are still taking place. Existing landforms, such as shingle spits, move and change over time; there is the creation of new land by deposition and erosion of existing features.

The construction of existing coast protection works and other barriers to sediment transport, such as harbours, has already interfered with these natural processes. The indirect effect of such constructions on the geomorphological value (and its associated ecological value) of downdrift sites is often considerable, depriving beaches and other accreting features of sediment. These indirect 'costs' may be greater than the benefits to be obtained by protection. A balance must be reached between allowing the continuation of coastal processes and the benefits of protection.

For an archaeological site, protection would, in general, be the most favoured option. However, coastal erosion can be beneficial, although most such factors cannot be predicted in advance: unknown sites can be exposed by erosion.

In aesthetic terms, concrete sea walls, rock revetments or groynes do not represent a 'natural' coastal landscape. Careful consideration should be given to ways in which project design will blend in with existing characteristics of a site. It may be that most consideration should be given to 'soft' engineering options, such as beach recharge, in situations where it is important to maintain as natural-looking a coastline as possible. On the other hand, access along the coast may be improved by the construction of vehicular routes for repairs to be carried out and by creating good walking conditions along berms.

Ecologists, in particular, have been thrown into a defensive posture as regards ecological sites and there may be disagreements as to priorities and policy between different interest groups and between those with different attitudes to protection. Their energies have been directed at protecting sites from development, for which they feel that legislative protection is inadequate. Natural and semi-natural areas of land are constantly being reduced in size by industrial, housing and recreational developments, making protection for those that remain an increasingly high priority.

At the coast, however, a dichotomy may arise. A sea defence or coast protection project can interfere with important natural ecological coastal processes which are part of the dynamic nature of coastal sites. For example, stages in vegetation succession which would naturally occur can be interrupted and coastline stabilisation can eliminate early stages of colonisation. Changes in sedimentation processes can affect the value of the site itself as well as those downdrift. The engineering works themselves can also cause direct damage to aspects of value at a site. On the other hand, projects can also protect an ecological site. For example, cliff-top habitats can be protected by the halting of the cliff erosion; damaging seawater flooding of freshwater habitats may be prevented.

It is easier to predict which sites will be lost through continued erosion or flooding than to assess whether changes to a site will result in overall ecological gain or loss. If a rare species were to be lost as a result of the destruction of its habitat, for example, it would be impossible to predict with certainty that an equally rare, or even a more valuable, species characteristic of the new habitat formed would inevitably colonise the site. Some ecologists would prefer protection of the features of known value at a site. Others would prefer the 'do-nothing' scenario which would preserve natural processes and could result in the development of habitats with potentially greater significance than those already existing.

The destruction of habitats may escalate in the future with sea-level rise and many important habitats may be lost or changed beyond recognition. In some areas it may be possible for sea defences to move back to earlier sea defence lines, so that land reclaimed in the past which may now be agricultural land can revert to semi-natural habitats. But in many areas the presence of existing development means that land cannot be abandoned to revert by a process of natural recolonisation and so the protection of existing sites will be a priority in such locations. Thus, even within the ecological perspective there is frequently division between those who regard the protection of vulnerable habitats as the greater priority at the coast and those who wish to maintain natural dynamic processes.

Therefore, at any one coastal site, there is not necessarily any one environmental priority that should dominate and hence it is for this reason, as well as others discussed in Chapter 2, that it would be very difficult, if not impossible, to allocate a single figure for environmental 'value'. An overall gain/loss equation cannot be devised but rather an assessment is possible of the different impacts and changes that are taking place and will, or could, take place both with and without a coast protection or sea defence project.

6.2.5 Why environmental effects should be considered at an early stage in project design

It is important that environmental effects are taken into account at the preliminary stages in the planning of any coast protection or sea defence project. As can be seen, it may not be in the best interests of the site from an environmental point of view for protection to occur at all. An environmentally sensitive site will require input from all the environmental groups involved, it may require an Environmental Impact Assessment (EIA), and may require specialist environmental consultancy advice in terms of design, siting and impacts.

There are several good reasons why the environmental effects should be assessed at as early a stage as possible in the promulgation of a project, as follows:

1. *Resources* For grant aid to be given, any project must not be environmentally damaging. Rather than having to redesign a project after much expense has already been incurred on an environmentally unsuitable option, the most environmentally favourable proposal should, if possible, be chosen at the outset. Considerable expense can be spared by the initial recognition of the most suitable option. All further outlay, such as for the benefit-cost analysis, can then be directed at the project which is most likely to be accepted for grant-aid purposes.

2. *Time* Time can be saved by early identification of the most environmentally sound option. Further research may need to be undertaken and this should be identified and included in the time budget. An EIA may be identified as a requirement and may take some time to carry out. At the later stages a planned project may be called in for public inquiry at a sensitive site, however careful the evaluation has been to take into account all the available environmental information. Time (and expense) can be saved by having all the environmental data marshalled and arguments already prepared.

3. *A 'green' image* A firm or authority which can show that it does take into account environmental effects can only gain in terms of public esteem. By demonstrating a willingness to co-operate with environmental bodies, relationships can be smoothed and any trade-offs or mitigating effects can be suggested in the knowledge that environmental effects have been given serious thought. Future projects and plans will be looked at in a more favourable light if a reputation is gained for positive co-operation.

At the present time, there is no agreement on how environmental values should be included in site assessments. Environmental specialists may value the environment differently from the public in general, both in terms of the use they make of it and the inherent values they see in it. **It is accepted, and recommended here, that given the present state of the art of environmental evaluation this should not be by the explicit calculation of monetary values.** Not only is the methodology ill-defined, but the interpretation of gains, losses and change are not easily amenable to economic analysis. It may also be the case that economic evaluation of these environmental values should not be attempted as a matter of principle.

As has already been indicated, however, it is important that full consideration of environmental effects should be incorporated in site assessment. The recommendation here is that the solution to the problem of valuation is **procedural**, and involves incorporating environmental issues at the preliminary stages of site evaluation as part of the established and agreed project appraisal procedure.

6.3 The state of the art of ecological and environmental evaluation

Before an ecological site can be valued, the elements of value at that site must be classified before prioritisation can occur. In this respect research into different species, communities and habitats has produced a large body of knowledge upon which to draw.

However, this research is not complete for all species groups, habitats and communities – particular topic areas may be under-represented. Also, when known information is considered from a particular site, the mobility of individuals means that information already obtained from earlier survey work may not be completely accurate. Bearing in mind these limitations, there is a well-developed system of site classification which is being used in the designation of SSSIs by English Nature (Nature Conservancy Council, 1989) for all terrestrial habitats, including coastal ones. This system is based on the classification of different habitat selection units, by community type, under the National Vegetation Classification (NVC) which characterises habitats and sub-habitats by the different species which habitually grow together. Ecological criteria are applied to these communities, and to additional species information, by which prioritisation for designation is made.

6.3.1 Types of habitat

At the coast, the habitats present can be broadly defined as terrestrial, coast-specific, and marine. Terrestrial habitats include woodland, grassland, heath and freshwater and are not confined to the coast, although there may be some maritime influence on their character where they occur at the coast. Coastal habitats are confined to the coast, including estuaries. The level of influence of the sea covers a spectrum from mean low water to above the level of the highest spring tides. The marine environment can be classified as that below mean low water, but there may also be considerable marine significance in the inter-tidal zone.

The complexities of the ecological character of any one site can be shown by an example for coastal habitats: sand-dune systems (Table 6.2). As well as the seven sub-habitats which can be present, there may also be gradations of change from dune grassland and dune heath into communities of grassland, heathland and fen. In addition, criteria should also apply to definitions of species assemblages. Groups include the following: vascular and non-vascular plants (e.g. lichens and bryophytes), mammals, birds, reptiles and amphibians, fish, butterflies, dragonflies and other invertebrates.

Table 6.2
Sand-dune systems

Sub-habitats	Description
1.	Strandline, within which are four NVC communities.
2.	Yellow dune, within which are four NVC communities, subdivided into eleven sub-communities.
3.	Dune grassland, with two NVC communities divided into six sub-communities.
4.	Acid dry dune grassland, with four NVC communities divided into nine sub-communities.
5.	Dune heath – several different heath communities may be represented.
6.	Dune slack, with two NVC communities divided into four sub-communities.
7.	Dune scrub, with one NVC community divided into four sub-communities.

The information available about some species groups may be inadequate to make any kind of judgement about their distribution and numbers, especially with regard to invertebrates. The numbers of different macro-invertebrate species in Britain include approximately 22,500 insect species and 7,500 species within other groups such as spiders, molluscs, millipedes, centipedes, and crustaceans. Micro-invertebrates, such as the 2,000 mite species and 1,000 nematode species, are not normally quantified (Nature Conservancy Council, 1989).

Marine habitats, which can be directly and indirectly affected by coast protection works, cannot so easily be classified into distinct habitats. Although there has been a considerable body of research in recent years with regard to marine habitats, there has been no generally accepted method of classification. Habitats can be defined in terms of their position on the shore (for inter-tidal habitats), their physical substrate, by the degree of exposure to waves, by the effects of tides and currents, and by community structure. Their position in relation to biogeographical trends can be of particular importance. Where these biogeographical boundaries overlap the community structure will contain a diversity of species from both adjacent areas.

The physical influences on community structure are interlinked, but would generally be considered separately, providing a wide range of possible combinations. Insufficient information may be available for a habitat classification to have been made, so an assessment of value will, in turn, be difficult. The effects of engineering works on the marine environment have often, in the past, not been sufficiently taken into account. The level of interest in marine ecosystems at the present time is high, so that the difficulties of obtaining information, and lack of knowledge about changes which can occur, are likely to be reduced in the future as the marine environment comes to be regarded as of greater value than hitherto.

Changes which can occur in patterns of sediment movement and changes in sediment type can result in extreme changes to the sea bed. Fine sediments released by coastal engineering works can smother immobile bottom-living species which, in turn, provide shelter and food for other, more mobile species, including fish.

The range of climatic, biogeographical and physicochemical influences mean that the possible combinations of conditions and species present are virtually infinite. In practice it is inevitable that a subjective judgement will have to be made about which aspects of site value are considered as a basis upon which to make a decision about the evaluation of a site: it would be very rare for every possible research dimension to have been fully assessed at any site.

The importance of any one site will relate to the total

amount of all habitats, sub-habitats and variants present, as well as the significance of individual species. An 'objective' scientific assessment of the absolute value of an ecological site is an impossibility in terms of its relationship to the total complement of species which could be present. Inevitably, a judgement has to be made about the classification of a site in terms of available information at that point in time.

6.3.2 Criteria for evaluation

After the collection of information about a site and the classification of its main features of significance, an assessment of the value of that site can then be made, by the application of specific criteria. Criteria are chosen with a particular objective in mind – usually for site designation or management purposes – and frequently by bodies with different nature conservation interests. As the reasons for the choice of criteria vary, so do the criteria which have been used; standardisation of criteria is unlikely to occur. Margules and Usher (1981), found that in seventeen studies they examined, a total of twenty-four biological/ecological criteria were used. The four in most frequent use were area/size, rarity (of habitat or species), diversity (of habitat or species), and naturalness.

The criteria used by English Nature were introduced in the Nature Conservation Review (Ratcliffe, 1977). The English Nature criteria are used for all habitat types, and are as follows:

1. **Primary criteria:** size, diversity, naturalness, rarity, fragility, and typicalness; and
2. **Secondary criteria:** recorded history, position in an ecological/geographical unit, potential value, and intrinsic appeal.

There may be problems in the use of these criteria for assessing the value of all ecological sites at the coast for the purposes required by the coastal engineer or planner. First, other attributes of importance, not just the strictly scientific, have been omitted. This is particularly true at sites of local importance, where reasons for placing a high value on a site may include use value for the local community and particular cultural/historic associations.

Secondly, not all the criteria apply equally to each habitat type. For example, naturalness is less applicable as a criterion in a semi-natural habitat; diversity is not applicable to a habitat, such as a mud-flat, which by its nature is not diverse. Thirdly, an engineer requires to know what effect the engineering works will have on the ecological characteristics of a site – the value or importance of the site is not the only information that is required. As the criteria are designed for designation or management reasons they do not take into account human induced change, or, frequently, natural changes.

6.3.3 Site scoring systems

Attempts have been made to summarise the total value of a site by assigning a numerical score to the results of the criteria assessment but this raises several issues, as follows:

1. Basing value on a total score figure would result in a loss of information obscuring the features of significance at a site. The effects of erosion or flooding, or of a coast protection project, will rarely affect an entire site. Each constituent habitat and community should be evaluated separately; the value of the criteria cannot be combined.

2. The use of numbers in themselves implies objectivity. Evaluation is frequently of a subjective nature as absolute scores cannot be given to every attribute. Where a scale of values is involved, different observers may arrive at different values, making any calculation of a numerical score subjective.

3. If a numerical score is given, there is a danger of mathematical manipulation in a way that is not appropriate for ecological parameters. For example, the loss of part of a site may diminish the value of the entire site by more than the value of the portion lost, and also affect the value of adjacent sites. This decrease in value could not be expressed adequately by a simple numerical calculation.

4. In a particular situation it may be desirable to give particular weight to one or more of the criteria. The function of a salt-marsh as a pollution dispersant will be more important adjacent to an industrial area, for example, although this could have the effect of reducing the numbers and variety of species present. A subjective change in the application of the criteria would result in a lack of standardisation and comparability.

5. Whether particular criteria should be given added weight as being 'more important' would not only involve a subjective judgement, but would also raise mathematical problems, as all scores should be standardised to the same scale even though they are measured in different units.

6. Multi-criteria indices, which add scores for each attribute, run into other mathematical difficulties. By the interactional nature of ecological value,

many criteria overlap, and so the addition of scores could be regarded as double-counting. To overcome this, some ecologists have suggested that criteria indices could be multiplied. However those criteria which do not overlap would have to be added together. The combinations of criteria involved mean that a simple formula is not possible.

7. Single criteria have been used on their own as a measure of site value. This simplifies the methodology and reduces the quantification required from site surveys. The main disadvantage is in the amount of information which is not taken into account; there may be many reasons for a site to be of wildlife significance.

6.3.4 Rarity

One of the most frequently cited criteria to be used on its own is that of rarity. This is popular with the public, and often produces a strong emotive reaction which can act as a catalyst for species protection. There is considerable political weight behind the concept of rarity which has resulted in legislative requirements for protection, for example in the Wildlife and Countryside Act 1981. The EEC Habitat Directive (OJ 1988 C247/3) (European Council, 1989) contains not only habitats which are considered to require protection but also rare species lists.

However, there are several problems involved in using this criterion on its own:

1. Species are omitted from rare species lists. These are usually those which have attracted least public or research significance, so extra value is given to known taxa.

2. There are different levels of rarity. In international terms, an endemic species (i.e. one restricted to a particular locality) should merit greatest protection. In local terms, species that are rare in that particular locality can be of particular significance, although they may be common elsewhere. For example, birdwatchers often show a high level of interest in regionally or nationally rare species in their own country even though they may be common elsewhere.

3. The accident of a rare species happening to occur at a particular site does not necessarily mean that there are any other features of significance. Conversely, it is more likely that sites which demonstrate a range of factors of significance will produce conditions in which rare species are more likely to colonise.

4. Any species is dependent on the ecosystem in which it lives. This includes the following: the physical habitat for shelter, sleeping and feeding; other species which are part of the food chain; and the wider countryside to provide a gene bank for a healthy population and a natural flow of species into and out of an area. Suitable habitats are required for long-term survival: no rare species can exist on its own.

5. By concentrating on politically popular rare species, which are frequently mammals and the top predators, the ability of an ecosystem to support itself can become unbalanced. Species may have to be culled to maintain a sufficient food supply for a healthy population.

6.3.5 Area and species diversity

Area is another criterion which could be considered on its own. In general terms, a larger site should be more valuable and more likely to contain rare species, species groups of significance and a variety of communities. However, each habitat has to be considered individually as some habitats have a limited distribution so that even a small area is of importance. Sub-habitats and particular features of significance may be localised at small sites.

Species diversity can be considered to be one of the most important attributes of ecological sites. Problems arise at the coast in particular as recently formed habitats, which are by their nature not diverse, are still of ecological significance. With an increase in diversity at such sites, the pioneer species become overshadowed by more successful competitors.

The range of species present and the numbers in which they are found can be represented in mathematical form as a diversity index. These appear to be an attractive evaluation system, but there are several problems in their use. For example, problems arise in the mathematics of the index calculations, as follows:

1. In indices based on multiplication, less weight may be given to species which occur in smaller numbers, other indices may be weighted in favour of species which occur in smaller numbers by the use of logarithms.

2. Practical problems can occur during calculation due to lack of standardisation. For example, Simpson's index can be expressed as D, as 1-D, or as 1/D. The logarithm base for the Shannon-Weaver index has not been standardised: base 10, base 2, and base e

have all been used (Washington, 1984).

3. The scales of indices may be non-linear. This means, for example, that a decline of diversity from 4.5 to 4.3 may be more significant than a decline from 1.5 to 1.3.

4. Diversity indices measure two components, the evenness and variety or numbers of species. Information is required for both. However, as both aspects are being taken into account when the index is being calculated, either factor may be the main influence on the total. For example, the diversity index can increase in value when species numbers decline, if the evenness of their distribution increases.

Problems of comparison arise as the indices are usually calculated for a sample of a population. This should be standardised over the number of sampling points, and time, area, etc. As this is frequently not the case it should be overcome by standardising the inputs, which would involve using the smallest sampling level, with obvious loss of data.

Biological problems also arise in the use of indices. First, it may be desirable to exclude certain species from the index depending on the purpose of the calculation. These include, for example, 'incidentals', which are species that just happen to occur at the time of sampling in what would normally be unlikely or adverse conditions, or species planted or introduced by humans, which would not be desirable as a measure of 'natural' diversity. The treatment of these possible exceptions has not been standardised.

Secondly, if the index is only calculated for flowering plant species diversity, which is most usual, this may not correspond with the value of a site for other species groups. Certainly vegetation diversity does not necessarily correspond with bird diversity, especially at the coast, where many species may feed on mud-flats of extremely low botanical diversity, dominated by one species.

A diversity index alone is not considered appropriate for birds, because of loss of information. The importance of sites may be based on the large size of flocks of one or a few species which congregate for roosting or feeding, frequently on a highly seasonal basis. The normal densities of different bird species vary, so information is required on the numbers and distribution of each important or rare species (Fuller and Langslow, 1986; Gotmark et al., 1986).

Thirdly, the scientific basis for the choice of sites of greatest value would be undermined if it were based on the use of one species group alone. This is especially true at the coast, where the pioneering species are frequently lichens or bryophytes, and the value of many sites may be ascribed to invertebrates. Unfortunately it would not be possible to use diversity indices for all species at a site, partly due to lack of knowledge of the distribution and occurrence of many species groups and also due to the time required for site surveys, which would be prohibitive. This suggests that habitat criteria, which could indicate sufficient diversity to cover the requirements of as many taxa as possible, would be of greater validity in the assessment of value of sites.

Fourthly, diversity in unstable and in developing habitats is likely to be low, but not necessarily of lower value than a habitat of high diversity. A diversity index for a site could then be high or low without this giving any indication of its comparative or even absolute value. It would be more usual to use the index to compare habitats of the same type, so as to compare like with like.

Several authors, including Usher (1986), agree that species richness is a better measure of diversity than a diversity index. Washington (1984) concludes that the continued use of the most popular index, the Shannon-Weaver index, is due to its 'entrenched nature' rather than to any biological relevance.

6.3.6 Other approaches and approaches in other countries

Omitted from most of the biological systems of valuation are those criteria which reflect the social and economic functions of wildlife. These functions could be said to be part of any rational basis for the assessment of value of any natural area, unless only the biological aspects are to be considered. They include educational and aesthetic values, ethical or moral values, and research value.

However, most of these values cannot be considered as site specific, but are rather part of the general value of wildlife as a whole (Chapter 2). Of particular relevance at the coast are the potential agricultural and medical values of species which can prove of benefit to humankind. Some maritime species may also have potential economic value (e.g. the wild ancestor of commercial sugar beet grows at the coast). Wild maritime species may provide gene banks for the development of such attributes as disease-resistance or salt tolerance.

Other countries make assessments of site value by different methods. In the United States, individual States operate their own systems of evaluation which reflect their own particular circumstances. Nationally, the United States Fish and Wildlife Service recognise three methods, all of which are computer-based (Brack et al., 1987). Sites can be analysed by several approaches, as follows:

1. Energy flow. This is only possible for a small area and also assumes that a site is self-contained, which is not necessarily the case.
2. Population estimates of particular species. This results in many of the same problems as in the United Kingdom: lack of data for all populations; deciding which populations to choose; the problems of time required for measurement; as well as natural fluctuations and cost.
3. The Habitat Evaluation Procedure (HEP). Habitat-based quantification is based on the idea that criteria can be quantified, qualified, enumerated and compared. Again there are problems of missing data: any computer model is only as good as the information that is entered. This method is also most time-consuming and requires training and certification.

The United States also gives protection to rare species through the United States Endangered Species Act – more political and economic weight is given to particular species than in the United Kingdom. Projects, such as dams and other construction works, have been halted by the requirements to protect species. The focus on larger mammals and top predators has caused problems, affecting natural diversity, which would be best preserved by focusing on biotic communities and habitats (Beiswanger, 1986).

In The Netherlands, the approach is one of integrated management. In order to achieve this, it was realised that measures are also required in other areas including land use planning, environmental policy, nature conservation and development and agricultural policy (Ministry of Transport and Public Works, 1989). The third National Policy Document on Water Management, 1989, was therefore co-signed by the Ministers of Housing, Physical Planning and Environment, and Agriculture, Nature Management and Fisheries.

The system is based on the functions of water systems as a whole. For coastal areas these functions would include coastal defence, nature conservation, recreation and freshwater abstraction. They may be integrated functions, or one or more may be given priority. All functions have to be considered for any project. It is expected that the relevant interests will co-operate in the preparation, execution and evaluation phases of coastal projects since a change in one aspect can affect any of the other functions. The decision-making process takes into account the following (de Jong and Visser, 1983):

1. The reasons why a project is being considered;
2. A good understanding of natural processes and their relationships;
3. A description of the future consequences of action; and
4. An assessment of available techniques and their likely success.

This means that interested bodies are not only consulted in the decision-making process but included as an integral part. Any plans that are drawn up are done so with input from all interested parties. The question of the 'value' of an area of ecological importance has been considered at a policy level in the evaluation of functional importance, rather than at a site level.

6.3.7 Assessment

To conclude, ecological evaluation is, by the interactive nature of ecological systems, an exercise which cannot be carried out without an element of subjectivity. Even with adequate site information, the choice of criteria for valuation is subjective in itself. A system of values, in which each aspect of value can be totalled and standardised, and which can also be used for comparative purposes, is not possible now, and may never be possible.

6.4 Archaeological, geological and landscape values

The components of value of archaeological sites, geological sites and landscape at the coast may appear to be less complex than for sites of ecological value. Fewer separate interests are involved, and the approach to evaluation can be more clearly defined. For archaeological sites, however, it should be borne in mind that once damaged or destroyed the information lost cannot be recreated, as it might be, for example, with a damaged ecological site.

6.4.1 The value of archaeological sites at the coast

An accepted criterion for a coastal archaeological monument is one which lies within 1 kilometre of the median high tide mark (Darvill, 1987).

Coastal environments encompass a wide variety of geomorphological forms, from estuaries to sand-dunes to 'hard' cliff coastlines. Partly as a result of this diversity and partly because the coast acts as a natural 'break-point' in human occupancy and communication, archaeology at the coast often displays a special significance. In addition to such coastal-specific sites there is, of course, the continuation to the coast of those sites that have a wider distribution inland.

Archaeological features at the coast have value for

the following reasons:

1. With Britain being an island, the coast has always had a special place in human occupancy, and attracted communities for specific activities, for example, as a provider of food (on-shore and off-shore), as a source of raw materials, and as a breakpoint for trade and defence. These sites are coast-specific whose functions are not recognised elsewhere; they are therefore unique features.

2. Waterlogged conditions clearly prevail at the coast. Where these occur in association with peat and marsh accumulation they can provide the ideal environment for the preservation of pollen, which is invaluable for archaeological dating purposes. Such anaerobic conditions are also suitable for the preservation of organic materials (e.g. textiles, leather, wood). Conversely, some dry sites, notably sand-dunes, are increasingly recognised as suitable for the survival of early mollusca remains, which are also used for dating/environmental purposes.

3. In many places the sea has encroached over parts of the coastline formerly used for human occupancy. Depending upon the prevailing conditions, this can be either destructive or preservative. The same consideration applies to the archaeology of shipwrecks on the foreshore, which is a growing area of archaeological significance and expertise. In other areas, windblown sand, shingle and alluvial accretion can have a similar preservation effect.

4. Much of the coastline is still undisturbed by urban and industrial use and the widespread, largely informal, leisure use has done relatively little to damage archaeological sites. Over half the coastline is under some form of protective status or ownership (e.g. the National Trust) and this generally holds good prospects for the future of archaeological remains. However, it should be noted that none of these agencies has a formal protective remit for archaeological sites.

In terms of the problems of identification and quantification, there are two issues to be addressed here. First, there is the issue of what proportion of the total archaeological sites have so far been identified? Secondly, what is the future potential of an area for further discoveries?

Despite systematic survey programmes, the collection of archaeological data on buried archaeological sites, in coastal areas as elsewhere, is often fortuitous and accidental. In material terms there may be some 400–600,000 archaeological sites in England. They range from standing monuments to earthworks to buried features and stretch over a vast period from the early Palaeolithic (circa 30,000 BC) to the very recent past, with Second World War structures protected under ancient monuments legislation. It is the buried features that represent the greatest area of uncertainty as to their total coverage; experience suggests that current knowledge covers only a small proportion of the potential sites. To complicate the picture further, some monuments stand alone while others make up part of a palimpsest or relic landscape where numerous features of several periods lie together.

Only a minute proportion of the archaeological resource is protected by scheduling under the ancient monument legislation. The sites protected are those which are judged to be of 'national importance'. There are currently some 13,000 scheduled monuments, encompassing some 18,000 sites, although it is estimated that this is only a quarter of the total that should be protected. In coastal areas of England a mere 140 sites are scheduled, nineteen of which are additionally under state guardianship. The majority of these are surface features, medieval or post-medieval in date and, perhaps not surprising, have a fair proportion of defensive sites among them.

It is clear, therefore, that current scheduling is unrepresentative through its small sample (albeit covering sites that may be of national or international importance) but some general criteria can be established as to the importance of certain categories of coastal archaeological sites, as follows:

1. Prehistoric sites are well represented at the coast but prehistoric habitation sites may be more distinctive than some other features (e.g. burials, which will also have a wider distribution further inland);

2. From the Late Iron Age to the present century the construction of coastal defences accords a high distinctiveness/uniqueness to some features;

3. Temporary and permanent sites in association with the collection of on-shore and off-shore materials, including food, make such coastal sites very important (e.g. from Mesolithic shell middens to salt-making and fish-traps);

4. Archaeological sites connected with coastal/overseas trade should be accorded a high priority. Such places often served as gateway communities for a wide inland area, and contain an unusually rich and varied assemblage of finds.

6.4.2. *The value of coastal geological sites*

Sites of geological and geomorphological interest can

be considered together as representing the earth science heritage of the country and are therefore valuable for a number of reasons:

1. Geological science started in Britain and coastal sites have played a significant role in the history of its development.
2. Some coastal geomorphological features are unique and should therefore be preserved in their own right.
3. Some geological strata only outcrop at the coast and therefore cannot be studied easily in any other locations.
4. Opportunities to study the earth's history through its rocks, fossils and minerals are disappearing as sites are being lost through development.

Overall then, there are sound scientific and educational reasons why such sites are of value. They form a range of types of site between two end points which are defined according to their conservation management needs. 'Exposure' sites, such as cliffs and foreshore exposures, are those whose value is in providing accessible exposures of a particular deposit. 'Integrity' sites are of value because they contain landforms or restricted deposits which are irreplaceable if destroyed, for example shingle bars and spits. Many of these sites also have wildlife and archaeological interest and cliff sites, in particular, may be considered an integral part of the scenery of a coastal area as a whole since they cannot be separated from the landscape which lies behind them by any clear dividing line.

In addition to the direct scientific and educational value of such sites they have an indirect value as part of a coastal sedimentation system or 'cell'. Shingle and sand features are the beneficiaries of erosion from elsewhere along the coast whilst the erosion of rock features provides material which may protect land downdrift from erosion. Proposed protection works updrift may therefore result in disbenefits through loss to the downdrift feature where this is of shingle or sand. Similarly, protection of a soft rock area may also have disbenefits through loss of sediment elsewhere.

It can therefore be seen that maintenance of the value of these two types of site differs. In the case of exposure sites, it depends on the maintenance of a clean exposure or the creation of a new one where maintenance of the original one is not feasible. Whereas in the case of integrity sites the creation of an alternative feature is not feasible and so the maintenance of the sedimentation patterns on which the existing feature depends is essential in order to retain its value.

Even where the sediment regime is relatively undisturbed, erosional changes to both soft rock and shingle or sand features may be undesirable on grounds of affecting the scientific interest of the site. Although, for soft rock deposits, some erosion is desirable as it keeps rock faces clear of vegetation and thus allows the strata to be clearly seen in the case of shingle or sand features, it is the changes in the dynamic processes of deposition and erosion that are of scientific interest as much as the feature itself. Hence any artificial stabilisation of that pattern at a particular point in time is undesirable.

In the case of exposure sites of palaeontological significance, some erosion is also desirable as it leads to fresh exposures. Since the importance of such sites is the potential for the discovery of fossils which, from a scientific perspective, substantiate or fill gaps in the existing fossil record, the value of erosion is partly dependent on the abundance of fossil hunters as material may otherwise be removed by the sea and lost.

6.4.3 Coastal landscapes: their significance and categorisation

Coastal scenery is highly valued: witness the large numbers of people who visit the coast each weekend to sit in their cars and look at the view. It is worthwhile for society to protect, therefore, these amenity values.

Coastal landscapes can be degraded by floods and erosion, and by coast protection and sea defence projects. Many of the controversial coast protection and sea defence projects in the past have centred around the loss of visual amenity from hard structures which impose themselves and disrupt the view (Parker and Penning-Rowsell, 1982). Moreover it is the 'soft' coastlines that are prone to erosion that also have the landscapes that are more vulnerable and more likely to be disturbed by coastal protection and sea defence works. This is because those works may feature as large elements in an otherwise flat coastal landscape. On the other hand, hard coastlines with cliffs, embayments and rocky foreshores are unlikely to need coast defence works to spoil the grandeur of their scenery.

But very little systematic research has been undertaken on landscape preferences at the coast, in contrast to other rural landscapes (Penning-Rowsell, 1981; 1989; Uzzell, 1989). Evaluating those research results for 'normal' rural landscapes, however, shows that the factors that promote this prediction of landscape preferences include landscape coherence, legibility, complexity and mystery; it is the last that is found, empirically, to have the greatest explanatory power, and complexity to have the least significance (Kaplan and Kaplan, 1990).

Familiarity is also important (Penning-Rowsell, 1982). Direct experience and knowledge of a place can clearly affect preference, although it is less clear, or predictable, what the effect may be. But knowledge of a local scene does not ensure increased preference, and

134 THE ECONOMICS OF COASTAL MANAGEMENT

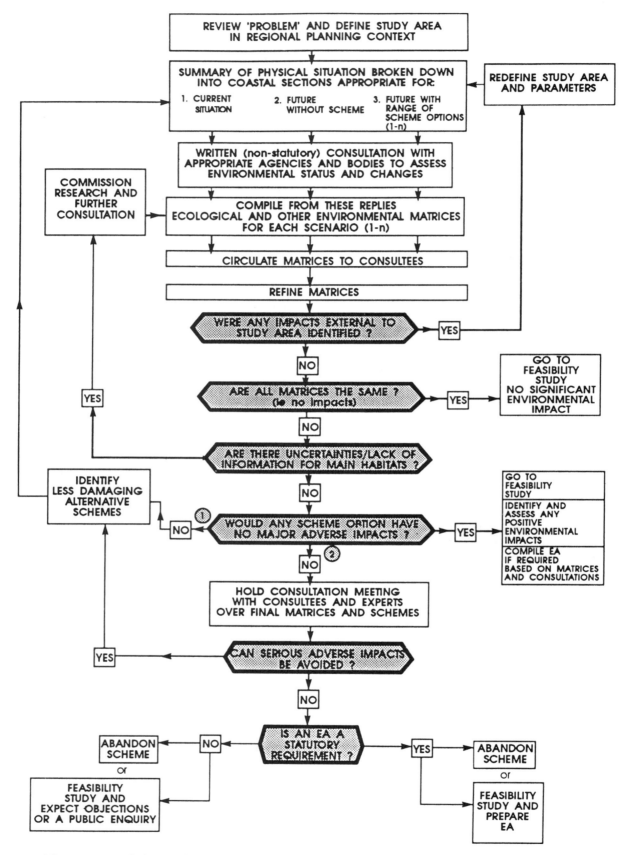

Figure 6.2 The recommended procedure for consultation on the environmental effects of proposed coast protection and sea defence projects

the novelty of foreign places may enhance it. The more elderly prefer pastoral scenes while adolescents prefer landscape with action potential: shops and buses! Those with specialist knowledge of landscape features have markedly different preferences and perceptions than those without such knowledge, and landscape professionals differ in their perceptions from lay people, but usually do not recognise those differences.

Natural environments are often those that are aesthetically preferred, because the content of nature contributes aesthetic experiences, and because the patterns and rhythms of nature promote attraction. Natural settings often provide mystery, which is also attractive, and also coherence, which gives satisfaction. The aesthetic factors in natural landscapes give pleasure, therefore, and they also support human functioning. They provide a context in which people can manage information more effectively, and they promote exploration with confidence and comfort and foster the recovery from mental fatigue.

6.5 The recommended procedure for incorporating environmental values into decision-making

Because of the wide range of environmental interests that may be affected by coast protection and sea defence projects, it is important to have a systematic way of bringing these interests and aspects together in order to assess the environmental effects of a project.

6.5.1 Steps in the evaluation process

Figure 6.2 gives the recommended procedure for consultation on the environmental effects of proposed coast protection and sea defence projects. This consultation should include consideration of all potential components of environmental value of a site. The procedure should be implemented in every case of a project proposal, whereas statutory Environmental Impact Assessment (EIA) procedures are only relevant in those cases which fall within the requirements of the EIA guidelines.

This consultation procedure should be initiated at the pre-feasibility study stage, for two reasons. First, there is the need for full liaison with those representing environmental interests from an early stage, and, secondly, there is the likelihood that inadequate environmental information will exist on which to base an evaluation of project effects, thus necessitating further information gathering. The steps in the procedure are outlined below.

1. The first step in the procedure is the identification of the problem and definition of the study area. The procedure may be applied both to a project proposal at a specific location or in the context of the need to develop a regional or sub-regional strategy for coast protection and/or sea defence. The nature of the problem and/or the scale of study area are first defined in engineering terms by the responsible Coast Protection Authority (usually the Local Government Authority or the National Rivers Authority (NRA) region).

2. An engineering survey should summarise the physical characteristics of the coastal site under three headings, as follows:

    - the current situation – there may or may not be existing coastal protection works in place;
    - the future situation with no coast protection or sea defence project – there will be continued environmental change resulting in erosion/accretion/flooding. These may be considered desirable or undesirable from different environmental viewpoints; and
    - the future situation – there will be a range of possible project options.

3. At this stage, and before proceeding further with outline project proposals in engineering terms, it is important to identify the various components of environmental value which should be considered, and this may require the advice of the appropriate environmental specialists. The next step, in any case, is to consult specialists to obtain advice in order to assess the current status of the area and its possible vulnerability to change.

4. The clearest way to summarise this potentially complex and diverse information is in the form of three, or more, environmental matrices for each of the components of environmental value, as in the example of Hengistbury Head (Section 8.6). This should either be done by in-house environmental expertise of the proposing agency or, more usually, by specialists in each field.

5. This stage should highlight any areas of missing information which prevent the drawing up of matrices. It is then necessary to consult more widely to ascertain whether such information exists or to commission further studies to enable data gathering from empirical sources, where the absence of such information is potentially significant.

6. The matrices should then be circulated for comment and refinement among all relevant consultees.

7. Environmental advisors will also be able to indicate whether there are any impacts external to the study area (i.e. 'knock-on' effects to other sites, for example as a result of a reduction in sediment supply). If

this is the case, it will then be necessary to redefine the physical parameters of the study area before proceeding further.

8. The matrices should then be compared to see whether or not there are any environmental impacts with the proposed project. The same questions should be asked of all environmental advisors in parallel concerning any relevant components of environmental significance (e.g. archaeology, geology and ecology).

9. In the unlikely case of all matrices being the same (i.e. a project with no environmental impacts), then the next stage is to proceed to a feasibility study of the project. Where environmental impacts occur of such a minor nature as to be judged unimportant then the study can also proceed to the feasibility stage. This clearly raises the issue of what constitutes a significant or minor project impact and who should be responsible for making such judgements. This can only be resolved through negotiation with the appropriate interests in each individual case.

10. A more likely situation would be that the matrices will show some environmental changes resulting from the project. It is then necessary to identify whether all or any of the project variants would have major adverse impacts.

11. Where a comparison of matrices shows uncertainty concerning the prediction of change for any of the main habitats, the question of how to handle a lack of information is raised. If the inadequacies of existing information are such as to make the prediction of project effects impossible, the next step must be to carry out further ecological research. It is clear that the need for this should be identified early on in the appraisal process, to save delays at a later stage, especially where seasonal considerations constrain survey work. The scope and scale of such research will depend, in practice, on the scale of the coastal protection works proposed. The larger the project and the more extensive the area, the greater will be the necessary lead-in time for the project.

The stage indicated as 'further research' may involve no more than collecting together all the existing information about the area from all those agencies and organisations with data and observations to contribute. For example, in the case of Hengistbury Head, the timescale given by the Local Authority was such that two site visits, consultation with the Nature Conservancy Council (NCC), the local wildlife Trust, a Natural History Society, a local ornithological group and the warden of the Local Nature Reserve were all that could be carried out. Predictions had to be made on the best informed estimates, and from available secondary source descriptive material.

In a larger project or one where the ecological importance of the area was more clearly recognised at the outset, further research could involve more detailed investigations, primary fieldwork, and predictive modelling of ecological impacts. In any case, further investigation would enable the matrices to be completed in a more detailed way, and ideally would ensure that predictions of change reflected the sum of best available information at that point in time.

12. With this more comprehensive information, it should then be possible to identify whether any project variant would be without major adverse impacts. If so, a feasibility study of that specific project variant would follow. At that stage it is necessary to ask whether the project could also have a positive ecological impact, such as the creation of new habitats by altered deposition and, if so, to evaluate these changes as benefits of the project.

13. In the event of all proposed projects having adverse impacts, it is important for environmentalists and engineers together to attempt to identify less damaging alternatives. This would apply either to the coast protection or sea defence strategy overall, or to ways of implementing any existing project proposals.

If such an alternative can be identified, this would require the procedure to be followed through for a second time when, ideally, a project could be produced which would avoid any serious impacts. The main project feasibility study could then be undertaken with confidence that any environmental impacts had been fully recognised and taken into account at the outset.

14. A more complex situation arises where different project alternatives differ in their impacts on the various components of environmental value. It may be impossible to identify which is the least damaging project, in environmental terms, in the absence of a common numeraire. The only solution available here is to attempt to reach agreement between the representatives of the environmental interests concerned, through discussion and trade-offs. This is, in essence, a process of consensual evaluation.

15. Whichever project alternative is finally selected, if serious impacts to any component of environmental value appear to be unavoidable despite negotiation, and the coastal protection authority does not wish to abandon plans for a project, all that may be possible is to note the likelihood of contention over the project and the possible objections or need for a

public inquiry before proceeding.

A consultation procedure such as the one given above should meet the objectives of the need for a co-ordinated environmental input, in a summary form, easily comprehended by the non-specialist and at an early stage in the decision-making process.

6.5.2 Ecological bodies to be consulted

Statutory consultees, under the Wildlife and Countryside Act 1981, are English Nature and the Countryside Commission, for England, and the Countryside Council for Wales (before April 1991 these were known as the Nature Conservancy Council and the Countryside Commission). The regional officer would normally be consulted.

The County Ecologist, if there is one, should also be consulted at any site. Information about local designations and the incorporation of sites within local plans, as well as local interest groups involved at a site, can usually be obtained. An alternative title for the officer responsible for ecological matters may be Countryside Officer or Conservation Officer (although this can refer to responsibility for historic buildings). The Council department concerned may be Planning, Parks and Recreation, Countryside, Museums and Conservation or Leisure Services: there is frequently no distinction made between responsibility for informal/formal recreation and nature conservation. District Councils may also employ 'countryside' staff.

Other bodies which may be involved at a site, as owners or managers, include:

Public landowners (including government departments) and private landowners;
The Royal Society for the Protection of Birds (RSPB).
The National Trust;
The Wildfowl Trust;
The Marine Conservation Society;
The County Naturalists/Conservation/Wildlife Trusts, which are represented collectively by the Royal Society for Nature Conservation (RSNC);
Local Naturalists Clubs or Natural History Societies, not necessarily affiliated to any national organisation;
The British Trust for Conservation Volunteers (BTCV); its affiliated groups may be involved in voluntary site management; and
Other local unaffiliated voluntary management groups.

Information about the local involvement of the various societies should be obtainable from the statutory consultees, from the Local Authority ecologist or Countryside department, or from local library and information services.

6.5.3 The procedure for the assessment of archaeological value

The archaeological resource is diverse and complex and the assessment of archaeological sites must continue to rely on the consistent application of professional archaeological judgement. It is difficult to see how a benefit-cost procedure can be applied to archaeological sites, any of which may contain unique information.

In order to identify existing sites and future potential, reference should be made to one or more of the following:

1. English Heritage (formerly, the Historic Buildings and Monuments Commission). The Conservation Group of the body maintains a record of scheduled monuments.

2. The Royal Commission on Historical Monuments. This body maintains a National Archaeological Record but detailed queries, especially in terms of cultural resource management, should normally be referred to the County Archaeological Officer. RCHM can also supply useful aerial photographs of many coastal sites.

3. The County Archaeological officer. Most counties now employ an archaeologist (or a team of archaeologists), either within their Planning or Conservation sections or as part of an established Archaeological Unit. Each has responsibility for a Sites and Monuments record in their area.

Since 1988, county archaeologists have been adopting a methodology developed for internal use by the Monuments Protection Programme of English Heritage (Section 6.6.2). This uses a point scoring system applied to a standardised set of assessment criteria and aimed at improving the consistency of professional judgements.

6.5.4 Assessing geological value

Key geological sites are identified in the Geological Conservation Review (GCR) which identifies and assesses all the most significant earth science sites in Great Britain. These sites are either already designated as SSSIs or it is anticipated they will be by 1996. It is necessary to be aware, therefore, that not all sites of geological significance will be currently designated SSSIs for similar reasons to those identified for ecological sites (Section 6.2.3). The GCR classified sites according to which of 97 subject areas they fall into (Nature Conservancy Council, 1991). These areas are defined according to stratigraphic time periods and subject divisions such as igneous and metamorphic and structural geology, palaeontology, mineralogy and geomorphology. The coastal geomorphology subject area is divided into four blocks: coastal geomorphology of England, Scotland, and Wales, and saltmarsh geomorphology.

The criteria used to assess the value of these sites are explained in detail in the introduction to the Geological Conservation Review (in press).

There is also a network of Regionally Important Geological and Geomorphological Sites (RIGS) which are non-statutory sites. These are important for their educational role and are also used for amateur study and for leisure activities. Such sites are generally widely accessible and so have a significant value in stimulating general interest in earth sciences as well as their role as a local amenity. They have a different function from the SSSIs and should not be regarded as second tier sites in any sense.

Information about the value of earth science SSSIs in England is available, in the first instance, from the appropriate office of English Nature (successor to the Nature Conservancy Council in England). The precise justification for the designation is set out in the notification statement. Similar information is held by local offices of the Countryside Council for Wales. Site information may be available in the form of a computerised database of the National Scheme for Geological Site Documentation (NSGSD).

For sites which are not of national importance, but where local RIGS groups exist, they will keep information about sites of local importance.

Other sources of specialist information are:

- the Geologists Association;
- the Geological Society and the Geological Curators Group; and
- the British Geological Survey.

For other sites, if data is not already available, a survey and documentation of the site will be necessary.

6.5.5 Approaches to the assessment of landscape values

Landscape is important to the public, but landscape evaluation remains controversial (Lowenthal and Penning-Rowsell, 1986; Countryside Commission, 1987) because it is essentially subjective, and replication levels tend to be low. Hedonic methods have been attempted (Li and Brown, 1980), but these have not resulted in rigorous or practical methods of assessing landscape values. Nevertheless there are some techniques that do assist in the recording of landscape features that people find attractive and important in landscape terms.

In addition to considering the use of these techniques, the Countryside Commission for either England or Wales should be consulted about national and local landscape designations in the areas where there may be landscape changes as a result of coast protection or sea defence projects (e.g. Areas of Outstanding Natural Beauty; Areas of Great Landscape Value). Other sources of opinion on these matters are the many interest groups concerned with landscape, of which the most important is probably the Council for the Protection of Rural England.

The most common approach used to assess landscape values in the past has been to devise measurement techniques that assess the attractions that landscapes have using surrogate combinations of variables such as land use and topography. As such these approaches are more concerned with the appearance of landscapes than with their historical or cultural features and associations, and therefore they cannot be considered either complete or comprehensive. In addition, the techniques tend to emphasise the importance of landscape diversity, and this may not be fully appropriate or relevant in coastal situations where the landscapes that are most vulnerable tend to be somewhat monotonous (for example, saltmarsh and estuarine landscapes).

Nevertheless these approaches can be used to focus attention on the landscape features that may change or disappear with coast protection projects, and also be used as part of visual displays to assist in public consultations. In essence they are a guide to systematic description, rather than useful as good predictors of public landscape preferences. However, this description of likely change in landscapes with coast protection and sea defence projects is a useful first step in the evaluation process.

The range of techniques that is available has been developed for rural landscapes (Penning-Rowsell, 1981; 1989) and these could be used, with adaption, for assessing the aesthetic value of coastal landscapes (Section 6.6.4). The simpler techniques involve field recording and the ranking of the importance of landscape features, while the most complex relate these features in different ways to public preferences for particular sample landscapes.

Another approach is to undertake questionnaire surveys of the public to gauge directly their valuation of the landscapes concerned (Penning-Rowsell, 1982). These surveys could determine what landscape features they value, and what would be the impact of flooding and erosion, and of coast protection and sea defence projects. As such these techniques are part of the participative approach to project feasibility study, and gives a direct input to the decision making of bodies acting on behalf of the public.

6.6 Methods and techniques

6.6.1 A matrix method for assessing ecological value

A multi-criteria and habitat matrix is the simplest

ENVIRONMENTAL GAINS AND LOSSES

	HABITATS													
	Marine[1]	Maritime[2]								Terrestrial[2]				
CRITERIA		Mud-flats	Saltmarsh			Vegetated shingle		Dunes			Lagoons	Cliffs	Fresh water	Grass-land	Heath	Wood-land
Size of habitat[3]																
Size of sub-habitats[4]			L	H	B	P	H	M	F	S						
Combinatory value[5]																
Physico-chemical specialisation[6]																
Naturalness[7]																
Rare species[8]																
Populations of Interest[9]																
Mammals																
Birds																
Other vertebrates																
Invertebrates																
Flora: flowering																
non-flowering																
Specialist research interest[10]																
Rare/declining habitats[11]																

L = Low   P = Pioneer   M = Mobile
H = High   H = Shingle heath   F = Fixed
B = Brackish   S = Slacks

Key to Habitat-Criteria Matrix

[1]Marine: The marine habitat may be defined as below low water mark and other inter-tidal habitats not included in categories of maritime habitats. Classification of marine habitats is more complex than can be represented on the matrix. For details consult: *A Coastal Directory for Marine Conservation* (Gubbay, 1988) and/or the Marine Conservation Society.

[2]Maritime and Terrestrial: Habitat divisions based on Ratcliffe (1977). Consult English Nature/Countryside Council for Wales, Local Authority ecologist and any management body for the site. These organisations should have information on any other bodies which may be involved at the site.

[3]Size of habitats: Habitat size category does not depend on actual extent at the site but is assessed relative to the size range of that habitat in general. Guidance on this for maritime and terrestrial habitats is given in Ranwell's semi-quantitative index (Ratcliffe, 1986). Other habitats are more complex, e.g. lagoons have several alternative categorisations. (Gubbay, 1988; Barnes, 1989).

[4]Size of sub-habitats: Where applicable, habitats may be separated into sub-habitats, if the information is available. NB: Habitats which by their nature have no sub-habitats are not necessarily of less value.

[5]Combinatory value: Applicable where site is adjacent to other similar habitats, either immediately adjacent or as part of a sequence with interbreeding populations or migratory species movements.

[6]Physico-chemical specialisation: indicates rare or unusual conditions of geology, soils, sediment, hydrology or geomorphology.

[7]Naturalness: Based on a visual, on-site, assessment of relative degree of disturbance.

[8]Rare species: Determined according to criteria for rare species at International, National, Regional and Local scale.

[9]Important populations of interest: Determined according to International, National, Regional and Local importance. NB: Information may not be available for all species groups.

[10]Specialist research interest: Related to amount of recorded history of past research and amount of on-going research.

[11]Rare/Declining habitats: Determined according to extent of decline; whether International, National, Regional or Local.

Figure 6.3 Habitat-criteria matrix for evaluating ecological status of project affected area (with and without proposed projects)

method of summarising ecological site value. The information required by the engineer and planner for coast protection and sea defence projects needs to be presented in a form in which the value of a site is summarised in language easily understood by the non-ecologist. Inevitably a level of ecological detail will be lost, but the matrix is not intended to preclude a full ecological study, if one is required. A blank example of the matrix along with definitions of the categories used in the cells is given in Figure 6.3.

In order to highlight features of particular site value, whilst also including other aspects of value, visual indicators are used in the cells, where applicable (Section 8.6.5). The visual indicators represent a three point scale: for example, large, medium, and small; or international/national, regional and local. As already discussed (Section 6.3), the use of a numerical scale is not appropriate in view of the different biological attributes of the different habitats and numerical difficulties.

Subjective opinions will still arise in any assessment, but the standing of many species and habitats can be recognised at the above three levels and a larger number of categories would introduce a spurious appearance of precision. In addition, knowing the relative scale of importance can be of practical value to engineers and planners for assessing site value, and also for assessing the implications of possible action. Although features of local value at an undesignated site could be regarded by ecologists as of less 'value', there may well still be a high level of significance and concern for a particular site and the effects of any changes on them should still be taken into account in a systematic way.

The matrix can also act as a checklist of information by indicating where data is not available or is in insufficient detail for the matrix to be completed. If this is the case, further data will need to be obtained and an ecological survey may need to be carried out.

Information on whether a site is designated (or not) has not been included in the matrix because this is information which would be obtained in the normal course of site investigation. Moreover, as non-designation does not mean a site is without ecological value, this factor should not be used as a criterion of itself. If a site is already designated then the information on its ecological significance should be readily obtainable. If it is not designated then it is likely that local naturalists groups may still know about or be involved in keeping records at a site (Section 6.5.1).

Habitat categories The matrix has been designed to include the various habitats which may be present at any site. Marine habitats may appear to be affected by a comparatively limited number of projects, for example those involving dredging, or offshore constructions. But these are only direct effects. Indirect effects by changed patterns of sediment supply could also affect a marine site, so consultation with the Marine Conservation Society is always desirable.

Lagoons may occupy an ambiguous position. Whether some lagoons should be considered as a marine habitat is dependent on which authorities are used, as Barnes' (1989) list of 41 lagoons does not agree with other interpretations of what is meant by a lagoon (e.g. Gubbay, 1988). They have been placed in the category of maritime habitats, with a note that other intertidal water bodies should be included in the marine category.

The maritime habitats listed in the matrix follow those of Ratcliffe (1977) in the Nature Conservation Review as being the simplest and most generally known and accepted classification. Although there may be boundary or descriptive problems with categories of habitat, a more complex system would be unacceptable in terms of the uses to which the matrix will be put (being interpreted by engineers and planners, rather than ecologists).

Terrestrial habitats are listed in four categories, although it is recognised that there may be variants within the categories, and gradients between them.

Criteria As a single score or criterion for site assessment is inadequate as an indication of value (Section 6.3), several commonly accepted criteria have been used. These are discussed below.

Using the absolute **size** of each habitat can only be appropriate where comparisons are being made within single habitat types. Otherwise it is likely that habitats which are by their nature small could be regarded as less important even though any one site may be one of the largest of its habitat type.

The size grades developed by Ranwell (Ratcliffe, 1986) have been reduced to fit the matrix scale, but these are only suggested for some maritime habitats. Size categories for lagoons depend on the lagoon sub-type – some very small lagoons may be of particular significance.

Habitat **diversity** is not included as a separate criterion, but where several habitats and sub-habitats are present this will be indicated by the larger numbers of visual indicators present in the top row of the matrix.

The number of **sub-habitats** included has been reduced to a minimum for simplicity. The system used for Sites of Special Scientific Interest (SSSI) assessment based on the National Vegetation Classification (Nature Conservancy Council, 1989) has too many subdivisions for this simplified summary of value, but could be used for a full-scale ecological survey.

In some instances the examples of succession provided by within habitat changes can be one of the main points of significance at a coastal site. However, for other habitats, such as mudflats, their homogeneity may be one of their main characteristics. It must always, therefore, be stressed that sub-habitats are only of value where they might be expected to be present, and a habitat without different sub-habitats is not necessarily of lesser value.

The **combinatory value** of a site, that is its ecological and spatial relationship to other sites of significance, is of particular importance at the coast where fragmentation of sites due to development is a common occurrence. It should be pointed out here, however, that sites isolated by urban and other development may be of great value locally, and that consequently such a site could give rise to strong public feelings apparently out of proportion to its ecological value.

A site's **physico-chemical specialisation** may be due to several factors, and these should be itemised and summarised, where present, in the text or as footnotes. This aspect of a site's value may be especially liable to change at the coast – for example, changing the hydrographical regime from more specialised brackish to salt as a result of seawater flooding, or to freshwater after the construction of impermeable coast protection works.

Site **naturalness** is a measure of the level of disturbance at a site. Lack of disturbance directly by people, rather than by past human activities, has been stressed so that unimproved grazed marshes could be included in the top category. A site degraded, for example, by overgrazing or overtrampling by animals would obviously be in the lower category. The word 'structures' was used in the definition of visual indicators, rather than buildings, so that, for example, telephone or electricity pylons, which would require regular access for maintenance, would be included.

This criterion has not been used here in the same sense as in the Nature Conservation Review (Ratcliffe, 1977) as unmodified vegetation not altered by human activities. The less restrictive definition has been adopted for several reasons. First, this criterion would be rated differently according to the area concerned. Secondly, it may also be difficult to assess the degree of past human intervention. Finally, patterns of grazing, drainage modifications, or even old borrow pits for coast protection works, can increase diversity of habitat and species and produce habitats of significance and value.

Any **rare species** may have legal implications under the Wildlife and Countryside Act 1981. It is also one of the main questions asked by the general public about a wildlife site. The species concerned should be itemised separately in the text accompanying the matrix.

Important **populations or communities of significance** have been grouped into broad categories, which could be added to if required. Mammals can be land and/or marine based. Non-flowering plants are important at the coast and include marine algae, and also pioneering lichens and bryophytes which can disappear where a site is stabilised by coast protection works.

A complete list of taxonomic groups is not essential as it obscures the important issues. A more comprehensive list would be drawn up for an ecological survey.

A site's **specialist research significance** was certainly held by Ratcliffe (1977) to enhance its ecological value. Since one of the major features of coastal sites is change, then knowledge of the changes and processes of change over time can only increase a site's value. The destruction or damage of a studied site could therefore be considered to cause a greater loss than a similar loss at an unresearched site.

Here, we are recommending that educational use of a site, for example by school and college students, should not be considered in the same category as research. Educational use could be better described as a use value (and possibly valued by other methods). Also educational use, by encouraging trampling and the collecting of specimens, may well be detrimental to a site's ecological value.

Any **rare and declining habitats** are an important factor when considering the value of habitats. Some rare habitats may be protected by legislation under the European Community's Habitats Directive. The *Guidelines for Selection of Biological SSSIs* (Nature Conservancy Council, 1989) gives some parameters for rare habitats at a national level, and considers that any habitat with a total area under 10,000 hectares (ha) can be regarded as rare. They would include at present the following: coastal heath over 10 ha, although even smaller fragments may be of national value if they also have high physico-chemical specialisation; lagoons over 0.5 ha; larger shingle features over 25 ha. Unimproved grazing marsh is among the most rapidly declining habitats.

Application of the matrices The matrices should be used, first, to identify the components of ecological significance at a site under current conditions, secondly to indicate their significance and, thirdly, to highlight the changes which may be expected to occur with erosion/flooding in the absence of a project and with a coastal protection project (or range of project scenarios). Thus three or more matrices will normally be produced; the first showing the features of ecological significance under current conditions, the second showing the ecological effects of future changes without a project such as those that are predicted to occur with continued erosion/flooding, and the third (or further matrices) showing the changes with the proposed project or project variants.

The occurrence of features of significance is indicated by the presence of a visual indicator in the appropriate cell of the matrix. The significance of the feature is represented by the size of indicator used and the predicted changes with and without a project can be highlighted by using coloured overlays where appropriate or by shading the cells affected.

Comparison of these matrices enables a relative evaluation to be made of the changes in ecological significance under different conditions. An example of the use of the matrices is given in Section 8.6 in the case study of Hengistbury Head (Figures 8.6.4–8.6.6). The three basic matrices are, as follows:

1. A matrix of ecological significance of the site under existing conditions. This should be completed by, or with the advice of, a specialist coastal ecologist. Cells which cannot be completed because of missing data should be clearly shown as such, as this may indicate where further ecological information is required;
2. A matrix of changes in ecological significance without a coastal protection project. The main changes which would occur need to be predicted by a coastal ecologist. In the event of a lack of agreement over predictions, different scenarios could be presented using matrices in parallel; and
3. A matrix of changes in ecological significance with a coast protection project. These should be highlighted by changes in the visual indicators in the appropriate cells. Losses or gains may be represented.

The use of visual indicators in the matrix gives a way of presenting information about the site as a whole in a simple visual form in which it can be easily assimilated. It is important to stress that the value of the matrix is for *comparative* purposes only, to enable the significant changes to be clearly identified in order to make a judgement about the overall vulnerability of the area to changes, with and without a project.

In one sense the matrix is only as useful as the accuracy of the information which is used to compile it. In particular, accurate predictions of ecological change under different conditions are particularly difficult to make.

In addition, however, the matrix method enables gaps in available information to be identified and rectified where possible. Where cells cannot be completed due to lack of available information it is suggested that a convention is adopted to indicate this so that areas of uncertainty are clearly identified. Ideally, such missing data would be collected before the completion of the matrices, but in reality this may not always be possible due to seasonality, etc. An important advance with the use of these matrices is that, if necessary, decisions can still be made on inadequate information but at least the limitations of the decision will have been made explicit.

6.6.2 A method of archaeological assessment

The monument evaluation method devised by English Heritage uses a simple accumulative scoring system applied to a number of set criteria, as outlined below. This forms part of the judgement process of comparing sites belonging to the same monument class and its prime purpose is to help define a threshold of national importance for particular monument classes (and hence provide further objectivity in scheduling). The overall procedure, however, continues to rely heavily on professional judgement and other criteria (e.g. rarity) are considered when comparing different classes of monument. The criteria work effectively for single monument evaluation but a satisfactory system for relict landscapes remains problematic, and the criteria should not be applied to areas with combinations of monuments.

Each criterion is rated on a simple 1 to 3 scale, 1 being 'below-average' for the class, 2 being 'average', and 3 being 'above-average'. Each monument is assessed against the characteristics of the class as a whole, at a county or regional level. In order to give greater dispersion of the figures each criterion value is then squared before a final total is made. The higher the total, the higher the value of the particular monument.

All criteria are rated 1 to 3, except for number 6, below (Group value, clustering). This is only rated 1 or 2, being regarded as less significant. The standard criteria are as follows:

1. *Group Value (association)*. Any monument will have an enhanced value if it is associated with other monuments of other classes;
2. *Survival*. The relative abundance of a feature in its above or below ground state may be an important factor in signifying importance;
3. *Potential*. Many sites are poorly explored and have considerable potential for further analysis. Greatest potential equals greatest value;
4. *Documentation (archaeological)*. Some sites will already have had the benefit of some small scale excavation, indicating further potential;
5. *Documentation (documentary)*. This is normally only important for monuments of recent origin although some sites may have been recorded by antiquarians;
6. *Group value (clustering)*. Within each class some monuments will have a tendency for clustering or dispersal. Variation from the norm will give the higher score (i.e. 2);
7. *Diversity (features)*. Most monuments have a number of component parts or features. Some sites will contain a higher diversity than normal, thus enhancing value;
8. *Amenity value*. This relates to the value of a monument to the community, for example in an educative role, through its accessibility, or because of its contribution to archaeological knowledge.

6.6.3 Techniques for assessing the value of geological features

'Integrity' sites are the most difficult to assess since they are, by definition, finite and irreplaceable and it is unrealistic to attempt to put a value on their 'uniqueness'. The cost of maintaining them will relate to the cost of preserving the deposit intact by restricting changes due to human activities and, in the case of landforms, the cost of preventing a change in erosion or sedimentation patterns in the vicinity (Section 6.5.4).

In the case of 'exposure' sites where the maintenance of an exposure is required, all geological and geomorphological sites can have considerable use value for educational purposes. In addition to specialist training and fieldwork for higher education, the National Curriculum for schools is anticipated to generate several million visits by school children to such sites. For a particular site, some idea of its value can be obtained by evaluating the loss of its geological significance in economic terms. This can be done indirectly by taking the additional cost per educational user of undertaking a visit to a site of equivalent geological significance. The replaceability of the site will clearly vary depending on the level of specialisation of the educational user.

In the case of rock strata subject to erosion, protection strategies are available which slow down erosion but still allow the maintenance of the clean exposure (Hydraulic Research, 1991.) Alternatively, it may be possible to maintain a clean exposure through other means such as an inspection chamber built into the protection works or a vertical shaft cut into the rock (Earth Science Conservation, 1987). The cost of such measures can be taken as a variant of the 'shadow project' approach to economic valuation.

In the case of 'type' exposures, a strategy to maintain the existing exposures is likely to be required as 'type' sites are the key reference sections for geologists. Unless such works are undertaken then strong opposition to the scheme can be expected on geological grounds. The cost of the works necessary to permit easy inspection of the existing exposure can be used for shadow project pricing.

In addition, any anticipated disbenefits of reducing sediment supply by slowing erosion must be considered, and the cost of supplying replacement sediment should be included in the benefit-cost analysis.

No such valuation strategy exists for shingle and sand features. Due to commercial sand and shingle extraction, together with other human activities, sites in a near natural condition are becoming rarer and consequently, application of the 'shadow project' approach is unlikely to be possible.

Palaeontological sites will also have a recreational value associated with amateur fossil hunters. This can be valued in the same way as other recreational uses (Section 4.5). In addition, recovered fossils have a market value. Thus, it is potentially possible to estimate the value of the reduction in supply of fossils recovered in consequence of a reduction in erosion at a given site. Such surrogate values will not, of course, reflect the whole geological value of any site.

6.6.4 Techniques for assessing the value of landscape features

The procedure common to most of the landscape evaluation techniques that have been devised (Penning-Rowsell, 1991) involves the following stages:

1. Divide the area liable to be affected by the coast protection or sea defence project into **landscape units.** These can either be topographic units (based on valleys, ridge lines or other features that alter and interrupt views) or grid squares;
2. Record the **land use**, topography and **landscape features** in these landscape units, including the characteristics of the view out from the units as well as the character of the landscape in each unit; and
3. Use a **rating scale** such as that devised by Tandy (1971) to assign values to the landscape in each unit. This is a subjective process and should not be seen as anything but that, although there is some evidence (Penning-Rowsell, 1982) that there is a fair measure of correlation between the values thus assigned and public preferences for particular landscape types.

The procedure should be repeated for the 'with project' and the 'without project' situations. It can also be used to complement public consultation exercises. More complex evaluations should involve questionnaire surveys of the local and visiting people likely to be affected by a project, using sketches and photographs, to gain an assessment of the public's evaluation of the likely landscape changes. This approach is clearly amenable to being incorporated into any user or residents surveys of recreation benefits (Chapter 4), in which case the scenarios presented in interviews should be drawn up to reflect wider landscape impacts (where these are predicted) and not just narrow views of seafronts.

However, we know that this type of landscape evaluation is relatively undeveloped and the results lack rigour. They should be used with caution, and more as an aid to focusing attention on landscape change rather than in producing definitive evaluations of the benefits or losses from those changes. In this way it is important that these losses or benefits should not be ignored just because the techniques available for their assessment are as yet rudimentary.

6.7 Summary, assessment and guidance checklist

This chapter has outlined the complexities of measuring gains and losses resulting from environmental change. It has highlighted the difficulties of developing acceptable methods for valuing, in economic terms, the environment generally, and the changes resulting from

**Table 6.3
Guidance checklist of steps in assessing environmental values**

1. Identify the nature of the project. A project proposed for a specific location or a regional strategy for coast protection or sea defence may be involved. The physical characteristics of the site: (a) in the current situation, (b) in the future with no protection project, and (c) in the future with protection, should each be considered. The implications of the 'do nothing' scenario should be given particularly careful attention.

2. Identify the components of environmental significance. The statutory consultees, national and local groups should be contacted to establish the presence of features of significance.

3. Consult all the relevant bodies representing environmental interests. A preliminary meeting to introduce the proposed project and discuss the nature of possible environmental effects would be useful at this stage.

4. Predict the changes that will result from the project as accurately as possible with the available information. This should include indirect changes to other sites, for example those areas 'downdrift' which may be deprived of sediment.

5. Assess the information in comparable forms, for example using matrices similar to those for the assessment of ecological value.

6. Assess the need for further research to enable more accurate prediction of change. This could be in the form of ecological surveys or other environmental research, or hydrological or other research in relation to the physical characteristics of the project.

7. Identify the gains and losses in relation to each of the environmental interests at the site.

8. Identify areas of possible conflict between different environmental interests where a gain in one aspect of environmental value can result in loss for another.

9. Meet with representatives of appropriate environmental interests to discuss and decide upon any trade-offs which can be made to help resolve conflicts. If consensus cannot be reached the project could either proceed in the knowledge that objections or a public inquiry may result, or it should be abandoned.

---

coast protection and sea defence projects specifically. The key steps recommended in this Chapter are given as a guidance checklist in Table 6.3.

The key problems identified are as follows:

1. A lack of appropriate **theoretical underpinning** to reflect society's values in any terms other than as the sum of individual valuations;
2. A lack of an appropriate **methodology** for expressing environmental values in economic terms;
3. A lack of **agreement** between environmental interests not only on how to proceed but indeed on whether we should be proceeding along this route of valuing the environment in economic terms at all.

6.7.1 Theoretical and methodological limitations on environmental valuation

If Contingent Valuation Methods were to be used to value environmental change, the grossing-up of such values would still not reflect the wider components of society's values such as non-use values. Progress on this issue is extremely problematic.

Moreover, Contingent Valuation Methods can only be applied satisfactorily where a recognised population can be identified. In the case of recreational users – both existing and potential – the CVM can be applied through the use of resident and user surveys. However, in the case of environmental valuation, it would be misleading to limit valuation to current users as this would take no account of values that relate to non-use or to option or bequest values for future generations. Little progress has been made on this aspect although the household survey referred to in Section 2.6.5 is one attempt to obtain valuations from a wider sample of the population.

6.7.2 Limitations arising from lack of agreement between environmental interests

These limitations create both practical and conceptual problems for those trying to value the effects of environmental change.

The need for a consensus between environmentalists over whether this type of evaluation should be attempted at all and, if so, how it should be carried out, can seriously limit attempts to obtain an environmental valuation. Through our research, however, progress has been made on developing and establishing a **proce-**

dure for consultation between project proponents and environmental interests over the effects of a project on the various components of environmental significance of an area. In view of the considerable problems which have emerged in relation to measuring environmental value in economic terms, a procedural response is the only one appropriate at present.

Attempts to value the environment in economic terms simply reflect what is easiest to measure at this point in time and not necessarily what it is most important to measure. Even if economic values were given to various components of the environment, different environmental interests would be unlikely to agree about the validity of any valuation system proposed.

The production of matrices, recommended here, is an important step forward in ensuring that environmental information is presented in a clear and comparable form for each of the various environmental interests affected by a coastal project. This approach also ensures that environmental values are made explicit.

The consultation procedure outlined here represents the process of trying to reach an implicit valuation by consensus of the impact of the changes resulting from a project proposal. It provides an acceptable compromise approach to environmental valuation at a time when no economic evaluation method can be recommended.

6.7.3 Archaeology

Where individual monuments occur at the coast, then importance will normally have been gauged using the criteria previously outlined (Section 6.6.2).

However, where the importance of a site lies in the diversity of its archaeological record, it is not adequate to evaluate it by applying these criteria and adding the scores obtained. The significance of the site in the context of the whole archaeological record must always be the subject of professional judgement and reference should be made initially to the appropriate County Archaeologist for advice. The lack of a quantitative evaluation system should not lead to the archaeological interest being overlooked.

6.7.4 Geology

Coast protection and sea defence works can have a considerable impact on geological and geomorphological sites at the coast, by making exposures inaccessible or preventing their continued erosion, and by cutting off or altering sediment supplies and hence changing the natural patterns of erosion and accretion. Various indirect methods of valuing geological sites exist for exposures which can be recreated elsewhere, including shadow pricing. For unique sites, the Travel Cost Method for education users could be taken as a surrogate value. However, it is important to be aware that these are surrogate values and do not represent the entire value of a site. For many sites which are unique, particularly geomorphological ones, there is no adequate method of evaluation and it is therefore necessary to compile as much information as possible to allow an informed judgement to be made on its significance in comparison with other scheme factors which have to be taken into account.

6.7.5 Landscape

We also know that the impact of coast protection and sea defence on the landscape at the coast can be important and some recommendations are contained in this chapter for its assessment (Section 6.5.5). However, this type of landscape evaluation is relatively undeveloped and the results lack rigour. Landscape assessment should be used with caution, and as an aid to focusing attention on landscape change, rather than in producing definitive evaluations of the benefits or losses from those changes. In this way it is important that these losses or benefits should not be ignored just because the techniques available for their assessment are as yet rudimentary.

6.7.6 Overview of environmental guidelines

Evaluating impacts of coast protection and sea defence on the 'environment' (ecology, geology, archaeology, and landscape as defined in this Chapter) is difficult. Both use and non-use values are involved.

User benefits can be both quantified and evaluated in economic terms, either as activities which have market prices or through the CVM enjoyment of visit technique recommended in Chapter 4. Where the latter is the case, the instruments used in the CVM should be designed to capture all the user impacts identified, such as landscape impacts for example. In addition there may be separate educational uses which would not be captured by the techniques of Chapter 4 (for example, school and university field courses), but which could be quantified and evaluated by similar methods.

However, ecology, geology and archaeology all have non-use components to their value to society and there is no generally acceptable means of placing an economic value on these aspects of coast protection impacts. Nevertheless these impacts are clearly very important and cannot be ignored. The methods recommended in this chapter, therefore, concentrate on a procedural approach. This is designed firstly to identify and quantify the importance of the site affected by flooding or erosion of the proposed project, and the potential environmental impacts of these changes, and secondly provides the basis for negotiating and averting any serious adverse impacts.

7 Integrated computer-based analysis

7.1 Summary of information

This chapter provides guidance on the analysis of data described elsewhere within this Manual, focusing on the computer programs that are available and those analytical methods that can be deployed using proprietary software packages. A further key point covers the phasing and integration of this analysis across the different areas of coast protection and sea defence covered in the preceding chapters of this Manual.

The case for computer data processing is simply that it is more accurate and reliable than 'hand' calculations, although the use of computers should not reduce the need for the results to be examined and interpreted critically to ensure that there are no major or minor errors that can be overlooked when computations are computerised.

Section 7.2 describes the need for different levels of analysis, and the requirement therefore for data processing models working at different levels of generality and accuracy. Section 7.3 describes and illustrates the available computer software for assessing the benefits of flood alleviation, and this section needs to be read in conjunction with Chapter 5. Section 7.4 describes and illustrates the use we have made of proprietary spreadsheets to devise an appraisal system for coast protection projects. This section needs to be read in conjunction with Chapters 3 and 4. Section 7.5 gives a checklist of points relevant to this chapter.

The case studies in Chapter 8 also contain references to computer based analysis, and also contain illustrations of the output from the different programs.

7.2 Multiple levels of analysis

7.2.1 Introduction: the range of needs

A wide range of coast protection and sea defence projects need to be appraised, including those just concerned with flooding, those just concerned with erosion, and projects which involve both flooding and erosion. For the last type, which can be very complex, some form of integrated analysis is needed.

In addition, there are several levels of analysis that are required (Penning-Rowsell *et al.*, 1988), from the most outline pre-feasibility analysis that is designed to guide decisions as to whether more detailed investigations are necessary, to the most detailed analyses where every step is undertaken with the maximum attention to detail and accuracy (Section 2.2).

Models – computational systems – are needed for all these circumstances and, ideally, they should be linked conceptually, or even operationally, so that data collected at low levels of detail and accuracy for an overview analysis is not entirely wasted when more detailed studies are subsequently shown to be necessary.

7.2.2 Planning levels

The three following levels of project planning are normal, and require somewhat different models for efficient analysis (see also Section 2.2):

1. *The investment priority-setting level* Assuming that programme allocation has been done, and therefore budgets are available, methods are required for project prioritisation, including both the following:

a) Methods to determine which projects out of a range of possible options – covering a large area such as an entire hydrographic unit (at a regional scale of analysis) or a major metropolitan area at the coast – are likely to generate the largest benefits in excess of costs, and be most socially desirable, and have minimum adverse environmental effects; and
b) Methods to determine whether *any* engineering solution to a flood or erosion problem in a particular area is likely to be economically viable, as opposed to other strategies for controlling flood damage and the loss of land to the sea by erosion.

Thus a 'high level' method is needed to 'scan' the various sea defence and coast protection problems, which are in turn investment possibilities, so as to identify those investment options that are likely to produce the greatest return from society's investment, and be within the bounds of environmental acceptability. That list will, in turn, provide the basis for devising subsequently a phased programme of more detailed studies concerned to construct a regional coastal management plan, or a capital works programme lasting many years, or both (preferably with the latter based on the former).

2. *The project feasibility investigation level* Assessment methods are required which focus on individual projects within the overall framework provided by the priority-setting exercise carried out as above. This is what we term here 'project appraisal', as discussed in Section 2.2. What is needed are methods for the following:

a) Methods to investigate in detail the coast protection and sea defence problems at a particular locality, and hence the likely benefits and costs of specific flood protection and erosion control proposals; and
b) Methods to evaluate alternative project designs and standards of flood protection and alternative standards of erosion control at a particular locality, one of these standards of service being in the form of the proposed life of the project.

Such methods for the project feasibility investigation level will need, first, to use more accurate and reliable economic and hydraulic/hydrographic data than at the investment priority setting stage. They will obviously also need, secondly, to be related to the decision-making procedures recommended in Chapter 6 concerned with assessing and gauging the likely environmental impacts of coast protection and sea defence projects. They will also have to take into consideration the timing of data collection suggested in Chapter 4 (Section 4.7), since this data collection (and that related to the material in Chapter 6) may need a long lead time.

3. *Full project feasibility investigation level* Once a decision has been made to pursue a project at a particular locality, further studies may be required to provide more detailed estimates of project impacts, benefits and costs, and in particular the phasing of the expenditure and the take-up of the resulting benefits.

This level of analysis is as much concerned with project management as with project appraisal, as the two activities become interconnected. The same points made in the paragraph above about linking data collection stages within a phased programme also apply here.

Table 7.1
Multi-level modelling methods available for evaluating the benefits and costs of urban flood protection projects

Requirement/ Purpose	Appraisal system	Data needs	Estimation accuracy	Relative study cost (per cent)
Priority-setting level	BOCDAM*	Low, limited	Low but reliable for purpose	10–20
Project feasibility level	ESTDAM** low	Relatively low for purpose	High and reliable	40–70
Full project feasibility level	ESTDAM** intensive	More data for purpose	High and reliable	100

* applicable to an entire catchment or urban area, or part thereof where there is a range of flood problems
** more applicable to individual flood protection projects

7.3 Models for assessing the benefits of flood alleviation

Research and development work at the Middlesex Polytechnic over the last twenty years has developed two systems for assessing the benefits of flood alleviation. Each has very different data requirements, and gives results with different levels of accuracy (Penning-Rowsell et al, 1988).

7.3.1 The 'high level' BOCDAM program

BOCDAM is a computer model for 286, 386 or 486 personal computers which calculates flood damage potential in river reaches, or specified coastal areas, for flood events of specified magnitudes (Green and N'Jai, 1987). The expected value of flood damages and net present value calculations are performed separately within a pre-formatted spreadsheet program (BENAL) using the flood event damage results from BOCDAM.

BOCDAM is designed for situations where there are limited resources for undertaking detailed hydrologic/hydrographic, hydraulic and engineering investigations of project feasibility, and where a preliminary estimate of the economic viability of flood protection is needed in a number of locations within a large catchment, coastal region or urban area. The model is particularly useful for large areas where limited resources generate the need to prioritise projects in order to protect some areas before others based on economic opti-

Table 7.2
Comparison of data source needs for BOCDAM and ESTDAM

Type of data	BOCDAM	ESTDAM
Land use		
Categories	Up to 10	10 to 500
Location	Allocated to nearest cross-section	Detailed grid reference
Property: area and number	Estimate from plans	Field survey plus plans
Property levels	Study area cross-sections then Monte Carlo technique to generate typical pattern of development relative to level	Property-by-property assessment based on maps and field survey
Damage data		
Depth-damage data (direct and indirect damages)	Per area of development, values for each category for 3-15 depths – standard data only	Per property/area values for each category for 5 to 15 depths – standard data plus site surveys of damage potential in selected major sites
Traffic disruption	Not assessed	SROAD.PRO program
Utility outages	Not assessed	Extra loss figures per non-flooded property affected
Hydrologic/hydraulic data		
Flood stages	Previous flood outlines then interpolate other events, resulting in flood elevations per cross-section	Detailed hydraulic study resulting in flood surface/profile
Probabilities	Expert estimate of likely exceedance probabilities of created events plus sensitivity analysis	Detailed hydrological assessment of exceedance probabilities – past events plus records

misation (Table 7.1), and where very large benefit areas are involved.

BOCDAM is therefore most useful as a 'high level' priority-setting technique which is used to devise a programme of capital investment, or to 'search' a region for areas where capital investment needs to be prioritised because the problems are particularly severe. Part of the value of BOCDAM lies in the ability of the method to determine whether it is likely to be worthwhile collecting further data necessary for project design, and whether it is worthwhile undertaking individual project-specific engineering and economic feasibility investigations, the latter using the ESTDAM or Micro-ESTDAM urban flood benefit assessment model (Section 7.3.2).

The results of a BOCDAM appraisal are generally less precise and contain greater uncertainties than appraisal using less generalised methods such as EST-DAM, but these results are sufficiently reliable both to indicate where further appraisals are worthwhile, and to identify on which areas and types of data the available resources should be concentrated within subsequent investigations.

BOCDAM has limited accuracy but also limited data input requirements (Table 7.2). The system uses readily-available cross-sectional topographic data for flood plains and Monte Carlo methods to generate synthetic patterns of development in terms of property type, location, area and floor level. Estimates taken from large scale plans of the percentage of the area which is developed for different land uses are used to generate patterns of development, while cross-section data for the benefit area is the basis for the generation of synthetic flood levels. Simplified sets of data on flood damage potential for the principal types of development are incorporated into the system.

The results are usually presented in the form of rankings of the sub-areas of the region studied in terms of their flood damage potential, either at a range of return periods or as estimated discounted annual benefits. In this way it can easily be seen which areas are liable to suffer the worst flooding and flood damage, and these can then be the subject of more detailed study (Green and N'Jai, 1987).

7.3.2 ESTDAM and Micro-ESTDAM

ESTDAM is a suite of computer programs (Chatterton and Penning-Rowsell, 1982; Penning-Rowsell et al., 1987) for estimating the benefits of urban flood protection projects in more detail and with greater accuracy than with BOCDAM. There is a mainframe version, running on IBM systems, which retains the ESTDAM title, and a Micro-ESTDAM version, running on 286, 386 or 486 personal computers. The database inherent in the system is retained at Middlesex University and is regularly updated (N'Jai et al., 1990).

This software uses all the data and techniques described in Chapter 5, and the other Manuals to which the reader is referred in that chapter. ESTDAM is most useful, therefore, in producing detailed estimates of benefits for comparing project benefits and costs (Table 7.1), and in medium sized flood-prone areas (i.e. up to 10,000 properties). ESTDAM incorporates a number of other programs, such as DDASS (Depth-Damage Data Assembly) designed to select, from a standard data base, relevant flood damage data in the form required by ESTDAM (Penning-Rowsell and Chatterton, 1982) (Figure 7.1). Recently VUEDAM has been developed to enhance the capability to present the results of ESTDAM in on-line spatial and graphical form for analytical, presentational and public consultation purposes.

The major difference between BOCDAM and EST-DAM is that the latter uses a three-dimensional model of the benefit area, analysing the width, depth and extent of flood prone areas, as a three dimension flood surface, rather than employing cross-sectional data.

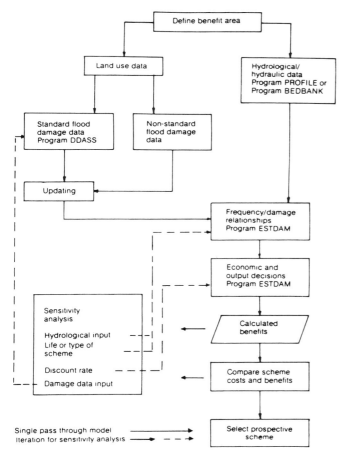

Figure 7.1 Main components of ESTDAM and Micro-ESTDAM

150　THE ECONOMICS OF COASTAL MANAGEMENT

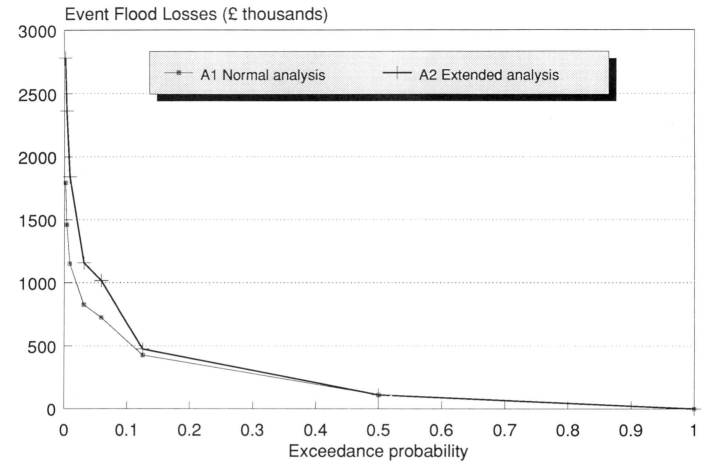

Figure 7.2　Example of a loss-probability relationship (Swalecliffe, Kent)

ESTDAM is therefore based on more detailed land use surveys, depth-damage data inputs and hydraulic/hydrographic data. The system also automatically computes annual average flood damage potential by incorporating flood probability data (Table 7.2).

Using hydraulic data inputs (flood frequency/flood height relationships), land use data for the project area, and the available depth-damage data banks (such as in Appendix 5.2), ESTDAM computes the likely flood damage in each property within a flood-prone area given different flood stages.

The total flood damage potential within a given area is thus derived by aggregating estimated flood damages in each property from a flood of given magnitude. From this data, and event probabilities, ESTDAM produces a loss-probability relationship (Figure 7.2), and the expected annual value of benefits is computed as the area under this curve up to any given protection standard. ESTDAM can be linked to a discounted cash flow program with which a stream of annual project benefits may be compared with project capital and maintenance costs.

Thus ESTDAM is a particularly versatile urban flood benefit assessment system and incorporates a wide range of options and sub-routines. For example, the effect on flood damages of embankments, depressions and basements can be modelled, as can traffic disruption avoidance benefits (SROAD.PRO programs, (Green, 1983)) and the damage reducing effects of flood warnings. By linking ESTDAM results with a discounted cash flow program, the impacts can be analysed of the common coastal flooding situation whereby infrequent overtopping of defences could be followed by a breach situation, within the discrete periods of which the flooding mechanism and character is quite different.

In its operation, ESTDAM may be used to produce both rapid desk-based benefit appraisals and more detailed full project feasibility analysis, depending on the user's needs. As described in Chapter 5, flood damage data is available in many different forms: either in an average, generalised form, or in a much more detailed property-specific form. Both can be used, if

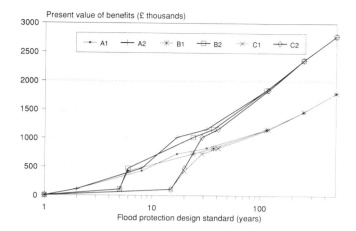

A1 MINIMUM SCHEME COSTS + 10%

A2 NATIONAL LOSS* (LONG DURATION)
FLOOD LEVELS + 0.1m

B1 MINIMUM SCHEME COSTS

B2 MINIMUM SCHEME COSTS - 10%

C1 NATIONAL LOSS* (LONG DURATION)
BEST ESTIMATE OF FLOOD LEVELS

C2 NATIONAL LOSS* (LONG DURATION)
FLOOD LEVELS - 0.1m

* DIRECT DAMAGES PLUS NATIONAL INDIRECT LOSSES

Figure 7.3 Sensitivity analysis of benefits and costs (Swalecliffe, Kent)

needed, to perform rapid 'hand' or spreadsheet-based calculations of benefit potential.

An example of the output from ESTDAM is given in Figure 5.6 for an appraisal at Swalecliffe, Kent. The Swalecliffe project was both complex and innovative (Parker, Penning-Rowsell and Green, 1983) and a large number of different flood scenarios were modelled in order to make the analysis as comprehensive as possible.

The output in Figure 5.6 gives a land use sector breakdown of potential flood damage for a range of floods of short to long return period, and from this range of events the annual average benefits and the discounted benefits of flood protection are computed (Figure 5.7). In this case the comparison of discounted project benefits and costs is aided by a sensitivity analysis designed to assess the impact of varying data inputs and project options (Figure 7.3).

7.4 Models for assessing the benefits of coast protection

No specialist computer programs have been designed specifically to compute the benefits of coast protection. MacKinder (1980) provided a series of proformas and tabulation sheets for evaluating these benefits, but these may be unnecessarily complex for most cases.

7.4.1 The Middlesex spreadsheet format

As part of the research project at Middlesex Polytechnic that led to this Manual we have devised a computer spreadsheet system to evaluate the benefits of coast protection. It can also be used to analyse the very complex situations of combined erosion and flooding problems. This system has been applied to the cases of Hengistbury Head (Thompson et al., 1987) and also Herne Bay (Section 8.5). These spreadsheets provide an ideal means of simplifying the evaluation, and make it much easier to undertake sensitivity analysis. They can easily be constructed using proprietary spreadsheets such as SUPERCALC-5 or Lotus 1-2-3.

The same computations can be carried out by 'hand' using the computation formulae in Chapter 3 and the data in the tables in Appendix 3.2. The disadvantage in hand calculation lies in coping with the many repetitive calculations and the complete reworking of the calculations as new data becomes available or sensitivity analysis is needed. The investment in setting up a comprehensive spreadsheet following the Middlesex format will be found to be well rewarded for all but the smallest and simplest project.

Figure 7.4 shows the recommended spreadsheet format for use in an overall appraisal of a coast protection project. Figure 7.4 therefore gives the recommended headings for use in reporting (and computing) the benefits of coast protection, and a version of this is described in more detail below (Section 7.4.2).

How detailed the assessment is depends on many factors. If different compartments or stretches of coast are being considered, it may be possible to identify stretches which appear to be worth protecting, and others which do not. But it is not enough just to consider the areas which are proposed to be protected. Care must also be taken to include or allow for impacts which are 'external' to the project area. If this is the case the appraisal needs to be re-worked to include modelling of the consequences of a range of different erosion rates in the different unprotected compartments following protection along other parts of the same coastline. This may entail multiple spreadsheets – interconnected or otherwise – to model these different scenarios.

It is important that the spreadsheet format allows for different measurement methods. In some cases a *capital* value for land permanently to be lost to the sea is likely to be available, in which case the treatment in one of the spreadsheet columns is as for property (Section 3.4). In other cases there may be an estimate of *annual*

152 THE ECONOMICS OF COASTAL MANAGEMENT

A. SPREADSHEET FOR EVALUATING CAPITAL LOSSES FROM EROSION PROTECTION

C1 Year p	C2 ELF for given s and r	C3 Domestic property number	C4 Total current value	C5 Business property number	C6 Total current value	C7 Investment number	C8 Infrastructure cost	C9 Total number property	C10 Total current value	C11 Present value of benefits
1	1							C3+C5	C4+C6+C8	C2xC10
2	$(1/(1+r)^{**}1)-(1/(1+r)^{**}1+s)$									
3	$(1/(1+r)^{**}2)-(1/(1+r)^{**}2+s)$									
4	.									
.	.									
Scheme life										
Time horizon										
Totals										

B. SPREADSHEET FOR EVALUATION OF ANNUAL BENEFITS FROM COAST PROTECTION AND SEA DEFENCE

C1 Year	C2 Discount factor D	Flooding ====>				Erosion ====>			Recreation ====>				Total
		C3 Average annual loss without project	C4 Annual residual loss with project	C5 Current value of flood protect. benefit	C6 PV of flood protect. benefit	C7 Annual agricul. benefit	C8 Annual disrup. benefit	C9 PV of erosion benefit	C10 Value enjoyment without project	C11 Value enjoyment with project	C12 Annual recreation benefit	C13 PV of recreation benefit	C14 Total PV of annual benefit
1	1			C3-C4	C5*C2			(C7+C8)*C2			C11-C10	C12*C2	C6+C9+C13
2	$1/(1+r)^{**}1$												
3	$1/(1+r)^{**}2$												
4	.												
Scheme life													
Time horizon													
Totals													

Note that annual benefits are evaluated over at least the same time horizon as for capital losses to allow for phasing in of changes after erosion recommences

Note spreadsheets may be modified/adapted to suit the features of a particular study area

Figure 7.4 Suggested spreadsheet frameworks for coast protection project appraisal

loss which is constant through time and is incurred annually once the land, or part of it, is to be regarded as written-off (a kind of annual average loss, akin to the annual average flood damage). This is likely to apply to the loss of agricultural land, where the loss will be quantified as the loss of the adjusted gross or net margins (Section 3.6.2). With erosion of environmental/recreational land it is likely that any estimated annual losses will *vary* over time, possibly from zero or minimal loss as erosion starts, to the full annual loss when the site is almost totally lost.

All of these measures can be incorporated into the spreadsheet format, but care must be taken to list the impacts separately, and to avoid double-counting by including impacts in several alternative evaluation columns (especially by counting both an annual sum and a capital sum for the same benefit).

7.4.2 A hypothetical example: 'Newton-on-cliff'

This example is of a project appraisal for a hypothetical coast protection project. Suitable proformas for pre-

senting the results of such a project appraisal are given by the Ministry of Agriculture, Fisheries and Food (1985a). However a different format from that used in flood alleviation is required in computing coast protection benefits. The purpose here, therefore, is to demonstrate the method recommended in Section 7.4.1, and

Table 7.3
Average property values in 'Newton-on-cliff' (January 1987 values)

LU code	Land use type	Market value (£)
11	Detached house	60,000/property
12	Semi-detached house	45,000/property
13	Terrace house	35,000/property
14	Bungalow	45,000/property
5	Retail and distribution	600/square metre
8	Manufacturing	250/square metre

the use of a suitable format for computing and reporting the erosion delay benefits of coast protection.

The hypothetical town, 'Newton-on-cliff', as a result of its location on soft substrata, faces the risk of loss of land and property due to erosion by the sea unless preventative action is taken. Land use surveys of the coastal zone have been carried out, and predicted erosion contours for a 100 year period are available. The local district valuers' office and local estate agents have provided estimates of the average market values of a variety of properties in the area which are not believed to be threatened by erosion (Table 7.3). One of the pro-

EXAMPLE OF SPREADSHEET EVALUATING EROSION DELAY BENEFITS WHERE PROPERTIES HAVE STANDARD VALUES

YEAR	ELF r = .06 s = 50	DETACHED VALUE £60,000 NUMBER	SEMI- DETACHED VALUE £45,000 NUMBER	BUNGALOW VALUE £45,000 NUMBER	RETAIL VALUE/SQM £600 AREA SQM	MANUFACTURING VALUE/SQM £250 AREA SQM	TOTAL VALUE	DISCOUNTED BENEFIT FROM EXTENSION OF LIFE
1	1						0	0
2	0.8921808						0	0
3	0.8416800						0	0
4	0.7940377						0	0
5	0.7490922						0	0
6	0.7066908						0	0
7	0.6666894						0	0
8	0.6289523						0	0
9	0.5933512						0	0
10	0.5597653	2					120000	67172
11	0.5280804		4				180000	95054
12	0.4981891						0	0
13	0.4699897			3			135000	63449
14	0.4433865	1			150		150000	66508
15	0.4182892						0	0
16	0.3946124						0	0
17	0.3722759						0	0
18	0.3512037						0	0
19	0.3313242	2		2			210000	69578
20	0.3125700						0	0
21	0.2948774						0	0
22	0.2781862		2			500	215000	59810
23	0.2624398						0	0
24	0.2475847						0	0
25	0.2335705	1					60000	14014
26	0.2203495						0	0
27	0.2078769						0	0
28	0.1961103			1			45000	8825
29	0.1850097				200		120000	22201
30	0.1745375		6				270000	47125
31	0.1646580						0	0
32	0.1553377						0	0
33	0.1465450						0	0
34	0.1382500						0	0
35	0.1304245						0	0
36	0.1230420						0	0
37	0.1160774						0	0
38	0.1095070						0	0
39	0.1033085						0	0
40	0.0974608	1	2	4			330000	32162
41	0.0919442						0	0
42	0.0867398	2	2				210000	18215
43	0.0818300				170		102000	8347
44	0.0771981						0	0
45	0.0728284					300	75000	5462
46	0.0687060						0	0
47	0.0648170						0	0
48	0.0611481						0	0
49	0.0576869						0	0
50	0.0544216						0	0
51	0.0513411						0	0
52	0.0484350			2			90000	4359
53	0.0456934			2			90000	4112
54	0.0431070						0	0
55	0.0406670						0	0
56	0.0383651						0	0
57	0.0361935						0	0
58	0.0341448						0	0
59	0.0322121						0	0
60	0.0303887						0	0
61	0.0286686						0	0
62	0.0270459						0	0
63	0.0255150						0	0
64	0.0240707						0	0
65	0.0227082						0	0
66	0.0214229						0	0
67	0.0202102						0	0
68	0.0190663						0	0
69	0.0179870						0	0
70	0.0169689						0	0
71	0.0160084						0	0
72	0.0151023						0	0
73	0.0142474						0	0
74	0.0134410						0	0
75	0.0126802						0	0
76	0.0119624						0	0
77	0.0112853	6	4	5			765000	8633
78	0.0106465						0	0
79	0.0100439	2					120000	1205
80	0.0094754		2		400		330000	3127
81	0.0089390						0	0
82	0.0084330	1		3			195000	1644
83	0.0079557						0	0
84	0.0075054			1			45000	338
85	0.0070805						0	0
86	0.0066797		4		150		270000	1804
87	0.0063017	1		6			330000	2080
88	0.0059450					300	75000	446
89	0.0056084	4	6				510000	2860
90	0.0052910						0	0
91	0.0049915					1000	250000	1248
92	0.0047090	1		10			510000	2402
93	0.0044424						0	0
94	0.0041910				400		240000	1006
95	0.0039537		2				90000	356
96	0.0037299			3		2000	635000	2369
97	0.0035188						0	0
98	0.0033196	2			100		180000	598
99	0.0031317						0	0
100	0.0029545			1			45000	133
		26	34	43	1570	4100	6992000	616641

Figure 7.5 Derivation of discounted erosion delay benefit using the Middlesex spreadsheet format

posed coast protection projects would involve structural works with an expected useful life of 50 years; this example evaluates the benefits of this project, implemented in year 1, compared with the 'do nothing' option of no protection.

Figure 7.5 illustrates the relative contributions to the project's benefits of extending the life of property close to the current cliff line compared with properties that are estimated to be at risk of erosion in the next 100 years. In this respect only 44 of the properties would be eroded before year 76, and the remaining 71 properties (62 per cent of the total) would only be eroded after year 76. Although over 60 per cent of the current value of erosion-prone property relates to these properties with over 50 years of usable life without coast protection, the discounted benefit of works done now to extend its life is, naturally, low (i.e. £616,641 minus £577,921 = £38,720).

Hence the present value of a protection project evaluated over 50 years is £0.57 millions (i.e. £569,450), but rises to only £0.62 millions (£616,641) when the same project is evaluated over a 100 year period. In fact 64 per cent of the final discounted value of benefits is achieved by protecting only 14 residential properties and one business property at risk of erosion within the first 19 years (cumulated discounted benefit of £361,761).

The project analysed in Figure 7.5 was contrasted with the same project evaluated on the assumption that implementation would be delayed until year 8 (this re-analysis is not shown in Figure 7.5). This can be achieved by reassigning the current year 8 as year 1, which preserves the functional relationships between elements in the spreadsheet. Effectively in a hand calculation the total current value of property moves up the table relative to the benefit factors. Ignoring properties more than 100 years worth of erosion away from the current cliff, the discounted benefit of this delayed project would be approximately £1.0 millions.

At Newton-on-cliff the present value of the costs of protecting the coast for 50 years proved to be £0.90 millions. This includes both discounted capital works costs, and a discounted annual maintenance provision which would allow for repair of damage to the works due to major storms (based on previous local experience). Hence a protection project was worthwhile and implemented, and none of the properties at risk of erosion were in fact lost. However, the coast protection authority delayed its works programme so that it was completed eight years after the initial economic appraisal. Although some land was lost in the process this had negligible economic value (and was not environmentally important), and as a result the works had a benefit-cost ratio of 1.05:1.

While the type of table in Figure 7.5 can be used in calculations, it can be compressed for presentation by omitting years with no loss, grouping years, and grouping property types. We recommend that any 'Derivation of discounted erosion delay benefits' tables include the following headings:

- Year (from operation of project);
- Extension of life factor;
- Residential property, businesses, agricultural land, other land use (for each of these four sub-headings give the number of properties or area lost, and the total market value of this type of loss where it can be estimate);
- Total current value of all losses (per year);
- Discounted benefit of protection (for each year);
- Cumulative discounted benefit of protecting land and property at risk from year 1 to year x.

In this way the user can see clearly the key points in the analysis, and yet not be overburdened by information which is important only within the calculations.

7.5 Summary and checklist

This chapter has emphasised the need to match the computer software used in project appraisals to the needs of that appraisal, which in turn reflect the relevant stage within the appraisal process (i.e. pre-feasibility, full feasibility, or priority setting). A range of software is available to match these different needs.

In undertaking an appraisal of coast protection or sea defence investment (or a combination of these, in an integrated project) care should be taken with a number of key elements, as follows:

1. Ensure that **time** is available for all the phases, and particularly that data that is seasonal in nature (particularly recreational activities and environmental characteristics) is collected in good time to be included in the analysis (see Chapters 1, 4 and 6);
2. Ensure that **field data collection** is organised such that repeat surveys to collect data in more detail are avoided and that the information gathered for the project appraisal matches that collected for the design process (if engineering works are envisaged);
3. Use the most appropriate **computer systems** for the purpose of the analysis (i.e. BOCDAM or ESTDAM, or use of the Middlesex spreadsheet format at appropriate levels of detail).
4. Ensure that **data collected on floor heights of properties**, and flood predictions, are as accurate as possible, since this is one of the variables that has most effect on calculated benefits for flood

alleviation projects;
5. Ensure that **erosion contours** are calculated and drawn with great care, since this is one of the variables that has the most effect on the calculated benefits of coast protection projects;
6. Evaluate a range of **options** at a range of **design standards**;
7. Ensure that **sensitivity analysis** is undertaken to assess whether there is the need to collect more data or model flooding and erosion more accurately if these inputs to the appraisals – which are inherently estimates – so dominate the results as to throw doubt on their reliability.

By following these general guidance notes, and those contained at the end of each of the preceding chapters, the user should be able to take the most efficient path through a coast protection or sea defence project appraisal, consistent with obtaining reliable results.

8 Case studies

8.1 Introduction

In this chapter we present five case studies designed to illustrate the range of benefit assessment issues in the field of coast protection and sea defence. The characteristics of the case studies are outlined in Table 8.1.1 and their location is indicated in Figure 8.1.1.

These case studies have either been pursued in the course of the research leading to this Manual, or they have been undertaken as research and consultancy projects by the Flood Hazard Research Centre between 1986 and 1992. As such they inevitably use and illustrate the data and techniques that were considered to be the best available at the time, and some of these sources and techniques have been improved since the particular case study was finished.

This point about the techniques and data used in these case studies is particularly true of the Hengistbury Head case study, which was the first conducted during the research period and which was itself the foundation of the recommendations concerning the appraisal procedure that we present here in Chapter 6. However, we have drawn out the lessons learned from that case study in Section 8.6.9, and we have indicated there that were this type of study to be repeated elsewhere then it would need to be structured somewhat differently.

In other respects the case studies are provided here only as guides to what may be needed elsewhere. Assessing the benefits of coast protection and sea defence is highly site-specific, and while techniques and data may be transferable from elsewhere this must always be done with caution. These case studies also generally describe major coastal projects (the exception being Peacehaven), and more day-to-day projects on a small scale will have different emphases and analytical needs. A 'horses-for-courses' approach is central to the guidance throughout this Manual, and these case studies should be viewed in this light.

Also, policies may change in the future with regard to important parameters in these studies, such as the relationship between land and sea levels if climatic warming leads to significant sea level rise. Finally, the projects that these case studies appraise are all engineering projects, yet it may well be the case elsewhere that this is not an appropriate response to the particular sea defence or coast protection problem to be found there. In particular it may, in certain circumstances, be more appropriate to abandon property and land threatened by erosion, and not to implement engineering schemes to alleviation flooding, but to adopt some other form or forms of non-structural response. In general, a mix of strategies is likely to be a wise policy to pursue, and to use the precautionary principle in seeking a sustainable policy base.

In several of these cases the benefit assessment formed the basis of the project appraisal leading to a scheme or project being implemented. In the case of Hastings, a scheme was already being implemented, which was not based on these results. But in the cases of Hengistbury Head and Herne Bay the projects being assessed were implemented, based generally on the benefit assessment results described below. In the case of Peacehaven the benefit assessment showed that the proposed cliff protection project was not worthwhile, and it was not undertaken. For Fairlight Cove, the work of the Fairlight Coastal Preservation Society – assisted by the Flood Hazard Research Centre – led to a coast protection scheme being implemented to protect the cliff and the cliff-top properties at risk.

Table 8.1.1
Summary of characteristics of case studies

ASPECT	HASTINGS	PEACEHAVEN	FAIRLIGHT COVE	HERNE BAY	HENGISTBURY HEAD
Type of sites	Beach	Cliff top	Cliff top	Seafront	Varied coastal habitats
Type of project	Beach and promenade reinstatement	Prevention of cliff erosion	Prevention of cliff erosion	New form of seafront protection	Erosion protection of headland
Main beneficiaries	Seafront users – informal recreation	1. Local householders 2. Local informal users of cliff top amenity area	Local householders	1. Existing seafront users – informal recreation 2. Potential new users – water based activities	1. Land and water based recreational users 2. Householders 3. Holiday chalet owners 4. Environmental interests: ecology, geology, archaeology
Type of survey	Site survey of seafront users: local/staying/day visitors	Residents survey	Land use survey	1. Site survey of seafront users 2. Residents survey 3. Land use survey	1. Site survey of users 2. Environmental site surveys 3. Land use survey
Date of survey	Summer 1988	Autumn 1988	Autumn 1988	Summer 1990	Summer 1986
Method of valuation and aggregation	Mean enjoyment per visit valuation	Mean willingness to pay times proportion of user population willing to pay	Erosion delay benefits	Mean enjoyment per visit times user population Erosion delay benefits Inundation protection benefits Recurrent flood protection benefits	Various (see text)

8.2 Hastings

8.2.1 Introduction

This case study demonstrates how recreation benefits can be assessed, using the Contingent Valuation Method and the enjoyment per visit technique to value the change in user enjoyment resulting from coast protection. Hastings represents a relatively straightforward example of assessing user benefits at a popular resort, where the main recreational activities are informal use of the beach and promenade. In this case, however, the construction of the coast protection works resulted in severe disruption to the use of the seafront, but the restrictions only operated over a short period of time.

The scheme extends over 2.5 kms of seafront, cost £6.8 millions in 1987, and took three years to construct. The scheme was justified in terms of the benefits to property protected and no sum for recreation benefits was included in the original benefit-cost analysis. However, the scheme provided a good opportunity to investigate the users' perceptions of an existing scheme, where the details of the improvements were already available and could be accurately portrayed in the drawings. It was, therefore, used in our research to develop the methodology of valuing user enjoyment, using drawings to represent the seafront under various conditions of erosion and protection. By using the methods developed, it would have been possible, in principle, to have calculated the recreational user benefits of the scheme as a component of the total scheme benefits.

8.2.2 Characteristics of the study area

Hastings is a popular South Coast resort with a range

158 THE ECONOMICS OF COASTAL MANAGEMENT

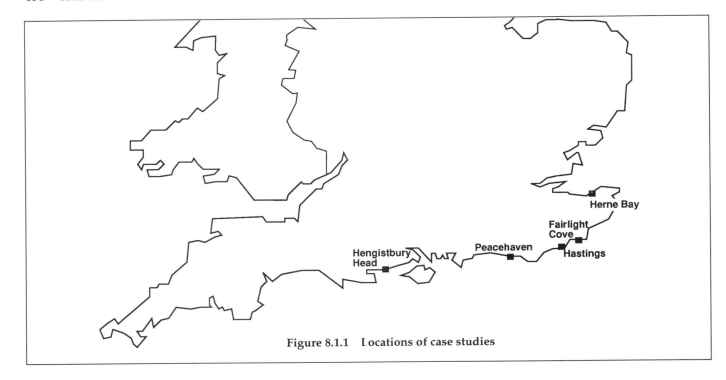

Figure 8.1.1 Locations of case studies

of facilities which attract both day users and holiday visitors, and include shopping, exhibitions, leisure facilities and the seafront. It had a population of over 76,000 in 1981, and has an estimated 240,000 staying visitors and 2.2 million day visitors annually together with an estimated 600,000 visitors staying with or visiting relatives.

The town has a built up seafront extending for 3 kms

Figure 8.2.1 Removal of beach material by longshore drift, Hastings

from the West Marina to the harbour in the Old Town at the eastern end. The character of the seafront varies considerably along its length with traditional fishing buildings in the Old Town, amenity flower gardens, level grassy areas and roadside parking. There is a continuous built-up promenade with a lower walkway in the central section.

Some of the earliest coast protection works, dating from 1830, form part of the lower promenade. The beach has numerous groynes and is heterogeneous in nature, with some sections showing signs of severe erosion while in others accretion has been taking place. The seafront has been subject to coast protection measures, with a seawall supporting a promenade and stonework groynes to protect against the effects of erosion, since the 1830s. Further works were carried out in the 1890s, and 1920s and 1930s.

Since the 1930s, erosion has been gradually affecting the beach, groynes and seawall in a number of ways. Beach material has been removed by longshore drift (Figure 8.2.1), which operates from west to east, and natural replacement has been largely prevented by coast protection measures to the west. The beach level has declined, in some sections little beach is left above the high tide level at the base of the promenade.

The effectiveness of the groynes in holding beach material has decreased as their condition has deteriorated. As the volume of beach material has decreased, so its effectiveness in protecting the base of the seawall from wave action has reduced. By the 1980s the seafront was suffering from quite serious erosion in some stretches due to failure and overtopping of the groynes (Figure 8.2.2). The seawall below the promenade was showing signs of cracking and hence there was the danger of instability with the possibility of the promenade being undermined by a severe storm. For a number of years Hastings Borough Council had managed the seafront by moving up to 12,000 cubic metres of shingle annually from the east end where it accretes against the harbour arm and replenishing the beaches at the west end of the Borough. This was not thought to be a satisfactory solution in the long term.

Without measures to improve the existing coast protection works it was anticipated that continued erosion would cause serious deterioration to the condition of the seafront. These changes would affect the enjoyment of recreational users. As the volume and composition of beach material changed, the beach level was expected to decline and the beach profile to flatten out. This would reduce the amount of beach accessible above high tide to the extent that in some stretches there would be none at all. Because the finer grained material is removed first, further erosion would create a deterioration in the quality of the beach material (Figure 8.2.3), with the remaining beach material comprising large pebbles and cobbles with no sandy material left at all on some sections of beach. Continued erosion would also result in the removal of so much beach material that the underlying parent material would be exposed in places, giving the beach an uneven profile.

Further erosion would also create problems in the form of dangers to beach access in the most heavily eroded sections, where the original steps from the promenade to the beach would no longer be long enough to reach the beach. The promenade itself could be seriously undermined in the event of a serious storm, creating potential access problems along some sections. The groynes would also continue to deteriorate and would have become dangerous, in addition to allowing longshore drift to take place unhindered.

These physical changes to the seafront could affect recreational use in two ways. Firstly, the overall level of use might be reduced, either due to users transferring to alternative locations or by a reduction in the level of use over all. Local users are likely to be affected to a greater extent than other users as they have fewer alternative choices without incurring greater costs. Secondly, the level of enjoyment of remaining users is likely to be reduced as the amenity of the seafront declines in quality.

Figure 8.2.2 Overtopping of groyne due to longshore drift, Hastings

160 THE ECONOMICS OF COASTAL MANAGEMENT

Figure 8.2.3 Deteriorating groynes and beach quality, Hastings

8.2.3 The nature of the coast protection project implemented

The present coast protection works were started in autumn 1987 and carried out in six stages over the following three years (Figure 8.2.4). They consist of extensive redevelopment of the existing seafront west of the pier, involving four types of work. Along most of the seafront the seawall has been reconstructed and rein-

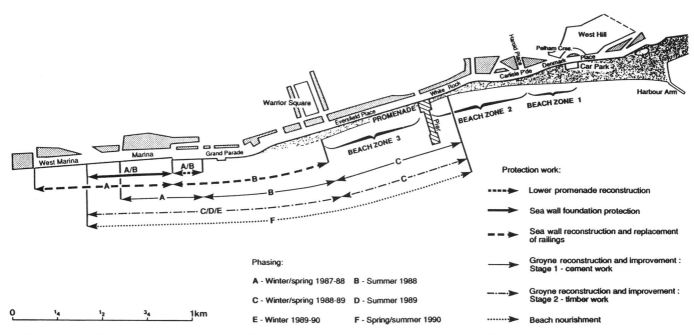

Figure 8.2.4 Hastings seafront showing phasing of coast protection works and sampling zones for user surveys

Figure 8.2.5 Seawall construction with 'bullnose' wave return profile, Hastings

forced in individual sections with a 'bullnose wave return profile' to reduce the impact of wave action (Figure 8.2.5). In one section the seawall foundation has been protected with masonry blocks (Figure 8.2.6). The third type of work consists of reconstruction of concrete groynes (Figure 8.2.7) to an improved design, including removable timberwork in the upper sections to facilitate the control of the movement of beach material. The fourth type of works involved reconstruction of the lower promenade. Finally, beach nourishment was carried out along the whole section of the works in summer 1990 using 0.25 million cubic metres of shingle dredged from the Hours Bank off Worthing and pumped ashore.

It was necessary to restrict access to some sections of beach while work was being undertaken. Throughout the period of the scheme, informative site notices and a series of leaflets explaining the nature of, and reasons for, the work in progress were produced for the public by Hastings Borough Council.

8.2.4 The methodology for assessment of the recreation benefits

The first stage in determining user benefits was to identify the extent of the area to be affected by the proposed scheme. This was a section of the seafront about 2.5 kms in lateral extent from West Marina to just east of the pier (Figure 8.2.4).

The scheme extended from the bottom of the groynes at mean low water mark to the landward side of the promenade. It was clear from familiarity with the area that the main users to be affected were informal users of the beach and promenade.

It was also necessary to identify the characteristics of the scheme in order to establish the extent to which it would affect recreational use. As the scheme had already been approved, it was possible to identify its characteristics down to the level of the design of individual groynes from information provided by the Engineers' Department of the Borough Council. This was invaluable in producing drawings of the seafront under different conditions, as in Figure 8.2.8.

The second step was to investigate existing sources of

162 THE ECONOMICS OF COASTAL MANAGEMENT

Figure 8.2.6 Construction of seawall foundations, Hastings

data on recreational use. A number of studies had been carried out by the tourism section of Hastings Borough Council (1985) and by the South East Tourist Board (1976). These studies provided reasonably up-to-date estimates of the level of use by staying and day visitors. Day visitors to Hastings in 1985 were estimated as 2.1 millions and staying visitors as 0.25 million. However, as none of these studies provided information on the activities of visitors, there was no basis on which to calculate the proportion of visitors using the seafront for recreation, as opposed to other types of activities such as visiting leisure and amusement facilities.

A site survey of informal recreational users of the seafront was therefore carried out employing the enjoyment per visit technique (Contingent Valuation Method) in order to quantify user enjoyment of informal recreation activities on the seafront. In order to ascertain how visit enjoyment would be affected by the coast protection scheme, drawings and descriptions were used to show the seafront under different conditions of erosion and protection (Figure 8.2.8).

The questionnaire employed separate sections in order to obtain the following information:

1. Category of user: to identify whether local (living within 3 miles of the seafront), day visitor (living more than 3 miles from the seafront but not staying away from home overnight), or staying visitor (spending more than one night away from home).
2. Characteristics of user and type and frequency of use of beach and promenade.
3. Characteristics of beach and promenade at Hastings which users value.
4. User perception of changes to the seafront brought about by coastal erosion and protection.
5. A monetary evaluation of enjoyment per visit to the seafront under different conditions of erosion and protection in order to determine the change in enjoyment resulting from this erosion and protection.
6. A monetary evaluation in terms of respondents' willingness-to-pay for sea defence and coastal protection schemes through 'increased rates and taxes', by comparison with current mean annual expenditure on rates and taxes per household.

8.2.5 Operationalisation of the survey

Due to the complexity of the Contingent Valuation Method questions, experienced professional interviewers were employed to carry out the survey. For reasons of cost effectiveness and also due to time constraints, it was only possible to conduct the survey during the summer months.

On the basis of previous knowledge of the distribution of beach users from a survey carried out by Hastings Borough Council in 1975, it was anticipated that different categories of user were not evenly distributed along the seafront. Visitors using the commercial attractions would be more likely to congregate at the eastern end of the resort. This would also be true of day visitors, particularly those arriving by coach, whereas locals and those seeking a quieter beach would be found at the western end.

In order to obtain a representative sample of visitors to the resort, the seafront was therefore divided into three sampling zones for interviews with beach users (Figure 8.2.4). Promenade users were also interviewed along the length of the promenade between Warrior Square and Pelham Crescent. On the basis of visual observation of user distributions, two thirds of the interviews were conducted on the beach and one third on the promenade.

Three drawings of a typical section of the seafront showing different scenarios were prepared (Figure 8.2.8). The first, illustrating current characteristics, was prepared from field sketches. Further drawings of the same section with continued erosion and coast protec-

Figure 8.2.7 Groyne reconstruction, Hastings

tion were developed from information provided by the Engineers' Department of Hastings Borough Council. A written description to emphasise the main characteristics of each drawing was also prepared (Figure 8.2.8).

The CVM enjoyment per visit technique required the respondents to put a monetary value on their enjoyment of that day's visit to the seafront by means of comparison with activities having a known cost (e.g. a visit to a leisure centre or theme park) (Section 4.3.1). They were then shown the drawings of the seafront under different conditions and asked to give valuations on a comparable basis.

8.2.6 Results

The importance of the beach and promenade in relation to other attributes of the resort overall is shown in Table 8.2.1. Responses to the question: 'What factors were important in your choice to come to Hastings?' showed clearly that 'enjoyment of previous visits', 'the historical interest of the town', and 'ease of access to the resort from home' were the main reasons cited. Other characteristics such as 'seafront facilities' and 'views from the front' were all more frequently mentioned than the actual characteristics of the beach itself, although of these, 'attractiveness', 'spaciousness', 'cleanliness' and 'good walking' were mentioned by 25 per cent or more of respondents.

In order to assess the relative importance attached to different beach characteristics, respondents were asked to rate these on a scale from –5 to +5 (Table 8.2.2). The results indicate that factors such as access and ease of walking on the beach were given a higher rating than physical attributes of the beach although, not surprisingly, safety figured quite highly.

In answer to the valuation questions, 82 per cent of respondents were willing to put a monetary value on their enjoyment of their seafront visit (Table 8.2.3), by comparing it with other activities which gave a similar amount of enjoyment and had a known cost. Some 14 per cent overall were unwilling to value their enjoyment in this way and 4 per cent did not know. There were, however, some differences in the response obtained from locals and visitors, with locals being less willing to put a monetary value on their enjoyment. This difference may be a reflection of age differences between these user groups.

164 THE ECONOMICS OF COASTAL MANAGEMENT

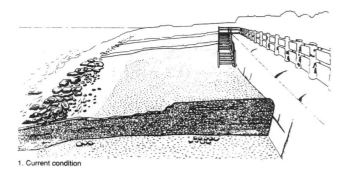

1. Current condition

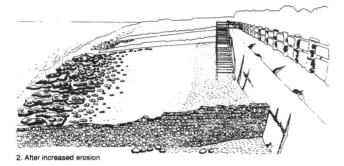

2. After increased erosion

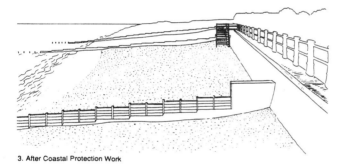

3. After Coastal Protection Work

Figure 8.2.8 Drawings of the same section of Hastings beach under different conditions

Drawing "1"

The picture shows the beach as it is now. There is no beach above high tide level, the tide comes up to the bottom of the sea wall. The beach is reached by a flight of steps from the promenade. The main part of the beach is made up of shingle and at low tide there is sand beyond the end of the breakwater with some rocks showing through. The breakwater is old and showing some signs of decay, the sea wall below the promenade is also old and has some cracks.

Drawing "2"

The picture shows the beach as it would be in a few years time if nothing were to be done to stop erosion. The beach level would be much lower so that the beach would be uncovered for less time at low tide. The drop from the promenade down to the beach would be much greater so the flight of steps would have to be longer. Some of the shingle at the top of the beach would have been washed away so that the sand would be visible further up the beach. There would be more rocks showing through higher up the beach, making the beach more uneven. The breakwater would appear to be higher due to the fall in level of the beach. It would also be more uneven and showing clear signs of decay with pieces falling off on to the beach. The sea wall below the promenade would also be deeply cracked with holes in places. Because the beach level would be lowered, the sea could begin to wash in underneath and the promenade could become in a dangerous state or even collapse as a result. The railings are rusted and in poor condition.

Drawing "3"

The picture shows the beach as it will be in a two years time when the coastal protection works are finished. The beach level will be much higher. There will no longer be such a big drop from the promenade down to the beach so the flight of steps will be shorter. There will be a bank of shingle left uncovered at high tide below the promenade. The breakwater will be longer than at present. There will be more shingle on the beach and it will be gently shelving down to the end of the breakwater. The breakwater will not appear to be so high, it will have wooden planking along the top and the concrete part will only be visible from one side. The sea wall below the promenade will be faced with new material. There will be new railings along the promenade.

Table 8.2.1
Reasons for choice of Hastings for visit

Statement	% day and staying visitors
I wanted to visit the coast	73
I wanted to visit somewhere new	21
I have liked it when I have been here before	58
To make visits to friends or relatives	13
To attend a conference or to carry out some other business	4
The enjoyment of the journey from home	40
The overall cost of the trip/holiday	22
The ease of access from home	60
The abundance of places to visit	38
The attractive setting of the beach	29
Good beach for children	13
The beach is good for taking walks on	25
The prom is a good place for walking or sitting	43
The type of beach (ie. sand, pebbles etc)	14
The spaciousness of the beach	28
The attractiveness of the town itself	39
The quietness, not too many people	39
The variety of evening entertainments, pubs and restaurants	13
The quality of the shopping, sports and leisure facilities in the area	22
The quality of accommodation	24
The sea is clean and good for swimming and paddling	17
The seafront has good facilities (e.g. car parks, deckchairs, toilets)	29
Views from the seafront	33
The beach is clean	25
The historic interest of the town	55
Other (specify)	25

Table 8.2.2
Perception of beach characteristics at Hastings
[from strongly disagree (–5) to strongly agree (+5)]

Statement	Mean score
Beach is clean	1.6
Plenty of beach above high tide	2
Breakwaters unattractive	0.08
Easy access to beach	3.9
Beach easy to walk on at low tide	3.2
Breakwaters decaying and becoming unsafe	1.6
Attractive mixture of shingle, sand and rocks on beach	2.3
Seawall and promenade maintained in safe condition	2.7
Unsafe for children due to difference in level on either side of breakwaters	2.0
Beach unattractive overall	–2.6

Table 8.2.3
Valuation of beach enjoyment, Hastings (percentages)

Response	Overall	Locals	Day visitors	Staying visitors
Valuation given of enjoyment in money terms	82	68	91	85
'Can't value in money terms'	14	24	8	11
'Don't know'	4	8	1	4

Table 8.2.4
Valuation of beach enjoyment based on drawings of alternative beach scenarios, Hastings

Scenario	Mean value (£)
'Today's visit'	7.33
Existing beach (drawing A)	5.41
Eroded beach (drawing B)	2.59
Protected beach (drawing C)	9.34

Note: See Figure 8.2.8 for scenario drawings

Table 8.2.5
Willingness to pay for coastal protection at Hastings through 'increased rates and taxes'

	Overall	Locals	Day visitors	Staying visitors
Percentage willing-to-pay	77	65	84	81

Excluding a single 'outlier' valuation (Section 4.6.2) from the analysis of values of enjoyment of the current beach gave a mean value of £7.33 per visit (Table 8.2.4). The values put on the different beaches shown in the drawings indicated that respondents consistently valued the eroded beach lower and the protected beach higher than the drawing of the existing beach showing its current condition. Differences between the value given to 'today's visit' and to the drawings of the existing beach can be accounted for by the fact that the drawing could only represent one section of beach, whereas the survey was carried out over a stretch with varying characteristics.

The other Contingent Valuation Method technique in the survey, using willingness-to-pay through 'increased rates and taxes', gave a mean willingness-to-pay figure of £5.58 per annum with 77 per cent of respondents being willing to pay in this way. However, the technique should not be expected to give results comparable with those from the enjoyment per visit approach for two reasons (Section 4.5.3). Firstly, willingness to pay is constrained by ability to pay (Section 4.2.3) and the results from this survey do indicate that locals appear to be less willing to pay for coastal protection than visitors (Table 8.2.5). This may be explicable in terms of age-related income, since a higher proportion of locals than other groups are in the 65+ years category. Secondly, willingness to pay contains an unquantifiable element of non-use benefits such as bequest values which tend to increase the valuation whereas the enjoyment per visit method is clearly a user valuation and hence the preferred method when estimating the recreation benefits of a scheme.

In order to obtain an indication of the order of magnitude of recreational benefits which could be expected with the coast protection scheme, it is first necessary to gross up these individual user valuations (Section 4.6.5). In the case of Hastings, although statistics are available for visitors to the resort, overall it would be necessary to carry out a general visitor survey in order to ascertain the proportion of visits which include recreational use of the seafront. Similarly, for holidaymakers, a weighting factor needs to be ascertained in order to determine the relative proportion of the holiday which is spent in informal recreation activities on the seafront. In both these cases the figures are likely to be specific to the resort in question since the range of alternative attractions will vary considerably from one coastal resort to another.

In allowing for local use levels when grossing up the valuation figures, it is necessary to carry out a residents' survey as was done at Herne Bay (Section 8.5) in order to obtain an accurate representation of seasonal patterns of use by locals. This is because locals may be under-represented in user surveys conducted in the summer months and, in any case, such site user surveys provide little information about patterns of use throughout the rest of the year, which could be quite considerable for locals.

8.2.7 Conclusions and lessons learned from the study

The study shows how recreational user benefits of a

scheme can be assessed for informal seafront activities at a resort site. In particular, it illustrates the problems of how to gross up individual user valuations to obtain overall scheme benefits.

In this case, the scheme had already been justified and it was not necessary to gross up the individual valuation figures. However, this case study shows the importance of identifying available user information at an early stage in scheme preparation since, if grossing up had been required, it would have been necessary to collect further information about patterns of visitor behaviour in order to do so. It would also be necessary to ensure that the user data was representative of user patterns throughout the whole year and not just the peak summer months. This is particularly important in the case of local users who give lower valuations for individual visits but whose visits are likely to be distributed throughout the whole year.

If the information were being collected for a scheme benefit assessment, a larger sample of 500–600 users would be required (Section 4.5.3), rather than the experimental sample size of 247 used in this case. As valuations of visit enjoyment differ between the three user groups – locals, day and staying visitors – the sampling should reflect the relative number of each user category. In this experimental study it was considered important to obtain an equal number of respondents from each of the three groups in order to identify whether differences in valuation existed.

One limitation of the approach in general is that it can only represent the valuations of current site users. At sites where serious deterioration in the quality of the seafront has already occurred this could introduce a bias: some former users may have already moved to alternative sites in response to the deterioration, and their valuations will not be reflected in the survey results. In this particular case study such an effect may have introduced a further bias, since the user survey was undertaken after the protection works had begun and access to certain sections of the seafront was not possible. Therefore, the sample could not include any former users who had ceased to use the resort altogether due to the engineering works rather than simply being displaced further along the beach.

8.3 The case of Peacehaven

8.3.1 Introduction

The Peacehaven study was undertaken after the 1988 beach surveys but before the final development of the survey instrument used in the 1989 surveys. The methodology used then was not fully consistent with that which is now recommended. Additionally, because of the timing of the proposed works, the survey had to be undertaken in October and November.

The cliff top at Peacehaven in East Sussex comprises a level grassy amenity area which extends for about 2

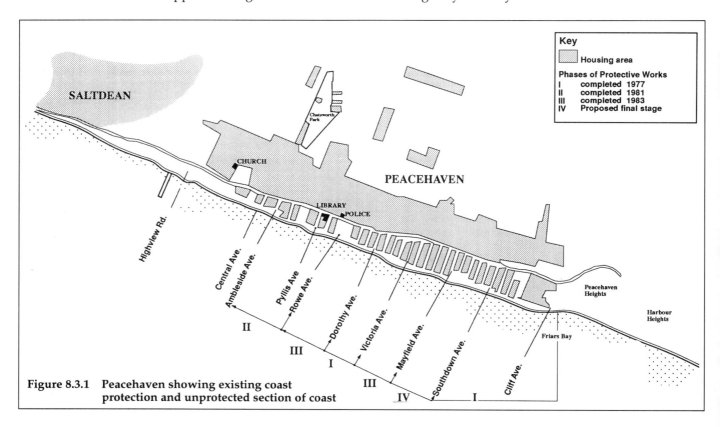

Figure 8.3.1 Peacehaven showing existing coast protection and unprotected section of coast

kms (Figure 8.3.1). The width of this strip is decreasing due to undercutting of the chalk cliff. The annual amount of erosion is highly variable, the mean annual rate being 0.5 to 1 metre (Posford Duvivier, 1988). The grassy expanse is currently about 25 metres wide to the nearest property, but it would be necessary to restrict recreational use before the whole area were lost, due to local authority restrictions on access for safety reasons.

Erosion of the soft chalk cliffs at this location was sufficiently severe to require major coast protection works of the most rapidly eroding sections of the cliff before 1983. By 1988 it became clear that protection works for the remaining unprotected stretch of cliff were required in order to prevent the cliff line from receding and forming a small bay at this point.

It was anticipated that the recreational benefits of cliff protection would be an important component of the total scheme benefits, and an evaluation of these recreational benefits was, therefore, carried out as part of the feasibility study for the scheme.

8.3.2 The characteristics of the study area

The area comprises a level stretch of grassy promenade about 25 m in width situated on the top of cliffs composed of a Cretaceous chalk deposit averaging about 800 metres in depth. The exposed Upper Chalk varies considerably in hardness and composition and the random pattern of vertical and horizontal fault lines makes projection of erosion rates difficult. The chalk is overlain with clay, varying from 150mm–8 metres in depth and subject to weathering by rain and frost and also to shrinkage by drought.

The foreshore beneath the cliffs consists of a small amount of shingle in natural embayments. Longshore drift in the area is from west to east but the movement of material is restricted by previous coast protection works, hence there is little new material entering the frontage. Any measures to increase the accumulation of beach material are desirable as the shingle is an essential barrier to wave action and helps to prevent scouring and undermining of coast protection structures such as seawalls. Its presence therefore reduces maintenance costs.

8.3.3 The process of erosion and the consequent problems

Erosion takes place in two ways at Peacehaven: from erosion of the cliff base by the sea, leading to undercutting and hence cliff falls, and secondly by weathering of the cliff top and upper faces. Both forms of erosion are accentuated in areas where major fault planes occur. Although the mean rate of recession at the study site is 0.5 metres per year, this occurs mainly in single large falls which have reduced the top width by as much as 7–8 metres in one collapse. These falls can occur at any time.

The major losses from erosion would be to property located within 50 metres of the cliff margin. In addition, there would be significant recreational and amenity losses resulting from erosion through the reduction in area of the grassy cliff-top promenade. The beach below is not sufficiently significant as an amenity for the effects of cliff falls on its use to be considered.

8.3.4 The coast protection project

Previous protection work to the cliffs had been carried out in three phases between 1977 and 1983 (Figure 8.3.1). This work was carried out to protect cliff-top properties under threat from continued erosion. The area under consideration in this study was a stretch of approximately 330 metres in length which would join the existing protection works and which had been identified as the fourth and final stage of the coast protection works for the area. A total of 23 properties had been identified as vulnerable to erosion and it was estimated that up to 12 of these would have to be abandoned within 50 years should protection work not be carried out. The main component of the benefits of the scheme would come from protecting these properties.

The proposed scheme was to be similar to the previous phases and consisted, first of cliff stabilisation, to be achieved by trimming the cliffs back to a stable angle of between 71.5° and 75°. Secondly, the construction was envisaged of a concrete sea wall with a reinforced concrete deck approximately six metres wide with concrete cliff facing three metres high behind it. Thirdly, two or three concrete groynes would be constructed incorporating beach steps.

The cliff trimming would result in a reduction of the grassed cliff top area available for recreation, by between 13 and 17 metres, including a two-metre safety zone.

8.3.5 The assessment of recreational benefits

No systematic data was available for the existing use of the area and from preliminary investigation it was clear that the current recreational use of the strip was of an informal nature, mainly by local residents. In addition, the area forms part of a coastal walk used by long distance walkers but although the level of recreational benefit was likely to be high for each individual walker, a relatively low level of use existed which indicated that it was not worthwhile to survey this user group.

Due to the predominantly local use of the area, it was

168 THE ECONOMICS OF COASTAL MANAGEMENT

considered that a residents survey would be the most appropriate method of determining the type and amount of existing recreational use. The survey population was restricted to the electors of the four wards of Peacehaven, as defined in the electoral register (a total of 10,556 people), and 214 residents interviews were carried out, representing a 1 in 50 sample of electors. Sampling was carried out in proportion to the number of electors in each of the streets of Peacehaven.

The questionnaire was based on the standard survey instrument described in Section 4.5.3 and incorporated a map of the area under consideration (Figure 8.3.1). It was structured in five sections:

1. The first was designed to obtain general information about the advantages and disadvantages of various characteristics of the local area, including cliff-top walks.
2. The second section asked about the amount and type of recreational use of the cliff-top at Peacehaven and other locations. In designing the survey questions to elicit levels of site use it was necessary to ensure that respondents could identify the effects of seasonal differences on the frequency of their use.
3. The third section utilised the Contingent Valuation Method (Section 4.3) in which respondents were asked to value their enjoyment of a single visit to the cliff-top for recreation in money terms.
4. In addition to using the survey to quantify and value current recreational use, the next two sections were used to assess the views of the local residents on the proposed project and its effect on their future recreational use. A map was used to indicate where the cliff line would probably be in fifty years' time if no protection project were undertaken. To illustrate the effects on the cliff of implementation of a protection project three drawings were also used: one showing the present cliff face, the second showing the cliff face without protection after fifty years of erosion and the third showing the cliff face after the proposed protection project (Figure 8.3.2). Each respondent was also given a series of statements detailing the effects of erosion and protection.

First, respondents were asked their views about erosion and how it would affect their level of visit enjoyment, their frequency of visit and the type of recreational activity which they carried out. Respondents were also asked if they would go elsewhere for their recreational activity with continued erosion and loss of access, and if so, whether it would cost them more or less or the same amount.

Next, respondents were asked similar ques-

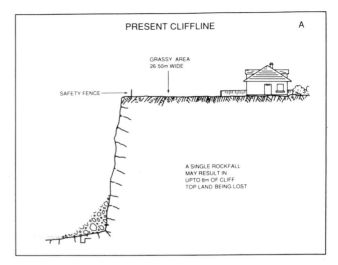

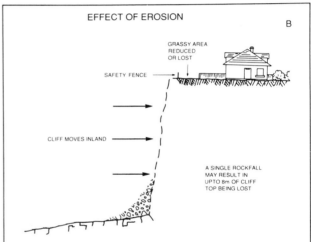

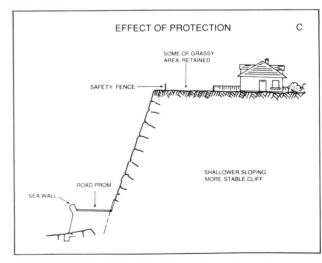

Figure 8.3.2 Scenarios of cliff erosion at Peacehaven used in residents surveys

tions about the effects of coast protection. Finally they were asked whether or not they were in favour of the Council carrying out protection works along that stretch of coast.

5. Fifthly, respondents were asked about their willingness to pay increased 'rates and taxes' in order to fund the protection of the eroding section of cliff. In designing the willingness-to-pay questions, it was important to be aware that respondents' estimates of willingness-to-pay can be influenced by the starting point value selected for the bidding process (Section 4.3.2). In order to obtain a realistic range of values two different starting points were used in the survey, these being £0.50 and £1.00.

One problem encountered was that although respondents were asked to consider the Peacehaven cliff-top specifically as a recreational resource in reporting their willingness-to-pay, other factors associated with the issues of erosion and protection were unavoidably considered. Respondents were only given information as to how erosion would affect the recreational use of the cliff-top, yet many mentioned the threat to property when reporting the degree to which they favoured protection.

Overall, the survey showed that 97 per cent of respondents favoured the proposed protection work being carried out on the unprotected section of cliff.

8.3.6 Calculation of recreational benefits

Just over 32 per cent of respondents said that they would visit less frequently with continued erosion, the large majority stating that they would visit at least as often as now. It is not surprising, and is certainly consistent, that those who said that they would continue to visit the cliffs tended to expect to suffer less loss of enjoyment after erosion than did those whose intention was to visit elsewhere. Those who would continue to visit the site reported that they would experience little loss of enjoyment. However, there is no significant difference between the two groups in terms of the value of enjoyment from visits at present (Table 8.3.1).

Of the 52 respondents who reported that they would visit somewhere else after erosion, slightly over half reported that they would gain at least the same amount of enjoyment at that site. Only 14 reported that it would cost more to visit that site.

It is impossible to estimate the loss per visit in the way set out in Chapter 4, because the survey did not ask for the values of enjoyment after erosion and the value of enjoyment at any other site that would be visited instead.

In the absence of a site count of users, the number of local users attracted to such a site might be first approximated by using those for the average local park. This gives a figure of 30,000 to 60,000 per annum (London Borough of Merton, 1984). In the case of Peacehaven, the average number of visits to the affected stretch was just over 18 per year per respondent (median = 3). The distribution suggests that the median figure is more representative of average visiting frequency than the mean. The maximum loss possible is then £3.59 x 3 x 10,556 = £113,688/year.

However, in this instance, it appears that only those who would transfer their visits to other sites are likely to experience any losses. The total number of movers who would experience any loss is also likely to be very small.

It is possible to derive an estimate of the annual loss to these respondents. Multiplying the additional travel costs to each respondent by the total number of visits that respondent reports gives a mean value of £21.11 per year. Removal of a single outlier reduces the mean

Table 8.3.1
Value of enjoyment of cliff top, Pacehaven

	Arithmetic mean (£)	Logarithmic mean	Logarithmic standard deviation
MOVERS			
value of enjoyment now	3.59	0.57	0.28
STAYERS			
enjoyment now	3.30	0.52	0.29

loss to £10.56. Multiplying this value by 0.32 (proportion of respondents who are movers), the proportion of movers who reported an additional travel cost (0.50), and the population which was sampled (10,556), gives: £17,766/per annum, equivalent to £0.59 per visit.

However, a full benefit estimation would require an estimate of the rate of erosion of the site, and then a phased introduction of recreational benefits into the cash flow. Since cliff falls are unpredictable and of variable size, this estimation of an erosion rate is very difficult.

8.3.7 Limitations of the study and lessons to be learnt

1. The analysis tends to confirm previous expectations that the recreational losses from cliff erosion will tend to be small, particularly where visitors are predominantly local.

   As an initial approximation in similar situations, it is recommended that the mean value of £0.59 is used as the loss per visit and the average loss to a local visitor from the beach survey data (Table 4.3) is used as the upper bound to the loss. Similarly, the average visitor number for a local park (London Borough of Merton, 1984) should be used as the first approximation to the number of visits which are made to the site.

2. Ideally, a site survey of users to give representative information on usage throughout the year should have been carried out to clarify the proportion of total users who were local residents.

3. One of the main problems in calculating recreation benefits in this study was that it was not possible to judge when the benefits would come on stream. This was because the occurrence of cliff falls is unpredictable. Therefore it was not possible to determine precisely at what stage in the future the prohibition of access, on safety grounds by the local authority, would occur. As a result the recreation benefit figures were expressed as per annum benefits of coast protection once the resource was rendered unusable through erosion.

4. Although willingness to pay questions were used in this case study, it was concluded that they did

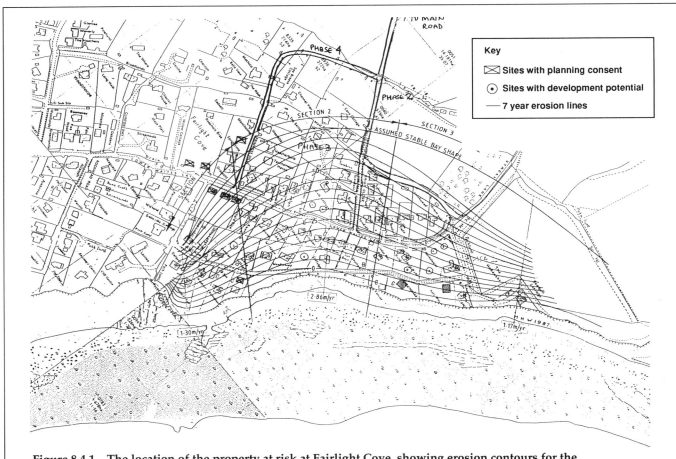

Figure 8.4.1 The location of the property at risk at Fairlight Cove, showing erosion contours for the 7 year erosion rate and the alternative road layout following erosion: Working diagram

not measure recreation benefits alone and are likely to be biased by the form of questioning. Hence the 'enjoyment of visit' CVM has been preferred.

8.3.8 Conclusion

This case study illustrates a relatively straightforward situation where the main recreational users are local residents.

The recreational use was low and the residents predominantly local with alternative recreation locations. Overall, the recreation benefits of the proposed scheme were insufficient when combined with the benefits of property protection, to justify the scheme being built.

8.4 Fairlight Cove, Sussex

8.4.1 Introduction: the context

The village of Fairlight, Sussex, is located at the top of eroding cliffs composed of unconsolidated Hastings Beds (Figure 8.4.1). The rate of erosion of the cliff has apparently increased in recent years, probably as a result of a diminished supply of shingle arriving at the base of the cliff.

This diminished supply of shingle may be attributable to the construction of harbour and coast protection works up-drift to the west, and possibly is related to the harbour works at Hastings – some 5 kilometres to the west – behind which has accumulated a huge bank of shingle. Whatever the cause, houses and gardens are now very close to the edge of the cliff at Fairlight, and there have been property losses in the recent past. Without coastal protection works further losses would have been inevitable (Figure 8.4.2).

The protection of Fairlight has been controversial (Penning-Rowsell *et al.*, 1989). Many possible schemes have been reviewed by a number of different consulting engineers, but there has been some reluctance on the part of the local District Council to promote a scheme. A range of benefit figures were produced for the scheme in the late 1980s, reflecting different assumptions about what could be protected, the value of the property there, and the rates of erosion that are likely to occur in the future.

The matter was resolved in 1990 when Rother District Council promoted a coast protection scheme and the Ministry of Agriculture, Fisheries and Food (MAFF) agreed to support their proposal with grant aid. This decision followed many years of controversy (Weaver, 1987) about the desirability of a scheme, its cost-effectiveness, and the basis of the benefit-cost analysis.

Figure 8.4.2 Residential property at risk from erosion at Fairlight (Source: Weaver 1987)

172 THE ECONOMICS OF COASTAL MANAGEMENT

8.4.2 The benefit assessment: scope and assumptions

The benefit assessment that was finally accepted as justifying works at Fairlight was relatively simple, and used the methods of assessing the value of the delayed erosion of property that are contained in Chapter 3 of this Manual (Figure 8.4.3).

Earlier assessments – which were not accepted by MAFF – had relied on other loss calculations, including those incorporating the loss of the capital value of the properties concerned. There were also assessments which included the loss of plots within the village which either had or did not have planning permission and which had values that reflected the market for residential building land in the area, despite the fact that the houses that might be built would eventually fall into the sea.

The basis of the final benefit assessment was the estimation of erosion contours, for each five-year period, reflecting the loss of the cliff-top and properties over a 100-year period. This was not the anticipated life of the proposed scheme (which was 75 years), but was chosen as the appraisal period because significant numbers of properties beyond the scheme life contour would also have their loss delayed by the scheme.

The rate of erosion at Fairlight was not known definitively for the analysis, and there was evidence that the rate was increasing. Thus two sets of erosion contours were drawn up reflecting the rate of erosion over the last 7 years (1981–1988) and the last 15 years

```
FAIRLIGHT CLIFFS BCA TDR 6%  7YR EROSION RATE SCHEME 0%PA HOUSE PRICE RISE
```

Year	75 Year erosion factors	Houses lost	House value 0% inflation	Present value of erosion benefit (0%)	Road loss	Sewer loss	Other infrastructure loss	Infrastructure disruption benefit	Sum of discounted benefits
1	0.9314631		130000	0			0	0	0
2	0.8787388		130000	0		25000	0	22250	22250
3	0.8289988	6	130000	646619	27889		117650	122197	768816
4	0.7820744		130000	0				0	0
5	0.7378060		130000	0				0	0
6	0.6960434		130000	0			2650	1868	1868
7	0.6566447		130000	0				0	0
8	0.6194762	2	130000	161064				0	161064
9	0.5844115		130000	0				1569	1569
10	0.5513316		130000	0		38000		21219	21219
11	0.5201241		130000	0				0	0
12	0.4906831		130000	0			2650	1317	1317
13	0.4629086	8	130000	481425				0	481425
14	0.4367062		130000	0				0	0
15	0.4119870		130000	0	384384	300000	106000	329800	329800
16	0.3886670		130000	0		4000		1575	1575
17	0.3666670		130000	0		4000		1485	1485
18	0.3459123	3	130000	134906		4000		1401	136307
19	0.3263323		130000	0		4000		1322	1322
20	0.3078607		130000	0		4000		1247	1247
21	0.2904346		130000	0		4000		1177	1177
22	0.2739949		130000	0		4000		1110	1110
23	0.2584858	4	130000	134413		4000		1047	135460
24	0.2438545		130000	0		4000		988	988
25	0.2300514		130000	0	109079	4000		26347	26347
80	0.0093326		130000	0		4000		38	38
81	0.0088043		130000	0		4000		36	36
82	0.0083060		130000	0		4000		34	34
83	0.0078358		130000	0		4000		32	32
84	0.0073923		130000	0		4000		30	30
85	0.0069739		130000	0		4000		28	28
86	0.0065791		130000	0		4000		27	27
87	0.0062067		130000	0		4000		25	25
88	0.0058554		130000	0		4000		24	24
89	0.0055239		130000	0		4000		22	22
90	0.0052113		130000	0		4000		21	21
91	0.0049163		130000	0		4000		20	20
92	0.0046380		130000	0		4000		19	19
93	0.0043755		130000	0		4000		18	18
94	0.0041278		130000	0		4000		17	17
95	0.0038942		130000	0		4000		16	16
96	0.0036737		130000	0		4000		15	15
97	0.0034658		130000	0		4000		14	14
98	0.0032696		130000	0		4000		13	13
99	0.0030645		130000	0		4000		12	12
100	0.0029099		130000	0		4000		12	12
Totals		57		1908005	783835	703000	251600	562666	2470671

Figure 8.4.3 Part of the Middlesex spreadsheet used in the Fairlight benefit-cost appraisal

(1973–1988). The benefit assessment was completed for both scenarios.

A 10-metre safety zone between the cliff top and the nearest part of each property was used before it was assumed that the property became uninhabitable owing to the danger of cliff falls. Over the next 100 years with no coast protection it was estimated that some 55 properties would be lost through the continuation of the erosion, including three development sites with planning permission. A single average was used for the value of each of the properties at risk, based on an assessment of the value of an 'erosion free' property in that area (£100,000 in 1988, £130,000 in 1989).

By cutting access road and utility provisions, erosion would make houses uninhabitable before they would themselves be eroded. The methods detailed in Section 3.5.4 were followed by including the least cost adjustments in a 'do nothing' scenario as a benefit from coast protection.

The costs of upgrading alternative access roads, creating new roads, and of demolishing garages and rebuilding them at set-back locations were phased into the benefit stream according to the erosion contour which would make these actions necessary. Associated with the new road proposals would be new fences that demarcate the new spur road that would be necessary as erosion proceeds (Figure 8.4.1). The costs of the 'do nothing' adjustments would be less than the loss due to earlier abandonment of houses.

One other effect of erosion would be to remove the infrastructure that currently services the residential properties: in this case the water supply, the sewerage system, and electricity (but not gas) power supplies. The assessment of the 'do nothing' situation included the costs of reinstating the services to the properties concerned, based on estimates provided by the utility suppliers themselves.

Re-siting sewers would be necessary to provide the required service for the properties that remain on the cliff top but would be cut off from the sewage works because the sewers followed the lines of the existing roads that would by then be cut. A new sewage pumping station would be required in year 15 (7-year erosion lines) or year 20 (15-year erosion lines). This new pumping system would cost £300,000 as a capital item and £4,000 per annum thereafter as running costs, since the naturally draining system could not by then be operational.

Similar calculations were undertaken for the water and electricity services, and the costs were based on estimates supplied by the utility company concerned; for water this was Southern Water. In the case of electricity a new substation would be required (at a cost of £20,000) so that the electricity service could be provided uninterrupted for the properties that would not be eroded.

8.4.3 The Fairlight Coastal Preservation Society results

A number of schemes have been appraised during the lifetime of the Fairlight controversy. Eventually the main options were for a single coast protection construction, with a length of 500 m, using rock armouring of the cliff top. One scheme originally designed by Sir William Halcrow and Partners was costed at £2.5 millions, and an alternative scheme proposed by Dobbie and Partners, on the same principles, was assumed to cost £1.72 millions (all costs in 1989 prices).

The total cost of the scheme promoted by the Fairlight Coastal Preservation Society was estimated at £1.54 millions, including recurrent maintenance. Various sensitivity analyses were carried out – for example on the maintenance costs – but these did not show significant diversions from the Fairlight Coastal Preservation Society base case.

The results of this benefit assessment, as produced by the Fairlight Coastal Preservation Society, are given in Table 8.4.1. The base case shows that the benefits vary between £2.16 millions and £1.84 millions depending on the rate of erosion taken from the 7-year record or the 15-year record, respectively. Obviously the second figure is lower because the erosion rate over the longer time period is lower.

Table 8.4.1
Summary of the benefit-cost analysis
for the Fairlight Cove coast protection scheme

Real house price inflation rate projected over scheme life	Sum of discounted benefits (75 year scheme life; 6% discount rate)[1]	Benefit-cost ratio[2]
7-year erosion lines		
0.0	£2,163,488	1.26
1.0	£2,473,622	1.44
2.5	£3,179,135	1.84
3.0	£3,511,122	2.04
15-year erosion lines		
0.0	£1,841,819	1.07
1.0	£2,125,444	1.23
2.5	£2,805,813	1.63
3.0	£3,142,284	1.82

Notes: 1. Original calculations used an average house value of £100,000 (1988), subsequently an assessment using an average value of £130,000 was also used (as shown in Figure 8.4.3).
2. Assumes present value of costs of £1.72 million.

With these benefits (which do not include intangible effects of the erosion threat on the population involved) the benefit-cost ratio for the Fairlight Coastal Preservation Society £1.54m scheme was 1.41:1 and 1.20:1 respectively. A conservative approach was taken to certain assumptions, such that no value for the residual life of the scheme was assumed after the 75-year scheme life. However, the discounted value of that benefit would be small.

8.4.4 The question of rising house prices

One interesting topic raised in the Fairlight case is the question of how to treat the rise (and fall) in house prices over time. The methodology recommended in Chapter 3 does not take the depressed value of property caused by the presence of the erosion threat, because we wish to value the use of the house type in question. But there is another matter, which is less clear-cut at first sight, and that is the real secular rise in house prices in Britain and whether these rises can be projected into the future and used to justify the protection of that property.

The Fairlight Coastal Preservation Society made a case that suggested that these real rises in house prices into the future might be used. They demonstrated that house prices had risen in real terms by between 2.5 per cent and 3.0 per cent between 1946 and 1986. The impact of projecting real rises in house prices of those properties to be protected by the scheme in the future is to reduce the effective discount rate. Table 8.4.1 shows the effect that this has on the benefits: they rise substantially, as does the benefit-cost ratio.

This practice of using past trends in house prices to predict future real prices has not been accepted by MAFF and, indeed, the Fairlight coast protection scheme was considered to be worthwhile without counting the projected real rise in house prices.

Our view is that it is important to understand why house prices have risen in the past and may rise over time in the future, rather than simply take it for granted that they will. Part of the real rise in value of houses is because the quality of houses has improved over time, for example the structures are now on average bigger – through extension – than they were in 1946, and they also have fixtures, fittings and decorations that are more valuable. It should be noted here that the rise of contents value is not important here since it is assumed that contents are removed before the building is abandoned and falls off the cliff.

If further quality improvements can be predicted, then this part of the rising value of housing might be a legitimate benefit to take into consideration when assessing the benefits of coast protection, but this is perhaps only half of the real percentage increases cited above. The remainder of the real increase is a reflection of the higher real incomes of those buying houses, which leads them to be willing to pay more for those properties. This income growth is arguably a function of macro-economic factors that create a mixture of real growth and inflation which makes investment in housing relatively attractive as a hedge against inflation. Experience in 1990–92 suggests that the real growth trend can be reversed when recession occurs. Future uncertainties mean that this potential element of the real rise in house prices in the future is not an admissible element in national economic efficiency benefit-cost analysis.

8.4.5 Assessment

The Fairlight example is interesting in highlighting a number of issues in assessing the benefits of coast protection.

First, the case makes it clear how important are the rates of erosion to the final results, and indicates the attention that is needed in determining these rates as accurately as possible. It is, of course, the future rate of erosion that is important, and in that respect rates observed in the past will have to be studied in detail and yet these may only be a guide to what may happen in the future.

Secondly, the case illustrates the problem of infrastructure support being cut to houses which are not themselves at risk from erosion at the same time. The consequences of this type of sequence of events must be determined and the least cost solution modelled (whether it is to abandon the houses or to reconnect the services, including the means and costs of reconnecting the services). Each stage needs to be appraised in sequence, and thereby determine what choices would be made at each point in time as the erosion affects both the houses and the infrastructure.

Thirdly, the case shows the effect that future real house price inflation might have on the benefit calculation. This aspect of the benefit results is not accepted by MAFF, but our view (Section 8.4.4) suggests that there is some merit in part of the argument in theory. However, a predictive model of part of the change in house prices into the distant future is unlikely to be easily developed.

Finally what the case shows is that it is generally only when houses are very near to the edges of cliffs, and are there in some number, that coast protection will be justified. The structures necessary for effective coast protection are expensive, and discounting property use values that will be lost through erosion many decades into the future substantially reduces the value of their protection.

8.5 Herne Bay case study

8.5.1 Background

Herne Bay is located on the North Kent Coast of England. It is one of a series of small towns which line this coast and are characterised by declining tourist industries and a mixture of local residential areas and light industrial development. The central section of the seafront (the sea defence wall and promenade) was found to be decaying and to have a high risk of failure (breaching) which would consequently result in flooding and erosion of the town. Hence in 1990 a benefit-cost analysis was undertaken for sea defence and coast protection works proposed by Canterbury City Council.

Figure 8.5.1 shows the location of Herne Bay, the area at risk from erosion over 100 years if no action were taken, and the area at risk from a 1-in-1000 year flood with no investment in new protection. The assessment concentrated on assessing the losses to the economy from doing nothing (these losses would be averted by the alternative schemes considered and hence are the main potential benefits), and on assessing the recreational gains from alternative designs all intended to offer the same level of protection. Although impacts on the local economy (the area served by the City Council in this case) are not normally the same as national economic impacts and are not included in benefit-cost calculations, they are nevertheless important in making local planning decisions, hence the local economic impacts (multiplier effects) of changed recreational use were also estimated.

This benefit assessment is important as an example of integrating the potential benefits from: erosion protection, flood protection, recreational impacts, and local economic aspects. Two particular issues are illustrated by the case study. Firstly the trade-off and sequencing between recurrent flood losses, permanent loss of use due to frequent flooding, and permanent loss due to erosion; and secondly the differences in recreational value assessment between users and local residents.

8.5.2 Flood alleviation and coast protection benefits

Estimation of direct economic benefits from sea defence and coast protection works used both the EST-DAM and spreadsheet appraisal systems to estimate potential losses over a 100-year period.

Without new protection works it was estimated that the sea front would fail after about 15 years, and thereafter would erode steadily. As the defences fail, properties would be eroded and flooding would become more frequent and more severe. However, at later stages during the process the tidal flooding would be able to flow out more quickly through a larger breach, while the mean sea level would be slightly higher. Hence a complex set of return periods (provided by Canterbury City Council) for the areas north and south of the High Road for each ten year period in the next 100 years of doing nothing were used in estimating benefits. Additionally, properties could become uninhabitable and 'written off' either because the erosion line is within the minimum safety margin for the property or because they would be flooded too frequently to be habitable. Hence a different annual average flood loss estimate (for damages to properties which continue in use) was calculated for each of these bands of years using the ESTDAM model.

For normal flood damages the standard fresh water flood damage data sets developed by FHRC were used (N'Jai et al., 1991) with an additional factor of 1.15 times to reflect additional damages from salt contamination (the data in Appendix 5.2 are later updates of this type of data). No indirect losses were included, other than extra heating costs and emergency service

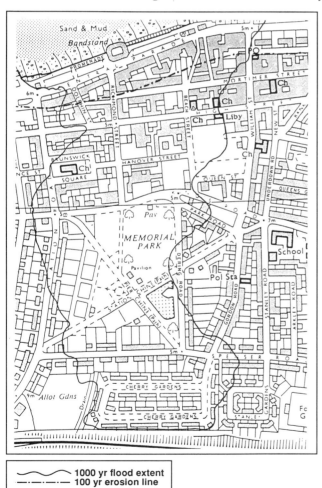

Figure 8.5.1 Herne Bay study area showing 100 year erosion line and 1000 year flood extent

costs which would be avoided when damages are prevented. The businesses which would be affected by flooding are mostly small service sector businesses whose loss of value added would be taken up by other businesses in the area. Flood levels for a given return period were estimated to be the same within the two flood compartments (north and south of the High Road), but would differ between the two areas, and over time for a given return period, due to secular sea level changes (assumed to be a rise of 1.5 mm a year) and the sequential erosion of the sea front. Around 500 properties are at risk from flooding: the figure is approximate since several are also at risk from erosion. Recurrent flood losses could not be estimated without first modelling erosion losses to properties and secondly comparing property floor levels with water levels in frequent floods. There is a lack of empirical evidence on the acceptable maximum frequency of flooding within houses in Britain; for the purposes of this assessment properties were assumed to be abandoned if they were flooded once in three years or more frequently (although the implications of abandonment up to a 1-in-2 year flood risk were also assessed).

Erosion 'contours' for five year intervals were provided by Canterbury City Council for the next 100 years of erosion assuming no new protection works. These were plotted on maps of the properties at risk which had been coded and valued following site surveys (with approximate freehold market values estimated assuming no erosion risk). Values reflected the state of repair and use of properties in mid-1990 and hence showed considerable variation between properties which appear similar on the map. The method does not allow for changes in use or quality over the next 100 years. The same individual property values were used for properties which would be 'written off' due to frequent flooding.

A 5-metre safety margin was allowed in identifying properties which would be lost from erosion in each band of years. However, the erosion contours were also superimposed on maps of the utilities serving the area (gas, electricity, water and sewerage). If an individual property's services would be cut before the 5 metre margin was reached then that property was written off at the earlier date. Additionally for the 'do nothing' scenario least cost alternatives were assumed rather than writing off properties in two cases (Section 3.5.4). In year 30 access to the Pier Pavilion would be eroded although the structure itself would still be intact – the cost of building a bridge from a point which would be safe in year 100 was considerably less than the estimated value of the Pavilion itself and the smaller cost was taken to be the loss avoided by a protection scheme. Likewise in that year the sewerage system serving much of the coastal part of the town would be cut. The cost of linking the properties affected to a

Table 8.5.1
Summary of erosion and flood protection benefit estimates, Herne Bay

Parameter	Inundation loss from 1-in-2 year flood	Inundation loss from 1-in-3 year flood
Total properties lost to erosion by year 2089	138	129
PV of erosion delay benefits	£0.957 million	£0.897 million
Total properties lost to inundation (frequent flooding) by year 2089	50	75
PV of inundation loss	£0.069 million	£0.466 million
No properties affected by 1-in-100 year flood in years 1–14	408	408
Event loss 1-in-100 year flood in years 1–14	£3.82 millions	£3.82 millions
Annual average flood loss in years 1–14	£0.179 million	£0.179 million
No. properties affected by 1-in-100 year flood in years 20-29	464	439
Event loss 1-in-100 year flood in years 20–29	£4.32 millions	£3.83 millions
Annual average flood loss in years 20–29	£0.622 million	£0.481 million
Net PV of recurrent flood losses	£6.28 millions	£5.398 millions
Present value of erosion and flood protection benefits	£7.31 millions	£6.76 millions

modified network with pumping would be much less than the market value of the properties affected.

8.5.3 Sequence of calculations in flood and erosion protection benefits

A combined land use file for properties at risk of erosion and flooding, ranked by floor level, was compiled. Properties identified to be lost due to erosion in a band of years were deleted from the file for that band of years and added to the Middlesex spreadsheet for erosion delay benefits (least cost adjustments were treated in the same way). Property floor levels were compared with the 1-in-3 year flood level for that band of years. If the floor was below that flood level the property was deleted from the relevant land use file for that band of years and its market value added into the Middlesex spreadsheet as a loss due to frequent flooding. The normal ESTDAM computations were made for the remaining land use file to give an annual average loss for each of the years in that band.

Hence the process was cumulative – removing properties from successive land use files for each band of years for the normal ESTDAM flood loss calculations, and adding the same properties and their market values into the Middlesex calculations to estimate present values of permanent losses which would be delayed by proposed protection works. In addition, estimates of residual flood losses with the proposed schemes were made – they are designed to protect up to a 1-in-100 year standard but in events of 1-in-100 year and greater return periods some properties would be affected (181 in a 1-in-1000 year event). Residual losses would increase over time (due to rising sea levels) but by a very small amount – from an estimated annual average loss of £10,728 in 1990 to £11,843 in 2089 (1990 prices).

The results are quite sensitive to the floor levels of properties and to the assumed tolerable frequency of over-floor flooding in properties. Table 8.5.1 summarises results given in Middlesex Polytechnic Flood Hazard Research Centre (1990), and shows that assuming inundation loss ('writing off') in a 1-in-2 year flood the total present value of benefits from flood and erosion protection would be £7.31 millions, whereas if properties are written off by less frequent flooding (1-in-3 years) the present value of benefits would be £6.76 millions. This apparently illogical result is because substantial flood losses would be incurred frequently in the few properties with floor levels between the 1-in-2 and 1-in-3 year levels. It is economically more efficient for these properties to be abandoned earlier than for occupants to continue to use them as normal in the face of frequent flooding (the adjustments which people might make in these circumstances and the associated loss of utility are not known and cannot be observed).

Overall, recurrent flood losses make by far the largest contribution to these benefit calculations because considerable average annual losses are predicted relatively early in the 100-year period of assessment, whereas the main erosion losses would not occur until year 30 and after and hence are heavily discounted.

8.5.4 Recreation benefits

While flood and erosion protection benefits to properties depend on the risk reduction from the scheme but not on the type of scheme involved, recreation benefits depend on the attributes of the sea front in a 'do nothing' scenario and with possible alternative schemes. The method adopted for estimating and valuing recreation benefits was the Contingent Valuation Method (CVM), (see Section 2.5.5). Recreation benefit assessment and the CVM depend on estimates of the visitor use under different scenarios and the value to the users of those recreational experiences. There was a lack of available data on recreational use of Herne Bay seafront, and the area of concern in the assessment is relatively small and primarily used by local people rather than tourists. For Herne Bay two separate recreation surveys were carried out: a sample of 189 residents was interviewed based on the electoral register, and a sample of 143 seafront users was interviewed during July–October 1990. The two types of survey are

Figure 8.5.2 Herne Bay drawing A: the beach in current condition

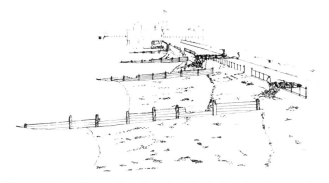

Figure 8.5.3 Herne Bay drawing B: the beach after erosion

Figure 8.5.4 Herne Bay drawing C scheme A version 1: the reef without watercraft

Figure 8.5.5 Herne Bay drawing D scheme A version 2: the reef with watercraft

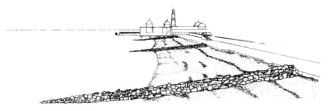

Figure 8.5.6 Herne Bay drawing E scheme B: rock groynes and no reef

a check on one another and are complementary. User surveys will catch non-local users and the most regular of local users, while resident surveys provide a more representative sample of the actual and potential local users.

Seafront user's valuations of a day's enjoyment of the seafront were obtained with the help of a series of drawings (Figures 8.5.2–8.5.6), supported by verbal descriptions, representing the existing beach and seafront (A), the situation after doing nothing for several years (B), and three alternative schemes (an offshore reef with (D) and without (C) moorings and boat launching facilities, and a series of rock groynes (E)). Additionally, respondents were asked about their expected frequency of visits to the seafront in the 'do nothing' scenario, and with alternative protection schemes, and if they would visit an alternative site following the changes to the seafront in each of the scenarios and the economic loss involved in doing this.

The method used depends on, first, comparing the seafront experience with other recreational experiences for which expenses are incurred to estimate the value of the enjoyment, and secondly on the difference between enjoyment values reported under present conditions and with the alternative scenarios, rather than the absolute valuations reported (Section 4.6.3). Overall the mean valuation of today's visit was quite high at Herne Bay at £12.34 (across all users), compared with other values in a variety of beach/seafront situations (Table 4.1) and compared with a much lower mean enjoyment per visit of £3.59 for residents (Table 8.5.2).

It is notable that while residents and users would suffer a serious loss of enjoyment if the seafront eroded, one of the scheme alternatives (rock groynes) would have provided less enjoyment than the seafront in its current (1990) condition, and was also rated as less desirable on separate enjoyment scales. This option was discarded during the scheme decision making process.

An issue which arises among the residents in particular is the rationality of their seafront use. Most residents (79 per cent) stated that they would visit the

Table 8.5.2
Mean value (£ in 1990) of enjoyment of a visit to Herne Bay seafront in different scenarios

Scenario	Residents	Local users	Day visitors	Staying visitors
Current seafront (drawing A)	3.59	6.46	11.09	13.11
Eroded seafront (drawing B)	0.50	1.73	3.53	2.45
Offshore reef no boats (drawing C)	6.12	12.29	13.81	17.73
Offshore reef with boats (drawing D)	7.44	13.70	13.19	15.19
Rock groynes (drawing E)	2.94	6.05	10.65	11.56
Sample (varies between scenarios)	151–158	27–29	46–49	49–50

seafront less often in its eroded state and would go elsewhere, but there were some 32 per cent of 'irrationals' in the sample. These are people for whom the enjoyment of the alternative site, less the difference in travel costs, was more than the enjoyment at Herne Bay (and hence who would achieve a negative loss from switching to an alternative site). It is hypothesised that the method adopted does not allow for multiple purpose visits – residents may visit the seafront as a secondary activity and hence accept a 'second best' recreational experience as a matter of convenience.

This problem may arise in other coastal sites. If the method does not measure the loss of value which these 'irrationals' would actually experience, the alternatives are to set their loss equal to zero, or to omit them from the analysis and still to gross up by the number of residents/visitors as appropriate. The latter will give a higher benefit but assumes that those people whose loss could not be calculated from the survey instrument would actually lose on average the same as those whose loss could be calculated but whose recreational behaviour is different. The more conservative assumption is that the loss of other benefits from multipurpose visits would cancel out the recreational gain from alternative sites, and hence that the net value of the change for these users is zero.

Table 8.5.3 shows the alternative estimates of losses and gains to recreational users for different scenarios. Since the resident sample is larger, and believed to be more representative, the value estimates from this sample are preferred to those derived from local people interviewed on the seafront (locals). Confidence intervals for the mean values were also calculated; however these have not been reported since the valuations do not show distributions approximating to a normal distribution.

The differences in mean value of enjoyment have to be multiplied by the number of visits ('grossing up') in order to estimate annual benefits and losses from different scenarios. This was a problem in the case of Herne Bay since there are no reliable figures for the number of visitors to the central section of the seafront.

Infra-red counters were installed to monitor visitor numbers, but had only been installed for two months when the benefit assessment was needed.

Hence data from the residents survey was used to estimate their visit frequency and could be multiplied up for the four different areas surveyed by the number of electors – this gives an estimate of 207,000 visits by locals each year. Some data on tourist visitor rates was available from a tourism survey (Canterbury City Council, 1986) and from the study of the local tourist economy (Section 8.5.5). This gave a general estimate of 300,000 day and staying visitor days at Herne Bay each year. However, not all of these visitors are likely to use the central section of seafront, and it was assumed that two thirds (200,000) of visitor days a year were spent in the central section (comprising 49.5 per cent day visitors and 50.5 per cent staying visitors).

Multiplying the value estimates by numbers of users for each category gives the estimated annual benefits/losses from alternative strategies in Table 8.5.4. These estimates are economic benefits – for example the benefit from having the offshore reef (drawing C) instead of an eroded seafront is £1.819 millions a year. The lower 95 per cent confidence intervals would still give considerable annual benefits to recreation and were used in the original benefit-cost assessment on the basis that they represent a more conservative benefit estimate. The rock groynes were not considered further in the analysis. The benefit cash flows allowed for the addition in recreational benefit over 1990 conditions from the expected completion of the scheme, but only included the loss averted by preventing erosion of the seafront once this was expected to occur (after 10 years).

A further limitation on the recreation benefit assessment is that only gains in enjoyment to current users are estimated. Coast protection may bring further recreational gains: in terms of more frequent visits by existing users, by attracting users from alternative sites, and by generating entirely new recreational visits. To estimate these additional gains would require much more complex interview surveys over a wider area.

Table 8.5.3
Mean gain or loss (£ in 1990) in enjoyment per visit to Herne Bay seafront between present and future scenarios

Scenario	Residents	Locals	Day visitors	Staying visitors
Loss with erosion	1.33[1]	1.94	1.82	7.56
Additional gain with drawing C	2.26	2.06	1.77	3.94
Additional gain with drawing D	3.10	1.95	1.33	1.23
Additional gain with drawing E	–3.99	–1.24	–1.77	–2.00

Note: 1. Weighted mean of 'stayers' and 'movers', the latter treating 'irrationals' as experiencing zero loss

Table 8.5.4
Annual incremental benefits from preventing erosion and additional recreational attributes of alternative schemes for Herne Bay

Scenario	Total annual benefit (£ in 1990)
Do nothing	–£1.219 millions
Offshore reef (drawing C) additional benefit	£0.620 million
Offshore reef (drawing D) additional benefit	£0.898 million
Rock groynes (drawing E) additional benefit	–£1.203 millions

8.5.5 Local economic impacts

Local economic impacts, as opposed to impacts on the national economy, are not allowed within the benefit-cost calculations for government grant aid towards coast protection and sea defence works. However, Canterbury City Council in this case wished to assess the possible additional impacts on the local tourist economy (in terms of degeneration without a scheme and regeneration with a scheme) since this would affect local economic planning and the local objective of regenerating the tourist industry in Herne Bay.

A purpose built income multiplier model for the Herne Bay economy was developed based on data from Canterbury City Council (1986). The model calculated local income and turnover generated by tourist expenditure (first round effect) and a subsequent round of local income generation induced by the first round. The results of this model were critically dependent on the estimated number of visitors to Herne Bay – the lower estimate of 300,000 a year was used in the assessment (Section 8.5.4). The model estimated annual tourist expenditure of about £3 millions, and that for every £1.00 spent by tourists some £0.35 of local income is generated.

Based on the declining infrastructure estimated over time from erosion and frequent flooding (Section 8.5.2) a decline in the number of tourists was estimated in the 'do nothing' scenario. With the basic scheme (drawing C) a 10 per cent increase in tourist visits was estimated, and with additional recreational facilities (drawing D) an increase of 25-33 per cent was assumed. It was estimated that the present value of lost tourist related expenditure (local income) would be £2.22 millions over a 100 year period. Scheme construction would generate £0.70 million of additional local income, while for the two scheme scenarios the additional local income generated would be £2.4 millions or £5.34 millions respectively. The relatively low value of induced income growth compared with the construction related income generation reflects the discounting of benefits which arise gradually over a long period (a 6 per cent discount rate, the same as the test discount rate, was assumed).

8.5.6 Conclusions

Overall the benefit assessment gave estimates of the present value of benefits from the offshore reef, without additional recreational investment, of £6.76 millions for protection of property and £29.06 millions for recreational gains (or £5.58 millions for the lower 95 per cent confidence interval for recreational gains – which was used in the scheme justification). A local economic gain of £2.4 millions (present value) was also estimated, but should not be added to the other benefits when calculating national economic benefits.

The Herne Bay assessment is important in illustrating the following points in coast protection benefit assessment:

1. Benefits from preventing erosion or breaching of a seafront are likely to depend on a mixture of flood and erosion related losses. It is important to model the changes in land uses at risk of flooding as erosion progresses, and to determine the use of properties which would become subject to frequent flooding. Based on this case study it seems reasonable to assume that properties which are flooded above floor level once in three years, or more frequently, would be uninhabitable and that their market value should be taken as a once-and-for-all loss once they become subject to frequent flooding (in the same way as erosion losses are assessed). Consequently annual average flood losses are not constant over the period of assessment and must be modelled separately for each band of years when either occupied land uses would be different or hydrological conditions and flood risks would be different.
2. When modelling 'permanent' losses from erosion or flooding care needs to be taken to estimate the

least cost 'do nothing' strategy – affecting continued use of flood-prone properties and the provision of utilities and services to properties which continue to be habitable.

3. Recreational benefits may be large compared with property protection benefits, even where urban areas are at risk. The seafront is the first land use to be affected by erosion, and the potential for enhanced benefits for many users mean that large benefits to society may be achieved (although they are not revealed in the marketplace).

4. Where there are many local people using a seafront the CVM which has been developed for estimating recreational impacts may mispecify the impact on some people. So called 'irrationals' would appear to achieve a net welfare gain by going to alternative sites instead of their preferred site. Where only a small part of the sample appear to behave in this way it may reflect measurement error and the mean value of enjoyment excluding these cases may be an appropriate measure. However, if a large part of the sample appear to be irrational this is likely to mean that they gain other benefits from visiting the preferred site (such as multipurpose visits to shops and seafront), and that this unmeasured benefit may be less in the alternative site, hence it is more appropriate to set the change in enjoyment of such people to zero.

5. Local economic impacts are unlikely to be very large, but may still be considerable. These impacts do not change the national economy where the resort competes for domestic visitors. However, they may be important locally and might justify locally funded investment to enhance a coast protection or sea defence scheme.

8.6 The case of Hengistbury Head

8.6.1 Introduction

The purpose of this case study is to demonstrate how

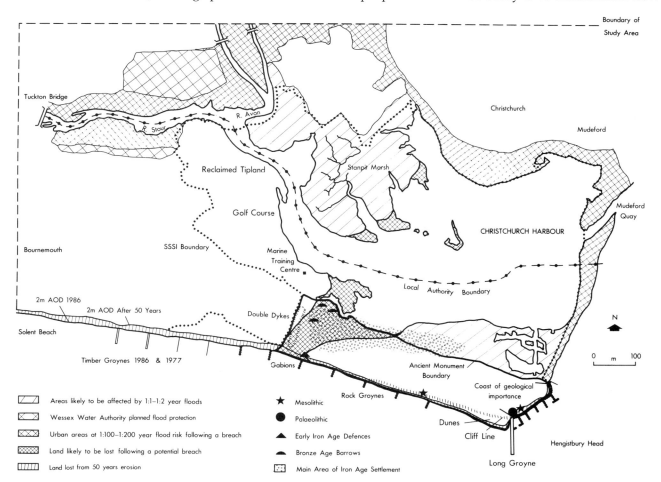

Figure 8.6.1 Hengistbury Head and Christchurch Harbour showing main features, existing coast protection, and works proposed in the scheme appraisal

an evaluation of a site of environmental value should be undertaken using Hengistbury Head and Christchurch Harbour as the example. Coastal erosion in this area was severe enough to warrant a major coast protection project in the late 1980s, and the evaluation process discussed here relates to the pre-project situation (Thompson et al., 1987; Parker and Thompson, 1988).

In this example the major environmental damage would be caused by the losses consequent upon continued coastal erosion, rather than by the project itself, although the adverse impacts of the project also needed consideration in order to modify the project to reduce these impacts. This differs from the situation at many coastal sites where it is the coast protection project itself which causes the main environmental impact.

8.6.2 The characteristics of the case study area

Hengistbury Head and the neck of land which connects it to Bournemouth protect Christchurch Harbour from the south westerly winds and currents which prevail on this part of the south coast (Figure 8.6.1). The Head rises to 36 metres and is the highest point in the immediate coastal zone, and is managed as a public open space. The sheltered harbour has lower water levels under severe storm conditions than does the open sea to the south of the Head and a lower tidal range. This unique set of physical features has produced a multiplicity of human and environmental resources.

The Harbour provides a sheltered anchorage for fishing boats and yachts, and safe water for dinghy sailing and for sailboarding, but is too shallow for major commercial use. A great deal of the low-lying land on the north side of the Harbour in Mudeford and Christchurch has been developed for residential and commercial use. The Harbour-Headland complex is a major tourist attraction and in the early 1980s at least 740,000 day-visits per year were made to the Headland alone (May and Osborne, 1981). The Harbour and Quay areas attract additional visitors. Part of the recreational use of the area is centred on some 350 beach huts which are located on the sand spit to the north of the Head (Figure 8.6.2.).

The Head and Harbour contain, in a very small area, a great diversity of habitats. These habitats have value not only intrinsically but also for teaching and research

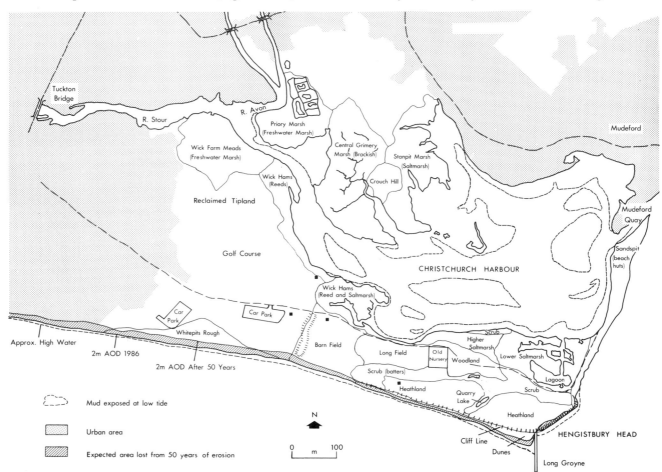

Figure 8.6.2 Main ecological divisions of the Hengistbury Head-Christchurch Harbour complex in 1986

and as an attractive open space. The area contains a Local Nature Reserve designated in 1964, and the whole area was rescheduled in 1986 as a Site of Special Scientific Interest both on ecological and geological grounds. Cliff exposures of geological importance exist to the east of the Long Groyne.

In the past the Headland's unique location on the south coast of England made it a major settlement and port. Major archaeological sites, ranging from Palaeolithic to Iron Age, now cover most of the Head (Cunliffe, 1987). A large part of the coastline, including the Headland, is scheduled as an Ancient Monument.

8.6.3 The process of erosion and the consequent problems

The coastal frontage of Bournemouth has been subject to a series of coast protection works over a number of years. Protection works have included the construction of sea walls, promenades and extensive groyne fields. In addition, dredged beach material has been transferred in large quantities to Bournemouth beaches which have been extensively renourished (Wilmington, 1983). There have also been minor works and many proposals of major projects to protect the coast between Solent Beach and Hengistbury Head.

However, prior to 1986 the only major engineering works comprised the Long Groyne built between 1937 and 1939 (Figure 8.6.1). Hence the coastline between Solent Beach and the Long Groyne was largely unprotected from erosion by the sea. Since the mid-nineteenth century, when naturally occurring ironstone boulders which protected the beach were removed by a mining company, erosion has been rapid along this stretch of coastline. Although the Long Groyne was successful in slowing erosion to its immediate west, and has created an incipient dune system there, up to the late 1980s when Bournemouth Borough Council implemented a coast protection project, erosion remained a serious problem at two points.

First, immediately to the east of the Long Groyne erosion was progressing at a rate of up to 0.8 metres per annum, and there were large cliff falls in 1984–5. This revealed geological sections of national importance, but was causing the loss of wind-trimmed heathland and important archaeological sites at the top of the Head. If erosion rates had increased, as had been predicted, then in the long term the Head could have become dissected.

Secondly, and more significantly, erosion at up to 1.25 metres per year was taking place near Double Dykes – an Iron Age settlement defence system. This is the narrowest point in the 'landbridge' connecting Hengistbury Head and Bournemouth, and is also a low point. Here the coastal cliff is the highest point, but is still only some 4 metres high and is comprised of soft, sandy gravel. A severe storm overtopping this cliff could be powerful enough to erode a channel northwards into the Harbour.

Hydraulics Research Ltd (1986) indicated that such a breach would become a permanent entrance to the Harbour if it ever occurred, and that the channel could be 100–150 metres wide. Not only would much more land be lost in a breach than from high current erosion rates, but a breach would radically alter the tidal regime of the Harbour. Low tides would become significantly lower and high tides slightly higher, but under severe conditions a harbour open to the full force of south-westerly gales and wave set-up would suffer occasionally much higher water levels (Parker and Thompson, 1988). Hence, flooding around the harbourside would become much more severe, affecting both low-lying properties and the ecological character of the marshland.

In summary, coastal erosion threatened to create a sequence of impacts on the Hengistbury Head area in the following order:

1. A breach of the coastline at Double Dykes would open the harbour to wave action, eliminating or severely reducing the area of safe water.
2. Low-lying urban areas would become much more flood-prone than at present. More properties would become flood-prone; flooding would be deeper in currently flood-prone properties; and the frequency of flooding would increase. Some properties could be flooded once per annum or more.
3. The Headland would become an island, making access difficult for visitors.
4. Similarly, the only land access to the beach huts would be removed, and some beach huts would become flood-prone.
5. Major archaeological sites would be lost in a breach, and were already being lost to erosion to the east of the Long Groyne.
6. The environment and ecological diversity of the Site of Special Scientific Interest would be altered in a complex way; some valuable areas would be lost.

8.6.4 The coast protection project

Following extensive hydraulic and economic investigations, and consideration of the various options, Bournemouth Borough Council implemented a coast protection project in the late 1980s comprising a series of timber and rock groynes (Figure 8.6.1) with beach replenishment. In addition a sea wall has been constructed to the east of the Long Groyne to protect the cliffs at that point.

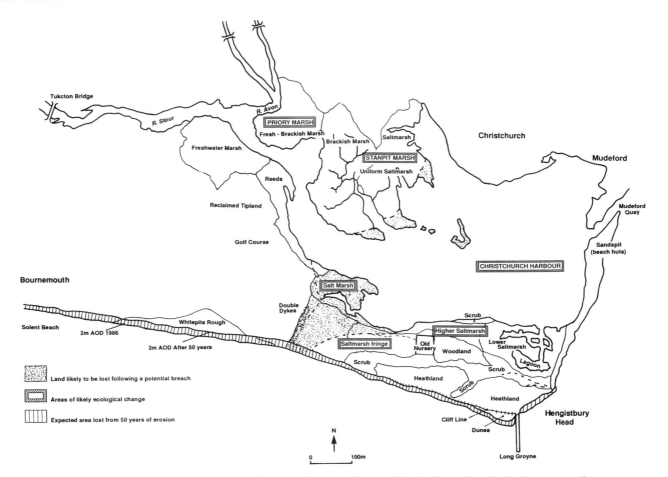

Figure 8.6.3 Main ecological changes predicted in the Hengistbury Head-Christchurch Harbour complex following a breach at Double Dykes

8.6.5 The assessment of ecological impacts

The project appraisal approach adopted in 1986 to evaluate this Bournemouth Borough Council project (Thompson et al., 1987) combined qualitative and non-monetary quantitative evaluation of the likely ecological impacts of (a) continued coastal erosion, and (b) protecting the coast from further erosion.

An ecological 'audit' of the study area was undertaken during 1986, in consultation with the Nature Conservancy Council and local environmental organisations. This determined the existing major ecological divisions (using a classification of habitats) (Figure 8.6.2) which would be retained if the coastline were to be protected from erosion, and the most likely ecological divisions should erosion be permitted to continue and following a breach (Figure 8.6.3). The major habitat types were distinguished and fifteen ecological divisions were identified which are discrete areas supporting specific communities.

Appraisal method and limitations This initial ecological assessment has been re-evaluated in the form of matrices (Section 6.6.1) which show the relative importance of each of the components of ecological value in a systematic visual form which can be easily comprehended by those without specialist ecological knowledge (Figures 8.6.4 to 8.6.6). These three matrices summarise the relative ecological value of each principal site under: existing conditions (Figure 8.6.4), under conditions following the implementation of a coast protection project (Figure 8.6.5), and under conditions in which erosion is allowed to continue (Figure 8.6.6) – without a coast protection project.

The ecological interest of the study area is mainly related to the high diversity of habitats within a small area rather than to the outstanding ecological importance of any individual habitat or species. The interest and value of this area in ecological terms is reflected in its designation as a Site of Special Scientific Interest and in the designation of part of the area, known as Stanpit marshes, as a Local Nature Reserve.

The initial evaluation failed to consider the ecological importance of the harbour itself, and this should be considered as an additional habitat, comprising a further division of the matrix. Some authorities, for example Barnes (1989), would classify it as an extreme type of lagoon and it has been shown as such in the matrix.

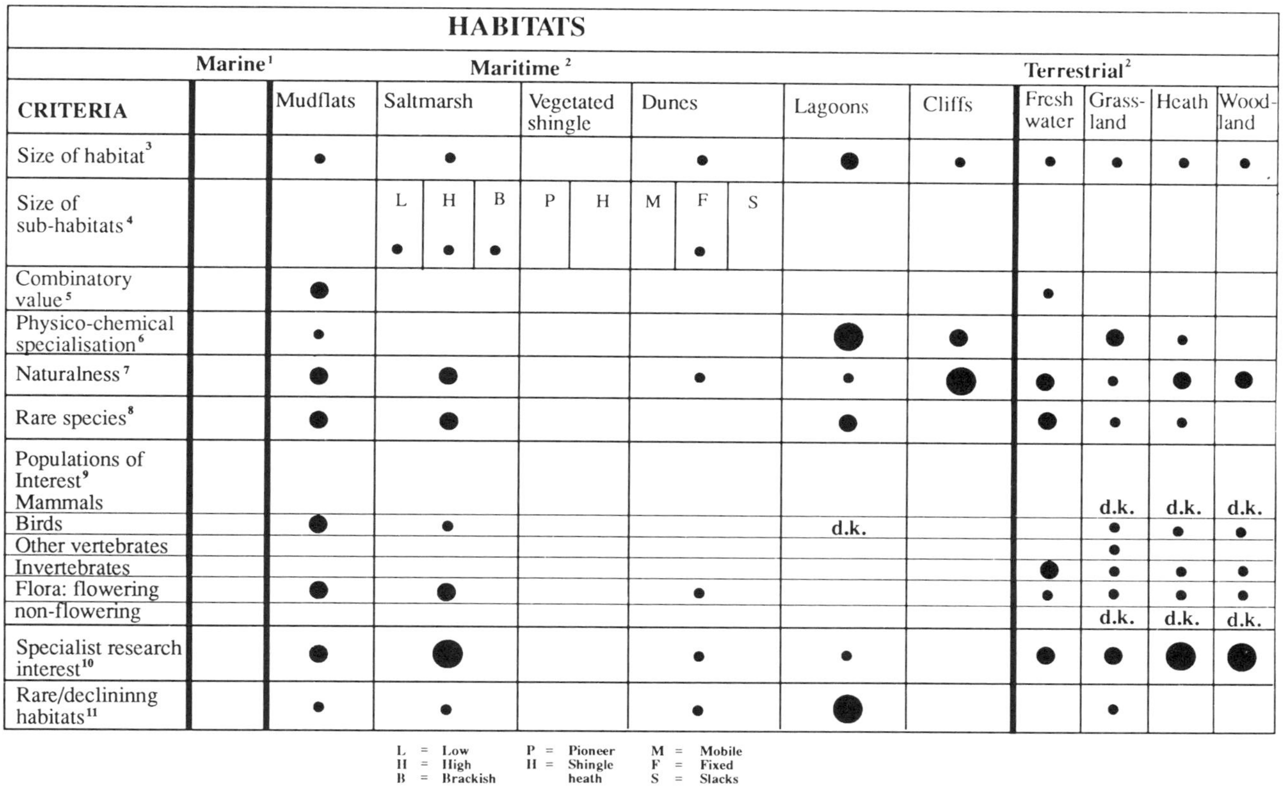

CASE STUDIES 185

CRITERIA	Marine[1]	Maritime[2]						Terrestrial[2]			
		Mudflats	Saltmarsh	Vegetated shingle	Dunes	Lagoons	Cliffs	Fresh water	Grass-land	Heath	Wood-land
Size of habitat[3]		•	•		•	●	•	•	•	•	•
Size of sub-habitats[4]			L • / H • / B •	P • / H •	M / F • / S						
Combinatory value[5]	●							•			
Physico-chemical specialisation[6]	•					●	•		•	•	
Naturalness[7]	●	•			•		•	●	•	●	●
Rare species[8]	●	•				•		●	•	•	
Populations of Interest[9] Mammals									d.k.	d.k.	d.k.
Birds	●	•				d.k.			•	•	•
Other vertebrates									•		
Invertebrates								●	•	•	•
Flora: flowering	●	•		•				•	•	•	•
non-flowering									d.k.	d.k.	d.k.
Specialist research interest[10]	•	●			•	•		•	•	●	●
Rare/declining habitats[11]	•	•		•		●			•		

L = Low P = Pioneer M = Mobile
H = High H = Shingle heath F = Fixed
B = Brackish S = Slacks

Figure 8.6.4 Ecological summary matrix for current conditions (1986) at Hengistbury Head-Christchurch Harbour

CRITERIA	Marine[1]	Maritime[2]						Terrestrial[2]			
		Mudflats	Saltmarsh	Vegetated shingle	Dunes	Lagoons	Cliffs	Fresh water	Grass-land	Heath	Wood-land
Size of habitat[3]		•	•		•	●	•	•	•	•	•
Size of sub-habitats[4]			L • / H • / B •	P • / H •	M / F • / S						
Combinatory value[5]	●							•			
Physico-chemical specialisation[6]	•					●	•		•	•	
Naturalness[7]	●	•			•		●	●	•	●	●
Rare species[8]	●	•				•	▨	●	•	•	
Populations of Interest[9] Mammals									d.k.	d.k.	d.k.
Birds	●	•				d.k.			•	•	•
Other vertebrates									•		
Invertebrates								●	•	•	•
Flora: flowering	●	•		•			▨	•	•	•	•
non-flowering									d.k.	d.k.	d.k.
Specialist research interest[10]	•	●			•	•		•	•	●	●
Rare/declining habitats[11]	•	•		•		●			•		

L = Low P = Pioneer M = Mobile
H = High H = Shingle heath F = Fixed
B = Brackish S = Slacks

▨ Ecological change

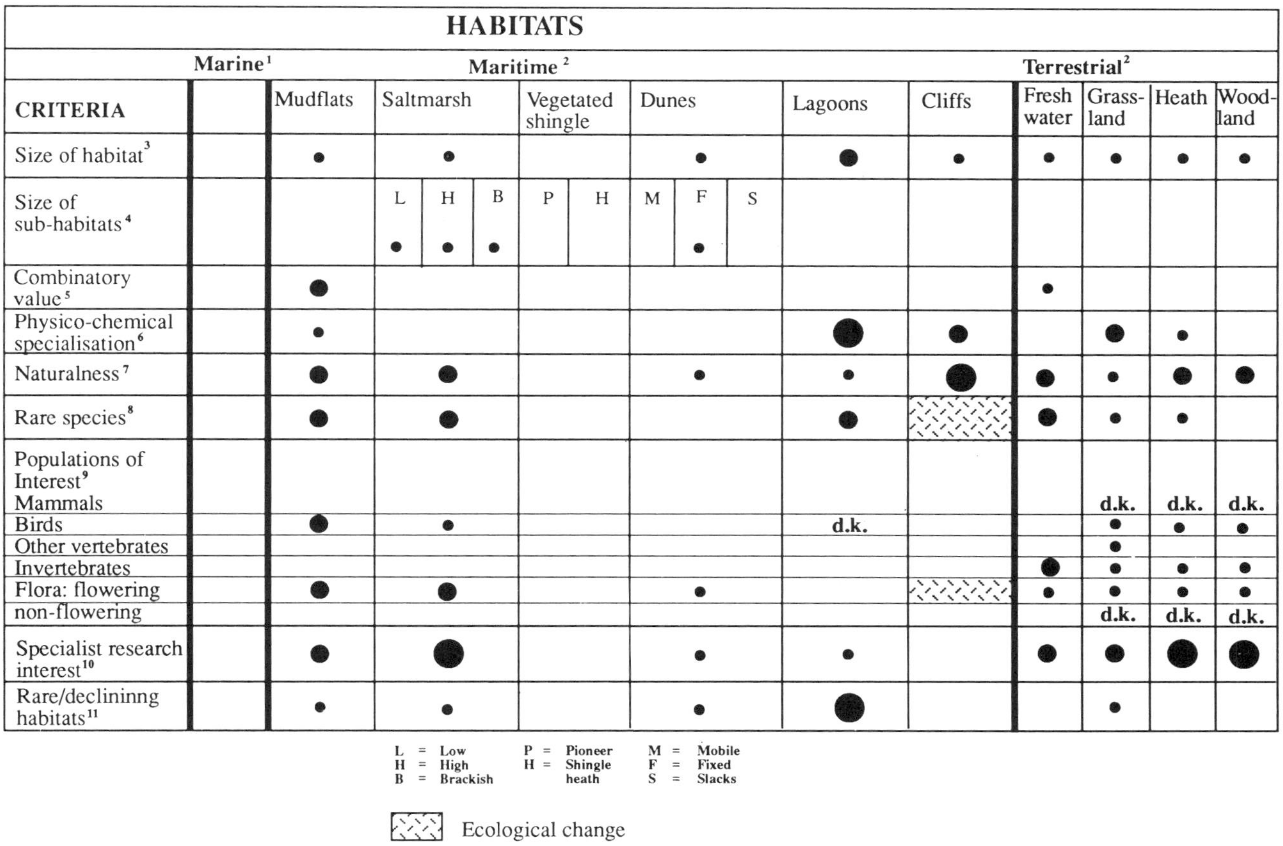

Figure 8.6.5 Ecological summary matrix for future conditions with coast protection at Hengistbury Head-Christchurch Harbour

186   THE ECONOMICS OF COASTAL MANAGEMENT

CRITERIA	Marine[1]	Maritime[2]									Terrestrial[2]			
		Mudflats	Saltmarsh			Vegetated shingle	Dunes			Lagoons	Cliffs	Fresh water	Grass-land	Heath	Wood-land
Size of habitat[3]		●	●				●			▨		●	●	●	●
Size of sub-habitats[4]			L ●	H ●	B ▨	P	H	M	F ●	S					
Combinatory value[5]	●											▨			
Physico-chemical specialisation[6]		▨								▨	●		▨	●	
Naturalness[7]		●	●				●				⬤	●		●	●
Rare species[8]		▨	▨							▨		⬤	●	●	
Populations of Interest[9] Mammals												d.k.	d.k.	d.k.	
Birds		●	●							d.k.		●	●	●	
Other vertebrates													●		
Invertebrates												●	●	●	●
Flora: flowering		▨	▨				●					●	●	●	●
non-flowering												d.k.	d.k.	d.k.	
Specialist research interest[10]		●	⬤				●			▨		●	▨	⬤	⬤
Rare/declininng habitats[11]		●	●				●			▨			●		

L = Low   P = Pioneer   M = Mobile
H = High   H = Shingle heath F = Fixed
B = Brackish   S = Slacks

▨ Ecological change

Key to the Habitat - Criteria matrix

Where information about site is unknown, this should be indicated on the matrix (for example, by DK).

[1]Marine: The marine habitat may be defined as below low water mark and other inter-tidal habitats not included in categories of maritime habitats. Classification of marine habitats is more complex than can be represented on the matrix. For details consult: *A Coastal Directory for Marine Conservation* (Gubbay, 1988) and/or the Marine Conservation Society.

[2]Maritime and Terrestrial: Habitat divisions based on Ratcliffe (1977). Consult English Nature/Countryside Council for Wales, Local Authority ecologist and any management body for the site. These organisations should have information on any other bodies which may be involved at the site.

[3]Size of habitats:

    large
    medium
    small

Habitat size category does not depend on actual extent at the site but is assessed relative to the size range of that habitat in general.

Examples:

Mudflats

Area (ha)
≤ 1600
400-1599
≥ 400

Saltmarsh and sand dunes

Area (ha)
≤ 400
80-399
≥ 80

Vegetated Shingle

Area (ha)
≤ 80
20-79
≥ 20

Cliffs

Undistributed run of cliff length (km)
≤ 40
8-39
≥ 8

(Based on Ranwell's semi-quantitative index (Ratcliffe 1986)

Other habitats are more complex, e.g. lagoons have several alternative categorisations (Gubbay 1988; Barnes 1989).

[4]Size of sub-habitats: Where applicable, habitats may be separated in sub-habitats, if the information is available.

    large
    medium
    small

NB: Habitats which by their nature have no sub-habitats are not necessarily of less value.

[5]Combinatory value: i.e. whether adjacent to other similar habitat sites, either immediately adjacent or as part of a sequence with interbreeding populations or migratory species movements.

    Adjacent to another habitat site International/National value
    Adjacent to another habitat site of Regional value
    Adjacent to another habitat site of Local value

[6]Physico-chemical specialisation: rare or unusual conditions of geology, soils, sediment, hydrology or geomorphology.

    Internationally/Nationally rare
    Regionally rare
    Locally rare

[7]Naturalness: a visual assessment on-site.

    Undisturbed by people
    Some pathways, some litter
    Structures, tracks, trampling, polluted water

[8]Rare species:

    Wildlife and Countryside Act protected species, International and British Red Data Book species
    Regionally rare species, nationally rare species not included in the above
    Locally rare species

[9]Important populations of interest:

    Internationally/nationally important populations
    Regionally/Locally important populations
    Representative interest

NB: Information may not be available for all species groups

[10]Specialist research interest:

    Recorded history of past research and on-going research
    Recorded history or future research plans
    Some records

[11]Rare/declining habitats:

    Internationally/nationally rare/declining (defined in general)
    Regionally rare/declining (defined by region)
    Locally rare/declining (defined by site)

Figure 8.6.6   Ecological summary matrix for future conditions without coast protection at Hengistbury Head-Christchurch Harbour

A further limitation is that the shoreline at the base of the cliffs and the marine area on the seaward side of Hengistbury Head was not considered in the original survey and an evaluation of its significance has not been included in the matrix although it is a habitat which could be affected by both coastal erosion and the protection project. This illustrates the significance of a systematic checklist in the assessment process.

There are two important steps in completing the matrices: allocating values to the matrix cells and combining values to give an assessment of overall performance.

Allocating values to the matrix cells The matrices show that the value of the site comprises four main components. The first is its high diversity of habitat types. Every coastal habitat except stabilised shingle is represented, and the sub-habitats of saltmarsh include low, high and brackish marsh, although areas of each are small in size in comparison to the national size range of these habitats. Secondly, the physico-chemical characteristics of the lagoon area and its national rarity as a habitat contribute to the overall site value. The third important factor is the naturalness of the cliffs on the headland. Fourthly, the research data available for the saltmarsh and wind-trimmed heath vegetation, and for the woodland in relation to its ornithological records, make a significant contribution to the site value.

The 'combinatory role' of the mudflats (i.e. their ecological role in a wider context than that of this site alone) is given an intermediate sized indicator in the matrix because of the inter-relationship with other sites of regional value, for example as a feeding ground for bird species using adjacent sites such as Poole Harbour.

Assessing overall importance of the site It is not possible to combine the values in the cells of the matrix in order to give a simple summary of the overall importance of the site (Coker and Richards, 1992). It would be unrealistic to attempt to obtain a 'mean value' from the visual indicators in the rows or columns of the matrix, neither should the maximum value in the row be taken as an overall maximum. It is therefore important that a matrix is always completed by a professional ecologist using the best available ecological expertise in order to ensure comparability as far as it is possible to do so.

It should be stressed that the function of the matrix is for comparative purposes only, to enable the significant changes which would occur both with and without a coastal protection project to be clearly identified in order to make a judgement about the overall vulnerability of the area to changes. It should not be seen as an attempt to evaluate a site overall in order to enable comparison with other sites. Therefore any difference between the professional judgements of individual ecologists is unlikely to be significant in this context.

Change in ecological interest of the site without coast protection The likely impacts on the salt marshes of a breach near Double Dykes are complex (Figures 8.6.2 and 8.6.3). High tides and higher storm tides would maintain saline influence, but lower tides would increase the drying out of the salt marshes, tending to benefit the plants found on higher salt marsh, which could spread to the detriment of the more desirable lower salt marsh. Furthermore, some areas of salt marsh would be lost, either directly in the breach forming event or consequently from wave action through the breach. At risk from this, for example, is a colony of Dwarf Spike-rush *(Eleocharis parvula)* on an area of brackish mud due north of the likely breach. This species is categorised as 'nationally scarce' by English Nature.

The areas most directly at risk from a breach are two fields immediately east of Double Dykes. These areas comprise unimproved maritime grassland, with a wide variety of herbs, and tussocks of heather *(Calluna vulgaris)*. This area is an important amenity area for the Head, but a breach would leave only some 40 per cent of the existing area as fields.

Although the woodland and sanctuary areas (Old Nursery) on the Head are not directly at risk from a breach, these areas are in fact low-lying; much is below 1.55 metres AOD. Hence large parts of these areas would be subject to saline flooding at least once a year if a breach occurred, which would seriously affect their ecological character. Although mature trees would probably survive, provided the groundwater remains fresh, regeneration and the existing ground flora would be affected, diminishing the long term interest of the area. Furthermore, these changes would limit the opportunities for improving the ecological value of the area, such as by the creation of dragonfly ponds in the Old Nursery, as saline flooding would reduce the ecological diversity.

Overall, there would be reduction from 15 to 11 units with a breach. This is represented in Figure 8.6.6. as a loss of lagoon features, a loss of the sub-habitat of brackish marsh and a loss in rare species from mudflats. These changes can all be considered as ecological losses although there might also be some gains which have not been identifiable.

Change in ecological interest of the site with coastal protection At Hengistbury Head the changes caused by the proposed coastal protection project are restricted in scope but would result in losses to the cliff area which would be an ecological cost of the project (Figure 8.6.5). In addition, the building of groynes would be likely to alter the seashore and marine habitat in ways which are more difficult to predict and are not fully represented in the matrix. However, it should not be assumed that all ecological change is necessarily detrimental, as the creation of rock groynes, in particular, could provide a new substrate for marine organisms and enhance the marine ecological interest overall.

Species diversity A further method available for assessment of the ecological impacts of coastal erosion and protection is the use of a diversity index, of which there are several (Section 6.3.5). The index devised by Shannon and Weaver (1949) – H – (Section 6.3.5) was employed in the initial assessment to measure habitat diversity (Thompson *et al.*, 1987). The value for H in 1986 (existing conditions and without coast protection) in this case was 2.75, but would be only 2.11 following a breach.

8.6.6 The assessment of archaeological impacts

A whole range of archaeological sites are found on Hengistbury Head (Figure 8.6.7) covering almost every period from the Upper Palaeolithic to the Roman Iron Age as well as industrial archaeological sites from the nineteenth century. Visually attractive features such as the Double Dykes and the Bronze Age tumuli have attracted visitors for many years, and visits to the area by the archaeologically aware are apparently growing.

Over the past 200 years the coastline of Hengistbury Head has advanced about 100 metres northwards, considerably reducing the Head. Cliff erosion has caused a loss of archaeological sites and continued erosion would accelerate these losses. The impact of a possible breach across the headland in the Double Dykes area would be even more serious and there would be drastic consequences for the archaeology of the area as the predicted course of the breach channel would pass through one of the most important areas of prehistoric settlement as well as seriously threatening the Dykes themselves. It was not envisaged that the proposed coast protection works would have any detrimental impact upon these archaeological sites: coast protection was judged to benefit archaeological interests.

The initial assessment of archaeological impacts reported in the 1987 study (Thompson *et al.*, 1987, Table 9.1) focused upon the principal archaeological periods present (for example the Palaeolithic, Mesolithic, and Early Iron Age); their apparent archaeological status; the impact of erosion; and the impact of a breach. This initial evaluation has been re-assessed in the light of the approach developed by English Heritage (Chapter 6.5.3). However, their scoring system using a standardised set of assessment criteria has

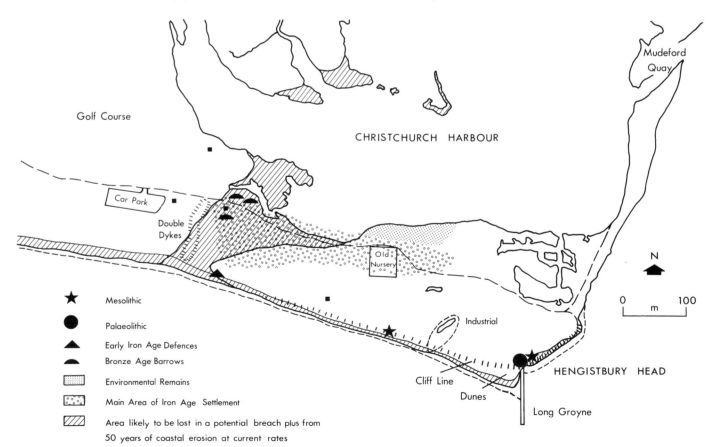

Figure 8.6.7 Features of archaeological importance threatened by erosion at Hengistbury Head

been developed in order to assess individual monuments and sites. It is not appropriate to apply it to an area such as Hengistbury, which contains a wide variety of records from different periods and whose interest lies partly in the overall diversity of the archaeological record at the location and partly in the significance of the individual sites. Of particular significance in an international context is the evidence of the Late Iron Age indicating an 'urban overseas trading centre'. Hengistbury is now recognised as a site of crucial importance in understanding the Iron Age of southern Britain. In the first century Hengistbury Head became a port and entrepot with well developed contacts with Italy and north-western France. Excavations conducted earlier this century in Long Field and Barn Field uncovered considerable evidence of this settlement, and the site contained a great wealth of pottery, coins and goods which were traded. Further archaeological work is planned for this site and many years of excavation and scientific investigation will be required before any detailed interpretation of this major site could be made.

Further work is needed on a methodology to enable evaluation of sites with a complex archaeological record.

8.6.7 The assessment of geological impacts

The initial assessment of geological impacts within the 1987 study was incomplete (Thompson *et al.*, 1987). A more complete assessment using the methods outlined in Section 6.6.3 has now been undertaken.

Hengistbury Head provides a unique geological exposure of the Hengistbury Beds of Eocene or middle Tertiary age. These contain a variant of the Barton Beds, and the Boscombe Sands formation of the Bracklesham Beds, and are of significance for comparison with deposits at other locations along the coastline. There are two features important to the site: an unusual 'marginal' variety of the Barton Beds in the upper cliff; and in the base of the cliff Boscombe sands which contain a unique type of bituminous sand important in studies of the late Auversian (Upper Bracklesham) period. The exposure site is a Site of Special Scientific Interest on geological grounds.

Current erosion in the section immediately east of the Long Groyne maintains the geological exposures in pristine condition and makes them available for both research and teaching. However, rapid erosion of 0.8 metres per annum or more is undesirable in the very long term as far as geological interest is concerned since this could eventually lead to the loss of the cliffs.

The coast protection works proposed by Bournemouth Borough Council would have an adverse effect on the areas of geological value as the cliffs would self grade and vegetate. This would result in less and less of the exposure being available for research and teaching.

Following consultation with the Nature Conservancy Council which initially objected to Bournemouth Borough Council's project because of its adverse effect of preventing active erosion, the Council agreed to the signing of a Section 15 agreement under the Countryside Act of 1968. This agreement now enables English Nature to manage the SSSI with the Council acting as their agents to clear eroding material from the base of the cliff, thus permitting erosion to continue and the cliffs to remain in pristine condition.

8.6.8 The assessment of recreational impacts

The likely recreational impacts of a breach of the coastline at Double Dykes are assessed in this section, as are the likely effects of allowing coastal erosion to continue along the Head. Hydraulic analyses revealed that, should a breach occur at Double Dykes, Hengistbury Head would probably become an island (although it was also considered possible that the existing harbour entrance might close up, introducing the possibility of a new landbridge to the Head).

After careful consideration of the particular circumstances at Hengistbury Head, variations of both the Clawson Travel Cost Method and the Contingent Valuation Method were applied in 1986, and this section reports the results of the analysis subsequently undertaken (Thompson *et al.*, 1987).

Evaluation of existing data A search of existing data sources revealed that an attraction site survey was undertaken by Dorset Planning Department in the summer of 1974. This survey investigated 19 sites in south-east Dorset, one of which was Hengistbury Head. The data was evaluated to determine their possible suitability for use in applying the Clawson Travel Cost Method approach. Unfortunately, the appraisal of the data revealed that it was insufficiently detailed on the origin of journeys; there was little indication of trip expenditures and the data was out of date. Subsequently, further information from a survey undertaken in 1980 (May and Osborne, 1981) became available and provided corroborative evidence when placed together with the results from the special survey described below.

Survey of recreational use of Hengistbury Head A special survey of recreational users of the Head was undertaken in 1986. The survey was designed to provide information on the number, type and sources of visitors to the Head area during the summer months.

A questionnaire was constructed in three parts. Most of the initial questions were of a factual nature, while other parts involved the respondent making value judgements. The questionnaire was designed in such a way that variants of both the Travel Cost Method and Contingent Valuation Method could be used in estimating the benefits of protecting the Head area (Thompson et al. 1987, pp. 106–111).

The main data items within the questionnaire were as follows:

1. Visitor origin;
2. Purpose of visit;
3. Travel mode;
4. Group size;
5. Length of stay at the Head;
6. Total expenditure at the Head;
7. Next best alternative if access to the Head were not possible; and
8. Willingness-to-pay for journeys by boat if the Head became an island.

A sample of 397 groups (comprising 1,528 people) was interviewed using a questionnaire of approximately 15 minutes duration. Interviews were undertaken at nine predetermined sites over the Head and at the Head car park. The non-response rate was 2 per cent. The full results of the survey are summarised in Thompson et al. (1987, pp. 66–75). Three principal types of visitor were identifiable: local visitors; day visitors; and tourists.

Two approaches, based on scenarios, were adopted to evaluate in economic terms the value to society of preserving the current access to the Head against the risk of a possible breach. A variation of the Travel Cost Method was used to estimate the loss of consumer surplus from not being able to visit the Head area.

The 'no access' scenario Respondents were asked to state their next best alternative recreational destination if no access were possible to Hengistbury Head. For the three main visitor types the difference between the distance travelled between their origin and the Head and the distance between their origin and the next best alternative destination was calculated. A weighted mean of the change in distance was used to reflect the relative 'strength' of the answers provided, as indicated by the total number of visits per year which it was estimated that the questionnaire represented.

To evaluate the travel costs, resource and travel time costs were calculated using methods explained in Thompson et al., (1987, pp. 69–81). Time values were calculated as leisure time values (Department of Transport, 1986). The economic loss from not being able to visit Hengistbury Head and having to use alternative sites is summarised in Table 8.6.1. The weighted mean method is preferred because it overcomes some of the biases of treating all questionnaires equally irrespective of the number of visits they represent.

A qualification of the results is that tourists may gain some benefit from the journeys which they make which could counterbalance the costs involved. The method assumes that the journeys lost by people who would visit their alternative site less often are counterbalanced by the extra journeys of people who would visit their alternatives more often. The estimated loss takes no account of the loss of utility due to not having access to the unique features of Hengistbury Head – clearly the Head has certain intrinsic merits, otherwise on the evidence from the response of local visitors, people would visit alternative sites which are closer to their homes and thereby save money.

Access by boat scenario The second method used to estimate the value of protecting the Head was directly to ask respondents their willingness-to-pay for boat trips should the Head become an island.

A weighted mean was used to calculate the willingness-to-pay for a boat service to the Head area if it became an island. A factor was used to take into account the number of visits lost because people deemed a boat trip as an inconvenience, hence the

Table 8.6.1
Aggregate travel cost results, Hengistbury Head
(£ value per year)

Type of visitor	Extra travel costs (£ January 1986)	
	Using weighted means	Without weighted means
Locals (gain)	−44,690	−64,141
Other daytrippers	9,385	53,599
Tourists	602,284	1,340,374
TOTAL	566,989	1,329,831

value estimated was only for people who would continue to visit the head by boat. The method resulted in a weighted mean willingness-to-pay, per return boat journey, for local people of £0.62 and for tourists of £0.72. From these figures a total willingness-to-pay to maintain access to the Head (and the consumer surplus thereby achieved) of £191,901 per year was estimated.

It is important to recognise that this result ignores the loss to visitors who would not be prepared to use the boat service, and underestimates the full loss for those who would use it less often than they currently visit the Head.

However, it is apparent that although willingness-to-pay for a boat trip was interpreted as an opportunity cost avoided by the scheme, the way in which visit enjoyment would be affected by a boat trip should also have been assessed. It may be that a boat trip would have increased the enjoyment of a visit for some people, in which case willingness-to-pay for a boat trip would include an element of recreational benefit, rather than a loss. At the time of the analysis, this interpretation was not taken into account.

It is, therefore, important when using Contingent Valuation Methods to ensure that the change in enjoyment under different scenarios is adequately valued.

8.6.9 Conclusions and lessons learnt from case study

1. The study is one of the earliest examples of an attempt to incorporate evaluation of environmental components into the assessment of coastal scheme benefits and was used to develop the methods and approaches which are now recommended. This means that the approaches adopted differ in a number of ways from the methods recommended in this Manual.
2. Ultimately, the coast protection scheme at Hengistbury Head was justified in economic terms without the need to put a value on the environmental components. The study identified that these impacts were important but also found that these components cannot easily be valued in economic terms. This creates difficulties where environmental evaluation is critical to the benefit-cost calculation.
3. At the outset of a scheme proposal, judgements need to be made as to which aspects of the environment are important. In the case of Hengistbury Head, it was clear that agricultural use was not a significant factor whereas ecological, archaeological, geological interest and recreational use were all important.
4. In any scheme proposal, a pre-feasibility assessment is essential in order to assess the order of magnitude of the environmental effects. In this case, it was clear that a breach would significantly affect the ecological, archaeological, and recreational use whereas the scheme proposals would affect the geological interest. Thus a more detailed environmental appraisal could focus on these aspects.
5. Sufficient lead-in time should always be allowed for the environmental appraisal to identify the locations which will be both directly and indirectly affected by the proposed scheme, and for collection of information about the environment, whether from primary or secondary sources. This was inadequate in the initial Hengistbury Head study.
6. Specialist opinion should always be sought on the ecological, archaeological and geological interest of the area. The case study helped to identify the professionals who should be consulted (Sections 6.5.2, 6.5.3 and 6.5.4 respectively).
7. In the case of ecological interest, the available information should now be compiled in the form of the recommended matrices to facilitate comparisons (Section 6.6.1), this approach was developed after this case study. The use of the matrices enables a systematic approach to be adopted which focuses on the evaluation of **relative** impacts rather than evaluating the ecological interest *per se* and would have improved the initial Hengistbury Head case study.
8. Archaeological information is difficult to present meaningfully in the form of a matrix (Sections 6.2.2), particularly where the archaeological importance of a location is derived from the presence of a number of sites of different periods, as was the case at Hengistbury Head. In particular, this difficulty may be compounded by a lack of information about the individual sites.
9. Most coast protection schemes will affect different components of the environment in different ways. The Hengistbury case study emphasises the need for a formal mechanism for discussing trade-offs, based on the presentation of the ecological matrices and the available archaeological and geological opinion, and provides an example of the negotiation of a compromise solution.
10. Currently, it is not possible to make an environmental evaluation, whether of aspects of archaeological, ecological, or geological importance, in economic terms. Even indirect methods of economic ('monetary') valuation are not sufficiently developed to be used in relation to these environmental components.
11. The assessment of recreational benefits used surrogate measures which are not regarded as being

as reliable as the 'enjoyment of visit' CVM approach subsequently developed. Also the assessment ignored the potential impacts on the Harbour-Headland landscape which might well have been important and might have been included in the scenarios presented to users had the CVM been used.

References

Aachen, C. H., *Interpreting and Using Regression*, Beverly Hills, Sage, 1982

Abelson, P., *Cost Benefit Analysis and Environmental Problems*, Aldershot, Gower, 1979

Abt, S. R., Wittler R. J., Taylor A. and Love D. J., 'Human stability in a high flood hazard zone', *Water Resources Bulletin*, 1989, vol. 25, no. 4, pp. 881–890

Akerlof, G. A., 'The market for lemons: qualitative uncertainty and the market mechanism', *Quarterly Journal of Economics*, 1970, vol. 84, no. 4, pp. 488–500

American Psychological Association, *Technical Recommendations for Psychological Tests and Diagnostic Techniques*, Washington, American Psychological Association, 1954

Antle, L. G., 'Analysis of property values to determine the impact of flooding' in *Proceedings of the Social Scientists Conference*, Fort Belvoir, Va., Institute for Water Resources, 1977

Arrow, K. J. and Fisher, A. C., 'Environmental preservation, uncertainty, and irreversibility', *Quarterly Journal of Economics*, 1974, vol. 88, pp. 312–319

Arrow, K. J. and Lind, R. C., 'Uncertainty and the evaluation of public investment decisions', *American Economic Review*, 1970, vol. 160, pp. 364–378

Association of Directors of Social Services, *Major Disasters: The Social Services Contribution*, Reading, Association of Directors of Social Services, 1987

Barnes, R. S. K., 'The coastal lagoons of Britain: an overview and conservation appraisal', *Biological Conservation*, 1989, vol. 49, pp. 295–313

Bateman, I., Green, C. H., Tunstall, S. M. and Turner, R. K., *The Contingent Valuation Method*, Report to the Transport and Road Research Laboratory, Enfield, London, Middlesex Polytechnic Flood Hazard Research Centre, 1991

Beiswanger, R. E., 'An endangered species, the Wyoming Toad *Bufo hemiophsys baxteri* – the importance of an early warning system', *Biological Conservation*, 1986, vol. 37, pp. 59–71

Bisset, R., 'Quantification, decision-making and environmental impact assessment in the United Kingdom', *Journal of Environmental Management*, 1978, vol. 7, pp. 43–58

Black, C. J. and Bowers, J. K., 'The level of protection of UK agriculture', *Oxford Bulletin of Economics and Statistics*, 1984, vol. 46, no. 4, pp. 291–310

Blockley, D., *The Nature of Structural Design and Safety*, Chichester, Ellis Horwood, 1980

Bowers, J., *Economics of the Environment: the Conservationists' Response to the Pearce Report*, Telford, British Association for Nature Conservation, 1990

Brack Jr, V., Holmes, V. R., Cable, T. T. and Hess, G. K., 'A wetland habitat assessment using birds', in Magoon, O.T. (ed.), *Coastal Zone*, vol. 1, American Society of Civil Engineers, 1987

Brennan, A., *Thinking About Nature: An Investigation of Nature, Value and Ecology*, London, Routledge, 1988

Brookshire, D. S., Eubanks, L. S. and Sorg, C. F., 'Existence values and normative economics: implications for valuing water resources', *Water Resources Research*, 1986, vol. 22, no. 11, pp. 1509–1518

Brookshire, D. S., Thayer, M. A., Schulze, W. D. and d'Arge, R. C., 'Valuing public goods: a comparison of survey and hedonic approaches', *American Economic Review*, 1982, vol. 72, no. 1, pp. 165–176

Brown, T. C., 'The concept of value in resource allocation', *Land Economics*, 1984, vol. 60, no. 3, pp. 213–246

Callicott, J. B., 'Intrinsic value, quantum theory, and environmental ethics', *Environmental Ethics*, 1985, vol. 7, pp. 257–75

Canterbury City Council, *Canterbury Tourism Survey 1985*, Canterbury, Kent, Canterbury City Council, 1986

Caulkins, P. P., Bishop, R. C. and Bouwes, N. W., 'The travel cost model for lake recreation: a comparison of two methods for incorporating site quality and substitution effects', *American Journal of Agricultural Economics*, 1986, pp. 291–297

Central Statistical Office, *Regional Trends 26*, London, HMSO, 1991

Central Statistical Office, *Economic Trends*, London, HMSO, 1992

Chatterton, J. B. and Penning-Rowsell, E. C., Computer modelling of flood alleviation benefits, *Proceedings of the American Society of Civil Engineers*, 1982, vol. 107 (WR2), pp. 533–547.

Cheshire, P. C. and Stabler, M. J., 'Joint consumption benefits in recreational site "surplus": an empirical estimate', *Regional Studies*, 1976, vol. 10, pp. 343–351

Chevas, J. P., Stoll, J. and Sellar, C., 'On the commodity value of travel time in recreational activities', *Applied Economics*, 1989, vol. 21, pp. 711–722

Clausen, L., *Potential Dam Failure: Estimation of Consequences and Implications for Planning*, unpublished M.Phil. thesis,

Enfield, London, Middlesex Polytechnic, 1989

Clawson, M., *Methods of Measuring the Demand for Outdoor Recreation*, Reprint No. 10, Washington DC, Resources for the Future, 1959

Clawson, M. and Knetsch, J., *Economics of Outdoor Recreation*, Baltimore, Johns Hopkins, 1966

Coker, A. and Richards, C., *Valuing the Environment: Economic Approaches to Environmental Evaluation*, London, Belhaven Press, 1992

Cole, G. and Penning-Rowsell, E. C., 'The place of economic evaluation in determining the scale of flood alleviation works', in *Flood Studies Report – 5 years on*, London, Institution of Civil Engineers, 1981

Commission on the Third London Airport, *Commission on the Third London Airport Report*, HMSO, London, 1970

Common, M. S., 'A note on the use of the Clawson method for the evaluation of recreation site benefits', *Regional Studies*, 1973, vol. 7, pp. 401–406

Construction Industry Research and Information Association, *Seawall Design Guidelines*, London, CIRIA, 1991

Coopers and Lybrand Associates, *Review of Investment Appraisal Techniques in the Water Industry*, Swindon, Water Research Centre, 1986

Countryside Commission, *The 1986 Countryside Recreation Survey*, Cheltenham, Countryside Commission, 1986

Countryside Commission, *Landscape Assessment: A Countryside Commission Approach*, CCD 18, Cheltenham, Countryside Commission, 1987

Cummings, R. G., Brookshire, D. S. and Schulze, W. D., *Valuing Environmental Goods: An Assessment of the Contingent Valuation Method*, Totowa, Rowman and Allanheld, 1986

Cunliffe, B., *Hengistbury Head, Dorset. Vol. 1 The Prehistoric and Roman Settlement, 3500 BC–AD50*, Oxford University, Comm. for Archaeology, Monograph 13, Oxford, UCA, 1987

Daly, H. E. and Cobb, J. B., *For The Common Good*, London, Green Print, 1990

Dalvi M. Q., *The Value of Life: a Search for a Consensus Estimate*, London, Department of Transport, 1988

Darvill, T., *Ancient Monuments in the Countryside: an Archaeological Management Review*, English Heritage, Arch. Report No. 5, London, Historic Buildings and Monuments Commission for England, 1987

Dalvill, T., *Monuments Protection Programme: Monument Evaluation Manual Part 1, Release 01*, London, London Heritage, 1988.

Darvill, T., 'A question of national importance: approaches to the evaluation of ancient monuments for the Monuments Protection Programme in England', *Antiquity*, 1987, vol. 61, pp. 393–408

Davies, C. D., 'Wirral scheme', in *Coastal Management*, proceedings of the conference organised by the Maritime Engineering Board of the Institute of Civil Engineers and held in Bournemouth on 9–11 May 1989, pp. 293–307, London, Thomas Telford, 1989

de Jong, G. J. and Visser, J., 'Environmental aspects of reinforcements in coastal dune areas', *Water Science and Technology*, 1983, vol. 16, pp. 377–386

Department of Employment, *Employment Gazette*, (weekly), London, HMSO, 1990

Department of the Environment, *This Common Inheritance: Britain's Environmental Strategy*, London, HMSO, 1990

Department of the Environment and Welsh Office, *Environmental Assessment: a Guide to Procedures*, London, HMSO, 1990

Department of Transport, *COBA Manual*, London, Department of Transport, 1971

Department of Transport, *COBA 9 Manual*, London, Department of Transport, 1981

Department of Transport, *Highways Economics Note 2*, London, Department of Transport, 1986

Dicks, M. J., 'The housing market', *Bank of England Quarterly Review*, February, 1989, pp. 66–77

Doizy, A., *Evaluation of the Non-Monetary Impacts of Flooding on Households*, Enfield, London, Middlesex Polytechnic, Flood Hazard Research Centre, 1991

Donelly, W. A., 'Hedonic price analysis of the effect of a floodplain on property values', *Water Resources Bulletin*, 1989, vol. 25, no. 3, pp. 581–585

Dreze, J., 'L'utilité sociale d'une vie humaine', *Revue Française de Recherche Opérationnel*, 1962, vol. 6, no. 23, p. 93

Duckworth, D. H., 'Psychological problems arising from disaster work', *Stress Medicine*, 1986, vol. 2, pp. 315–323

Duffield, J., 'Travel cost and contingent valuation: a comparative analysis', *Advances in Applied Micro-Economics*, 1984, vol. 3, pp. 67–87

European Council, *Working Document of the Proposal for the Directive on the Protection of Natural and Semi-Natural Habitats of Wild Flora and Fauna*, Report 10039/89, Brussels, European Community, 1989

Fairlight Coastal Preservation Society, *Fairlight Cove, Fairlight, East Sussex: Benefit-Cost Analysis*, Fairlight, Sussex, Fairlight Coastal Preservation Society, 1988

Fisher, A. C. and Krutilla, J. V., 'Resource conservation, environmental preservation, and the rate of discount', *Quarterly Journal of Economics*, 1978, vol. 139, pp. 358–370

Fox, W., 'Anthropocentric and nonanthropocentric foundations of environmental decision-making', paper given at the *First International Conference on Ethics and Environmental Policies*, Borca di Cadore, 1990

Freeman, A. M., *The Benefits of Environmental Improvement: Theory and Practice*, Baltimore, Johns Hopkins, 1979

Friends of the Earth, *Beyond Rhetoric*, Discussion Paper No. 2, London, Friends of the Earth, 1990

Foster, H. D., *Disaster Planning: The Preservation of Life and Property*, Berlin, Springer-Verlag, 1980

Fouquet, M.-P., Green, C. H. and Tunstall, S. M., *Hurst Spit: an assessment of the benefits of coast protection*, Enfield, London, Middlesex Polytechnic Flood Hazard Research Centre, 1991

Fuller, R. J. and Langslow, D. R., 'Ornithological evaluation for wildlife conservation', in Usher, M. B. (ed.), *Wildlife Conservation Evaluation*, London, Chapman and Hall, 1986

Gibson, S., 'Recreational land use', in Pearce, D. W. (ed.), *Evaluation of Social Cost*, London, Allen and Unwin, 1978

Goicoechea, A., Hansen, D. R. and Duckstein, L., *Multiobjective Decision Analysis with Engineering and Business Applications*, New York, John Wiley, 1982

Goicoechea, A., Krouse, M. and Antle, L. G., 'An approach to risk and uncertainty in benefit-cost analyses of water resources projects', *Water Resources Research*, 1982, vol. 18, no. 4, pp. 791–799

Goodin, R., 'Discounting discounting', *Journal of Public Policy*, 1986, vol. 2, pp. 53–72

Gotmark, F., Ahlund, M. and Eriksson, M. O. G., 'Are indices reliable for assessing the conservation value of natural areas? An avian case study', *Biological Conservation*, 1986, vol. 38, pp. 55–73

Government Statistical Service, *Housing Construction Statistics 1978–1988 Great Britain*, London, HMSO, 1989

Government Statistical Service, *Housing and Construction Statistics Great Britain*, September quarter 1991, no. 47, part 1, London, HMSO, 1991

Green, C. H., *Evaluating Road Traffic Disruption from flooding*, Enfield, London, Middlesex Polytechnic Geography and Planning Paper no. 11, 1983

Green, C. H., 'The relationship between the magnitude of flooding, stress and health', paper presented at the Annual Meeting of the British Psychology Society, London, 1988

Green, C. H., 'Investment appraisal in the water industry', in *IWEM 90: Design and Construction of Works for Water and Environmental Management*, Glasgow, 1990

Green, C. H., Emery, P. J., Penning-Rowsell, E. C. and Parker, D. J., *The Health Effects of Flooding: a Survey at Uphill*, Avon, Enfield, London, Middlesex Polytechnic Flood Hazard Research Centre, 1985

Green, C. H. and N'Jai, A., *Thames Overview Pre-Feasibility Study*, Enfield, London, Middlesex Polytechnic Flood Hazard Research Centre, 1987

Green, C. H. and Penning-Rowsell, E. C., 'Flooding and the quantification of "intangibles"', *Journal of the Institution of Water and Environmental Management*, 1989, vol. 3, no. 1, pp. 27–30

Green, C. H. and Tunstall, S. M., 'Is the economic evaluation of environmental goods possible?', *Journal of Environmental Management*, 1991a, vol. 33, pp. 123–141

Green, C. H. and Tunstall, S. M., 'The evaluation of river water quality improvements by the contingent valuation method', *Applied Economics*, 1991b, vol. 23, pp. 1135–1146

Green, C. H. and Tunstall, S. M., 'Theoretical bridges and empirical trains: economics and environmental decision making', paper given at the Annual Conference of the Regional Studies Association, Oxford, 1991c

Green, C. H., Tunstall, S. M. and Fordham M., 'Perceptions of the risks of flooding', paper given at the Conference of the Society for Applied Anthropology, York, 1990

Green, C. H., Tunstall, S. M. and House, M. A., 'Investment appraisal for sewerage schemes: benefit assessment', in Laikari, V. (ed.), *River Basin Management – V*, Oxford, Pergamon, 1989

Green, C. H., Tunstall, S. M. and House, M. A., 'Evaluating the benefits of river water quality improvement', in van der Staal, P. M. and van Vught, F. A. (eds), *Impact Forecasting and Assessment: Methods, Results, Experiences*, pp. 171–180, Delft, Delft University Press, 1989

Green, C. H., Tunstall, S. M., N'Jai, A. and Rogers, A., 'Economic evaluation of environmental goods', *Project Appraisal*, 1990, vol. 5, no. 2, pp. 70–82

Green, C., Tunstall, S. and Turner, K., (eds), *Workshop on the Methodological and Theoretical Issues of the CVM: Conclusions and Recommendations*, Enfield, London, Middlesex Polytechnic Flood Hazard Research Centre, 1991

Gubbay, S., (ed.) *A Coastal Directory for Marine Nature Conservation*, Ross-on-Wye, Marine Conservation Society, 1988

H. M. Treasury, *Investment Appraisal in the Public Sector: A Technical Guide for Government Departments*, London, HMSO, 1991

Haggett, P., *Locational Analysis in Human Geography*, London, Edward Arnold, 1965

Harrison, A. J. M. and Stabler, M. J., 'An analysis of journeys for canal-based recreation', *Regional Studies*, 1981, vol. 15, no. 5, pp. 345–358

Hastings Borough Council, *Strategy for the Resort 2000*, Hastings, East Sussex, Hastings Borough Council, 1985

Health and Safety Executive, *Canvey: an Investigation of Potential Hazards from Operations in the Canvey Island/Thurrock Area*, London, HMSO, 1979

Henry, C., 'Investment decisions under uncertainty: the irreversibility effect', *American Economic Review*, 1974, vol. 64, no. 6, pp. 1006–1012

Hess, T. M. and Morris, J., 'A computer model for agricultural land drainage scheme appraisal', paper to the Ministry of Agriculture, Fisheries and Food Conference of River Engineers, Cranfield, 1985

Hoinville, G. and Jowell, J. R., *Survey Research Practice*, London, Heinemann, 1982

Hopkinson, P. G., Bowers, J. and Nash, C. A., *The Treatment of Nature Conservation in the Appraisal of Trunk Roads*, Peterborough, Nature Conservancy Council, 1990

Hourthan, K., 'Context-dependent models of residential satisfaction', *Environment and Behavior*, 1984, vol. 16, no. 3, pp. 369–393

Hydraulics Research Ltd, *A Guide to the Selection of Appropriate Coast Protection Works for Geological SSSIs*, Report to the Nature Conservancy Council, Wallingford, Hydraulics Research, 1991

Hydraulics Research Ltd, *Hengistbury Head Coast Protection Study, Summary and Technical Reports EX1459 and EX1460*, Wallingford, Hydraulics Research Ltd, 1986

Jones-Lee, M., *The Value of Life: An Economic Analysis*, London, Martin Robertson, 1976

Kaplan, S. and Kaplan, R. *The Experience of Nature: a Psychological Perspective*, Cambridge, Cambridge University Press, 1990

King, A. and Wathern, P., *Environmental Assessment Methodology*, National Rivers Authority Report 214/1/T, Aberystwyth, University College, Wales, 1991

Kish, L., *Survey Sampling*, New York, Wiley, 1965

Kosmin, R., 'A paper on house prices and the retail price index', in Fairlight Coastal Preservation Association, *Fairlight Cove Benefit-cost Analysis*, Fairlight, East Sussex, Fairlight Coastal Preservation Association, 1988

Krutilla, J. V., 'Conservation reconsidered', *American Economic Review*, 1967, vol. 57, no. 4, pp. 77–86

Krutilla, J. V. and Fisher, A. C., *The Economics of Natural Environments*, Washington DC, Resources for the Future, 1975

Lammers, C. J., *Studies in Holland Flood Disaster 1953: Volume II Survey of Evacuation Problems and Disaster Experiences*, Amsterdam, Instituut voor Sociaal Onderzoek van het Nederlandse Volk, 1955

Langbein, L. L. and Lichman, A. J., *Ecological Inference*, Sage, Beverly Hills, 1978

Leopold, L. B., Clarke, F. E., Hanshaw, B. B. and Balsey, J. R., *A Procedure for Evaluating Environmental Impact*, US Geological Survey Circular 645, Springfield Va., NTIS, 1971

Li, M. M. and Brown, H. J., 'Micro-neighborhood externalities and hedonic housing prices', *Land Economics*, 1980, vol. 56, no. 2, pp. 125–141

Lichfield, N., 'Cost-benefit analysis in plan evaluation', *Town Planning Review*, 1964, vol. 35, no. 2, pp. 159–169

Lind, R.C., (ed) *Discounting for Time and Risk in Energy Policy*, Washington DC, Resources for the Future, 1982

Little, I. M. D. and Mirrlees, J. A., *Project Appraisal and Planning for Developing Countries*, London, Heinemann, 1974

London Borough of Merton, *Parks User Survey*, London, London Borough of Merton, 1984

Lowenthal, D. and Penning-Rowsell, E. C., *Landscape Meanings and Values*, London, George Allen and Unwin, 1986

Lutz, M. A. and Lux, K., *Humanistic Economics: The New Challenge*, New York, The Bootstrap Press, 1988

Lystad, M. (ed.), *Innovations in Mental Health Services to Disaster Victims*, Rockville, National Institute of Mental Health, 1985

Mackinder, I. H., *The Economics of Coast Protection*, Reading, Local Government Operational Research Unit, 1980

Madriaga, B. and McConnell, K. E., 'Exploring existence value', *Water Resources Research*, 1987, vol. 23, no. 5, pp. 936–942

Margules, C. and Usher, M. B., 'Criteria used in assessing wildlife conservation potential: a review', *Biological Conservation*, 1981, vol. 21, pp. 79–109

Markandya, A. and Pearce, D. W., *Environmental Considerations and the Choice of the Discount Rate in Developing Countries*, Environment Department Working Paper No 3, Washington DC, World Bank, 1988

May, V. J. and Osborne, K. A., *Recreation Use of Hengistbury Head*, Christchurch, Dorset, Dorset Institute of Higher Education, 1981

Middlesex Polytechnic Flood Hazard Research Centre, *Herne Bay Coastal Protection Benefit Assessment*, 4 volumes, London, Enfield, Middlesex Polytechnic Flood Hazard Research Centre, 1990

Milne, G., 'Cyclone Tracy: some consequences of the evacuation for adult victims', *Australian Psychologist*, 1977, vol. 12, no. 1, pp. 39–54

Ministry of Agriculture, Fisheries and Food, *Investment Appraisal of Arterial Drainage, Flood alleviation and Sea Defence Schemes – Guidelines for Drainage Authorities*, Land and water Service River and Coastal Engineering Group note AD1AAK, London, MAFF, 1985

Ministry of Agriculture, Fisheries and Food, *Investment Appraisal of Arterial Drainage Flood Protection and Sea Defence Schemes: Agricultural Output Valuation Adjustments, Economics (RU) Note*, London, MAFF, 1985

Ministry of Agriculture, Fisheries and Food, *Prices of Agricultural Land in England, Three Months ended 30 June 1990. No 187/90*, London, Government Statistical Service, 1990

Ministry of Transport and Public Works, *Water in the Netherlands: a Time for Action*, The Hague, Ministry of Transport and Public Works, 1989

Mishan, E. J., *Cost-Benefit Analysis*, London, George Allen Unwin, 1971

Mitchell, R. C. and Carson, R. T., *Using Surveys to Value Public Goods: The Contingent Valuation Method*, Washington, Resources for the Future, 1989

Mitchell, R. C. and Carson, R. T., *How Far Along the Learning Curve is the Contingent Valuation Method?*, Discussion Paper QE 87-0, Washington DC, Resources for the Future, 1987

Moore, P. G. and Thomas, H., *The Anatomy of Decisions*, London, Penguin, 1988

Morris, J. and Hess, T. M., 'Drainage benefits and farmer uptake', paper to the Ministry of Agriculture, Fisheries and Food Conference of River Engineers, Cranfield, 1984

Morris, J., Hess, T. M., Ryan, A. M. and Leeds-Harrison, P. B., *Drainage Benefits and Farmer Uptake*, Report to Severn-Trent Water Authority, Silsoe, Silsoe College, 1984

Moser, C. A. and Kalton, G., *Survey Methods in Social Investigation*, London, Heinemann, 1971

Moser, D. A. and Dunning, C. M., *A Guide for Using the Contingent Value Methodology in Recreation Studies*, National Economic Development Procedures Manual – Recreation, Vol 2, IWR report 86-R-5m, Fort Belvoir, Va., Institute for Water Resources, 1986

Munro, M. and Lamont, D. 'Neighbouring perception, preference and household mobility in the Glasgow private housing market', *Environment and Planning*, 1985, vol. 17, pp. 1331–1350

Murphy, M. C., *Report on Farming in the Eastern Counties of England*, Cambridge, Agriculture Economics Unit, Department of Land Economics, University of Cambridge, 1989

Nash, C., Pearce, D. W. and Stanley, J., 'Criteria for evaluating project evaluation techniques', *Journal of the American Institute of Planners*, March 1975, vol. 41, pp. 83–89

Nash, G. A., 'Future generations and the social rate of discount', *Environment and Planning A*, 1973, vol. 5, no 5, pp. 611–617

National Institute of Mental Health, *Crisis Intervention Programs for Disaster Victims in Smaller Communities*, Washington DC, US Department of Health and Human Services, 1979

Nature Conservancy Council, *Earth Science Conservation no. 23*, Peterborough, Nature Conservancy Council, 1987

Nature Conservancy Council, *Guidelines for the Selection of Biological SSSIs*, Peterborough, Nature Conservancy Council, 1989

Nature Conservancy Council, *19th Annual Report*, Peterborough, Nature Conservancy Council, 1990

Nature Conservancy Council, *Earth Science Conservation in Great Britain: A Strategy*, Peterborough, Nature Conservancy Council, 1991

Neal, J. and Parker, D. J., *Flood Plain Encroachment: A Case Study of Datchet, UK*, Enfield, London, Middlesex Polytechnic School of Geography and Planning, Paper no. 22, 1988

Neal, J. and Parker, D. J., *Flood Warnings in the Severn-Trent Water Authority Area: an Investigation of Standards of Service, Effectiveness and Customer Satisfaction*, Enfield, London, Middlesex Polytechnic School of Geography and Planning, Paper no. 23, 1989

Nelson, J., 'Airport noise, location, rent and the market for residential amenities', *Journal of Environmental Economics and Management*, 1979, vol. 6, pp. 320–331

Nelson, J. P., 'Highway noise and property values: a survey of recent evidence', *Journal of Transport Economics and Policy*, 1982, vol. XVI, pp. 117–130

Newsome, D. W. and Green, C. H., *Economic Value of Improvements to the Water Environment*, NRA R&D Note, Bristol, National Rivers Authority, 1991

Nix, J., *The Farm Management Pocketbook*, Wye, Kent, Farm Business Unit, Wye College, 1990

N'Jai, A., Tapsell, S. M., Taylor, D., Thompson, P. M. and Witts, R.C., *FLAIR 1990 (Flood Loss Assessment Information Report)*, Enfield, London, Middlesex Polytechnic Flood Hazard Research Centre, 1990

Novick, D. (ed), *Program Budgeting: Program Analysis and the Federal Budget*, Cambridge, Mass., Harvard University Press, 1967

Page, T., 'Equitable use of the resource base', *Environment and Planning A*, 1977, vol. 9, pp. 15–22

Parker, D. J., 'Principles of urban flood alleviation benefit-cost appraisal', paper given at a Seminar on Flood Protection and Land Drainage, Wallingford, Hydraulics Research Ltd, 1987

Parker, D. J., *The Damage Reducing Effects of Flood Warnings*, Enfield, London, Middlesex Polytechnic Flood Hazard Research Centre, 1991

Parker, D. J., Green, C. H., Thompson, P. M. and Penning-Rowsell, E. C., *Flood Alleviation Benefit-Cost Analysis: Cranfield Conference Papers*, London, Middlesex Polytechnic Flood Hazard Research Centre, 1983

Parker, D. J., Green, C. H. and Thompson, P. M., *Urban Flood Protection Benefits: A Project Appraisal Guide*, Aldershot, Hampshire, Gower Technical Press, 1987

Parker, D. J. and Handmer, J. W., *Hazard Management and Emergency Planning: Perspectives on Britain*, London, James and James, 1992

Parker, D. J. and Penning-Rowsell, E. C., *Whitstable Central Area Coast Protection Scheme: Benefit Assessment*, Enfield, London, Middlesex Polytechnic Flood Hazard Research Centre, 1981

Parker, D. J. and Penning-Rowsell, E. C., 'Flood risk in the urban environment', in Hurbert, D. H. and Johnston, R. J. (eds), *Geography and the Urban Environment*, Chichester, Wiley, 1982

Parker, D. J., Penning-Rowsell, E. C. and Green, C. H., *Swalecliffe Coast Protection Proposals*, Enfield, London, Middlesex Polytechnic Flood Hazard Research Centre, 1983

Parker, D. J., Penning-Rowsell, E. C. and Green, C. H., *Whitstable Central Area Sea Defence Scheme: Evaluation of Potential Benefits*, Enfield, London, Middlesex Polytechnic Flood Hazard Research Centre, 1984

Parker, D. J. and Thompson, P. M., 'An "extended" economic appraisal of coast protection works: a case study of Hengistbury Head, England', *Ocean and Shoreline Management*, 1988, vol. 11, no. 1, pp. 45–72

Pearce, D. W., 'Noise nuisance', in Pearce, D. W. (ed.) *The Valuation of Social Cost*, London, George Allen and Unwin, 1978

Pearce, D. W., *Cost-Benefit Analysis*, London, Macmillan, 1984

Pearce, D. W., Barbier, E. and Markandya, A., *Sustainable Development*, London, Earthscan, 1990

Pearce, D. W. and Markandya, A., *Environmental Policy, Benefits, Monetary Evaluation*, Paris, OECD, 1989

Pearce, D. W., Markandya, A. and Barbier, E., *Blueprint for a Green Economy*, London, Earthscan, 1989

Penning-Rowsell, E. C., *The Effect of Salt Contamination on Flood Damage to Residential Property*, Enfield, Middlesex Polytechnic Flood Hazard Research Centre, 1978

Penning-Rowsell, E. C., 'Fluctuating fortunes in gauging landscape value', *Progress in Human Geography*, 1981, vol. 5, no. 1, pp. 25–41

Penning-Rowsell, E. C., 'A public reference evaluation of landscape quality', *Regional Studies*, 1982, vol. 16, pp. 97–112

Penning-Rowsell, E. C., 'Landscape evaluation in practice: a survey of local authorities', *Landscape Research*, 1989, vol. 14, no. 2, pp. 35–37

Penning-Rowsell, E. C. and Chatterton, J. B., *The Benefits of Flood Alleviation: A Manual of Assessment Techniques*, Aldershot, Gower, 1977

Penning-Rowsell, E. C., Chatterton, J. B., Day, H. J., Ford, D. T., Greenaway, M. A., Smith, D. I. and Witts, R. C., 'Comparative aspects of computerized floodplain data management', *Journal of Water Resources Planning and Management, American Society of Civil Engineers*, 1987, vol. 113, no. 6, pp. 725–744

Penning-Rowsell, E. C., Coker, A. M., N'Jai, A., Parker, D. J. and Tunstall, S. M. 'Scheme worthwhileness', in *Coastal Management*, proceedings of the conference organised by the Maritime Engineering Board of the Institute of Civil Engineers and held in Bournemouth on 9–11 May 1989, pp. 227–241, London, Thomas Telford, 1989

Penning-Rowsell, E. C. and Parker, D. J., *Chesil Sea Defence Scheme: Benefit Assessment*, Enfield, London, Middlesex Polytechnic Flood Hazard Research Centre, 1980

Penning-Rowsell, E. C. and Parker, D. J., 'The indirect effects of floods and benefits of flood alleviation: evaluating the Chesil sea defence scheme' *Applied Geography*, 1987, vol. 7, pp. 263–288

Penning-Rowsell, E. C. Parker, D. J., Thompson, P. M. and Green, C. H., 'Flood loss data and models for appraising flood alleviation investment: explanation and critical evaluation', paper presented to the 39th Meeting of the International Commission on Irrigation and Drainage, Dubrovnik, Yugoslavia, 1988

Petersen, G. L., Driver, B. L. and Gregory, R. (eds), *Amenity Resource Valuation*, State College, Venture Publishing, 1988

Popper, K. R., *The Logic of Scientific Discovery*, London, Hutchinson, 1968

Posford Duvivier, *Peacehaven and Telscombe Coast Protection Scheme: Report on Phase IV*, Lewes, East Sussex, Lewes District Council, 1988

Raiffa, H., *Decision Analysis*, Reading, Mass., Addison-Wesley, 1968

Randall, A., Hoen, J. P and Brookshire, D., 'Contingent valuation surveys for valuing environmental assets', *Natural Resources Journal*, 1983, vol. 23, no. 3, pp. 635–48

Randall, A., Ives, B. and Eastman, C., 'Bidding games for aesthetic environmental improvements', *Journal of Environmental Economics and Management*, 1974, vol. 1, pp. 132–149

Ratcliffe, D. A. (ed.), *A Nature Conservation Review*, vol. 1, Cambridge, Cambridge University Press, 1977

Ratcliffe, D. A., 'Selection of important areas for wildlife conservation in Great Britain: the Nature Conservancy Council's approach', in Usher, M. B. (ed.) *Wildlife Conservation Evaluation*, pp. 135–159, London, Chapman and Hall, 1986

Regan, T., 'The nature and possibility of an environmental ethic', *Environmental Ethics*, 1981, vol. 3, pp. 19–34

Rolston, H.,'Valuing wildlands', *Environmental Ethics*, 1985, vol. 7, pp. 23–48

Rosen, S., 'Hedonic prices and implicit markets: product differentiation in Pure Competition', *Journal of Political Economy*, 1974, vol. 82, pp. 34–55

Rosenthal, D. H., Loomis, J. B. and Peterson, G. L., *The Travel Cost Model: Concepts and Applications*, USDA Forest Service General Technical Report RM-109, Fort Collins, Rocky Mountain Forest and Range Experimental Station, 1984

Rubinstein, D., 'A statistician's view of NRC statistics', paper given to the American Statistical Association Ad Hoc Committee on Nuclear Regulatory Research, Washington DC, 1981

Ryan, M. S. and Bonfield, H. E., 'The Fishbein extended model and consumer behavior', *Journal of Consumer Research*, 1975, vol. 2, no. 2, pp. 118–136

Sagoff, M., *The Economy of the Earth*, Cambridge, University Press, 1988

Schelling, T. C., 'The life you save may be your own', in Chase S.B., (ed.), *Problems in Public Expenditure Analysis*, Washington, Brookings Institute, 1968

Schlaifer, R., *Probability and Statistics for Business Decisions: an Introduction to Managerial Economics under Uncertainty*, New York, McGraw-Hill, 1959

Schmalensee, R., 'Optimum demand and consumer surplus: valuing price changes under uncertainty', *American Economic Review*, 1972, vol. 62, pp. 813–824

Sen, A. K., 'Rational fools: a critique of the behavioral foundations of economic theory', *Philosophy and Public Affairs*, 1977, vol. 6, pp. 317–344

Sen, A. K., *On Ethics and Economics*, London, Basil Blackwell, 1987

Shabman, L. A. and Damianos, D. I., 'Flood-hazard effects on residential property values', *Journal of the American Society of Civil Engineers: Water Resources Planning and Management Division*, April 1976, pp. 151–162

Shannon, C. E. and Weaver, W., *The Mathematical Theory of Communication*, Urban, University of Illinois, 1949

Smith, D. I., 'Cost-effectiveness of flood warnings' in Smith D. I. and Handmer J. W. (eds) *Flood Warning in Australia*, Canberra, Centre for Resource and Environmental Studies, Australian National University, 1986

Smithson, M., 'Toward a social theory of ignorance', *Journal for the Theory of Social Behaviour*, 1985, vol. 15, pp. 151–172

Smithson M., 'The changing nature of ignorance', paper given at the Joint CRES/ACDC Workshop *Risk Perception in Australia*, Mount Macedon, Victoria, 1989

Sorensen, J. H., *Evaluation of Warning and Protective Action Implementation Times for Chemical Weapons Accidents*, ORNL/TM-10437, Oak Ridge, Oak Ridge National Laboratory, 1988

Sorensen, J. H. and Rogers, G. O., *Protective Actions for Extremely Toxic Chemical Accidents*, Oak Ridge, Oak Ridge National Laboratory, 1989

South-East Tourist Board, *Study of Tourism in Hastings*, Hastings, South-East Tourist Board, 1976

Spackman, M., *Discount Rates and Rates of Return in the Public Sector: Economic Issues*, Government Economic Service Working Paper No. 113 (Treasury Working Paper No. 58), London, H. M. Treasury, 1991

Squire, L. and van der Tak, H., *Economic Analysis of Projects*, Baltimore, Johns Hopkins, 1975

Stabler, M. J. and Ash, S., *The Amenity Demand for Inland Waterways*, Reading, Amenity Waterway Study Unit, 1978

Sugden, R. and Williams, A., *The Principles of Practical Cost-Benefit Analysis*, Oxford, Oxford University Press, 1978

Suleman, M., N'Jai, A., Green, C. H. and Penning-Rowsell, E. C., *Potential Flood Damage Data: a Major Update*, Enfield, London, Middlesex Polytechnic Flood Hazard Research Centre, 1988

Tandy, C. R. U., *Landscape Evaluation Technique*, Croydon, England, Land Use Consultants, 1971

Thompson, P. M., Parker, D. J., Coker, A., Grant, E., Penning-Rowsell, E. C. and Suleman, M., *The Economic and Environmental Impacts of Coast Erosion and Protection: A Case Study of Hengistbury Head and Christchurch Harbour, England*, Enfield, London, Middlesex Polytechnic Geography and Planning Paper No. 19, 1987

Thompson, P. M., Penning-Rowsell, E. C., Parker, D. J. and Hill, M. I., *Interim Guidelines for the Economic Evaluation of Coast Protection*, Enfield, London, Middlesex Polytechnic Flood Hazard Research Centre, 1987

Tourism and Recreation Research Unit, *Recreation Site Survey Manual*, London, Spon, 1983

Tukey, J. W., *Exploratory Data Analysis*, Reading, Addison Wesley, 1977

Turner, R. K., Bateman, I. and Brooke, J. S., 'Valuing the benefits of coastal defence: a case study of the Aldeburgh Sea Defence Scheme' in Coker, A. and Richards, C. (eds), *Valuing the Environment*, London, Belhaven Press, 1992

Turner, R. K. and Brooke, J., *Cost-Benefit Analysis of the Lower Bure, Halvergate Fleet and Acle Marshes I. D. B. Flood Protection Scheme*, Norwich, Environmental Appraisal Group, University of East Anglia, 1988

Tunstall, S. M., Green, C. H. and Lord, J., *The Contingent Valuation Method*, Enfield, London, Middlesex Polytechnic Flood Hazard Research Centre, 1987

Tunstall, S. M., Green, C. H. and Lord, J., *The Evaluation of Environmental Goods by the Contingent Valuation Method*, Enfield, London, Middlesex Polytechnic Flood Hazard Research Centre, 1988

Tunstall, S. M., Green, C. H., Lewin, P. and Coker, A. C., *Herne Bay Coast Protection Benefit Assessment: Report III The Recreational Benefit Assessment*, Enfield, London, Middlesex Polytechnic Flood Hazard Research Centre, 1990

Usher, M. B., *Wildlife Conservation Evaluation*, London, Chapman and Hall, 1986

US Water Resources Council, *Principles and Guidelines for Water and Related Land Resources Implementation Studies*, Washington DC, US Government Printing Office, 1983

Uzzell, D. L., *People, Nature and Landscape: an Environmental Psychological Perspective*, Report to the Landscape Research Group, Guildford, England, Department of Psychology, University of Surrey, 1989

Von Winterfeldt, D. and Edwards, W., *Decision Analysis and Behavioral Research*, Cambridge, Cambridge University Press, 1986

Washington, H. G., 'Diversity, biotic and similarity indices', *Water Research*, 1984, vol. 18, no. 6, pp. 653–694

Weaver, M., 'Within months this house will tumble into the sea ...' *Daily Telegraph*, August 26 1987, p. 11

Western J. S. and Milne G., 'Some social effects of a natural hazard: Darwin residents and Cyclone Tracy' in Heathcote, R. L. and Thom, B. G. (eds), *Natural Hazards in Australia*, Canberra, Australian Academy of Science, 1979

Wetherill, G. B., Duncombe, P., Kenward, M., Kollerstrom, J., Paul, S. R. and Vowden, B. J., *Regression Analysis with Applications*, London, Chapman and Hall, 1986

Willis, K. G., 'Valuing non-market wildlife commodities: an evaluation and comparison of benefits and costs', *Applied Economics*, 1990, vol. 22, pp. 13–30

Willis, K. G. and Benson, J. F., 'A comparison of user benefits and costs of nature conservation at three nature reserves', *Regional Studies*, 1987, vol. 22, no. 5, pp. 417–428

Willman, E. A., *External Cost of Coast Beach Pollution: An Hedonic Approach*, Washington, Resources for the Future, 1984

Wilmington, R. H., 'The renourishment of Bournemouth beaches 1974–75', in *Shoreline Protection*, pp. 157–162, Proceedings of the Institution of Civil Engineers Conference, London, Thomas Telford, 1983

Wilson, A. G., *Models in Urban Planning, a Synoptic Review of Recent Literature*, Working Paper No. 3, London, Centre for Environmental Studies, 1968

Wilson, A. A. G., *Assessing the Bushfire Hazard of Houses: A Quantitative Appraisal*, Rural Fire Research Committee Technical Paper no. 6, Victoria, RFRC, 1984

Wood, C. and Jones, C., *Monitoring Environmental Assessment and Planning*, London, HMSO, 1991

World Commission on Environment and Development, *Our Common Future*, London, Oxford University Press, 1987

Appendix 3.1 The derivation of the Extension of Life Factors (ELFs) and Formula 3.1

The market value of a property is dependent on a number of factors: the cost of construction, its state of repair, the opportunity cost (i.e. the alternative use) of the plot of land, the demand for property in that area, and the expected future life of the property. Normally, land has an infinite life, while properties in Britain have very long lives, perhaps not different from infinity since the buyer usually expects to be able to sell the property much later without any loss in its expected life since it is assumed to be maintained such that it would not lose its value.

Where a property is threatened by erosion by the sea at some time in the future its expected life will be finite, and its value to society will depend on when it can be expected to become uninhabitable because of this erosion. Ideally the value of a property with such a finite life could be assessed as its market value, and it would be possible then to take the market values for similar properties with different lives to estimate the benefit from delaying the loss of a property through protection from erosion.

However, as discussed in Section 3.4.4 the market is imperfect in these circumstances: no threat is perceived when the property is far from the sea and its remaining life is long (but finite), while the market value collapses when the threat is perceived (usually when loss is more or less imminent). Often there will be an implicit assumption that action will be taken to prevent erosion before it affects such property and hence market prices will not be adjusted to more remote dates of erosion. In these circumstances a shadow pricing method is needed which makes use of readily available and observable information.

An annual value for the services of a property can be estimated from its market value by assuming that this is the present value of a stream of constant annual values discounted over the remaining life of the property. The discount factor (D_n) which brings a value in year n to its present value is defined by Formula A3.1 where r is some discount rate:

$$D_n = 1/(1+r)^n \qquad \text{Formula A3.1}$$

The market value of the property is then the sum of a series of equal annual values discounted over the life of the property (which is a finite geometric progression). If MV is the market value of the property and the expected life is finite, then the annual value (A) is given by Formula A3.2:

$$A = MV \cdot r_1/(1-D_n) = MV \cdot r_1/(1-1/(1+r_1)^n) \qquad \text{Formula A3.2}$$

For an infinite property life, which is assumed in the following presentation, Formula A3.2 simplifies to Formula A3.3:

$$A = MV \cdot r_1 \qquad \text{Formula A3.3}$$

Here r_1 is some rate of interest (presumably the market rate of interest adjusted for constant prices rather than expectations of inflation).

Hence the annual value derived from a given market value assuming an infinite life is less than the annual value based on the same market value but assuming a finite life. Therefore assuming an infinite life is a conservative assumption in computing an annual property value; in practice the difference for lives of over 100 years is very small.

A coast protection scheme will extend the life of a property by the life of the protection works. Hence the benefit to society of that scheme is the difference between the present value of the property with its extended life (with the scheme) less the present value of the property without the scheme. If p is the number of years of life of the property without a scheme, then using Formula A3.3 to define the annual value of the property and substituting this in a rearranged Formula A3.2 gives Formula A3.4:

$$PV_P = MV.r_1.((1-1/(1+r_2)^p)/r_2) \qquad \text{Formula A3.4}$$

Here r_2 signifies the discount rate used to estimate the present value to society of the property with finite life due to erosion.

It follows that if s is the life of the coast protection scheme in years, then the present value to society of the property with the scheme is given by PV_p+_s in Formula A3.5:

$$PV_p+_s = MV.r_1.((1-1/(1+r_2)^{p+s})/r_2) \qquad \text{Formula A3.5}$$

Hence the present value of the benefit from extending the life of the property (PVB) is given by Formula A3.5 minus Formula A3.4 as given in Formula A3.6:

$$PVB = MV.r_1/r_2.\{[1-1/(1+r_2)^{p+s}] - [1-1/(1+r_2)^p]\}$$

$$= MV.r_1/r_2.[1/(1+r_2)^p - 1/(1+r_2)^{p+s}] \qquad \text{Formula A3.6}$$

The remaining question in evaluating this formula is: what should be the discount rates r_1 and r_2? The benefit assessment is undertaken to assess the value to society of protecting the property, and the method adopted in a normal appraisal is to discount using a Test Discount Rate. It can be argued that the calculation of the PVB should use the Test Discount Rate (TDR), that is r_2 = TDR; but the calculation of an annual value for the property need not use the same discount rate, and it might be argued that it should be a social discount rate (to calculate a social annual value for property) or the market discount rate.

However, to avoid inconsistencies, for simplicity, and on the basis that the TDR reflects the real rate of return on investments in the economy in the long term, it appears reasonable for assessing coast protection benefits to assume that r_1 is also equal to the TDR. Hence it is assumed that: $r_1 = r_2 = r$ = TDR. It then follows that Formula A3.6 reduces to Formula A3.7:

$$PVB = MV.[1/(1+r)^p - 1/(1+r)^{p+s}] \qquad \text{Formula A3.7}$$

We thus define the extension of life factor (ELF) by Formula 3.1 (Section 3.4.2), reproduced here:

$$PVB = MV.[1/(1+r)^p - 1/(1+r)^{p+s}]$$

$$= MV.ELF_{p+s} \qquad \textbf{Formula 3.1}$$

This calculation of PVB can be expressed as the discounted value of the property loss in the year of loss without the coast protection scheme, less the discounted value of the property loss in the year of loss with the scheme, where the property value is its market value as though there were no threat to its future life from erosion.

However, it is important to note that the ELF factor is based on calculating an annual value for the property and the equality of the two discount rates (one used in calculating an annual value, and the other used in estimating the benefit from extending that stream of annual values for the life of the proposed scheme).

Appendix 3.2 Discounted Extension of Life Factors (ELFs) for property losses

Extension of life (ELF) factors for TDR = 10%,   r =   0.1

p = Year of erosion; Dn = Discount factor; ELF for p given by row and s (scheme life) given by column

p = Year of erosion	Dn = Discount factor	s=20 yr	s=25 yr	s=30 yr	s=50 yr	s=75 yr	s=100 yr	s=125 yr
1	0.90909	0.77396	0.82519	0.85699	0.90135	0.90838	0.90902	0.90908
2	0.82645	0.70360	0.75017	0.77908	0.81941	0.82580	0.82639	0.82644
3	0.75131	0.63964	0.68197	0.70826	0.74491	0.75072	0.75126	0.75131
4	0.68301	0.58149	0.61997	0.64387	0.67720	0.68248	0.68296	0.68301
5	0.62092	0.52863	0.56361	0.58534	0.61563	0.62043	0.62088	0.62092
6	0.56447	0.48057	0.51238	0.53212	0.55967	0.56403	0.56443	0.56447
7	0.51316	0.43688	0.46580	0.48375	0.50879	0.51275	0.51312	0.51315
8	0.46651	0.39716	0.42345	0.43977	0.46253	0.46614	0.46647	0.46650
9	0.42410	0.36106	0.38496	0.39979	0.42048	0.42376	0.42407	0.42409
10	0.38554	0.32823	0.34996	0.36345	0.38226	0.38524	0.38552	0.38554
11	0.35049	0.29840	0.31814	0.33041	0.34751	0.35022	0.35047	0.35049
12	0.31863	0.27127	0.28922	0.30037	0.31592	0.31838	0.31861	0.31863
13	0.28966	0.24661	0.26293	0.27306	0.28720	0.28944	0.28964	0.28966
14	0.26333	0.22419	0.23903	0.24824	0.26109	0.26312	0.26331	0.26333
15	0.23939	0.20381	0.21730	0.22567	0.23735	0.23920	0.23937	0.23939
16	0.21763	0.18528	0.19754	0.20516	0.21578	0.21746	0.21761	0.21763
17	0.19784	0.16844	0.17958	0.18651	0.19616	0.19769	0.19783	0.19784
18	0.17986	0.15312	0.16326	0.16955	0.17833	0.17972	0.17985	0.17986
19	0.16351	0.13920	0.14842	0.15414	0.16212	0.16338	0.16350	0.16351
20	0.14864	0.12655	0.13492	0.14013	0.14738	0.14853	0.14863	0.14864
21	0.13513	0.11504	0.12266	0.12739	0.13398	0.13502	0.13512	0.13513
22	0.12285	0.10459	0.11151	0.11581	0.12180	0.12275	0.12284	0.12285
23	0.11168	0.09508	0.10137	0.10528	0.11073	0.11159	0.11167	0.11168
24	0.10153	0.08643	0.09216	0.09571	0.10066	0.10145	0.10152	0.10152
25	0.09230	0.07858	0.08378	0.08701	0.09151	0.09222	0.09229	0.09230
26	0.08391	0.07143	0.07616	0.07910	0.08319	0.08384	0.08390	0.08390
27	0.07628	0.06494	0.06924	0.07191	0.07563	0.07622	0.07627	0.07628
28	0.06934	0.05904	0.06294	0.06537	0.06875	0.06929	0.06934	0.06934
29	0.06304	0.05367	0.05722	0.05943	0.06250	0.06299	0.06303	0.06304
30	0.05731	0.04879	0.05202	0.05402	0.05682	0.05726	0.05730	0.05731
31	0.05210	0.04435	0.04729	0.04911	0.05165	0.05206	0.05209	0.05210
32	0.04736	0.04032	0.04299	0.04465	0.04696	0.04733	0.04736	0.04736
33	0.04306	0.03666	0.03908	0.04059	0.04269	0.04302	0.04305	0.04306
34	0.03914	0.03332	0.03553	0.03690	0.03881	0.03911	0.03914	0.03914
35	0.03558	0.03029	0.03230	0.03354	0.03528	0.03556	0.03558	0.03558
36	0.03235	0.02754	0.02936	0.03050	0.03207	0.03232	0.03235	0.03235
37	0.02941	0.02504	0.02669	0.02772	0.02916	0.02939	0.02941	0.02941
38	0.02673	0.02276	0.02427	0.02520	0.02651	0.02671	0.02673	0.02673
39	0.02430	0.02069	0.02206	0.02291	0.02410	0.02429	0.02430	0.02430
40	0.02209	0.01881	0.02006	0.02083	0.02191	0.02208	0.02209	0.02209
41	0.02009	0.01710	0.01823	0.01894	0.01992	0.02007	0.02008	0.02009
42	0.01826	0.01555	0.01657	0.01721	0.01810	0.01825	0.01826	0.01826
43	0.01660	0.01413	0.01507	0.01565	0.01646	0.01659	0.01660	0.01660
44	0.01509	0.01285	0.01370	0.01423	0.01496	0.01508	0.01509	0.01509
45	0.01372	0.01168	0.01245	0.01293	0.01360	0.01371	0.01372	0.01372
46	0.01247	0.01062	0.01132	0.01176	0.01237	0.01246	0.01247	0.01247
47	0.01134	0.00965	0.01029	0.01069	0.01124	0.01133	0.01134	0.01134
48	0.01031	0.00878	0.00936	0.00972	0.01022	0.01030	0.01031	0.01031
49	0.00937	0.00798	0.00851	0.00883	0.00929	0.00936	0.00937	0.00937
50	0.00852	0.00725	0.00773	0.00803	0.00845	0.00851	0.00852	0.00852
51	0.00774	0.00659	0.00703	0.00730	0.00768	0.00774	0.00774	0.00774
52	0.00704	0.00599	0.00639	0.00664	0.00698	0.00703	0.00704	0.00704

53	0.00640	0.00545	0.00581	0.00603	0.00635	0.00640	0.00640	0.00640
54	0.00582	0.00495	0.00528	0.00548	0.00577	0.00581	0.00582	0.00582
55	0.00529	0.00450	0.00480	0.00499	0.00524	0.00529	0.00529	0.00529
56	0.00481	0.00409	0.00436	0.00453	0.00477	0.00480	0.00481	0.00481
57	0.00437	0.00372	0.00397	0.00412	0.00433	0.00437	0.00437	0.00437
58	0.00397	0.00338	0.00361	0.00375	0.00394	0.00397	0.00397	0.00397
59	0.00361	0.00308	0.00328	0.00341	0.00358	0.00361	0.00361	0.00361
60	0.00328	0.00280	0.00298	0.00310	0.00326	0.00328	0.00328	0.00328
61	0.00299	0.00254	0.00271	0.00281	0.00296	0.00298	0.00299	0.00299
62	0.00271	0.00231	0.00246	0.00256	0.00269	0.00271	0.00271	0.00271
63	0.00247	0.00210	0.00224	0.00233	0.00245	0.00247	0.00247	0.00247
64	0.00224	0.00191	0.00204	0.00211	0.00222	0.00224	0.00224	0.00224
65	0.00204	0.00174	0.00185	0.00192	0.00202	0.00204	0.00204	0.00204
66	0.00185	0.00158	0.00168	0.00175	0.00184	0.00185	0.00185	0.00185
67	0.00169	0.00143	0.00153	0.00159	0.00167	0.00168	0.00169	0.00169
68	0.00153	0.00130	0.00139	0.00144	0.00152	0.00153	0.00153	0.00153
69	0.00139	0.00119	0.00126	0.00131	0.00138	0.00139	0.00139	0.00139
70	0.00127	0.00108	0.00115	0.00119	0.00126	0.00127	0.00127	0.00127
71	0.00115	0.00098	0.00104	0.00109	0.00114	0.00115	0.00115	0.00115
72	0.00105	0.00089	0.00095	0.00099	0.00104	0.00105	0.00105	0.00105
73	0.00095	0.00081	0.00086	0.00090	0.00094	0.00095	0.00095	0.00095
74	0.00086	0.00074	0.00079	0.00082	0.00086	0.00086	0.00086	0.00086
75	0.00079	0.00067	0.00071	0.00074	0.00078	0.00079	0.00079	0.00079
76	0.00071	0.00061	0.00065	0.00067	0.00071	0.00071	0.00071	0.00071
77	0.00065	0.00055	0.00059	0.00061	0.00064	0.00065	0.00065	0.00065
78	0.00059	0.00050	0.00054	0.00056	0.00059	0.00059	0.00059	0.00059
79	0.00054	0.00046	0.00049	0.00051	0.00053	0.00054	0.00054	0.00054
80	0.00049	0.00042	0.00044	0.00046	0.00048	0.00049	0.00049	0.00049
81	0.00044	0.00038	0.00040	0.00042	0.00044	0.00044	0.00044	0.00044
82	0.00040	0.00034	0.00037	0.00038	0.00040	0.00040	0.00040	0.00040
83	0.00037	0.00031	0.00033	0.00035	0.00036	0.00037	0.00037	0.00037
84	0.00033	0.00028	0.00030	0.00031	0.00033	0.00033	0.00033	0.00033
85	0.00030	0.00026	0.00028	0.00029	0.00030	0.00030	0.00030	0.00030
86	0.00028	0.00023	0.00025	0.00026	0.00027	0.00028	0.00028	0.00028
87	0.00025	0.00021	0.00023	0.00024	0.00025	0.00025	0.00025	0.00025
88	0.00023	0.00019	0.00021	0.00021	0.00023	0.00023	0.00023	0.00023
89	0.00021	0.00018	0.00019	0.00020	0.00021	0.00021	0.00021	0.00021
90	0.00019	0.00016	0.00017	0.00018	0.00019	0.00019	0.00019	0.00019
91	0.00017	0.00015	0.00016	0.00016	0.00017	0.00017	0.00017	0.00017
92	0.00016	0.00013	0.00014	0.00015	0.00015	0.00016	0.00016	0.00016
93	0.00014	0.00012	0.00013	0.00013	0.00014	0.00014	0.00014	0.00014
94	0.00013	0.00011	0.00012	0.00012	0.00013	0.00013	0.00013	0.00013
95	0.00012	0.00010	0.00011	0.00011	0.00012	0.00012	0.00012	0.00012
96	0.00011	0.00009	0.00010	0.00010	0.00011	0.00011	0.00011	0.00011
97	0.00010	0.00008	0.00009	0.00009	0.00010	0.00010	0.00010	0.00010
98	0.00009	0.00007	0.00008	0.00008	0.00009	0.00009	0.00009	0.00009
99	0.00008	0.00007	0.00007	0.00008	0.00008	0.00008	0.00008	0.00008
100	0.00007	0.00006	0.00007	0.00007	0.00007	0.00007	0.00007	0.00007

Extension of life (EL) factors for TDR = 6%, r = 0.06

p = Year of erosion	Dn = Discount factor	ELF for p given by row and s given by column						
		s=20 yr	s=25 yr	s=30 yr	s=50 yr	s=75 yr	s=100 yr	s=125 yr
1	0.94340	0.64924	0.72359	0.77914	0.89218	0.93146	0.94062	0.94275
2	0.89000	0.61249	0.68263	0.73504	0.84168	0.87874	0.88737	0.88939
3	0.83962	0.57782	0.64399	0.69343	0.79404	0.82900	0.83714	0.83904
4	0.79209	0.54512	0.60754	0.65418	0.74909	0.78207	0.78976	0.79155
5	0.74726	0.51426	0.57315	0.61715	0.70669	0.73781	0.74506	0.74675
6	0.70496	0.48515	0.54071	0.58222	0.66669	0.69604	0.70288	0.70448
7	0.66506	0.45769	0.51010	0.54926	0.62895	0.65664	0.66310	0.66460
8	0.62741	0.43178	0.48123	0.51817	0.59335	0.61948	0.62556	0.62698
9	0.59190	0.40734	0.45399	0.48884	0.55977	0.58441	0.59015	0.59149
10	0.55839	0.38428	0.42829	0.46117	0.52808	0.55133	0.55675	0.55801
11	0.52679	0.36253	0.40405	0.43507	0.49819	0.52012	0.52523	0.52643
12	0.49697	0.34201	0.38118	0.41044	0.46999	0.49068	0.49550	0.49663
13	0.46884	0.32265	0.35960	0.38721	0.44339	0.46291	0.46746	0.46852
14	0.44230	0.30439	0.33925	0.36529	0.41829	0.43671	0.44100	0.44200
15	0.41727	0.28716	0.32004	0.34461	0.39461	0.41199	0.41604	0.41698
16	0.39365	0.27091	0.30193	0.32511	0.37228	0.38867	0.39249	0.39338
17	0.37136	0.25557	0.28484	0.30671	0.35120	0.36667	0.37027	0.37111
18	0.35034	0.24110	0.26871	0.28935	0.33132	0.34591	0.34931	0.35010
19	0.33051	0.22746	0.25350	0.27297	0.31257	0.32633	0.32954	0.33029
20	0.31180	0.21458	0.23915	0.25752	0.29488	0.30786	0.31089	0.31159
21	0.29416	0.20244	0.22562	0.24294	0.27819	0.29043	0.29329	0.29395
22	0.27751	0.19098	0.21285	0.22919	0.26244	0.27399	0.27669	0.27731
23	0.26180	0.18017	0.20080	0.21622	0.24758	0.25849	0.26103	0.26162
24	0.24698	0.16997	0.18943	0.20398	0.23357	0.24385	0.24625	0.24681
25	0.23300	0.16035	0.17871	0.19243	0.22035	0.23005	0.23231	0.23284
26	0.21981	0.15127	0.16859	0.18154	0.20788	0.21703	0.21916	0.21966
27	0.20737	0.14271	0.15905	0.17126	0.19611	0.20474	0.20676	0.20723
28	0.19563	0.13463	0.15005	0.16157	0.18501	0.19316	0.19505	0.19550
29	0.18456	0.12701	0.14156	0.15242	0.17454	0.18222	0.18401	0.18443
30	0.17411	0.11982	0.13354	0.14380	0.16466	0.17191	0.17360	0.17399
31	0.16425	0.11304	0.12598	0.13566	0.15534	0.16218	0.16377	0.16414
32	0.15496	0.10664	0.11885	0.12798	0.14655	0.15300	0.15450	0.15485
33	0.14619	0.10060	0.11213	0.12073	0.13825	0.14434	0.14576	0.14609
34	0.13791	0.09491	0.10578	0.11390	0.13042	0.13617	0.13751	0.13782
35	0.13011	0.08954	0.09979	0.10745	0.12304	0.12846	0.12972	0.13002
36	0.12274	0.08447	0.09414	0.10137	0.11608	0.12119	0.12238	0.12266
37	0.11579	0.07969	0.08881	0.09563	0.10951	0.11433	0.11545	0.11571
38	0.10924	0.07518	0.08379	0.09022	0.10331	0.10786	0.10892	0.10916
39	0.10306	0.07092	0.07904	0.08511	0.09746	0.10175	0.10275	0.10298
40	0.09722	0.06691	0.07457	0.08029	0.09194	0.09599	0.09694	0.09716
41	0.09172	0.06312	0.07035	0.07575	0.08674	0.09056	0.09145	0.09166
42	0.08653	0.05955	0.06637	0.07146	0.08183	0.08543	0.08627	0.08647
43	0.08163	0.05618	0.06261	0.06742	0.07720	0.08060	0.08139	0.08157
44	0.07701	0.05300	0.05907	0.06360	0.07283	0.07603	0.07678	0.07696
45	0.07265	0.05000	0.05572	0.06000	0.06871	0.07173	0.07244	0.07260
46	0.06854	0.04717	0.05257	0.05660	0.06482	0.06767	0.06834	0.06849
47	0.06466	0.04450	0.04959	0.05340	0.06115	0.06384	0.06447	0.06461
48	0.06100	0.04198	0.04679	0.05038	0.05769	0.06023	0.06082	0.06096
49	0.05755	0.03960	0.04414	0.04753	0.05442	0.05682	0.05738	0.05751
50	0.05429	0.03736	0.04164	0.04484	0.05134	0.05360	0.05413	0.05425
51	0.05122	0.03525	0.03928	0.04230	0.04844	0.05057	0.05106	0.05118
52	0.04832	0.03325	0.03706	0.03990	0.04569	0.04771	0.04817	0.04828

53	0.04558	0.03137	0.03496	0.03765	0.04311	0.04500	0.04545	0.04555
54	0.04300	0.02959	0.03298	0.03551	0.04067	0.04246	0.04287	0.04297
55	0.04057	0.02792	0.03112	0.03350	0.03837	0.04005	0.04045	0.04054
56	0.03827	0.02634	0.02935	0.03161	0.03619	0.03779	0.03816	0.03824
57	0.03610	0.02485	0.02769	0.02982	0.03414	0.03565	0.03600	0.03608
58	0.03406	0.02344	0.02612	0.02813	0.03221	0.03363	0.03396	0.03404
59	0.03213	0.02211	0.02465	0.02654	0.03039	0.03173	0.03204	0.03211
60	0.03031	0.02086	0.02325	0.02504	0.02867	0.02993	0.03022	0.03029
61	0.02860	0.01968	0.02194	0.02362	0.02705	0.02824	0.02851	0.02858
62	0.02698	0.01857	0.02069	0.02228	0.02551	0.02664	0.02690	0.02696
63	0.02545	0.01752	0.01952	0.02102	0.02407	0.02513	0.02538	0.02544
64	0.02401	0.01652	0.01842	0.01983	0.02271	0.02371	0.02394	0.02400
65	0.02265	0.01559	0.01737	0.01871	0.02142	0.02237	0.02259	0.02264
66	0.02137	0.01471	0.01639	0.01765	0.02021	0.02110	0.02131	0.02136
67	0.02016	0.01387	0.01546	0.01665	0.01907	0.01991	0.02010	0.02015
68	0.01902	0.01309	0.01459	0.01571	0.01799	0.01878	0.01896	0.01901
69	0.01794	0.01235	0.01376	0.01482	0.01697	0.01772	0.01789	0.01793
70	0.01693	0.01165	0.01298	0.01398	0.01601	0.01671	0.01688	0.01692
71	0.01597	0.01099	0.01225	0.01319	0.01510	0.01577	0.01592	0.01596
72	0.01507	0.01037	0.01156	0.01244	0.01425	0.01487	0.01502	0.01505
73	0.01421	0.00978	0.01090	0.01174	0.01344	0.01403	0.01417	0.01420
74	0.01341	0.00923	0.01028	0.01107	0.01268	0.01324	0.01337	0.01340
75	0.01265	0.00871	0.00970	0.01045	0.01196	0.01249	0.01261	0.01264
76	0.01193	0.00821	0.00915	0.00986	0.01129	0.01178	0.01190	0.01192
77	0.01126	0.00775	0.00863	0.00930	0.01065	0.01112	0.01122	0.01125
78	0.01062	0.00731	0.00815	0.00877	0.01004	0.01049	0.01059	0.01061
79	0.01002	0.00690	0.00768	0.00827	0.00948	0.00989	0.00999	0.01001
80	0.00945	0.00650	0.00725	0.00781	0.00894	0.00933	0.00942	0.00945
81	0.00892	0.00614	0.00684	0.00736	0.00843	0.00880	0.00889	0.00891
82	0.00841	0.00579	0.00645	0.00695	0.00796	0.00831	0.00839	0.00841
83	0.00794	0.00546	0.00609	0.00655	0.00751	0.00784	0.00791	0.00793
84	0.00749	0.00515	0.00574	0.00618	0.00708	0.00739	0.00746	0.00748
85	0.00706	0.00486	0.00542	0.00583	0.00668	0.00697	0.00704	0.00706
86	0.00666	0.00459	0.00511	0.00550	0.00630	0.00658	0.00664	0.00666
87	0.00629	0.00433	0.00482	0.00519	0.00594	0.00621	0.00627	0.00628
88	0.00593	0.00408	0.00455	0.00490	0.00561	0.00586	0.00591	0.00593
89	0.00559	0.00385	0.00429	0.00462	0.00529	0.00552	0.00558	0.00559
90	0.00528	0.00363	0.00405	0.00436	0.00499	0.00521	0.00526	0.00527
91	0.00498	0.00343	0.00382	0.00411	0.00471	0.00492	0.00496	0.00498
92	0.00470	0.00323	0.00360	0.00388	0.00444	0.00464	0.00468	0.00469
93	0.00443	0.00305	0.00340	0.00366	0.00419	0.00438	0.00442	0.00443
94	0.00418	0.00288	0.00321	0.00345	0.00395	0.00413	0.00417	0.00418
95	0.00394	0.00271	0.00303	0.00326	0.00373	0.00389	0.00393	0.00394
96	0.00372	0.00256	0.00285	0.00307	0.00352	0.00367	0.00371	0.00372
97	0.00351	0.00242	0.00269	0.00290	0.00332	0.00347	0.00350	0.00351
98	0.00331	0.00228	0.00254	0.00273	0.00313	0.00327	0.00330	0.00331
99	0.00312	0.00215	0.00240	0.00258	0.00295	0.00308	0.00311	0.00312
100	0.00295	0.00203	0.00226	0.00243	0.00279	0.00291	0.00294	0.00295

Extension of life (ELF) factors for TDR = 5%,   r =   0.05

p = Year of erosion	Dn = Discount factor	ELF for p given by row and s given by column						
		s=20 yr	s=25 yr	s=30 yr	s=50 yr	s=75 yr	s=100 yr	s=125 yr
1	0.95238	0.59344	0.67114	0.73202	0.86933	0.92786	0.94514	0.95024
2	0.90703	0.56518	0.63918	0.69716	0.82793	0.88367	0.90013	0.90499
3	0.86384	0.53827	0.60874	0.66397	0.78851	0.84159	0.85727	0.86190
4	0.82270	0.51263	0.57976	0.63235	0.75096	0.80152	0.81645	0.82085
5	0.78353	0.48822	0.55215	0.60224	0.71520	0.76335	0.77757	0.78177
6	0.74622	0.46497	0.52586	0.57356	0.68114	0.72700	0.74054	0.74454
7	0.71068	0.44283	0.50082	0.54625	0.64871	0.69238	0.70528	0.70909
8	0.67684	0.42175	0.47697	0.52023	0.61782	0.65941	0.67169	0.67532
9	0.64461	0.40166	0.45425	0.49546	0.58840	0.62801	0.63971	0.64316
10	0.61391	0.38254	0.43262	0.47187	0.56038	0.59810	0.60924	0.61253
11	0.58468	0.36432	0.41202	0.44940	0.53369	0.56962	0.58023	0.58337
12	0.55684	0.34697	0.39240	0.42800	0.50828	0.54250	0.55260	0.55559
13	0.53032	0.33045	0.37372	0.40762	0.48408	0.51666	0.52629	0.52913
14	0.50507	0.31471	0.35592	0.38821	0.46102	0.49206	0.50123	0.50393
15	0.48102	0.29973	0.33897	0.36972	0.43907	0.46863	0.47736	0.47994
16	0.45811	0.28545	0.32283	0.35211	0.41816	0.44631	0.45463	0.45708
17	0.43630	0.27186	0.30746	0.33535	0.39825	0.42506	0.43298	0.43532
18	0.41552	0.25892	0.29282	0.31938	0.37929	0.40482	0.41236	0.41459
19	0.39573	0.24659	0.27887	0.30417	0.36122	0.38554	0.39272	0.39485
20	0.37689	0.23484	0.26559	0.28969	0.34402	0.36718	0.37402	0.37604
21	0.35894	0.22366	0.25295	0.27589	0.32764	0.34970	0.35621	0.35814
22	0.34185	0.21301	0.24090	0.26275	0.31204	0.33305	0.33925	0.34108
23	0.32557	0.20287	0.22943	0.25024	0.29718	0.31719	0.32310	0.32484
24	0.31007	0.19321	0.21850	0.23833	0.28303	0.30208	0.30771	0.30937
25	0.29530	0.18401	0.20810	0.22698	0.26955	0.28770	0.29306	0.29464
26	0.28124	0.17524	0.19819	0.21617	0.25672	0.27400	0.27910	0.28061
27	0.26785	0.16690	0.18875	0.20587	0.24449	0.26095	0.26581	0.26725
28	0.25509	0.15895	0.17976	0.19607	0.23285	0.24852	0.25315	0.25452
29	0.24295	0.15138	0.17120	0.18673	0.22176	0.23669	0.24110	0.24240
30	0.23138	0.14417	0.16305	0.17784	0.21120	0.22542	0.22962	0.23086
31	0.22036	0.13731	0.15529	0.16937	0.20114	0.21468	0.21868	0.21986
32	0.20987	0.13077	0.14789	0.16131	0.19157	0.20446	0.20827	0.20939
33	0.19987	0.12454	0.14085	0.15363	0.18244	0.19473	0.19835	0.19942
34	0.19035	0.11861	0.13414	0.14631	0.17376	0.18545	0.18891	0.18993
35	0.18129	0.11296	0.12775	0.13934	0.16548	0.17662	0.17991	0.18088
36	0.17266	0.10758	0.12167	0.13271	0.15760	0.16821	0.17134	0.17227
37	0.16444	0.10246	0.11588	0.12639	0.15010	0.16020	0.16319	0.16407
38	0.15661	0.09758	0.11036	0.12037	0.14295	0.15257	0.15541	0.15625
39	0.14915	0.09294	0.10510	0.11464	0.13614	0.14531	0.14801	0.14881
40	0.14205	0.08851	0.10010	0.10918	0.12966	0.13839	0.14097	0.14173
41	0.13528	0.08430	0.09533	0.10398	0.12348	0.13180	0.13425	0.13498
42	0.12884	0.08028	0.09079	0.09903	0.11760	0.12552	0.12786	0.12855
43	0.12270	0.07646	0.08647	0.09431	0.11200	0.11954	0.12177	0.12243
44	0.11686	0.07282	0.08235	0.08982	0.10667	0.11385	0.11597	0.11660
45	0.11130	0.06935	0.07843	0.08555	0.10159	0.10843	0.11045	0.11105
46	0.10600	0.06605	0.07470	0.08147	0.09675	0.10327	0.10519	0.10576
47	0.10095	0.06290	0.07114	0.07759	0.09215	0.09835	0.10018	0.10072
48	0.09614	0.05991	0.06775	0.07390	0.08776	0.09367	0.09541	0.09593
49	0.09156	0.05705	0.06452	0.07038	0.08358	0.08921	0.09087	0.09136
50	0.08720	0.05434	0.06145	0.06703	0.07960	0.08496	0.08654	0.08701
51	0.08305	0.05175	0.05853	0.06384	0.07581	0.08091	0.08242	0.08286
52	0.07910	0.04929	0.05574	0.06080	0.07220	0.07706	0.07849	0.07892

53	0.07533	0.04694	0.05308	0.05790	0.06876	0.07339	0.07476	0.07516
54	0.07174	0.04470	0.05056	0.05514	0.06549	0.06990	0.07120	0.07158
55	0.06833	0.04257	0.04815	0.05252	0.06237	0.06657	0.06781	0.06817
56	0.06507	0.04055	0.04586	0.05002	0.05940	0.06340	0.06458	0.06493
57	0.06197	0.03862	0.04367	0.04763	0.05657	0.06038	0.06150	0.06183
58	0.05902	0.03678	0.04159	0.04537	0.05388	0.05750	0.05857	0.05889
59	0.05621	0.03503	0.03961	0.04321	0.05131	0.05476	0.05578	0.05609
60	0.05354	0.03336	0.03773	0.04115	0.04887	0.05216	0.05313	0.05342
61	0.05099	0.03177	0.03593	0.03919	0.04654	0.04967	0.05060	0.05087
62	0.04856	0.03026	0.03422	0.03732	0.04432	0.04731	0.04819	0.04845
63	0.04625	0.02882	0.03259	0.03555	0.04221	0.04506	0.04589	0.04614
64	0.04404	0.02744	0.03104	0.03385	0.04020	0.04291	0.04371	0.04394
65	0.04195	0.02614	0.02956	0.03224	0.03829	0.04087	0.04163	0.04185
66	0.03995	0.02489	0.02815	0.03071	0.03647	0.03892	0.03965	0.03986
67	0.03805	0.02371	0.02681	0.02924	0.03473	0.03707	0.03776	0.03796
68	0.03623	0.02258	0.02553	0.02785	0.03308	0.03530	0.03596	0.03615
69	0.03451	0.02150	0.02432	0.02652	0.03150	0.03362	0.03425	0.03443
70	0.03287	0.02048	0.02316	0.02526	0.03000	0.03202	0.03262	0.03279
71	0.03130	0.01950	0.02206	0.02406	0.02857	0.03050	0.03106	0.03123
72	0.02981	0.01858	0.02101	0.02291	0.02721	0.02904	0.02958	0.02974
73	0.02839	0.01769	0.02001	0.02182	0.02592	0.02766	0.02818	0.02833
74	0.02704	0.01685	0.01905	0.02078	0.02468	0.02634	0.02683	0.02698
75	0.02575	0.01605	0.01815	0.01979	0.02351	0.02509	0.02556	0.02569
76	0.02453	0.01528	0.01728	0.01885	0.02239	0.02389	0.02434	0.02447
77	0.02336	0.01455	0.01646	0.01795	0.02132	0.02276	0.02318	0.02330
78	0.02225	0.01386	0.01568	0.01710	0.02031	0.02167	0.02208	0.02220
79	0.02119	0.01320	0.01493	0.01628	0.01934	0.02064	0.02102	0.02114
80	0.02018	0.01257	0.01422	0.01551	0.01842	0.01966	0.02002	0.02013
81	0.01922	0.01197	0.01354	0.01477	0.01754	0.01872	0.01907	0.01917
82	0.01830	0.01140	0.01290	0.01407	0.01671	0.01783	0.01816	0.01826
83	0.01743	0.01086	0.01228	0.01340	0.01591	0.01698	0.01730	0.01739
84	0.01660	0.01034	0.01170	0.01276	0.01515	0.01617	0.01647	0.01656
85	0.01581	0.00985	0.01114	0.01215	0.01443	0.01540	0.01569	0.01577
86	0.01506	0.00938	0.01061	0.01157	0.01374	0.01467	0.01494	0.01502
87	0.01434	0.00894	0.01010	0.01102	0.01309	0.01397	0.01423	0.01431
88	0.01366	0.00851	0.00962	0.01050	0.01247	0.01330	0.01355	0.01363
89	0.01301	0.00810	0.00917	0.73202	0.86933	0.92786	0.94514	0.95024
90	0.95238	0.59344	0.67114	0.73202	0.86933	0.92786	0.94514	0.95024
91	0.95238	0.59344	0.67114	0.73202	0.86933	0.92786	0.94514	0.95024
92	0.95238	0.59344	0.67114	0.73202	0.86933	0.92786	0.94514	0.95024
93	0.95238	0.59344	0.67114	0.73202	0.86933	0.92786	0.94514	0.95024
94	0.95238	0.59344	0.67114	0.73202	0.86933	0.92786	0.94514	0.95024
95	0.95238	0.59344	0.67114	0.73202	0.86933	0.92786	0.94514	0.95024
96	0.95238	0.59344	0.67114	0.73202	0.86933	0.92786	0.94514	0.95024
97	0.95238	0.59344	0.67114	0.73202	0.86933	0.92786	0.94514	0.95024
98	0.95238	0.59344	0.67114	0.73202	0.86933	0.92786	0.94514	0.95024
99	0.95238	0.59344	0.67114	0.73202	0.86933	0.92786	0.94514	0.95024
100	0.95238	0.59344	0.67114	0.73202	0.86933	0.92786	0.94514	0.95024

Extension of life (ELF) factors for TDR = 3%,   r =   0.03

p = Year of erosion	Dn = Discount factor	\multicolumn{7}{c}{ELF for p given by row and s (scheme life) given by column}						
		s=20 yr	s=25 yr	s=30 yr	s=50 yr	s=75 yr	s=100 yr	s=125 yr
1	0.97087	0.43332	0.50718	0.57089	0.74941	0.86510	0.92036	0.94675
2	0.94260	0.42070	0.49241	0.55426	0.72758	0.83990	0.89355	0.91917
3	0.91514	0.40845	0.47806	0.53812	0.70639	0.81544	0.86752	0.89240
4	0.88849	0.39655	0.46414	0.52244	0.68582	0.79169	0.84226	0.86641
5	0.86261	0.38500	0.45062	0.50723	0.66584	0.76863	0.81772	0.84117
6	0.83748	0.37379	0.43750	0.49245	0.64645	0.74624	0.79391	0.81667
7	0.81309	0.36290	0.42475	0.47811	0.62762	0.72451	0.77078	0.79289
8	0.78941	0.35233	0.41238	0.46418	0.60934	0.70341	0.74833	0.76979
9	0.76642	0.34207	0.40037	0.45066	0.59159	0.68292	0.72654	0.74737
10	0.74409	0.33211	0.38871	0.43754	0.57436	0.66303	0.70538	0.72560
11	0.72242	0.32243	0.37739	0.42479	0.55763	0.64372	0.68483	0.70447
12	0.70138	0.31304	0.36640	0.41242	0.54139	0.62497	0.66489	0.68395
13	0.68095	0.30393	0.35573	0.40041	0.52562	0.60676	0.64552	0.66403
14	0.66112	0.29507	0.34536	0.38875	0.51031	0.58909	0.62672	0.64469
15	0.64186	0.28648	0.33531	0.37742	0.49545	0.57193	0.60846	0.62591
16	0.62317	0.27813	0.32554	0.36643	0.48102	0.55528	0.59074	0.60768
17	0.60502	0.27003	0.31606	0.35576	0.46701	0.53910	0.57354	0.58998
18	0.58739	0.26217	0.30685	0.34540	0.45341	0.52340	0.55683	0.57280
19	0.57029	0.25453	0.29791	0.33534	0.44020	0.50816	0.54061	0.55611
20	0.55368	0.24712	0.28924	0.32557	0.42738	0.49336	0.52487	0.53992
21	0.53755	0.23992	0.28081	0.31609	0.41493	0.47899	0.50958	0.52419
22	0.52189	0.23293	0.27263	0.30688	0.40285	0.46503	0.49474	0.50892
23	0.50669	0.22615	0.26469	0.29794	0.39111	0.45149	0.48033	0.49410
24	0.49193	0.21956	0.25698	0.28926	0.37972	0.43834	0.46634	0.47971
25	0.47761	0.21317	0.24950	0.28084	0.36866	0.42557	0.45275	0.46574
26	0.46369	0.20696	0.24223	0.27266	0.35792	0.41318	0.43957	0.45217
27	0.45019	0.20093	0.23518	0.26472	0.34750	0.40114	0.42676	0.43900
28	0.43708	0.19508	0.22833	0.25701	0.33738	0.38946	0.41433	0.42621
29	0.42435	0.18940	0.22168	0.24952	0.32755	0.37812	0.40227	0.41380
30	0.41199	0.18388	0.21522	0.24225	0.31801	0.36710	0.39055	0.40175
31	0.39999	0.17852	0.20895	0.23520	0.30875	0.35641	0.37917	0.39005
32	0.38834	0.17332	0.20287	0.22835	0.29975	0.34603	0.36813	0.37869
33	0.37703	0.16828	0.19696	0.22170	0.29102	0.33595	0.35741	0.36766
34	0.36604	0.16337	0.19122	0.21524	0.28255	0.32617	0.34700	0.35695
35	0.35538	0.15862	0.18565	0.20897	0.27432	0.31667	0.33689	0.34655
36	0.34503	0.15400	0.18024	0.20288	0.26633	0.30744	0.32708	0.33646
37	0.33498	0.14951	0.17499	0.19697	0.25857	0.29849	0.31755	0.32666
38	0.32523	0.14516	0.16990	0.19124	0.25104	0.28979	0.30830	0.31714
39	0.31575	0.14093	0.16495	0.18567	0.24373	0.28135	0.29932	0.30791
40	0.30656	0.13682	0.16014	0.18026	0.23663	0.27316	0.29061	0.29894
41	0.29763	0.13284	0.15548	0.17501	0.22974	0.26520	0.28214	0.29023
42	0.28896	0.12897	0.15095	0.16991	0.22305	0.25748	0.27392	0.28178
43	0.28054	0.12521	0.14655	0.16496	0.21655	0.24998	0.26595	0.27357
44	0.27237	0.12157	0.14229	0.16016	0.21024	0.24270	0.25820	0.26560
45	0.26444	0.11803	0.13814	0.15549	0.20412	0.23563	0.25068	0.25787
46	0.25674	0.11459	0.13412	0.15096	0.19817	0.22877	0.24338	0.25036
47	0.24926	0.11125	0.13021	0.14657	0.19240	0.22210	0.23629	0.24306
48	0.24200	0.10801	0.12642	0.14230	0.18680	0.21563	0.22941	0.23598
49	0.23495	0.10486	0.12274	0.13815	0.18136	0.20935	0.22273	0.22911
50	0.22811	0.10181	0.11916	0.13413	0.17607	0.20326	0.21624	0.22244
51	0.22146	0.09884	0.11569	0.13022	0.17095	0.19734	0.20994	0.21596
52	0.21501	0.09597	0.11232	0.12643	0.16597	0.19159	0.20383	0.20967

53	0.20875	0.09317	0.10905	0.12275	0.16113	0.18601	0.19789	0.20356
54	0.20267	0.09046	0.10587	0.11917	0.15644	0.18059	0.19212	0.19763
55	0.19677	0.08782	0.10279	0.11570	0.15188	0.17533	0.18653	0.19188
56	0.19104	0.08526	0.09980	0.11233	0.14746	0.17022	0.18110	0.18629
57	0.18547	0.08278	0.09689	0.10906	0.14316	0.16527	0.17582	0.18086
58	0.18007	0.08037	0.09407	0.10588	0.13899	0.16045	0.17070	0.17559
59	0.17483	0.07803	0.09133	0.10280	0.13495	0.15578	0.16573	0.17048
60	0.16973	0.07576	0.08867	0.09981	0.13102	0.15124	0.16090	0.16552
61	0.16479	0.07355	0.08609	0.09690	0.12720	0.14684	0.15621	0.16069
62	0.15999	0.07141	0.08358	0.09408	0.12349	0.14256	0.15166	0.15601
63	0.15533	0.06933	0.08114	0.09134	0.11990	0.13841	0.14725	0.15147
64	0.15081	0.06731	0.07878	0.08868	0.11641	0.13438	0.14296	0.14706
65	0.14641	0.06535	0.07649	0.08609	0.11302	0.13046	0.13879	0.14277
66	0.14215	0.06344	0.07426	0.08359	0.10972	0.12666	0.13475	0.13862
67	0.13801	0.06160	0.07209	0.08115	0.10653	0.12297	0.13083	0.13458
68	0.13399	0.05980	0.07000	0.07879	0.10343	0.11939	0.12702	0.13066
69	0.13009	0.05806	0.06796	0.07649	0.10041	0.11591	0.12332	0.12685
70	0.12630	0.05637	0.06598	0.07426	0.09749	0.11254	0.11973	0.12316
71	0.12262	0.05473	0.06406	0.07210	0.09465	0.10926	0.11624	0.11957
72	0.11905	0.05313	0.06219	0.07000	0.09189	0.10608	0.11285	0.11609
73	0.11558	0.05159	0.06038	0.06796	0.08922	0.10299	0.10957	0.11271
74	0.11221	0.05008	0.05862	0.06598	0.08662	0.09999	0.10637	0.10942
75	0.10895	0.04862	0.05691	0.06406	0.08409	0.09708	0.10328	0.10624
76	0.10577	0.04721	0.05525	0.06220	0.08164	0.09425	0.10027	0.10314
77	0.10269	0.04583	0.05365	0.06038	0.07927	0.09150	0.09735	0.10014
78	0.09970	0.04450	0.05208	0.05863	0.07696	0.08884	0.09451	0.09722
79	0.09680	0.04320	0.05057	0.05692	0.07472	0.08625	0.09176	0.09439
80	0.09398	0.04194	0.04909	0.05526	0.07254	0.08374	0.08909	0.09164
81	0.09124	0.04072	0.04766	0.05365	0.07043	0.08130	0.08649	0.08897
82	0.08858	0.03954	0.04627	0.05209	0.06838	0.07893	0.08397	0.08638
83	0.08600	0.03838	0.04493	0.05057	0.06638	0.07663	0.08153	0.08387
84	0.08350	0.03727	0.04362	0.04910	0.06445	0.07440	0.07915	0.08142
85	0.08107	0.03618	0.04235	0.04767	0.06257	0.07223	0.07685	0.07905
86	0.07870	0.03513	0.04111	0.04628	0.06075	0.07013	0.07461	0.07675
87	0.07641	0.03410	0.03992	0.04493	0.05898	0.06809	0.07244	0.07451
88	0.07419	0.03311	0.03875	0.04362	0.05726	0.06610	0.07033	0.07234
89	0.07203	0.03215	0.03763	0.04235	0.05560	0.06418	0.06828	0.07024
90	0.06993	0.03121	0.03653	0.04112	0.05398	0.06231	0.06629	0.06819
91	0.06789	0.03030	0.03547	0.03992	0.05240	0.06049	0.06436	0.06620
92	0.06591	0.02942	0.03443	0.03876	0.05088	0.05873	0.06248	0.06428
93	0.06399	0.02856	0.03343	0.03763	0.04940	0.05702	0.06066	0.06240
94	0.06213	0.02773	0.03246	0.03653	0.04796	0.05536	0.05890	0.06059
95	0.06032	0.02692	0.03151	0.03547	0.04656	0.05375	0.05718	0.05882
96	0.05856	0.02614	0.03059	0.03444	0.04520	0.05218	0.05552	0.05711
97	0.05686	0.02538	0.02970	0.03343	0.04389	0.05066	0.05390	0.05544
98	0.05520	0.02464	0.02884	0.03246	0.04261	0.04919	0.05233	0.05383
99	0.05359	0.02392	0.02800	0.03151	0.04137	0.04776	0.05081	0.05226
100	0.05203	0.02322	0.02718	0.03060	0.04016	0.04636	0.04933	0.05074

Appendix 4.1 1989 Coastal recreation survey research questionnaire

This appendix reproduces the questionnaire schedule used in collecting the data reported in Chapter 4. However, it is a research questionnaire and it is not recommended for use in future assessments of the recreational benefits of coast protection schemes for which the questionnaire schedules in Appendix 4.2 (a) and (b) have been developed.

APPENDIX 4.1

MIDDLESEX POLYTECHNIC
FLOOD HAZARD RESEARCH CENTRE

CONFIDENTIAL

COAST PROTECTION: COASTAL RECREATION SURVEY (1989)

FOR OFFICE USE ONLY

Questionnaire number ....

Location	Bridlington = 1 Clacton = 2
	Hunstanton = 3 Morecambe = 4

Interviewer ....

Interview number ....

Date ..../..../1989

Day Mon = 1 Tue = 2 Wed = 3 Thu = 4
     Fri = 5 Sat = 6 Sun = 7

Time interview started (24 hour clock)..

Time interview finished (24 hour clock)..

Duration (minutes) ....

Weather conditions ....

(a) Sunny......... = 1 (c) Dry ....= 1
    Broken cloud.. = 2   Drizzle/showers = 2
    Overcast...... = 3   Persistent rain = 3

(b) Hot (>20 c)... = 1 (d) Calm........... = 1
    Warm (15-20 c) = 2   Breezy......... = 2
    Cool (<15 c).. = 3   Windy.......... = 3

SECTION A - applies to ALL visitors

1. Where is your normal place of residence?

    Town/Village......................
    County............................
    Post code.........................

2. Where did you stay last night (if different from place in 1)?

    Town..............................
    County............................

3. (a) Do you live within 3 miles of this point?

    Yes = 1
    No = 0

    IF YES, GO TO SECTION B

    IF NO, GO TO QUESTION 3(b)

    (b) Which of these statements best describes you? **(SHOW CARD A)**

    I live locally = 1
    I am on a part-day visit from home = 2
    I am on a day visit from home = 3

    I am on a touring holiday = 4
    I am staying away from home for 1 or 2 nights = 5
    I am staying away from home for more than 2 nights = 6

    IF ANSWER IS 1, 2 OR 3, GO TO SECTION C

    IF ANSWER IS 4, 5 OR 6, GO TO SECTION D

SECTION B - applies to LOCAL RESIDENTS only

☐ 1

4. How did you travel to this point today?
   PROMPT

   Car/motorcycle/van/dormobile = 1
   Public transport = 2
   Other = 3

5. How long did it take you to get here today?
   PROMPT

   0-15 minutes = 1
   16-30 minutes = 2
   31-60 minutes = 3
   Between 1 and 2 hours = 4
   Between 2 and 3 hours = 5
   Between 3 and 4 hours = 6
   > 4 hours = 7

6. (a) For how many years have you lived locally (ie. within 3 miles of this point)?
   CODE TO NEAREST YEAR

   Years.......

   (b) What are the advantages and disadvantages of living locally? (SHOW CARD B)

   | -5 | -4 | -3 | -2 | -1 | 0 | +1 | +2 | +3 | +4 | +5 | -9 |

   major disadvantage   not a consideration   major advantage   (N/A)

   (A) The town and its facilities ....
   (B) The beach ....
   (C) The seaside promenade ....
   (D) The coastal scenery ....
   (E) The scenery and places to visit in the surrounding area..
   (F) Other ....

(c) How often, on average do you come to this **seafront**?
PROMPT

At least daily = 1
At least 3 times a week = 2
Once or twice a week = 3
At least fortnightly = 4
At least once a month = 5
About 5 to 11 times a year = 6
About 2 to 4 times a year = 7
About once a year = 8
Less than once a year = 9
Never been before = 10

7. (a) Did you come to this seafront today alone or in a group?

Alone = 1
In group = 2

IF IN GROUP

(b) How many people including yourself are in the group?......

(c) How many of these people live locally
    (ie. live within 3 miles?) ....

8. Which of the following best describes your group?
   READ OUT

Adults with children aged under 10 (and older children) = 1
Adults with no children under 10 but children aged 10-15 = 2
Adults aged 24 and under only = 3
Adults aged between 25 and 65 only = 4
Adults aged 65 and over only = 5
Adults of mixed age groups without children = 6

9. What is the main reason for your outing today? Is it to visit the seafront, to make visits elsewhere, or are they of equal importance?

The main reason is to visit the seafront = 1
The visits are of equal importance = 2
The main reason is to make visits elsewhere = 3
Don't know = 8

10. (a) How will you spend your time at the seafront today?
READ OUT

$$\begin{aligned}\text{Only on the beach} &= 1\\ \text{Mainly on the beach} &= 2\\ \text{Equally, on both the beach and the promenade} &= 3\\ \text{Mainly on the promenade} &= 4\\ \text{Only on the promenade} &= 5\\ \text{Don't know} &= 8\end{aligned}$$

(b) How long do you intend to spend on **this beach** today?
PROMPT

$$\begin{aligned}\text{No time} &= 0\\ \text{0-15 minutes} &= 1\\ \text{16-30 minutes} &= 2\\ \text{31-60 minutes} &= 3\\ \text{Between 1 and 2 hours} &= 4\\ \text{Between 2 and 4 hours} &= 5\\ > \text{4 hours} &= 6\end{aligned}$$

11. What activities will you or any member of your group be taking part in during this visit to the seafront? **(SHOW CARD C)**

    Sitting/sunbathing/picnicking on the beach......Y / N / DK A
    Sitting/sunbathing/picnicking on the promenade..Y / N / DK B
    Playing with sand, stones or shells............ Y / N / DK C
    Swimming/paddling...............................Y / N / DK D
    Walking/strolling on the beach..................Y / N / DK E
    Walking/strolling on the promenade..............Y / N / DK F
    Long walks (2 miles or over)....................Y / N / DK G
    Games or sports on the beach....................Y / N / DK H
    Other (specify).................................Y / N / DK K

    (CODE YES = 1, NO = 0, DK = 8)

12. How important were each of the following in your choice to come to **this seafront today**? **(SHOW CARD D)**

    0   1   2   3   4   5   6   7   8   9   10   (-9)

    not at all   very   not
    important   important   applicable
    to my   to my   at this
    choice   choice   site

    (A) The quality of the seafront promenade ....
    (B) The characteristics of the beach ....

(C) The coastal scenery ....
(D) The suitability of the sea for swimming and paddling .....
(E) The convenience of the journey ....
(F) The costs of this journey ....

13. (a) The practical factors such as travel, time and cost considerations may have limited the range of places you could have visited today. In practical terms, how wide a choice of places to visit do you feel that you had?
    (SHOW CARD G)

    A very wide range (more than 10 places) = 1
    A fairly wide range (5-10 places) = 2
    A few places (2-4 places) = 3
    This was the only place we could consider visiting = 4

    (CODE DON'T KNOW = 8)

    (b) Who made the final decision to visit **this seafront today?**

    I made the decision = 1
    The decision was made jointly by myself and others = 2
    The decision was made by another member of the group = 3

APPENDIX 4.1 217

SECTION C - applies to DAY AND PART-DAY VISITORS only |2|

4. How did you travel to this point today?
   PROMPT

   Car/motorcycle/van/dormobile = 1
   Public transport = 2
   Other = 3

5. How long did it take you to get from home to **this point today?**
   PROMPT

   0-15 minutes = 1
   16-30 minutes = 2
   31-60 minutes = 3
   Between 1 and 2 hours = 4
   Between 2 and 3 hours = 5
   Between 3 and 4 hours = 6
   > 4 hours = 7

6. How often, on average do you come to **this seafront?**
   PROMPT

   At least daily = 1
   At least 3 times a week = 2
   Once or twice a week = 3
   At least fortnightly = 4
   At least once a month = 5
   About 5 to 11 times a year = 6
   About 2 to 4 times a year = 7
   About once a year = 8
   Less than once a year = 9
   Never been before = 10

7. (a) Did you come to this seafront alone or in a group?

   Alone = 1
   In group = 2

   <u>IF IN GROUP</u>

   (b) How many people including yourself are in the group?.......

   (c) How many of these people live locally
       (ie. live within 3 miles)? ....

8. Which of the following best describes your group?
   READ OUT

   Adults with children aged under 10 (and older children) = 1
   Adults with no children under 10 but children aged 10-15 = 2
   Adults aged 24 and under only = 3
   Adults aged between 25 and 65 only = 4
   Adults aged 65 and over only = 5
   Adults of mixed age groups without children = 6

9. What is the main reason for your outing today? Is it to visit the seafront, to make visits elsewhere, or are they of equal importance?

   The main reason is to visit the seafront = 1
   The visits are of equal importance = 2
   The main reason is to make visits elsewhere = 3
   Don't know = 8

10. (a) How will you spend your time at the seafront today?
    READ OUT

    Only on the beach = 1
    Mainly on the beach = 2
    Equally, on both the beach and the promenade = 3
    Mainly on the promenade = 4
    Only on the promenade = 5
    Don't know = 8

    (b) How long do you intend to spend on **this beach** today?
    PROMPT

    No time = 0
    0-15 minutes = 1
    16-30 minutes = 2
    31-60 minutes = 3
    Between 1 and 2 hours = 4
    Between 2 and 4 hours = 5
    > 4 hours = 6

11. What activities will you or any member of your group be taking part in during this visit to the seafront? (SHOW CARD C)

    Sitting/sunbathing/picnicking on the beach......Y / N / DK A
    Sitting/sunbathing/picnicking on the promenade..Y / N / DK B
    Playing with sand, stones or shells............ Y / N / DK C
    Swimming/paddling...............................Y / N / DK D
    Walking/strolling on the beach..................Y / N / DK E
    Walking/strolling on the promenade..............Y / N / DK F
    Long walks (2 miles or over)....................Y / N / DK G
    Games or sports on the beach....................Y / N / DK H
    Other (specify).................................Y / N / DK K

    (CODE YES = 1, NO = 0, DK = 8)

12. (a) How important were each of the following in your choice to come to .... today ? (SHOW CARD E)

    0   1   2   3   4   5   6   7   8   9   10   (-9)

    not at all   very   not
    important   important   applicable
    to my   to my   at this
    choice   choice   site

    (A) The town and its facilities ....
    (B) The quality of the seafront promenade ....
    (C) The characteristics of the beach ....
    (D) The coastal scenery ....
    (E) The scenery and places to visit in the surrounding area...
    (F) The suitability of the sea for swimming and paddling .....
    (G) The convenience of the journey ....
    (H) The cost of this trip ....

    (b) Are you here today for any of the following reasons?

    To visit friends and/or family = 1
    To attend a conference or to carry out some other business = 2
    Neither of the above = 3

13. (a) The practical factors such as travel, time and cost considerations may have limited the range of places you could have visited today. In practical terms, how wide a choice of places to visit do you feel that you had?
(SHOW CARD G)

$$\begin{aligned}
\text{A very wide range (more than 10 places)} &= 1 \\
\text{A fairly wide range (5-10 places)} &= 2 \\
\text{A few places (2-4 places)} &= 3 \\
\text{This was the only place we could consider visiting} &= 4
\end{aligned}$$

(CODE DON'T KNOW = 8)

(b) Who made the final decision to visit .... today?

$$\begin{aligned}
\text{I made the decision} &= 1 \\
\text{The decision was made jointly by myself and others} &= 2 \\
\text{The decision was made by another member of the group} &= 3
\end{aligned}$$

APPENDIX 4.1 221

SECTION D - applies to STAYING VISITORS only $\boxed{3}$

4. How did you travel to this point **today**?
    PROMPT

    Car/motorcycle/van/dormobile = 1
    Public transport = 2
    Other = 3

5. How long did it take you to get to **this point today** from where you are staying?
    PROMPT

    0-15 minutes = 1
    16-30 minutes = 2
    31-60 minutes = 3
    Between 1 and 2 hours = 4
    Between 2 and 3 hours = 5
    Between 3 and 4 hours = 6
    > 4 hours = 7

6. (a) For how many nights are you staying away from home?

        Nights ....

    (b) On how many days during this holiday do you expect to visit **this beach**?

        Days ....

    (c) How often, on average do you come to **this seafront**?
        PROMPT

        At least once a week = 3
        At least fortnightly = 4
        At least once a month = 5
        About 5 to 11 times a year = 6
        About 2 to 4 times a year = 7
        About once a year = 8
        Less than once a year = 9
        Never been before = 10

7. (a) Did you come to this seafront alone or in a group?

        Alone = 1
        In group = 2

IF IN GROUP

(b) How many people including yourself are in the group?.......

(c) How many of these people live locally
(ie. live within 3 miles)? ....

8. Which of the following best describes your group?
READ OUT

Adults with children aged under 10 (and older children) = 1
Adults with no children under 10 but children aged 10-15 = 2
Adults aged 24 and under only = 3
Adults aged between 25 and 65 only = 4
Adults aged 65 and over only = 5
Adults of mixed age groups without children = 6

9. What is the main reason for your outing today? Is it to visit the seafront, to make visits elsewhere, or are they of equal importance?

The main reason is to visit the seafront = 1
The visits are of equal importance = 2
The main reason is to make visits elsewhere = 3
Don't know = 8

10. (a) How will you spend your time at the seafront today?
READ OUT

Only on the beach = 1
Mainly on the beach = 2
Equally, on both the beach and the promenade = 3
Mainly on the promenade = 4
Only on the promenade = 5
Don't know = 8

(b) How long do you intend to spend on **this beach today?**
PROMPT

No time = 0
0-15 minutes = 1
16-30 minutes = 2
31-60 minutes = 3
Between 1 and 2 hours = 4
Between 2 and 4 hours = 5
> 4 hours = 6

11. What activities will you or any member of your group be taking part in during this visit to the seafront? **(SHOW CARD C)**

  Sitting/sunbathing/picnicking on the beach......Y / N / DK A
  Sitting/sunbathing/picnicking on the promenade..Y / N / DK B
  Playing with sand, stones or shells............ Y / N / DK C
  Swimming/paddling...............................Y / N / DK D
  Walking/strolling on the beach..................Y / N / DK E
  Walking/strolling on the promenade..............Y / N / DK F
  Long walks (2 miles or over)....................Y / N / DK G
  Games or sports on the beach....................Y / N / DK H
  Other (specify).................................Y / N / DK K

  (CODE YES = 1, NO = 0, DK = 8)

12. (a) How important were each of the following in your choice to come to .... ? **(SHOW CARD F)**

  0   1   2   3   4   5   6   7   8   9   10   (-9)

  not at all   very   not
  important   important   applicable
  to my   to my   at this
  choice   choice   site

  (A) The town and its facilities ....
  (B) The quality of the seafront promenade ....
  (C) The characteristics of the beach ....
  (D) The coastal scenery ....
  (E) The scenery and places to visit in the surrounding area...
  (F) The suitability of the sea for swimming and paddling .....
  (G) The convenience of the journey ....
  (H) The cost of this trip ....

  (b) Are you here **today** for any of the following reasons?

                         To visit friends and/or family = 1
    To attend a conference or to carry out some other business = 2
                                    Neither of the above = 3

13. (a) The practical factors such as travel, time and cost considerations may have limited the range of places you could have visited today. In practical terms, how wide a choice of places to visit do you feel that you had?
   (SHOW CARD G)

   A very wide range (more than 10 places) = 1
   A fairly wide range (5-10 places) = 2
   A few places (2-4 places) = 3
   This was the only place we could consider visiting = 4

   (CODE DON'T KNOW = 8)

   (b) Who made the final decision to visit ....?

   I made the decision = 1
   The decision was made jointly by myself and others = 2
   The decision was made by another member of the group = 3

SECTION E - applies to ALL respondents

14. How do you rate this **resort** overall as a place to visit?
    (SHOW CARD H)

    0   1   2   3   4   5   6   7   8   9   10

    very   about   very
    poor   average   good

15. (a) How would you rate this beach for walking on?
        (SHOW CARD H) ....

    (b) How would you rate this beach for sitting or lying on?
        (SHOW CARD H) ....

    (c) How would you rate this beach for children to play on?
        (SHOW CARD H) ....

16. How do you rate this **beach** overall as a place to visit?
    (SHOW CARD H) ....

17. How many years ago did you first visit this **seafront**?

    Have never been before = 0

    Years ....

IF NEVER BEEN BEFORE, GO TO Q20, ELSE GO TO Q18

18. Overall, during this time do you feel that **the beach** has changed for the better or for the worse or that it has not changed at all?

    Better = 1
    Same = 2
    Worse = 3
    Don't know = 8

19. Over the period you have visited **this seafront**, has **the beach** changed in any of the following ways? (SHOW CARD I)

For each of the pair of statements below ask the question in the following way:

"The beach is more sandy, the beach is less sandy or there has been no change?"

IF NO CHANGE, CODE AS 0

CODE THIS CHANGE AS 1	CODE THIS CHANGE AS 2	
Beach is more sandy = 1	Beach is less sandy = 2	☐
Beach is more crowded = 1	Beach is less crowded = 2	☐
Beach is steeper between high tide and low tide = 1	Beach is flatter between high tide and low tide = 2	☐
Beach is steeper above high tide = 1	Beach is flatter above high tide = 2	☐
Longer drop from seawall to beach = 1	Shorter drop from seawall to beach = 2	☐
The groynes have decayed = 1	The groynes have been improved = 2	☐
The sea is cleaner = 1	The sea is dirtier = 2	☐
There is more beach at high tide = 1	There is less beach at high tide = 2	☐
There is more beach at low tide = 1	There is less beach at low tide = 2	☐
Beach is cleaner = 1	Beach is dirtier = 2	☐
The promenade has decayed = 1	The promenade has been improved = 2	☐

READ OUT
We are trying to find out how much value you, **as an individual,** put on your enjoyment of this visit to **this beach today;** I MEAN IN £ AND PENCE.

Now, this is an unusual question to ask so let me explain it to you in this way:

Think of a visit or activity you have done in the past which gave you the same amount of enjoyment as your visit to this beach today. Here is a list of possibilities **(SHOW CARD J)**

(If not visiting the beach, probe as in interviewer instructions)

20. (a) What visit or activities gave you about the same enjoyment as your visit to this beach today?
    PROBE AND WRITE IN RESPONSE

    ....

Now, think about how much that visit (or other activities) cost you. Remember that the cost of a visit may include petrol and parking costs or bus or train fares as well as admission charges and any other costs.

You can use the costs of that visit (or other activities) as a guide to the value of your enjoyment of today's visit to this beach.

(b) So, now, what value do you put on your individual enjoyment of this visit to the **beach today?**

Value of today's beach visit   £......:......pence

Cannot value these things in money terms = -777.77
Don't know = -888.88
Other (please specify) ....
.... = -999.99

228   THE ECONOMICS OF COASTAL MANAGEMENT

READ OUT
   I am now going to show you a number of drawings which represent this beach in varying conditions.

21.   **CURRENT BEACH   (DRAWING A)**

The drawing shows that:

   (i)   There is no beach at high tide
   (ii) The beach is reached by sets of steps and ramps down from the promenade
   (iii) The beach is a mixture of sand and pebbles
   (iv) There are a lot of low groynes which are in a reasonably good condition
   (v)   The sea wall is in a reasonably good condition

21. (a) If you were making a visit (similar to today's visit) to the beach in this drawing, would you get more, less or the same amount of enjoyment from a visit compared to your enjoyment of today's visit? **(SHOW CARD K)**

-3	-2	-1	0	+1	+2	+3
Much less enjoyment			The same amount of enjoyment			Much more enjoyment

(CODE DON'T KNOW = -8)

IF DON'T KNOW, GO TO Q22

   (b) So, how much enjoyment would you get from your visit to the beach in this drawing?

£............

Cannot value these things in money terms = -777.77
Don't know = -888.88
Other (please specify) .... = -999.99

APPENDIX 4.1   229

22. ERODED BEACH (DRAWING B)

READ OUT
Now I would like to ask you what you would think if the beach changed to look like this.

Compared to DRAWING A:

   (i) There is no beach above high tide
  (ii) The drop from the promenade down to the beach is longer so the flights of steps are longer
 (iii) The beach is less sandy and more pebbly
  (iv) There are a lot of groynes and they show signs of decay
   (v) The sea wall is not in such a good condition and there are cracks in places

22. (a) If you were making a visit (similar to today's visit) to the beach in this drawing, would you get more, less or the same amount of enjoyment from a visit compared to your enjoyment of today's visit? (SHOW CARD K)

-3	-2	-1	0	+1	+2	+3
Much less enjoyment			The same amount of enjoyment			Much more enjoyment

(CODE DON'T KNOW = -8)

IF DON'T KNOW, GO TO Q22(c)

   (b) So, how much enjoyment would you get from your visit to the beach in this drawing?

£............

Cannot value these things in money terms = -777.77
Don't know = -888.88
Other (please specify) .... = -999.99

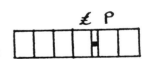

(c) If this beach looked like the beach in the drawing how often would you visit it? **(SHOW CARD L)**

-3	-2	-1	0	+1	+2	+3
Much less often than now			The same as now			Much more often than now

IF LESS GO TO (d) ELSE GO TO Q23

(d) Where would you go instead?

Town..................... County.....................

IF NO ANSWER GIVEN GO TO Q23

(e) Would you go there to visit a beach?

$$\text{Yes} = 1$$
$$\text{No} = 0$$
$$\text{Don't know} = 8$$

(f) How much enjoyment would you get from your visits to this alternative site?

£..............

Cannot value these things in money terms = -777.77
Don't know = -888.88
Other (please specify) .... = -999.99

(g) How much more or less would it cost you to go to this alternative site?

£............ More

£.............Less

Don't know = -888.88

23. IMPROVED BEACH (DRAWING C)

READ OUT
Now I would like to ask you what you would think if the beach changed to look like this.

Compared to DRAWING A:

(i) There is a wide beach at high tide and the high tide mark is shown in the drawing
(ii) The drop from the promenade down to the beach is shorter so the flights of steps are shorter
(iii) The beach is made up of sand and pebbles
(iv) There are no groynes visible
(v) The sea wall is in a good condition

23. (a) If you were making a visit (similar to today's visit) to the beach in this drawing, would you get more, less or the same amount of enjoyment from a visit compared to your enjoyment of today's visit? (**SHOW CARD M**)

| -3 | -2 | -1 | 0 | +1 | +2 | +3 |

Much less enjoyment   The same amount of enjoyment   Much more enjoyment

(CODE DON'T KNOW = -8)

IF DON'T KNOW, GO TO Q23(c)

(b) So, how much enjoyment would you get form your visit to the beach in this drawing?

£............

Cannot value these things in money terms = -777.77
Don't know = -888.88
Other (please specify) .... = -999.99

(c) If this beach looked like the beach in the drawing how often would you visit it? (SHOW CARD N).

| -3 | -2 | -1 | 0 | +1 | +2 | +3 |

Much less often than now

The same as now

Much more often than now

IF LESS, AND NO ALTERNATIVE SITE SPECIFIED IN Q22(d) ABOVE, GO TO (d)
ELSE GO TO Q24

(d) Where would you go instead?

Town...................... County......................

IF NO ANSWER GIVEN GO TO Q24

(e) Would you go there to visit a beach?

Yes = 1
No = 0
Don't know = 8

(f) How much enjoyment would you get from your visits to this alternative site?

£..............

Cannot value these things in money terms = -777.77
Don't know = -888.88
Other (please specify) .... = -999.99

(g) How much more or less would it cost you to go to this alternative site?

£............ More

£..............Less

Don't know = -888.88

READ OUT

24. We want you to think about how much extra (if anything) you would be willing to pay in increased rates and taxes so that more money can be spent on coastal protection. This may be a difficult question if you haven't thought about it in this way before. We hope these notes will help you to decide.
(SHOW CARD O)

(i) The average adult in England and Wales currently pays £890 a year in rates and taxes.

(ii) The total amount spent in England and Wales on coastal protection in 1987-1988 was £14.5 million. This is about 30p per adult per year.

(iii) In the current economic circumstances, you may not be able to afford to pay any more in rates and taxes to prevent erosion even if you would like to be able to. Also there may be other areas of public expenditure (such as education, law and order or health care) upon which you would prefer any extra money to be spent. Or you may prefer all public expenditure to be reduced so that rates and taxes are reduced.

(iv) If you decide you want to pay more then you will have to give up something - it may help you to think what would have been bought with the money you now want to spend on reducing coastal erosion.

24. (a) Would you as an individual be willing to pay increased rates and taxes so that more money can be spent to protect the coast from erosion?

> Yes = 1
> No = 2
> Don't know = 8

IF YES, ASK ALTERNATE RESPONDENTS PART (b) OR PART (c) AND THEN GO TO Q25

IF NO, GO TO Q26

IF DON'T KNOW GO TO Q27

234 THE ECONOMICS OF COASTAL MANAGEMENT

Follow the branches in the diagram below asking the respondent whether they would be willing to pay each of the amounts BOXED until you reach the end of a branch or box. Starting at 50 pence, the question to be asked for each amount is:

(b) "WOULD YOU BE WILLING TO PAY MORE OR LESS THAN .... PER YEAR **EXTRA** IN RATES AND TAXES?"
(SHOW CARD P)

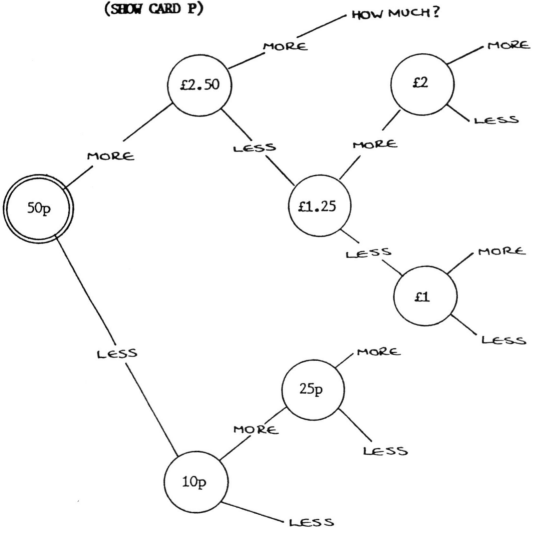

Start at 50p
Show direction of movement

Exact figure if given £..........

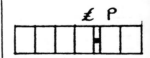

GO TO Q25

APPENDIX 4.1 235

Follow the branches in the diagram below asking the respondent whether they would be willing to pay each of the amounts BOXED until you reach the end of a branch or box. Starting at £2.50, the question to be asked for each amount is:

(c) "WOULD YOU BE WILLING TO PAY MORE OR LESS THAN .... PER YEAR **EXTRA** IN RATES AND TAXES?"
(SHOW CARD Q)

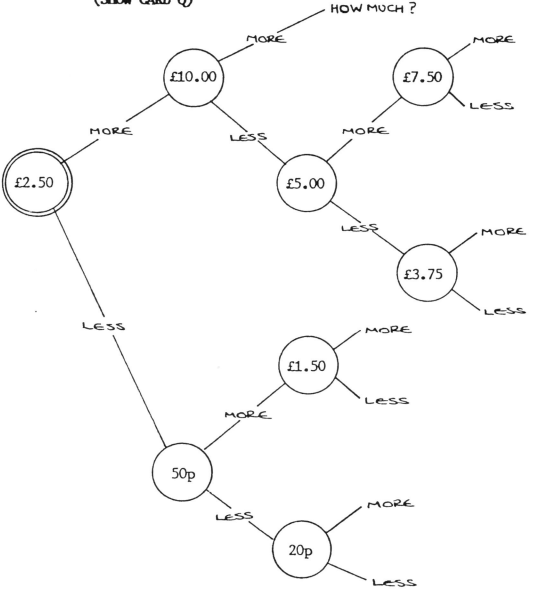

Start at £2.50
Show direction of movement

Exact figure if given £..........

GO TO Q25

IF WILLING TO PAY

25. In deciding how much you were willing to pay, did you think about any of the following? (SHOW CARD R)

CODE ALL THAT APPLY (YES = 1, NO = 0)

    (1). What I can afford to pay .... Y/N
    (2). What my household can afford to pay .... Y/N
    (3). What other people should pay .... Y/N
    (4). What it is fair for me to pay .... Y/N
    (5). The other things I would like to spend money on Y/N
    (6). The value to me of protecting the coast .... Y/N
    (7). How much the country is spending at the moment Y/N

    Other (specify) ....

GO TO Q27

IF NOT WILLING TO PAY

26. If you are not willing to pay increased rates and taxes to reduce coastal erosion, please indicate which of the following factors, if any, most closely describe your reasons and explain why. (SHOW CARD S)

CODE ALL THAT APPLY (YES = 1, NO = 0)

    (1). I can't afford it .... Y/N
    (2). I don't know enough about coastal erosion and protection Y/N
    (3). Rates and taxes are high enough already .... Y/N
    (4). Beach and coastal erosion is the problem of the individual property owner and taxpayers should not have to pay for the solution .... Y/N
    (5). We all have a right to protection from the sea .... Y/N
    (6). I have no desire to pay increased rates and taxes for something that doesn't benefit me personally .... Y/N

    Other (specify) ....

    EXPLANATION: ....

    ....

    ....

SECTION F - applies to ALL visitors

READ OUT
These following questions are standard for most questionnaires and are used for our own classification purposes: **Your answers are completely confidential**

27. Sex

    Female = 0
    Male = 1   ☐

28. To which of the following age categories do you belong?

    Under 18 = 1
    18-39 = 2
    40-65 = 3
    65+ = 4   ☐

29. Can you tell me at what age you completed full-time education?

    Still in full-time education = 1
    At age 15 years or under = 2
    16 years of age = 3
    17 years of age = 4
    18 years of age = 5
    19 years of age or older = 6   ☐

30. Which of the letters on this card represents the gross weekly income from **all sources** of your household?
    (SHOW CARD T)

    ....   ☐

31. We can't avoid or reduce some of our household's expenditure (eg on mortgages). How far do you feel that all your household's income has to be spent on necessities rather than you having a choice about how to spend it? (**SHOW CARD U**).   ☐

32. Do you or any other member of your household own or have access to a car, van etc?

    Yes = 1
    No = 0   ☐

ADDITIONAL NOTES:

Appendix 4.2 (a) Standard site user questionnaire
Appendix 4.2 (b) Standard residents questionnaire

These standard questionnaires are designed so that they can be photocopied directly from the Manual for use. Alternative question wording and spaces for site specific details to be pasted in have been provided.

It is intended that the Show cards presented in Appendix 4.3 should be used with the standard questionnaires.

Site specific drawings or photographic material and maps are required for use with the questionnaires.

Middlesex University

SITE USER QUESTIONNAIRE

CONFIDENTIAL

COAST PROTECTION: RECREATION SURVEY

NAME OF SITE ………………………………… YEAR OF SURVEY……………………

FOR OFFICE USE ONLY

Questionnaire serial number   ……………………………………………………………

Location of interviewer on site   ……………………………………………………………

Interviewer
(name, number, initials)   ……………………………………………………………

Interview number   ……………………………………………………………

Date (day/month/year)   ……/……/……

Day   Mon = 1   Tue = 2   Wed = 3   Thur = 4
              Fri = 5   Sat = 6   Sun = 7

Time interview started (24 hour clock)   ……………………………………………………………

Time interview finished (24 hour clock)   ……………………………………………………………

Duration (minutes)   ……………………………………………………………

TO COMPLETE - INTERVIEWERS PLEASE:
    **RING CODE NUMBER OR ENTER NUMBERS OR WORDS
    IN THE SPACES PROVIDED AS APPROPRIATE**

INTERVIEWERS' INSTRUCTIONS ARE PRINTED IN BOLD

APPENDIX 4.2 (a)

GOOD MORNING/AFTERNOON, MY NAME IS .... FROM .... SURVEY ORGANISATION. WE ARE CONDUCTING A SURVEY FOR ....

THE COAST/CLIFFS/BEACH HERE IS BEING ERODED BY THE SEA (AND THERE IS A RISK OF FLOODING IN THE FUTURE). THE AIM OF THE SURVEY IS TO FIND OUT HOW PEOPLE USE THE AREA AND HOW THEIR ENJOYMENT WILL BE AFFECTED IF THE EROSION IS ALLOWED TO CONTINUE.

..............., (name of coastal protection authority) IS CONSIDERING WHETHER WORKS TO PROTECT THE COAST/CLIFFS/BEACH ARE DESIRABLE AND ECONOMICALLY JUSTIFIED.

(Insert a brief description of the proposed coast protection works if desired)

THE SURVEY IS COMPLETELY CONFIDENTIAL. THE NAMES AND VIEWS OF INDIVIDUALS PARTICIPATING WILL NOT BE REVEALED TO ANYONE OUTSIDE THE SURVEY ORGANISATION. THE RESULTS WILL BE PUBLISHED IN STATISTICAL AND UNIDENTIFIABLE FORM ONLY.

SECTION A: INTRODUCTION - applies to ALL RESPONDENTS

1. Where is your normal place of residence?

    Town/Village...
    County...
    Post code...

    (If not UK, then terminate the interview)

2. (a) Do you live within 3 miles of this point (interviewer location)?

    Yes = 1   **GO TO SECTION B**
    No = 0   **GO TO Q2(b)**

    (b) Which of these statements best describes you? **(SHOW CARD A)**

    I live locally = 1   **GO TO**
    I am on a part-day visit from home = 2   **SECTION B**
    I am on a day visit from home = 3

    I am on a touring holiday = 4   **GO TO**
    I am staying away from home for 1 or 2 nights = 5   **SECTION C**
    I am staying away from home for more than 2 nights = 6

SECTION B - applies to LOCAL RESIDENTS, DAY AND PART DAY VISITORS only

3. How long did it take you to get from your home to **this point today**?
   PROMPT

   0-5 minutes = 1
   6-10 minutes = 2
   11-15 minutes = 3
   16-30 minutes = 4
   31-60 minutes = 5
   More than 1 hour - 2 hours = 6
   More than 2 hours - 3 hours = 7
   More than 3 hours - 4 hours = 8
   More than 4 hours = 9
   Don't know = -8

4. How often, on average, do you come to this seafront (beach/promenade/cliffs)?
   SHOW MAP DEFINING SEAFRONT EXTENT
   PROMPT

   At least daily = 1
   At least 3 times a week = 2
   Once or twice a week = 3
   At least fortnightly = 4
   At least once a month = 5
   About 5 to 11 times a year = 6
   About 2 to 4 times a year = 7
   About once a year = 8
   Less than once a year = 9
   Never been before = 10

5. (a) Did you come to this seafront today alone or in a group?

   Alone = 1 **GO TO SECTION D**
   In group = 2 **ASK Q5(b)+(c)**

   (b) How many of these people live locally (i.e. live within 3 miles?)

   ....

   (c) Does your group include children under the age of 16?

   Yes = 1
   No = 0

   GO TO SECTION D

APPENDIX 4.2(a) 243

SECTION C - applies to STAYING VISITORS only

3. How long did it take you to get to **this point today** from where you are staying?
    PROMPT

    $$
    \begin{aligned}
    \text{0-5 minutes} &= 1 \\
    \text{6-10 minutes} &= 2 \\
    \text{11-15 minutes} &= 3 \\
    \text{16-30 minutes} &= 4 \\
    \text{31-60 minutes} &= 5 \\
    \text{More than 1 hour - 2 hours} &= 6 \\
    \text{More than 2 hours - 3 hours} &= 7 \\
    \text{More than 3 hours - 4 hours} &= 8 \\
    \text{More than 4 hours} &= 9 \\
    \text{Don't know} &= -8
    \end{aligned}
    $$

4. (a) For how many nights are you staying away from home?

    .... Nights

    (b) Are you staying free with relatives or friends, or paying for your accommodation?

    Staying free = 1
    Paying for accommodation = 0

    (c) Have you ever visited this seafront (beach/prom./cliffs) before today, and if so, roughly how many times?
    SHOW MAP DEFINING SEAFRONT EXTENT

    Never = 0
    Once = 1
    2 to 5 times = 2
    6 to 10 times = 3
    more than 10 times = 4

5. (a) Did you come to this seafront alone or in a group?

    Alone = 1 **GO SECTION D**
    In group = 2 **GO TO Q5(b)**

    (b) Does your group include children under the age of 16?

    Yes = 1
    No = 0

    GO TO SECTION D

SECTION D - applies to ALL RESPONDENTS

6. What is the main reason for your outing today?
   READ OUT

   To visit this seafront (beach/promenade/cliffs) = 1
   To visit ....(name of town/resort) = 2
   To visit..............(named attraction if applicable) = 3
   To visit..............(named attraction if applicable) = 4
   To visit..............(named attraction if applicable) = 5
   To make visits elsewhere = 6
   Don't know = 8

7. How long do you intend to spend at this seafront today?
   PROMPT

   0-15 minutes = 1
   16-30 minutes = 2
   31-60 minutes = 3
   More than 1 hour - 2 hours = 4
   More than 2 hours - 3 hours = 5
   More than 3 hours - 4 hours = 6
   More than 4 hours = 7
   Don't know = 8

8. What activities will you or any member of your group be taking part in during this visit to the seafront? (**SHOW CARD B**)
   RING ALL CODES THAT APPLY

   A. Sitting/sunbathing/picnicking .... Y
   B. Walking... Y
   C. Swimming/paddling............................ Y
   D. Sail-boarding...................................... Y
   E. Games or sports................................. Y
   F. Bird watching..................................... Y
   G. Angling/fishing................................... Y
   H. Sailing/Boating................................... Y
   I. Visiting....................(named attraction) Y
   J. Visiting....................(named attraction) Y
   K. Visiting....................(named attraction) Y
   L. Other (specify)................................. Y
   ....

APPENDIX 4.2 (a) 245

SECTION E : VALUATION OF THE SEAFRONT - applies to ALL RESPONDENTS

9. Overall, how do you rate this seafront (beach/prom./cliffs not the town or resort) as a place to visit? **(RATE ON THE SCALE SHOWN BELOW)**
   (SHOW CARD C)

0	1	2	3	4	5	6	7	8	9	10
very poor					about average					very good

10. (a) How would you rate this seafront (beach/prom./cliffs) for walking on?
    (RATE Q10(a)-(e) ON THE SAME SCALE AS Q9)

    (SHOW CARD C) ....

    (b) How would you rate the seafront (beach/prom./cliffs) as a place for sitting, sunbathing and picnicking?

    (SHOW CARD C) ....

    (c) How would you rate this seafront (beach/prom./cliff) as a place for children to play?

    (SHOW CARD C) ....

    (Include if appropriate to seafront area)

    (d) How would you rate this seafront (beach/prom./cliff) as a place for..............(named specialist activity)?

    (SHOW CARD C) ....

    (e) How would you rate..........(named attraction) as a place to visit?

    (SHOW CARD C) ....

(Add further attractions/specialist activities etc to be rated as required)

READ OUT

11. We are trying to find out how much value you, **as an individual,** put on your enjoyment of this visit to **this seafront today;** I MEAN IN £ AND PENCE.

    Now, this is an unusual question to ask so let me explain it to you in this way:

    Think of a visit or activity you have done in the past which gave you the same amount of enjoyment as your visit to this seafront today. Here is a list of possibilities.
    (SHOW CARD D)

    (a) What visit or activities gave you about the same enjoyment as your visit to this seafront (beach/prom./cliff) today?

    WRITE IN RESPONSE

    ....

    ....

    ....

    ....

    Now, think about how much that visit (or other activities) cost you. Remember that the cost of a visit may include petrol and parking costs or bus or train fares as well as admission charges and any other costs.

    You can use the costs of that visit (or other activities) as a guide to the value of your enjoyment of today's visit to this seafront.

    (b) So, now, what value do you put on your **individual** enjoyment of this visit to **this seafront**?

    Value of today's visit   £.....:.....pence

    Cannot value these things in money terms = -777.77
    Don't know = -888.88
    Other (please specify) ....
    .... = -999.99

APPENDIX 4.2 (a) 247

READ OUT
I am now going to show you a number of drawings and maps which represent this seafront (beach/prom./cliffs) in varying conditions.

12. **ERODED SEAFRONT (DRAWING B)**

    READ OUT
    Now I would like to ask you what you would think if the seafront changed to look like this.

    SHOW DRAWING B AND READ OUT
    Drawing B shows that:

    (Insert a brief explanatory text describing the specific effects of erosive changes on the seafront as shown in the drawing)

13. (a) If you were making a visit (similar to today's visit) to the seafront in this drawing, would you get more, less or the same amount of enjoyment from a visit compared to your enjoyment of today's visit? (SHOW CARD E)
    PLEASE GIVE ME A FIGURE FROM THIS CARD.

-3	-2	-1	0	+1	+2	+3
Much less enjoyment			The same amount of enjoyment			Much more enjoyment

    (CODE DON'T KNOW = -8)

    IF DON'T KNOW, GO TO Q13(c)

    (b) So, how much enjoyment would you get from your visit to the seafront in this drawing in terms of pounds and pence?

    £.....:...... pence

    Cannot value these things in money terms = -777.77
    Don't know = -888.88
    Other (please specify).................. = -999.99

248   THE ECONOMICS OF COASTAL MANAGEMENT

(c)   If the seafront looked like this drawing (B), would you visit it more or less often than you do now? (SHOW CARD F)

$$\begin{aligned}
\text{Never visit} &= 0 \text{ GO TO Q13(d)} \\
\text{Visit much less often than now} &= 1 \text{ GO TO Q13(d)} \\
\text{Visit less often than now} &= 2 \text{ GO TO Q13(d)} \\
\text{Visit as often as now} &= 3 \text{ GO TO Q14} \\
\text{Visit more often than now} &= 4 \text{ GO TO Q14}
\end{aligned}$$

I do not intend ever to visit this seafront again - the change will make no difference = 5 GO TO Q14

(d)   Would you go somewhere else instead?

Yes = 1 GO TO Q13(e)
No = 0 GO TO Q14

(e)   How much enjoyment would you get from your visit to this alternative site in terms of pounds and pence?

£......:...... pence

Cannot value these things in money terms = -777.77
Don't know = -888.88
Other (please specify).................. = -999.99

(f)   How much more or less would it cost you to go to this alternative site?

£.........More

£.........Less

Don't know = -888.88

APPENDIX 4.2 (a) 249

14. **PROTECTED SEAFRONT (DRAWING C)**

    READ OUT
    The exact form of the proposed works to protect the seafront has not yet been finally determined. This decision will be taken after further investigation and consultation.

    At present, the most likely scheme/one scheme being considered/another scheme being considered will involve:
    SHOW DRAWING C AND READ OUT

    (Insert a brief explanatory text giving details of the scheme as shown in the drawing)

    (This section - PROTECTED SEAFRONT - Questions 14, 15, 16 can be repeated for each proposed scheme with drawings and explanatory text as appropriate)

15. (a) If you were making a visit (similar to today's visit) to the seafront in this drawing, would you get more, less or the same amount of enjoyment from a visit compared to your enjoyment of today's visit? **(SHOW CARD E)**
    PLEASE GIVE ME A FIGURE FROM THIS CARD.

-3	-2	-1	0	+1	+2	+3
Much less enjoyment			The same amount of enjoyment			Much more enjoyment

    (CODE DON'T KNOW = -8)

    IF DON'T KNOW, GO TO Q15(c)

    (b) So, how much enjoyment would you get from your visit to the seafront in this drawing in terms of pounds and pence?

    £.....:...... pence

    Cannot value these things in money terms = -777.77
    Don't know = -888.88
    Other (please specify).................. = -999.99

250   THE ECONOMICS OF COASTAL MANAGEMENT

(c)   If the seafront looked like this drawing (C), would you visit it more or less often than you do now? **(SHOW CARD F)**

Never visit = 0 **GO TO Q15(d)**
Visit much less often than now = 1 **GO TO Q15(d)**
Visit less often than now = 2 **GO TO Q15(d)**
Visit as often as now = 3 **GO TO Q16**
Visit more often than now = 4 **GO TO Q16**
I do not intend ever to visit this seafront
again - the change will make no difference = 5 **GO TO Q16**

(d)   Would you visit somewhere else instead ?

Yes = 1 **GO TO Q15(e)**
No = 0 **GO TO Q16**

(e)   How much enjoyment would you get from your visit to this alternative site in terms of pounds and pence?

£......:...... pence

Cannot value these things in money terms = -777.77
Don't know = -888.88
Other (please specify).................. = -999.99

(f)   How much more or less would it cost you to go to this alternative site?

£.........More

£..........Less

Don't know = -888.88

16.   Would you be in favour or against the coastal protection authority carrying out works to protect the seafront as shown in Drawing C ? **(SHOW CARD G)**
      PLEASE GIVE ME A FIGURE FROM THIS CARD.

-3	-2	-1	0	+1	+2	+3	
Strongly against			indifferent			Stongly in favour	Don't know = -8

APPENDIX 4.2 (a) 251

SECTION F - applies to ALL RESPONDENTS

17. Have you any comments you would like to make about the proposed works to protect the coast as shown in the Drawing(s)?
    PROBE AND WRITE IN COMMENTS VERBATIM

READ OUT
These following questions are standard for most questionnaires and are used for our classification purposes: **Your answers are completely confidential.**

18. Gender

    Female = 0
    Male = 1

19. To which of the following age categories do you belong?

    Under 18 = 1
    18-29 = 2
    30-49 = 3
    50-64 = 4
    65+ = 5

20. Can you tell me at what age you completed full-time education?

    Still in full-time education = 1
    At age 15 years or under = 2
    16 years of age = 3
    17 years of age = 4
    18 years of age = 5
    19 years of age or older = 6

21. Which of the letters on this card represents the gross (this is before tax and other deductions) weekly income from **all sources** of your household?
    (SHOW CARD H)   ....

THANK RESPONDENT AND COMPLETE ADDRESS SHEET ON NEXT PAGE AND FRONT PAGE

ADDRESS SHEET

PLEASE DETATCH THIS SHEET AND KEEP SEPARATE FROM QUESTIONNAIRE

So that the survey organisation can check that this interview has been carried out and that the interview was satisfactory, please would you sign this sheet and give me your name, address and telephone number. The information given in the questionnaire will be treated as strictly confidential and this sheet with your name and address will be detatched and kept separate from the questionnaire.

Respondent's name..

Respondent's address...

....

....

Respondent's telephone number..

Respondent's signature..

Interviewer's signature..

INTERVIEW NUMBER ....

OFFICE USE ONLY

QUESTIONNAIRE NUMBER ....

APPENDIX 4.2 (b)

Middlesex University

RESIDENTS QUESTIONNAIRE

CONFIDENTIAL

COAST PROTECTION: RECREATION SURVEY

NAME OF SEAFRONT.................................... **YEAR OF SURVEY**........................

FOR OFFICE USE ONLY

Location (name of seafront)..

Questionnaire serial number ....

Location of interview
(polling district, street) ....

Interviewer
(name, number, initials) ....

Interview number ....

Date (day/month/year)   ....../....../......

Day   Mon = 1   Tue = 2   Wed = 3   Thur = 4
                  Fri = 5   Sat = 6   Sun = 7

Time interview started (24 hour clock) ....

Time interview finished (24 hour clock) ....

Duration (minutes) ....

TO COMPLETE - INTERVIEWERS PLEASE:
   RING CODE NUMBER OR ENTER NUMBERS OR WORDS
   IN THE SPACE PROVIDED AS APPROPRIATE

INTERVIEWERS' INSTRUCTIONS ARE PRINTED IN BOLD

GOOD MORNING/AFTERNOON, MY NAME IS .... FROM .... SURVEY ORGANISATION. WE ARE CONDUCTING A SURVEY FOR ....

THE COAST/CLIFFS/BEACH (specify affected section)...
IS BEING ERODED BY THE SEA (AND THERE IS A RISK OF FLOODING IN THE FUTURE). THE AIM OF THE SURVEY IS TO FIND OUT HOW PEOPLE USE THIS AREA AND HOW THEIR ENJOYMENT WILL BE AFFECTED IF THE EROSION IS ALLOWED TO CONTINUE.

..............., (name of coastal protection authority) IS CONSIDERING WHETHER WORKS TO PROTECT THE COAST/CLIFFS/BEACH ARE DESIRABLE AND ECONOMICALLY JUSTIFIED.

(Insert a brief description of the proposed coast protection works if desired)

THE SURVEY IS COMPLETELY CONFIDENTIAL. A REPRESENTATIVE SAMPLE OF ADDRESSES HAS BEEN SELECTED FROM THE ELECTORAL REGISTER. THE NAMES AND VIEWS OF INDIVIDUALS PARTICIPATING WILL NOT BE REVEALED TO ANYONE OUTSIDE THE SURVEY ORGANISATION. THE RESULTS WILL BE PUBLISHED IN STATISTICAL AND UNIDENTIFIABLE FORM ONLY.

INTRODUCTION

---

1. How long have you lived in................(name of town/resort)?

    Years..................

2. Overall, how do you rate this town/resort as a place to live?
   (RATE ON THE SCALE SHOWN BELOW)
   (SHOW CARD C)

   0   1   2   3   4   5   6   7   8   9   10

   very   about   very
   poor   average   good

APPENDIX 4.2 (b)

RECREATIONAL USE OF THE SEAFRONT

3. (a) How often, on average, do you visit this seafront (beach/promenade/cliffs) for recreation in the Spring and Summer (April to September) ?
        SHOW MAP DEFINING SEAFRONT EXTENT
        PROMPT

        At least daily = 1
        At least 3 times a week = 2
        Once or twice a week = 3
        At least fortnightly = 4
        At least once a month = 5
        About 3 to 5 times = 6
        About once or twice = 7
        Less than once in Spring and Summer = 8
        Never visit in Spring and Summer = 9

    (b) How often, on average, do you visit this seafront (beach/promenade/cliffs) for recreation in the Autumn and Winter (October to March)?
        SHOW MAP DEFINING SEAFRONT EXTENT
        PROMPT

        At least daily = 1
        At least 3 times a week = 2
        Once or twice a week = 3
        At least fortnightly = 4
        At least once a month = 5
        About 3 to 5 times = 6
        About once or twice = 7
        Less than once in Autumn and Winter = 8
        Never visit in Autumn and Winter = 9

4. How long would it normally take you to get from your home to **this seafront**?
    SHOW MAP DEFINING SEAFRONT EXTENT
    PROMPT

    0-5 minutes = 1
    6-10 minutes = 2
    11-15 minutes = 3
    16-30 minutes = 4
    31-60 minutes = 5
    More than 1 hour - 2 hours = 6
    More than 2 hours - 3 hours = 7
    More than 3 hours - 4 hours = 8
    More than 4 hours = 9
    Don't know = -8

256 THE ECONOMICS OF COASTAL MANAGEMENT

IF NEVER VISIT AT ALL (Q3(a)=9 and Q3(b)=9) GO TO Q8

5. (a) Do you usually visit this seafront alone or in a group?

                                                Alone = 1 **GO TO Q6**
                                            In group = 2 **ASK Q5(b)**

   (b) Does the group usually include children under the age of 16?

                                                                 Yes = 1
                                                                 No = 0

6. What activities do you or members of your group usually take part in during visits to this seafront?
(SHOW CARD B)
RING ALL CODES THAT APPLY

          A. Sitting/sunbathing/picnicking .... Y
          B. Walking... Y
          C. Swimming/paddling........................... Y
          D. Sail-boarding..................................... Y
          E. Games or sports................................. Y
          F. Bird watching.................................... Y
          G. Angling/fishing................................. Y
          H. Sailing/Boating................................. Y
          I. Visiting....................(named attraction) Y
          J. Visiting....................(named attraction) Y
          K. Visiting....................(named attraction) Y
          L. Other (specify)................................. Y
              ....

7. How long do you usually spend at this seafront when you visit?
PROMPT

                                           0-15 minutes = 1
                                           16-30 minutes = 2
                                           31-60 minutes = 3
                        More than 1 hour - 2 hours = 4
                       More than 2 hours - 3 hours = 5
                       More than 3 hours - 4 hours = 6
                              More than 4 hours = 7
                                         Don't know = 8

8. Which local coast site do you visit most often for recreation: this seafront or another local coast site?
SHOW MAP DEFINING SEAFRONT EXTENT

                                               This seafront = 1
                             Another local coastal site = 2
                                    please specify:........................
                  Never visit local coast for recreation = 3

APPENDIX 4.2 (b)

VALUATION OF THE SEAFRONT

9.   Overall, how do you rate this seafront (beach/prom./cliffs not the town or resort) as a place to visit?
(RATE ON THE SCALE SHOWN BELOW)
(SHOW CARD C)

```
0     1     2     3     4     5     6     7     8     9    10

very                          about                        very
poor                          average                      good    Don't know = -8
```

10. (a)   How would you rate this seafront (beach/prom./cliffs) for walking on?
(RATE Q10(a)-(e) ON THE SAME SCALE AS Q9)

(SHOW CARD C) ....

(b)   How would you rate the seafront (beach/prom./cliffs) as a place for sitting, sunbathing and picnicking?

(SHOW CARD C) ....

(c)   How would you rate this seafront (beach/prom./cliff) as a place for children to play?

(SHOW CARD C) ....

(Include if appropriate to seafront area)

(d)   How would you rate this seafront (beach/prom./cliff) as a place for..............(named specialist activity)?

(SHOW CARD C) ....

(e)   How would you rate..........(named attraction) as a place to visit?

(SHOW CARD C) ....

(Add further attractions/specialist activities etc to be rated as required)

258   THE ECONOMICS OF COASTAL MANAGEMENT

READ OUT

I am now going to show you a number of drawings (and maps) which represent this seafront (beach/prom/cliffs) in varying conditions.

11.   **CURRENT SEAFRONT (DRAWING A)**
      SHOW DRAWING A AND READ OUT
      Drawing A shows that:

      (Insert a brief explanatory text describing the seafront in its current state)

READ OUT

We are trying to find out how much value you, **as an individual**, put on your enjoyment of an average visit to **this seafront in Drawing A;** I MEAN IN £ AND PENCE.

Now, this is an unusual question to ask so let me explain it to you in this way:

Think of a visit or activity you have done in the past which gave you the same amount of enjoyment as you would get from an average visit to the seafront in Drawing A. Here is a list of possibilities. **(SHOW CARD D)**

  (a)   What visit or activities gave you about the same enjoyment as a visit to the seafront in Drawing A?

                         WRITE IN RESPONSE

         ....

         ....

         ....

         ....

READ OUT

Now, think about how much that visit (or other activities) cost you. Remember that the cost of a visit may include petrol and parking costs or bus or train fares as well as admission charges and any other costs.

You can use the costs of that visit (or other activities) as a guide to the value of your enjoyment of a visit to this seafront in Drawing A.

SHOW DRAWING A

(b) So, now, what value do you put on your **individual** enjoyment of an average visit to **this seafront**?

                        Value of today's visit   £.....:.....pence

                        Cannot value these things in money terms = -777.77
                        Don't know = -888.88
                        Other (please specify) ....
                        .... = -999.99

12. **ERODED SEAFRONT (DRAWING B)**

    READ OUT
    Now I would like to ask you what you would think if the seafront changed to look like this.

    SHOW DRAWING B AND READ OUT
    Drawing B shows that:

    (Insert a brief explanatory text describing the specific effects of erosive changes on the seafront as shown in the drawing)

13. (a) If you were making an average visit to the seafront in this drawing (B), would you get more, less or the same amount of enjoyment from a visit compared to your enjoyment of a visit to the seafront in Drawing A? **(SHOW CARD E)**
    PLEASE GIVE ME A FIGURE FROM THIS CARD.

-3	-2	-1	0	+1	+2	+3
Much less enjoyment			The same amount of enjoyment			Much more enjoyment

    (CODE DON'T KNOW = -8)

    IF DON'T KNOW, GO TO Q13(c)

    (b) So, how much enjoyment would you get from your visit to the seafront in Drawing B in terms of pounds and pence?

    £.....:...... pence

    Cannot value these things in money terms = -777.77
    Don't know = -888.88
    Other (please specify).................. = -999.99

APPENDIX 4.2 (b)

(c) If the seafront looked like this drawing (B), would you visit it more or less often than you would visit the seafront in Drawing A ? **(SHOW CARD F)**

$$\begin{aligned}
\text{Never visit} &= 0 \textbf{ GO TO Q13(d)}\\
\text{Visit much less often} &= 1 \textbf{ GO TO Q13(d)}\\
\text{Visit less often} &= 2 \textbf{ GO TO Q13(d)}\\
\text{Visit as often} &= 3 \textbf{ GO TO Q14}\\
\text{Visit more often} &= 4 \textbf{ GO TO Q14}
\end{aligned}$$

I would not visit visit the seafront in
Drawing A - the change would make no difference = 5 **GO TO Q14**

(d) Would you go somewhere else instead?

Yes = 1 **GO TO Q13(e)**
No = 0 **GO TO Q14**

(e) How much enjoyment would you get from your visit to this alternative site in terms of pounds and pence?

£......:...... pence

Cannot value these things in money terms = -777.77
Don't know = -888.88
Other (please specify).................. = -999.99

(f) How much more or less would it cost you to go to this alternative site?

£.........More

£..........Less

Don't know = -888.88

14. **PROTECTED SEAFRONT (DRAWING C)**

    READ OUT
    The exact form of the proposed works to protect the seafront has not yet been finally determined. A decision will be taken after further investigation and consultation.

    At present, the most likely scheme/one scheme being considered/another scheme being considered will involve:
    SHOW DRAWING C AND READ OUT

    (Insert a brief explanatory text giving details of the scheme as shown in the drawing)

15. (a) If you were making an average visit to the seafront in this drawing (C), would you get more, less or the same amount of enjoyment from a visit compared to your enjoyment of a visit to the seafront in Drawing A ?
    (SHOW CARD E)
    PLEASE GIVE ME A FIGURE FROM THIS CARD.

-3	-2	-1	0	+1	+2	+3
Much less enjoyment			The same amount of enjoyment			Much more enjoyment

    (CODE DON'T KNOW = -8)

    IF DON'T KNOW, GO TO Q15(c)

    (b) So, how much enjoyment would you get from your visit to the seafront in Drawing C in terms of pounds and pence?

    £.....:..... pence

    Cannot value these things in money terms = -777.77
    Don't know = -888.88
    Other (please specify).................. = -999.99

(c) If the seafront looked like this drawing (C), would you visit it more or less often than you would visit the seafront in Drawing A ? **(SHOW CARD F)**

Never visit = 0 **GO TO Q15(d)**
Visit much less often = 1 **GO TO Q15(d)**
Visit less often = 2 **GO TO Q15(d)**
Visit as often = 3 **GO TO Q16**
Visit more often = 4 **GO TO Q16**
Would not visit the seafront in Drawing A and B = 5 **GO TO Q16**

(d) Would you visit somewhere else instead?

Yes = 1 **GO TO Q15(e)**
No = 0 **GO TO Q16**

(e) How much enjoyment would you get from your visit to this alternative site in terms of pounds and pence?

£......:...... pence

Cannot value these things in money terms = -777.77
Don't know = -888.88
Other (please specify).................. = -999.99

(f) How much more or less would it cost you to go to this alternative site?

£.........More

£.........Less

Don't know = -888.88

16. Would you be in favour or against the coastal protection authority carrying out works to protect the seafront as shown in Drawing C ? **(SHOW CARD G)**
PLEASE GIVE ME A FIGURE FROM THIS CARD.

-3	-2	-1	0	+1	+2	+3	
Strongly against			indifferent			Stongly in favour	Don't know = -8

17. Have you any comments you would like to make about the proposed works to protect the coast shown in the drawing(s)?
    PROBE AND WRITE IN COMMENTS VERBATIM

READ OUT
These following questions are standard for most questionnaires and are used for our classification purposes: **Your answers are completely confidential.**

18. Gender

    Female = 0
    Male = 1

19. To which of the following age categories do you belong?

    Under 18 = 1
    18-29 = 2
    30-49 = 3
    50-64 = 4
    65+ = 5

20. Can you tell me at what age you completed full-time education?

    Still in full-time education = 1
    At age 15 years or under = 2
    16 years of age = 3
    17 years of age = 4
    18 years of age = 5
    19 years of age or older = 6

21. Which of the letters on this card represents the gross (this is before tax and other deductions) weekly income from **all sources** of your household?
    (SHOW CARD H) ....

THANK RESPONDENT AND COMPLETE ADDRESS SHEET ON NEXT PAGE AND FRONT PAGE

APPENDIX 4.2 (b) 265

ADDRESS SHEET

PLEASE DETATCH THIS SHEET AND KEEP SEPARATE FROM QUESTIONNAIRE

So that the survey organisation can check that this interview has been carried out and that the interview was satisfactory, please would you sign this sheet and give me your name, address and telephone number. The information given in the questionnaire will be treated as strictly confidential and this sheet with your name and address will be detatched and kept separate from the questionnaire.

Respondent's name..

Respondent's address...

....

....

Respondent's telephone number...

Respondent's signature..

Interviewer's signature..

INTERVIEW NUMBER ....

OFFICE USE ONLY

QUESTIONNAIRE NUMBER ....

Appendix 4.3 Show cards

These show cards are intended for use with the standard seafront user and residents questionnaires presented in Appendix 4.2 (a) and (b). They are designed so that they can be photocopied and reproduced directly on to card for use in surveys.

The same cards are used for the site user survey and the residents survey with the exception of SHOW CARD F. A different version of SHOW CARD F is provided for the users and the residents survey.

SHOW CARD A

Q. Which of these statements best describes you ?

I live locally = 1

I am on a part-day visit from home = 2

I am on a day visit from home = 3

I am on a touring holiday = 4

I am staying away from home for 1 or 2 nights = 5

I am staying away from home for more than 2 nights = 6

SHOW CARD B

ACTIVITIES

READ OUT THE LETTERS WHICH APPLY

(A)   Sitting / Sunbathing / Picnicking

(B)   Walking

(C)   Swimming / paddling

(D)   Sailboarding

(E)   Games or sports

(F)   Birdwatching

(G)   Angling / fishing

(H)   Sailing / boating

(I)   Visiting................

(J)   Visiting................

(K)   Visiting................

(L)   Other - Please specify

SHOW CARD C

PLEASE RATE ON THIS SCALE:

0   1   2   3   4   5   6   7   8   9   10

very   about   very
poor   average   good

SHOW CARD D

A leisure centre

An art exhibition

A wildlife park

A nature reserve

A country house or country park

A local swimming pool

A concert

A cinema, theatre or Bingo

A cafe, pub or restaurant

SHOW CARD E

PLEASE GIVE ME A FIGURE FROM THIS CARD:

-3   -2   -1   0   +1   +2   +3

Much   The same   Much
less   amount of   more
enjoyment   enjoyment   enjoyment

SHOW CARD F (USER)

Never visit = 0

Visit much less often than now = 1

Visit less often than now = 2

Visit as often as now = 3

Visit more often than now = 4

I do not intend ever to visit this seafront
again - the change will make no difference   = 5

SHOW CARD F (RESIDENTS)

Never visit = 0

Visit much less often = 1

Visit less often = 2

Visit as often = 3

Visit more often = 4

I would not visit the seafront in
Drawing A - the change would make
no difference = 5

SHOW CARD G

Q. Would you be in favour or against the coastal protection authority carrying out works to protect the seafront as shown in Drawing C ?

PLEASE GIVE ME A FIGURE FROM THIS CARD.

-3   -2   -1   0   +1   +2   +3

Strongly   indifferent   Strongly
against   in favour

SHOW CARD H

GROSS HOUSEHOLD INCOME
(FROM ALL SOURCES BEFORE STOPPAGES)

	WEEKLY	MONTHLY	ANNUAL
(L)	Under £100	Under £415	Under £5,000
(Q)	£100-£149	£415-£624	£5,000-£7,499
(F)	£150-£199	£625-£834	£7,500-£9,999
(G)	£200-£249	£835-£1,039	£10,000-£12,499
(A)	£250-£299	£1,040-£1,249	£12,500-£14,999
(H)	£300-£349	£1,250-£1,459	£15,000-£17,499
(T)	£350-£399	£1,460-£1,669	£17,500-£19,999
(C)	£400-£499	£1,670-£2,084	£20,000-£24,999
(D)	£500 and +	£2,085 and +	£25,000 and +

Appendix 4.4　Visit count record sheet

APPENDIX 4.4 277

Visit Count Record Sheet

Sheet Number: ......

FOR USE WITH INFRA-RED COUNTERS (TO RECORD DATA FOR 10 ADULTS)

Name of Organisation .... Title/Site of Survey/Year ....

Date .... Recording period start time .... Recording period finish time ....

Location of count point/infra red counter ....

Name of recorder ....

ASK QUESTIONS 1-7 TO THE NEXT ADULT TO PASS	RECORD FOR EACH ADULT									
	1	2	3	4	5	6	7	8	9	10
RING ONE CODE Q1. Which of these statements best describes you? I live locally - within 3 miles I am on a day or part day visit from home I am on a touring holiday and/or staying away from home for 1 or more nights	1 2 3	1 2 3	1 2 3	1 2 3	1 2 3	1 2 3	1 2 3	1 2 3	1 2 3	1 2 3
RING ONE CODE Q2. How will you spend your time at the seafront today? Only on the beach Mainly on the beach Equally on the beach and promenade/ cliffs/dunes Mainly on the promenade/cliffs/dunes Only on the promenade/cliffs/dunes Don't know	1 2 3 4 5 -9	1 2 3 4 5 -9	1 2 3 4 5 -9	1 2 3 4 5 -9	1 2 3 4 5 -9	1 2 3 4 5 -9	1 2 3 4 5 -9	1 2 3 4 5 -9	1 2 3 4 5 -9	1 2 3 4 5 -9
WRITE IN NUMBER FOR QUESTIONS 3-7 Q3. How many times will you pass this point during today's visit? (include this time)										
Q4. How many times will you pass point X (2nd counter) during today's visit? (include this time)										
Q5. How many times will you pass point Y (3rd counter) during today's visit (include this time)										
Q6. How many times will you pass point Z (4th counter) during today's visit (include this time)										
Q7. How many people aged 6 - 18 are with you now?										

Appendix 4.5 Checklist of recreational uses

SPECIALIST

Water based

angling
board sailing
canoeing
cruiser sailing
dinghy sailing
jet biking
speedboating
surfing
water skiing

Shore based

bait digging
beach sports (informal)
nature study
sand yachting

GENERALIST

paddling
swimming

Shore based – beach

children's games
picnicking
strolling
sitting
sunbathing
walking

Shore based – promenade

roller skating
skate boarding

Appendix 5.1 Revised land use classification to be used in the appraisal of flood alleviation projects

This appendix comprises the revised hierarchical land use classification for all the MUFHRC land use sectors (1–9) which also appears in the FLAIR 1990 report (N'Jai *et al.*, 1990). This is a development of the land use classification originally developed in Penning-Rowsell and Chatterton (1977) and subsequently updated by Parker *et al.* (1987).

The Sector 1 (Residential) changes are:

Old name	New name
Post-Manual (post 1975)	Post 1975 (standard)
Post-Manual (timber-framed)	Post 1975 (timber-framed)
Post-Manual (garage/utility)	Post 1975 (garage/utility)

The garage/utility categories for flats and bungalows have now been excluded as it is not thought likely that these types of property will have garage/utilities on the ground floor.

Sector 5 has a few additions under the category Shops – Other non-food retailers (515). These are:

i) video libraries;
ii) pet shops;
iii) florists;
iv) wool shops; and
v) toy shops.

Under the General Stores category (516) there are two additions:

i) hypermarkets; and
ii) garden centres/nurseries.

The Services, Design Studio category (534) has one addition:

i) antiques.

No changes have been made to other Sectors.

Data sources given in this Appendix refer to the first documentation of the particular categories. These sources should be consulted for detailed information on the category concerned. The sources (referred to in this appendix) are abbreviations for earlier Flood Hazard Research Centre publications.

PC   refers to the 'Blue Manual' (Penning-Rowsell and Chatterton, 1977)
PGT   refers to the 'Red Manual' (Parker, Green and Thompson, 1987)
FHRC   refers to the 'Update' Report (Suleman *et al.*, 1988)
FLAIR   refers to the 'FLAIR' Report (N'Jai *et al.*, 1990)

For further details of the changes since the Red Manual consult the FLAIR report.

*   indicates storage of standard Blue Manual type depth-damage data on computer file at Middlesex University;
#   indicates storage of average depth-damage data from previous case studies on computer file at Middlesex University;
+   indicates storage of Red Manual type depth-damage data on computer file at Middlesex University.

Type	Category	Subcategory (Social class)	Depth/damage or loss data sources Appendix/ chapter/ section	Data on computer file at Middlesex University	Land use code Column number			
					21	22	23	24

Sector 1 Residential

Type	Category	Subcategory	Source	Data	21	22	23	24
Detached			PC A2.3	*	1	1		
	Pre-1918		PC A2.3	*	1	1	1	
		AB	PC A2.4	*	1	1	1	1
		C1	PC A2.4	*	1	1	1	2
		C2	PC A2.4	*	1	1	1	3
		DE	PC A2.4	*	1	1	1	4
	Inter-war (1918–38)		PC A2.3	*	1	1	2	
		AB	PC A2.4	*	1	1	2	1
		C1	PC A2.4	*	1	1	2	2
		C2	PC A2.4	*	1	1	2	3
		DE	PC A2.4	*	1	1	2	4
	Post-war (1939–65)		PC A2.3	*	1	1	3	
		AB	PC A2.4	*	1	1	3	1
		C1	PC A2.4	*	1	1	3	2
		C2	PC A2.4	*	1	1	3	3
		DE	PC A2.4	*	1	1	3	4
	Modern (post-1965)		PC A2.3	*	1	1	4	
		AB	PC A2.4	*	1	1	4	1
		C1	PC A2.4	*	1	1	4	2
		C2	PC A2.4	*	1	1	4	3
		DE	PC A2.4	*	1	1	4	4
	Post-1975 (standard)		FHRC A2.3	*	1	1	5	
		AB	FHRC A2.4	*	1	1	5	1
		C1	FHRC A2.4	*	1	1	5	2
		C2	FHRC A2.4	*	1	1	5	3
		DE	FHRC A2.4	*	1	1	5	4
	Post-1975 (timber-framed)				1	1	6	
		AB			1	1	6	1
		C1			1	1	6	2
		C2			1	1	6	3
		DE			1	1	6	4
	Post-1975 (garage/utility)		FLAIR 3.9		1	1	7	
		AB	FLAIR 3.9		1	1	7	1
		C1	FLAIR 3.9		1	1	7	2
		C2	FLAIR 3.9		1	1	7	3
		DE	FLAIR 3.9		1	1	7	4

APPENDIX 5.1

Type	Category	Subcategory (Social class)	Depth/damage or loss data sources Appendix/ chapter/ section	Data on computer file at Middlesex University	Land use code Column number			
					21	22	23	24
Semi-detached			PC A2.3	*	1	2		
	Pre-1918		PC A2.3	*	1	2	1	
		AB	PC A2.4	*	1	2	1	1
		C1	PC A2.4	*	1	2	1	2
		C2	PC A2.4	*	1	2	1	3
		DE	PC A2.4	*	1	2	1	4
	Inter-war (1918–38)		PC A2.3	*	1	2	2	
		AB	PC A2.4	*	1	2	2	1
		C1	PC A2.4	*	1	2	2	2
		C2	PC A2.4	*	1	2	2	3
		DE	PC A2.4	*	1	2	2	4
	Post-war (1939–65)		PC A2.3	*	1	2	3	
		AB	PC A2.4	*	1	2	3	1
		C1	PC A2.4	*	1	2	3	2
		C2	PC A2.4	*	1	2	3	3
		DE	PC A2.4	*	1	2	3	4
	Modern (post-1965)		PC A2.3	*	1	2	4	
		AB	PC A2.4	*	1	2	4	1
		C1	PC A2.4	*	1	2	4	2
		C2	PC A2.4	*	1	2	4	3
		DE	PC A2.4	*	1	2	4	4
	Post-1975 (standard)		FHRC A2.3	*	1	2	5	
		AB	FHRC A2.4	*	1	2	5	1
		C1	FHRC A2.4	*	1	2	5	2
		C2	FHRC A2.4	*	1	2	5	3
		DE	FHRC A2.4	*	1	2	5	4
	Post-1975 (timber-framed)				1	2	6	
		AB			1	2	6	1
		C1			1	2	6	2
		C2			1	2	6	3
		DE			1	2	6	4
	Post-1975 (garage/utility)		FLAIR 3.9		1	2	7	
		AB	FLAIR 3.9		1	2	7	1
		C1	FLAIR 3.9		1	2	7	2
		C2	FLAIR 3.9		1	2	7	3
		DE	FLAIR 3.9		1	2	7	4

Type	Category	Subcategory (Social class)	Depth/damage or loss data sources Appendix/ chapter/ section	Data on computer file at Middlesex University	Land use code Column number			
					21	22	23	24
Terrace			PC A2.3	*	1	3		
	Pre-1918		PC A2.3	*	1	3	1	
		AB	PC A2.4	*	1	3	1	1
		C1	PC A2.4	*	1	3	1	2
		C2	PC A2.4	*	1	3	1	3
		DE	PC A2.4	*	1	3	1	4
	Inter-war (1918–38)		PC A2.3	*	1	3	2	
		AB	PC A2.4	*	1	3	2	1
		C1	PC A2.4	*	1	3	2	2
		C2	PC A2.4	*	1	3	2	3
		DE	PC A2.4	*	1	3	2	4
	Post-war (1939–65)		PC A2.3	*	1	3	3	
		AB	PC A2.4	*	1	3	3	1
		C1	PC A2.4	*	1	3	3	2
		C2	PC A2.4	*	1	3	3	3
		DE	PC A2.4	*	1	3	3	4
	Modern (post-1965)		PC A2.3	*	1	3	4	
		AB	PC A2.4	*	1	3	4	1
		C1	PC A2.4	*	1	3	4	2
		C2	PC A2.4	*	1	3	4	3
		DE	PC A2.4	*	1	3	4	4
	Post-1975 (standard)		FHRC A2.3	*	1	3	5	
		AB	FHRC A2.4	*	1	3	5	1
		C1	FHRC A2.4	*	1	3	5	2
		C2	FHRC A2.4	*	1	3	5	3
		DE	FHRC A2.4	*	1	3	5	4
	Post-1975 (timber-framed)				1	3	6	
		AB			1	3	6	1
		C1			1	3	6	2
		C2			1	3	6	3
		DE			1	3	6	4
	Post-1975 (garage/utility)		FLAIR 3.9		1	3	7	
		AB	FLAIR 3.9		1	3	7	1
		C1	FLAIR 3.9		1	3	7	2
		C2	FLAIR 3.9		1	3	7	3
		DE	FLAIR 3.9		1	3	7	4

APPENDIX 5.1 283

Type	Category	Subcategory (Social class)	Depth/damage or loss data sources Appendix/ chapter/ section	Data on computer file at Middlesex University	Land use code Column number			
					21	22	23	24
Bungalow			PC A2.3	*	1	4		
	Pre-1918 (cottage)		PC A2.3	*	1	4	1	
		AB	PC A2.4	*	1	4	1	1
		C1	PC A2.4	*	1	4	1	2
		C2	PC A2.4	*	1	4	1	3
		DE	PC A2.4	*	1	4	1	4
	Inter-war (1918–38)		PC A2.3	*	1	4	2	
		AB	PC A2.4	*	1	4	2	1
		C1	PC A2.4	*	1	4	2	2
		C2	PC A2.4	*	1	4	2	3
		DE	PC A2.4	*	1	4	2	4
	Post-war		PC A2.3	*	1	4	3	
		AB	PC A2.4	*	1	4	3	1
		C1	PC A2.4	*	1	4	3	2
		C2	PC A2.4	*	1	4	3	3
		DE	PC A2.4	*	1	4	3	4
	Modern (post-1965)		PC A2.3	*	1	4	4	
		AB	PC A2.4	*	1	4	4	1
		C1	PC A2.4	*	1	4	4	2
		C2	PC A2.4	*	1	4	4	3
		DE	PC A2.4	*	1	4	4	4
	Post-1975 (post-Manual)		FHRC A2.3	*	1	4	5	
		AB	FHRC A2.4	*	1	4	5	1
		C1	FHRC A2.4	*	1	4	5	2
		C2	FHRC A2.4	*	1	4	5	3
		DE	FHRC A2.4	*	1	4	5	4
	Post-1975 (timber-framed)				1	4	6	
		AB			1	4	6	1
		C1			1	4	6	2
		C2			1	4	6	3
		DE			1	4	6	4
Flat			PC A2.3	*	1	5		
	Pre-1918 (tenement)		PC A2.3	*	1	5	1	
		AB	PC A2.4	*	1	5	1	1
		C1	PC A2.4	*	1	5	1	2
		C2	PC A2.4	*	1	5	1	3
		DE	PC A2.4	*	1	5	1	4

Type	Category	Subcategory (Social class)	Depth/damage or loss data sources Appendix/ chapter/ section	Data on computer file at Middlesex University	Land use code Column number			
					21	22	23	24
	Inter-war (1918–38)		PC A2.3	*	1	5	2	
		AB	PC A2.4	*	1	5	2	1
		C1	PC A2.4	*	1	5	2	2
		C2	PC A2.4	*	1	5	2	3
		DE	PC A2.4	*	1	5	2	4
	Post-war (maisonette)		PC A2.3	*	1	5	3	
		AB	PC A2.4	*	1	5	3	1
		C1	PC A2.4	*	1	5	3	2
		C2	PC A2.4	*	1	5	3	3
		DE	PC A2.4	*	1	5	3	4
	Modern (post-1965)		PC A2.3	*	1	5	4	
		AB	PC A2.4	*	1	5	4	1
		C1	PC A2.4	*	1	5	4	2
		C2	PC A2.4	*	1	5	4	3
		DE	PC A2.4	*	1	5	4	4
	Post-1975 (post-Manual)		FHRC A2.3	*	1	5	5	
		AB	FHRC A2.4	*	1	5	5	1
		C1	FHRC A2.4	*	1	5	5	2
		C2	FHRC A2.4	*	1	5	5	3
		DE	FHRC A2.4	*	1	5	5	4
	Post-1975 (timber-framed)				1	5	6	
		AB			1	5	6	1
		C1			1	5	6	2
		C2			1	5	6	3
		DE			1	5	6	4
Mobile home (caravan)					1	6		
Pre-fab			PC A2.3	*	1	7		
		AB	PC A2.4	*	1	7	3	1
		C1	PC A2.4	*	1	7	3	2
		C2	PC A2.4	*	1	7	3	3
		DE	PC A2.4	*	1	7	3	4

For all residential properties with an asterisk, standard depth/damage data is available on computer file at Middlesex University for two durations (less than 12 hours; more than 12 hours) and assuming three flood warning intervals (half-hour, 2 hours and 4 hours).

Type	Category	Depth/damage or loss data sources Appendix/chapter/section	Data on computer file at Middlesex University	Land use code Column number 21	22	23	24
Sectors 2–9							
Sector 2 Agricultural buildings							
Arable farm		PC A7.1	+	2	1		
Livestock farm		PC A7.1	+	2	3		
Dairy farm		PC A7.1	+	2	4		
Mixed farm		PC A7.1	+	2	4		

Agricultural buildings are not coded individually but collectively by broad farm type classification if known.

Type	Category	Depth/damage or loss data sources	Data on computer file	21	22	23	24
Sector 3 Non-built up land (non-agricultural)							
Recreational				3	1		
	Playing fields			3	1	1	
	Golf courses			3	1	2	
	Parks	PC Table 8.1		3	1	3	
	Stadia, race tracks, and courses			3	1	4	
	Children's playgrounds			3	1	5	
	Zoos and botanical gardens			3	1	6	
	Lidos and open air swimming pools			3	1	7	
	Picnic sites (designated)			3	1	8	
	Other non-specified public and private open spaces			3	1	9	
Cemeteries, crematoria etc.				3	2		
	Cemeteries and burial grounds			3	2	1	
	Crematoria			3	2	2	
	Memorial grounds (detached from parks)			3	2	3	
Sector 4 Non-domestic residential							
Accommodation ancillary to educational establishments				4	1		
	Halls of residence	PC 5.1		4	1	1	
	Student hostel (non-purpose built)	PC 5.1		4	1	2	
	School boarding house	PC 5.1		4	1	3	

Type	Category	Depth/damage or loss data sources		Land use code Column number			
		Appendix/ chapter/ section	Data on computer file at Middlesex University	21	22	23	24
Accommodation of health/welfare establishments				4	2		
	Nurses homes, staff hostels, etc.			4	2	1	
	Old people's homes			4	2	2	
	Children's homes			4	2	3	
	Family group homes			4	2	4	
	Homes for mentally subnormal adults			4	2	5	
	Homes for the blind			4	2	6	
Other special residential accommodation				4	3		
	Convents, monasteries, etc			4	3	1	
	Residential conference centres			4	3	2	
	Showmen's winter quarters			4	3	3	
	Itinerant's caravan sites			4	3	4	
Government residential establishments				4	4		
	Royal navy barracks			4	4	1	
	Army barracks			4	4	2	
	Royal air force barracks			4	4	3	
	Prisons, approved schools, borstals etc.			4	4	4	
	Other government establishments			4	4	5	
Hotel accommodation				4	5		
	Hostels and motels	PC5.1		4	5	1	
	Guest houses			4	5	2	
	Residential clubs			4	5	3	
	Other hotels including YMCA and YWCA			4	5	4	
Holiday establishments				4	6		
	Holiday camps			4	6	1	
	Camping sites			4	6	2	
	Holiday caravan sites			4	6	3	
	Holiday bungalows and chalets			4	6	4	
	Youth hostels			4	6	5	

APPENDIX 5.1 287

Type	Category	Depth/damage or loss data sources Appendix/ chapter/ section	Data on computer file at Middlesex University	Land use code Column number			
				21	22	23	24

Sector 5 Retail trading and related services

Type	Category	Source	Data	21	22	23	24
Shops		PC A3.1	*	5	1		
	Food	PC A3.1 PGT A5.1–5.3	*+	5	1	1	
	Grocers/deli., & provisions dealers	PC A3.1	*	5	1	1	1
	Dairymen	PC A3.1	*	5	1	1	2
	Butchers	PC A3.1	*	5	1	1	3
	Fishmongers, poulterers	PC A3.1	*	5	1	1	4
	Greengrocers, fruiterers	PC A3.1	*	5	1	1	5
	Bread and flour confectioners	PC A3.1	*	5	1	1	6
	Off licences	PC A3.1	*	5	1	1	7
	Supermarkets			5	1	1	8
	Confectioners, tobacconists and newsagents	PC A3.1 PGT A5.1–5.3	*+	5	1	2	
	Clothing & footwear	PC A3.1 PGT A5.1–5.3	*+	5	1	3	
	Boots and shoes	PC A3.1	*	5	1	3	1
	Men's and boy's wear	PC A3.1	*	5	1	3	2
	Women's and girl's wear	PC A3.1	*	5	1	3	3
	Household goods	PC A3.1 PGT A5.1–5.3	*+	5	1	4	
	Furniture and allied	PC A3.1	*	5	1	4	1
	Electrical and electronic goods	PC A3.1	*	5	1	4	2
	Radio and TV hire	PC A3.1	*	5	1	4	3
	Cycles and perambulators			5	1	4	4
	Hardware and china, wallpaper and paint	PC A3.1	*	5	1	4	5
	Electricity board showrooms			5	1	4	6
	Gas board showrooms	PC A3.1	*	5	1	4	7
	Other non-food retailers	PC A3.1 PGT A5.1–5.3	*+	5	1	5	

Type	Category	Depth/damage or loss data sources Appendix/chapter/section	Data on computer file at Middlesex University	Land use code Column number 21	22	23	24
	Books, stationers			5	1	5	1
	Chemists, photographic dealers	PC A3.1	*	5	1	5	2
	Jewellery, leather and sports	PC A3.1	*	5	1	5	3
	Other non-food			5	1	5	4
	Video Libraries			5	1	5	5
	Pet shops			5	1	5	6
	Florist			5	1	5	7
	Wool shop			5	1	5	8
	Toy shop			5	1	5	9
	General stores	PC A3.1 PGT A5.1–5.3	*+	5	1	6	
	Department stores	PC A3.1	*	5	1	6	1
	Variety and other general	PC A3.1	*	5	·1	6	2
	Hypermarkets			5	1	6	3
	Garden centres/nurseries			5	1	6	4
Vehicle services		PC A4.1 PGT A5.1–5.3	*+	5	2		
	Major garages with showrooms, repair depots and petrol filling facilities	PC A4.2	#	5	2	1	
	Petrol filling stations with or without service bay	PC Table 3.6	*	5	2	2	
	Tyre sales and servicing	PC A4.2	#	5	2	3	
	Car wash plant		#	5	2	4	
	Motorbike repairs and sales			5	2	5	
	Motor vehicle repairs	PC A4.2		5	2	6	
Services							
	Public houses	PC A3.2 PGT A5.1–5.3	*+	5	3	1	
	Public houses without cellars	PC A3.2	*	5	3	1	1
	Public houses with cellars	PC A3.2	*	5	3	1	2
	Cafes & restaurants	PC A3.2	*	5	3	2	
	Laundries, launderettes and dry cleaning plants (incl. linen hire)			5	3	3	
	Design studios, photographers' studios, picture framing, antiques			5	3	4	
	Take away food establishments			5	3	5	
	Repair establishments (e.g. TV/electrical goods)			5	3	6	

APPENDIX 5.1 289

Type	Category	Depth/damage or loss data sources Appendix/ chapter/ section	Data on computer file at Middlesex University	Land use code Column number 21 22 23 24
	Auction sales and markets			5 3 7
	Livestock sales and markets			5 3 8 1
	Other services	PC A3.2 PGT A5.1–5.3	*+	5 3 9
	Travel agencies	PGT A5.1–5.3	+	5 3 9 1
	Betting offices	PC A3.2	*	5 3 9 2
	Opticians and chiropodists	PC A3.2 PGT A5.1–5.3	*+	5 3 9 3
	Funeral services			5 3 9 4
	Barbers/ hairdressers, beauty parlours and slimming clinics	PC A3.2	*	5 3 9 5
	Car and taxi hire offices	PC A3.2	*	5 3 9 6
	Coach booking offices			5 3 9 7
	Driving school offices			5 3 9 8
	Printing shops			5 3 9 9
Contractors and other merchants and dealers		PC A4.1	#	5 4
	Builders' merchants contractors (electricians painters, plumbers, etc.)	PCA 4.2 PGT A5.1–5.3	#+	5 4 1
	Coal merchants and coal yards	PC A4.2 PGT A5.1–5.3	#+	5 4 2
	Scrap metal and rag and bone merchants	PC A4.2 PGT A5.1–5.3	#+	5 4 3
	Timber merchants	PC A4.2 PGT A5.1–5.3	#+	5 4 4
	Dealers in sand, gravel, bricks and tiles	PGT 5.1–5.6	+	5 4 5
	Construction (civil engineers, scaffolders & concrete contractors)	PC A4.2/Table 4.1	#	5 4 6
	Plant and crane hire	PC A4.2	#	5 4 7
	Waste rubber dealers and waste clearance and disposal merchants	PC A4.2	#	5 4 8
	Miscellaneous	PC A4.2 PGT A5.1–5.3	#+	5 4 9
Storage and wholesale establishments		PC A4.1	#	5 5
	Establishments engaged in the wholesale distribution of food products	PC A4.2 PGT A5.1–5.3	#+	5 5 1

Type	Category	Depth/damage or loss data sources Appendix/ chapter/ section	Data on computer file at Middlesex University	Land use code Column number			
				21	22	23	24
	Establishments engaged in the wholesale distribution of non-food products	PC A4.2 PGT A5.1–5.3	#+	5	5	2	
	Transit warehouses (including cash and carry depositories)	PC A4.2 PGT A5.1–5.3	#+	5	5	3	
	Bonded warehouses	PC A4.2	#	5	5	4	
	Post offices depots			5	5	5	
	Furniture and book depositories	PGT A5.1–5.3	+	5	5	6	
	Fuel storage tanks	PC A4.2	#	5	5	7	
	Establishments which do not suitably fit into any other category	PC A4.2 PGT A5.1–5.3	#+	5	5	8	

Sector 6 Professional and offices

Type	Category	Depth/damage or loss data sources	Data on computer file	21	22	23	24
Central government offices (including labour exchanges)		PC 3.6.1 PGT A5.1–5.3	+	6	1		
Local government offices (including rate and rent offices)		PC 3.6.1		6	2		
Embassies, consulates etc.				6	3		
Banks		PC A3.3 PGT A5.1–5.3	*+	6	4		
Commercial offices		PC A3.3 PGT A5.1–5.3	*+	6	5		
	Architects	PC A3.3	*	6	5	1	
	Building societies	PGT A5.1–5.3	+	6	5	2	
	Estate agents	PC A3.3 PGT A5.1–5.3	*+	6	5	3	
	Insurance brokers/ loss adjusters	PC A3.3	*	6	5	4	
	Surveyors/valuers	PC A3.3	*	6	5	5	
	Solicitors/accountants	PC A3.3	*	6	5	6	
	Other commercial offices, eg. secretarial, duplicating services etc.	PC A3.3 PGT A5.1–5.3	*+	6	5	7	
Surgeries			*	6	6		
	Dentists			6	6	1	
	Doctors			6	6	2	
	Osteopaths, etc.			6	6	3	
	Physiotherapists			6	6	4	
	Veterinary			6	6	5	
Institutional offices		PC 3.6.1		6	7		
	Hospital offices			6	7	1	
	University offices			6	7	2	
	Trade union offices			6	7	3	
	Other			6	7	4	
Research establishments				6	8		
	Research establishments			6	8	1	
	Observatories			6	8	2	
	Laboratories			6	8	3	

Type	Category	Depth/damage or loss data sources Appendix/ chapter/ section	Data on computer file at Middlesex University	Land use code Column number			
				21	22	23	24
Sector 7 Public Buildings and community services		PGT 7					
Central and Local government establishments and public corporations				7	1		
	Post offices			7	1	1	
	GPO sorting offices			7	1	2	
	Telephone exchanges	PC 5.4.1		7	1	3	
	Police stations	PC 5.1		7	1	4	
	Fire and ambulance stations			7	1	5	
	Radar and radio communications establishments			7	1	6	
	Law courts			7	1	7	
	Telecommunications and GPO depots			7	1	8	
	Armed forces offices			7	1	9	
Schools, colleges, etc.		PC 5.1		7	2		
	Nursery & kindergartens			7	2	1	
	Primary	PGT A8.1	+	7	2	2	
	Secondary	PGT A8.1	+	7	2	3	
	Technical, art, teacher training and FE colleges			7	2	4	
	Universities and polytechnics			7	2	5	
	Adult education centres			7	2	6	
	Schools for physically and mentally handicapped			7	2	7	
	Industrial training centres			7	2	8	
Health establishments				7	3		
	General and psychiatric hospitals			7	3	1	
	Convalescent and nursing homes			7	3	2	
	Clinics			7	3	3	
	Health centres			7	3	4	
	Rehabilitation centres			7	3	5	
Education and recreation centres							
	Major entertainment centres containing at least 2 of those listed below			7	4	1	
	Cinemas, theatres concert halls	PGT A5.1–5.3	+	7	4	2	
	Dance halls			7	4	3	
	Amusement parks and permanent fun fairs			7	4	4	
	Swimming baths			7	4	5	

292 THE ECONOMICS OF COASTAL MANAGEMENT

Type	Category	Depth/damage or loss data sources Appendix/ chapter/ section	Data on computer file at Middlesex University	Land use code Column number 21	22	23	24
	Sports halls and gymnasia	PGT A5.1–5.3	+	7	4	6	
	Skating rinks			7	4	7	
	Indoor games and Amusements			7	4	8	
	bingo halls	PGT A5.1–5.3		7	4	8	1
	billiard saloons			7	4	8	2
	amusement arcades			7	4	8	3
	casinos			7	4	8	4
	night clubs			7	4	8	5
Places of assembly				7	5		
	Public halls			7	5	1	
	Private halls (including church halls)	PGT A8.1	+	7	5	2	
	Community centres			7	5	3	
	Clubs (non-residential) and societies			7	5	4	
	Youth clubs and scout halls			7	5	5	
Places of culture and worship				7	6		
	Libraries			7	6	1	
	Museums			7	6	2	
	Art galleries			7	6	3	
	Exhibition halls			7	6	4	
	Ancient monuments			7	6	5	
	Historical buildings			7	6	6	
	Churches	PGT A8.1	+	7	6	7	
Other				7	7		
	Public conveniences	PGT A5.1–5.3	+	7	7	1	
	Public baths and wash houses			7	7	2	
	Turkish, foam and sauna baths			7	7	3	

APPENDIX 5.1

Type	Category	Appendix/ chapter/ section	Data on computer file at Middlesex University	Standard Industrial Classification Activity	Previous land use code	Revised land use code Column number			
						21	22	23	24

Sector 8 Manufacturing and extractive industries
PGT A4.2/Tables 4.6–4.7

Energy and extractive industry
				1100	89	8	1	0	0
1 Energy and extractive industry									
	Deep coal mines			1113	891	8	1	0	1
	Open cast coal mines			1114	891	8	1	0	2
	Solid fuel manufacture		+	1115	8211	8	1	0	3
	Coke ovens			1200	8211	8	1	0	4
	Oil and gas extraction			1300	894	8	1	1	1
	Mineral oil processing			1401	8221	8	1	1	2
	Other oil products			1402	8222	8	1	1	3
	Nuclear fuel production			1520	NA	8	1	2	1
	Extraction; metalliferous ores			2100	895	8	1	3	1
	Extraction stone, clay, sand etc.			2310	892/3	8	1	3	2
	Extraction of salt & refining			2330	895	8	1	3	3
	Other mineral extraction			2396	895	8	1	3	4

Metal and mineral manufacture
			+			8	2	0	0
2 Metal manufacture									
	Iron and steel manufacture			2210	8311	8	2	0	1
	Steel tubes			2220	8312	8	2	0	2
	Steel wire manufacture		+	2234	8365	8	2	0	3
	Other drawing and cold rolling of steel			2235	NA	8	2	0	4
	Aluminium manufacture (basic refining and semi-manufacture only)			2245	8314	8	2	0	5
	Copper, brass and other copper alloys (manufacture and semi processing)			2246	8315	8	2	0	6
	Other non-ferrous metal			2247	8316	8	2	0	7
3 Other mineral manufacture			+						
	Structural clay products (e.g. tiles)			2410	861	8	2	1	1
	Cement, lime, plaster			2420	864	8	2	1	2
	ready mixed concrete		+	2436	865	8	2	1	3
	Other concrete, cement, plaster, asbestos cement building products		+	2437	865	8	2	1	4
	Asbestos goods (except for cement and automotive gasket sets)		+	2440	865	8	2	1	5
	Stone, slate and other non-metallic mineral products		+	2450	865	8	2	1	6
	Abrasives (other than natural stone)			2460	865	8	2	1	7
	Flat glass		+	2471	863	8	2	2	1
	Glass containers			2478	863	8	2	2	2

Type	Category	Appendix/ chapter/ section	Depth/damage or loss data sources Data on computer file at Middlesex University	Standard Industrial Classification Activity	Previous land use code	Revised land use code Column number 21	22	23	24
	All other glass and glass fibre products (not optical glass or completed items with glass fibre parts)		+	2479	863	8	2	2	3
	Refractory goods (heat resistant blocks, linings etc.)			2481	861	8	2	3	1
	Ceramic goods		+	2489	862	8	2	3	2
Chemicals and derivatives									
	4 Chemicals and derivatives		+			8	3	0	0
	Inorganic chemical manufacture (excluding gases and finished products)			2511	8231	8	3	0	1
	Basic organic chemicals (excluding pharmaceutical chemicals & finished products)			2512	8231	8	3	0	2
	Artificial fertilizers			2513	8238	8	3	0	3
	Synthetic resins and plastic materials			2514	8236	8	3	0	4
	Synthetic rubber			2515	8236	8	3	0	5
	Dyestuffs and pigments			2516	8237	8	3	0	6
	Paints, varnishes and painters' fillings (not sealants)		+	2551	8234	8	3	1	1
	Printing ink			2552	8239	8	3	1	2
	Formulated adhesives and sealants			2562	8239	8	3	1	3
	Chemical treatment of oils and fats			2563	NA	8	3	1	4
	Essential oils and flavourings (not turpentine)			2564	NA	8	3	1	5
	Explosives, fireworks and matches (not ammunition)		+	2565	8239	8	3	1	6
	Miscellaneous industrial chemicals (industrial gases, tanning agents, waxes etc.)			2567	NA	8	3	1	7
	Formulated pesticides		+	2568	8239	8	3	1	8
	Adhesive film, cloth and foil (not paper)			2569	NA	8	3	1	9
	Pharmaceutical products		+	2570	8232	8	3	2	1
	Soaps and detergents			2581	8235	8	3	2	2
	Perfumes, cosmetics and toilet preparations		+	2582	8233	8	3	2	3
	Photographic materials and chemicals			2591	8239	8	3	2	4
	Chemical products not elsewhere specified (includes candles, waxes, food preservatives and unrecorded magnetic tape)		+	2599	8239	8	3	2	5

Type	Category	Appendix/ chapter/ section	Depth/damage or loss data sources Data on computer file at Middlesex University	Standard Industrial Classification Activity	Previous land use code	Revised land use code Column number 21	22	23	24
	Production of man-made fibres		+	2600	8511	8	3	3	1
Metal and electrical engineering etc.			+			8	4	0	0
5 Other metal goods									
	Ferrous metal foundries		+	3111	8311	8	4	0	1
	Non-ferrous metal foundries		+	3112	8314–6	8	4	0	2
	Forging, pressing and stamping			3120	8368	8	4	0	3
	Bolts, nuts, washers, springs, non-precision chains		+	3137	8364/8368	8	4	0	4
	Heat and surface treatment of metals			3138	8368	8	4	0	5
	Metal doors, windows etc.			3142	8368	8	4	0	6
	Hard tools and implements			3161	8362	8	4	1	1
	Cutlery, similar tableware & knives etc.			3162	8363	8	4	1	2
	Non-industrial metal storage vessels (e.g. cisterns, dustbins)			3163	NA	8	4	1	3
	All metal packaging (e.g. cans, drums, foil		+	3164	8366/8368	8	4	1	4
	Non-electrical domestic heaters and cookers (not central heating systems)			3165	8368	8	4	1	5
	Metal furniture and safes		+	3166	8368	8	4	1	6
	Domestic metal utensils (e.g. kettles, cookware)			3167	8368	8	4	1	7
	Other finished metal products (locks, needles, pins, metal fastenings)		+	3169	8368	8	4	1	8
6 Mechanical engineering									
	Fabricated constructional steelwork		+	3204	8331	8	4	2	1
	Boilers, process plant fabrications and other heavy steel fabrications (e.g. silos, bunkers etc. over 3mm thick)			3205	8331	8	4	2	2
	Agricultural machinery		+	3211	8411	8	4	2	3
	Wheeled tractors (agricultural)			3212	8411	8	4	2	4
	Metal-working machine tools (includes numerically controlled machine manufacture e.g. milling, bending etc. machines)			3221	8322	8	4	3	1
	Engineer's small tools (metal cutting tools and press tools e.g. dies, drills)		+	3222	8361	8	4	3	2
	Textile machinery			3230	8325	8	4	3	3

Type	Category	Appendix/ chapter/ section	Depth/damage or loss data sources Data on computer file at Middlesex University	Standard Industrial Classification Activity	Previous land use code	Revised land use code Column number			
						21	22	23	24
	Food, drink and tobacco machinery (processing, packaging and bottling)		+	3244	8329	8	4	3	4
	Chemical industry machinery; furnaces, kilns, gas, water and waste treatment plant		+	3245	NA	8	4	3	5
	Process engineering contractors (combining design, construction, assembly and commissioning of process industry plant)			3246	NA	8	4	3	6
	Mining machinery manufacture			3251	8329	8	4	3	7
	Construction and earth moving equipment			3254	8326	8	4	3	8
	Mechanical lifting and handling equipment (e.g. cranes, escalators, forklifts)		+	3255	8327	8	4	3	9
	Precision chains, plain bearings, gears, gearboxes and other mechanical power transmission equipment			3261	8333	8	4	4	1
	Ball, needle and roller bearings			3262	8333	8	4	4	2
	Machines working: wood, rubber, plastic, leather, footwear, paper, glass, brick, ceramics, and laundry and dry cleaning			3275	8329	8	4	4	3
	Printing, bookbinding and paper goods machinery		+	3276	8329	8	4	4	4
	Internal combustion engines (not for road vehicles, agricultural tractors or aircraft) including steam and gas turbines for any other prime movers, including ships			3281	8324	8	4	4	5
	Compressors and fluid power equipment (e.g. hydraulics and pneumatic control equipment)		+	3283	8323	8	4	4	6
	Refrigerating, air conditioning, ventilating and space heating equipment (not domestic gas or electric heaters)		+	3284	8329	8	4	4	7
	Scales, weighing machinery and portable power tools			3285	8329	8	4	4	8

APPENDIX 5.1

Type	Category	Appendix/ chapter/ section	Depth/damage or loss data sources Data on computer file at Middlesex University	Standard Industrial Classification Activity	Previous land use code	Revised land use code Column number 21	22	23	24
	Other industrial and commercial machinery (including non-electric test equipment, lawn mowers, machinery for foundries and rolling mills, other machines not specified elsewhere)			3286	NA	8	4	4	9
	Pumping manufacture (not for hydraulics or internal combustion engines)			3287	8323	8	4	5	1
	Industrial valve manufacture		+	3288	8323	8	4	5	2
	Mechanical, marine and precision engineering not elsewhere specified (including components common to a wide range of engines and machines, gas welding machinery, general subcontractors, establishments with very mixed engineering products, auxiliary marine machinery)		+	3289	NA	8	4	5	3
	Ordnance, small arms, ammunition and tracked military vehicles		+	3290	8332	8	4	5	4
7 Office machinery									
	Other machinery (not copiers, dictating machines)		+	3301	8328	8	4	6	1
	Electronic data processing equipment (computers, subassemblies and specialised computer peripherals)		+	3302	8356	8	4	6	2
8 Electrical and electronic engineering			+						
	Insulated wires and cables			3410	8352	8	4	7	1
	Basic electrical equipment (generators, power transmission, switch and control gear, electric motors and overhauling of electrical machinery)			3420	8351	8	4	7	2
	Batteries and accumulators			3432	8359	8	4	7	3
	Alarms and signalling equipment			3433	8357	8	4	7	4
	Electrical equipment for motor vehicles, cycles and aircraft (not railways or marine = 8472)			3434	8359	8	4	7	5

Type	Category	Appendix/ chapter/ section	Depth/damage or loss data sources Data on computer file at Middlesex University	Standard Industrial Classification Activity	Previous land use code	Revised land use code Column number 21	22	23	24
	Industrial electrical equipment not specified elsewhere (e.g. electric welding equipment)			3435	NA	8	4	7	6
	Telegraph and telephone equipment			3441	8353	8	4	8	1
	Electrical instrumental and control systems		+	3442	8344	8	4	8	2
	Radio and electronic capital goods (e.g. transmitters, TV cameras, radar, aerials, X-ray machines, electro-medical apparatus)		+	3443	8357	8	4	8	3
	Non-active electronic equipment components (e.g. resistors, capacitors, transformers, printed circuits)		+	3444	8354	8	4	8	4
	Gramophone records and prerecorded tapes			3452	8355	8	4	8	5
	Active components and electronic subassemblies (e.g. valves, transistors, integrated circuits, domestic aerials, loudspeakers, microphones record playing parts, tape decks, tuners, any subassemblies not specified elsewhere)			3453	8355	8	4	8	6
	Electronic consumer goods (e.g. domestic: radios, TVs, record players, tape recorders; public address systems etc. other electronic equipment not specified elsewhere)		+	3454	8355	8	4	8	7
	Domestic electric appliances (e.g. electric cookers, washing machines, refrigerators, toasters space heaters)		+	3450	8358	8	4	8	8
	Electric lamp bulbs and tubes and other electric lighting equipment (not for vehicles)			3470	8359	8	4	8	9
	Electrical equipment installation (by specialist non-manufacturers not wiring of buildings)			3480	NA	8	4	9	1

Type	Category	Appendix/ chapter/ section	Depth/damage or loss data sources Data on computer file at Middlesex University	Standard Industrial Classif- ication Activity	Previous land use code	Revised land use code Column number 21	22	23	24
Transport and instrument engineering			+			8	5	0	0
9 Motor vehicles			+						
	Manufacture of passenger cars, commercial vehicles and motor vehicle engines			3510	8412	8	5	0	1
	Motor vehicle body manufacture		+	3521	8412	8	5	0	2
	Trailers and semi-trailers			3522	8412	8	5	0	3
	Caravans			3523	8412	8	5	0	4
	Motor vehicles parts wholly or mostly metal		+	3530	8412	8	5	0	5
10 Other transport equipment			+						
	Shipbuilding and repairing (including inland and pleasure vessels)		+	3610	837	8	5	1	1
	Railway and tramway vehicles, rolling stock, parts and repairs			3620	8415/6	8	5	2	1
	Motor cycle manufacture (includes metal components not elsewhere specified)			3633	8413	8	5	3	1
	Pedal cycles and parts		+	3634	8413	8	5	3	2
	Aerospace equipment and repair (includes gliders, helicopters, guided weapons, launch vehicles, aero engines and power plants, hovercraft, balloons, parts but not instruments and electrical-electronic parts)			3640	8414	8	5	4	1
	Other vehicles (baby carriages, wheelchairs invalid carriages, animal drawn carts etc.)			3650	NA	8	5	5	1
11 Instrument engineering									
	Measuring and checking precision instruments (e.g. balances, thermometers, pressure gauges, laboratory apparatus, not electrical-electronic or optical)			3710	8344	8	5	6	1
	Medical, surgical and orthopaedic equipment (e.g. surgical cutlery, respirators, anaesthetic equipment, dental instruments, artificial parts)			3720	8343	8	5	6	2
	Spectacles and unmounted lenses			3731	8343	8	5	6	3

Type	Category	Appendix/ chapter/ section	Depth/damage or loss data sources		Previous land use code	Revised land use code Column number			
			Data on computer file at Middlesex University	Standard Industrial Classif- ication Activity		21	22	23	24
	Optical precision Instruments (e.g. microscopes, lasers; not cameras)			3732	8344	8	5	6	4
	Photographic and cinematographic equipment (cameras, photocopiers, mounted lenses)		+	3733	8341	8	5	6	5
	Clocks, watches and timing devices			3740	8342	8	5	6	6
Food, drink and tobacco manufacture			+						
12 Food, drink and tobacco manufacture									
	Margarine and compound cooking fats			4115	8119	8	6	0	1
	Processing organic oils and fats (not crude animal fats)			4116	8118	8	6	0	2
	Slaughterhouses (includes preparing fresh meat and freezing but not poultry)			4121	NA	8	6	1	1
	Bacon curing and meat processing (includes prepared meat products e.g. pies, sausages etc.)		+	4122	8113	8	6	1	2
	Poultry slaughter and processing			4123	8113	8	6	1	3
	Animal by-product processing (including slaughter and processing for non-human consumption)			4126	NA	8	6	1	4
	Preparation of milk and milk products			4130	8114	8	6	2	1
	Processing of fruit and vegetables (includes freezing, canning, pickling and jams etc.)			4147	8116	8	6	2	2
	Fish processing (includes shellfish, all means of preserving etc.)			4150	8113	8	6	2	3
	Grain milling (includes splitting and grinding pulses and producing uncooked breakfast cereals)			4160	8111	8	6	3	1
	Manufacture of starch			4180	8119	8	6	3	2
	Bread and flour confectionery manufacture		+	4196	8112	8	6	3	3
	Biscuit and crispbread manufacture			4197	8112	8	6	3	4
	Sugar and sugar by-products			4200	8115	8	6	4	1
	Ice-cream manufacture			4213	8114	8	6	4	2

APPENDIX 5.1 301

Type	Category	Appendix/ chapter/ section	Depth/damage or loss data sources		Previous land use code	Revised land use code Column number			
			Data on computer file at Middlesex University	Standard Industrial Classification Activity		21	22	23	24
	Cocoa, chocolate and sugar confectionery			4214	8115	8	6	4	3
	Compound animal feeds (includes protein concentrates)			4221	8117	8	6	5	1
	Pet foods and non-compound animal feeds, and supplements			4222	8117	8	6	5	2
	Miscellaneous foods (coffee, tea; crisps, snack products; infant and diabetic foods; starch and malt extract, puddings yeast; soups, sauces etc; pasta; ready-to-eat; breakfast cereals; any foods not elsewhere specified)		+	4239	8119	8	6	5	3
	Spirit distilling and compounding (includes producing raw ethyl alcohol and fermentation etc. of potable spirits)			4240	8123	8	6	6	1
	Production of wines bases on concentrates, cider, sherry, fruit wines, (including fresh grapes, but not on agricultural holdings)			4261	8123	8	6	6	2
	Brewing and malting		+	4270	8121	8	6	6	3
	Soft drinks, mineral waters and fruit juices			4283	8122	8	6	6	4
	Tobacco industry			4290	813	8	6	7	1
Textile and clothing industry									
13 Textile industry			+			8	7	0	0
	Woollen and worsted industry (preparation of wool and hair fibres, spinning and weaving, includes blankets but not carpets)		+	4310	8514	8	7	0	1
	Spinning and doubling on the cotton system (includes cotton, silk, and man-made if done using cotton system; production of yarns and finished thread)			4321	8512	8	7	0	2

Type	Category	Appendix/ chapter/ section	Depth/damage or loss data sources Data on computer file at Middlesex University	Standard Industrial Classification Activity	Previous land use code	Revised land use code Column number 21	22	23	24
	Weaving of cotton, silk and man-made fibre (of all sorts excluding for carpets, includes establishments both weaving and finishing)			4322	8513	8	7	0	3
	Throwing, texturing, crimping etc. continuous filament yarn (not where man-made fibres manufactured)			4336	NA	8	7	0	4
	Spinning and weaving of flax, hemp and ramie (includes establishments both weaving and finishing)			4340	8512	8	7	0	5
	Spinning, weaving and production of jute, fibrillated yarn and polypropylene fabric (not extrusion of man-made fibres or tapes)			4350	8515	8	7	0	6
	Hosiery and other weft knitted goods and fabrics (includes knitting and pile fabrics)			4363	8516	8	7	0	7
	Warp knitted goods (includes elastic fabrics, not making-up garments from purchased fabric)			4364	8516	8	7	0	8
	Textile finishing (includes bleaching, dyeing, printing etc. of all items listed above (8701– 8); if carried out by manufacturer assign to manufacturing)		+	4370	8518	8	7	0	9
	Pile carpets, carpeting and rugs (both woven and tufted)			4384	8517	8	7	1	1
	Needle and bounded carpets, rugs etc. and hard fibre matting (not jute)			4385	8517	8	7	1	2
	Lace (except for bleaching, dyeing etc. on commission = 8709)			4395	8516	8	7	1	3
	Rope, twine and net			4396	8515	8	7	1	4
	Narrow fabrics (elastics and elastometrics under 30cm wide, labels, ribbons tapes etc.)			4398	8519	8	7	1	5
	Miscellaneous textiles (felt textiles not elsewhere specified, kapok and hair fibre fillings)			4399	8514/8518	8	7	1	6
	Leather tanning and dressing and fell mongery			4410	8521	8	7	2	1

APPENDIX 5.1

Type	Category	Appendix/ chapter/ section	Depth/damage or loss data sources Data on computer file at Middlesex University	Standard Industrial Classification Activity	Previous land use code	Revised land use code Column number 21	22	23	24
	Leather goods (travel goods, belts, saddlery etc. and industrial leather goods)			4420	8522	8	7	2	2
14 Clothing and footwear			+						
	Footwear manufacture (not wooden, plastic or rubber component manufacture)		+	4510	8538	8	7	3	1
	Weatherproof outerwear			4531	8531	8	7	4	1
	Men's and boy's tailored outerwear (not jeans)			4532	8532	8	7	4	2
	Women's and girl's tailored outerwear (not jeans)			4533	8533	8	7	4	3
	Work clothing and men's and boy's jeans		+	4534	8534	8	7	4	4
	Men's and boy's shirts, underwear and nightwear (except for hosiery and knitted goods)		+	4535	8534	8	7	4	5
	Women's and girl's light outerwear, lingerie and infants' wear (includes jeans, dresses)		+	4536	8535	8	7	4	6
	Hat, caps and millinery (felt and other non-fur materials)			4537	8536	8	7	4	7
	Gloves (excludes knitted, rubber and plastic gloves and sportswear)			4538	8537	8	7	4	8
	Other dress industries (includes swimwear, foundation garments, umbrellas, handkerchiefs, neckties, scarves etc.)			4539	8537	8	7	4	9
	Soft furnishings (covered cushions, pillows etc.)			4555	NA	8	7	5	1
	Canvas goods, snacks, other made-up textile products			4556	8518	8	7	5	2
	Household textiles (e.g. quilts, sheets, towels)		+	4557	8518	8	7	5	3
	Fur goods (not sheepskin rugs, etc.)			4560	8523	8	7	6	1
Wood and paper industries			+			8	8	0	0
15 Timber and wooden furniture									
	Sawmilling, planing etc. (includes pit props, railway sleepers etc. skirting boards and flooring, but not parquet floors = 8803)		+	4610	8711	8	8	0	1

Type	Category	Appendix/ chapter/ section	Depth/damage or loss data sources Data on computer file at Middlesex University	Standard Industrial Classif- ication Activity	Previous land use code	Revised land use code Column number 21	22	23	24
	Semi-finished wood products, preservation and treatment of wood (e.g. veneers, plywoods, boards)			4620	8711	8	8	0	2
	Builders carpentry and joinery		+	4630	8711	8	8	0	3
	Wooden containers			4640	8715	8	8	1	1
	Other wooden articles (not furniture, e.g. beading, handles, wood for textile footwear, agricultural industries, wooden utensils)		+	4650	8716	8	8	1	2
	Brushes and brooms			4663	883	8	8	1	3
	Cork and cork articles, basketware, wickerware etc. (except for furniture)			4664	8716	8	8	1	4
	Wooden and upholstered furniture (both domestic and non-domestic, including cabinet work, wooden furniture components, beds and mattresses)		+	4671	8712/8713	8	8	2	1
	Shop and office fitting		+	4672	8714	8	8	2	2
16 Paper, printing and publishing			+						
	Manufacture of pulp, paper and board of all sorts (includes newsprint; coating and surface treatment by manufacturers)			4710	8721	8	8	3	1
	Wall coverings (wallpaper, paper backed vinyl and fabric wall coverings)			4721	8724	8	8	3	2
	Household and personal hygiene paper products based on paper (wrapping and toilet papers etc. = 8831)			4722	NA	8	8	3	3
	Stationery (paper and binders)			4723	8723	8	8	3	4
	Paper packaging products			4724	8722	8	8	3	5
	Board packaging products (fibre board, cartons etc.)		+	4725	8722	8	8	3	6
	Other paper and board products not specified elsewhere (not toys, playing cards or sensitised photographic paper)		+	4728	8724	8	8	3	7
	Printing and publishing of newspapers		+	4751	8725	8	8	4	1
	Printing and publishing of periodicals		+	4752	8726	8	8	4	2

Type	Category	Appendix/ chapter/ section	Depth/damage or loss data sources		Previous land use code	Revised land use code Column number			
			Data on computer file at Middlesex University	Standard Industrial Classification Activity		21	22	23	24
	Printing and publishing of books			4753	8727	8	8	4	3
	Other printing and publishing (includes security printing, cards, music, transfers, book binding and repairing, making printing plates, printing on metal, unspecialised printing works)		+	4754	8727	8	8	4	4
Rubber, plastics and miscellaneous manufacturing			+			8	9	0	0
17 Rubber and plastics			+						
	Rubber tyres and inner tubes (not retreading)			4811	881	8	9	0	1
	Other rubber products not elsewhere specified (includes rubber and plastic hose, tubing and belting; not complete rubber footwear = 8731 or adhesives)		+	4812	881	8	9	0	2
	Retreading and specialist repair of rubber tyres			4820	881	8	9	0	3
	Plastic coated textile fabric (not floorcoverings)			4831	NA	8	9	1	1
	Plastics semi-manufactures (e.g. film, sheet, tubes)			4832	886	8	9	1	2
	Plastic floorcoverings (also linoleum, felt base and woven plastic matting)			4833	882	8	9	1	3
	Plastic building products			4834	886	8	9	1	4
	Plastic packaging products		+	4835	886	8	9	1	5
	Plastic products not elsewhere specified (*not* complete items of footwear or clothing, brushes, mattresses, toys, sports goods, leather substitutes, fibre glass boats or mouldings for vehicles)		+	4836	886	8	9	1	6
18 Miscellaneous manufacturing			+						
	Jewellery and coins			4910	8367	8	9	2	1
	Musical instruments		+	4920	887	8	9	3	1
	Photographic and cinematographic processing laboratories			4930	NA	8	9	4	1
	Toys and games (not of rubber)			4941	884	8	9	5	1
	Sports goods		+	4942	884	8	9	5	2

Type	Category	Appendix/ chapter/ section	Depth/damage or loss data sources Data on computer file at Middlesex University	Standard Industrial Classif- ication Activity	Previous land use code	Revised land use code Column number 21	22	23	24
	Miscellaneous stationers' goods (pens, pencils, crayons etc. inks, duplicating materials, staplers etc.)		+	4954	885	8	9	6	1
	Other manufacturers not elsewhere specified (e.g. taxidermists, manufacturers of devotional ivory, bone etc. articles)		+	4959	887	8	9	7	1

APPENDIX 5.1 307

Type	Category	Subcategory	Depth/damage or loss data sources Appendix/ chapter/ section	Data on computer file at Middlesex University	Land use code Column number			
					21	22	23	24

Sector 9 Public utility and transportation

Transportation

	Category	Subcategory	App/ch/sec	Data	21	22	23	24
	Roads		PC 5.2.1 PGT 6		9	1		
		Road transport stations			9	1	1	1
	Bus and coach				9	1	2	
		Bus and coach depots	PC 8.3.4		9	1	2	1
		Road haulage depots			9	1	2	2
		Transport garages			9	1	2	3
		Taxi garages			9	1	2	4
	Car parks and garages				9	1	3	
		Multistorey car parks			9	1	3	1
		Lock-up garages (not attached to other property)	PC A2.1B		9	1	3	2
	Rail transport		PC 8.3.4		9	1	4	
		British Rail track and operational buildings	PC 5.2.2		9	1	4	1
		British Rail stations	PC 5.2.2		9	1	4	2
		Liner and goods depots			9	1	4	3
		Locomotive works and maintenance sheds			9	1	4	4
		National carriers etc.			9	1	4	5
	Air transport				9	1	5	
		Civil and commercial airports			9	1	5	1
		Private airfields			9	1	5	2
		Air terminals			9	1	5	3
		Heliports			9	1	5	4
	Water transport				9	1	6	
		Docks and other harbour installations			9	1	6	1
		Ocean passenger terminals			9	1	6	2
		Ocean freight terminals including transport warehouses			9	1	6	3
		Inland waterways and associated installations			9	1	6	4
		Boat hire installations			9	1	6	5
		Car ferry terminals			9	1	6	6

308 THE ECONOMICS OF COASTAL MANAGEMENT

Type	Category	Subcategory	Depth/damage or loss data sources Appendix/ chapter/ section	Data on computer file at Middlesex University	Land use code Column number			
					21	22	23	24
		Hovercraft terminals			9	1	6	7
Utilities			PGT 7		9	2		
	Local authority establishments				9	2	1	
		Sewage works	PC 5.7.1		9	2	1	1
		Sewage pumping stations	PC 9.2.2		9	2	1	2
		Local authority depots			9	2	1	3
	Central Electricity Generating Board & Area Electricity Board Establishment		PC 5.5		9	2	2	
		Power stations			9	2	2	1
		Transformer stations and substations	PC 5.5.1/9.2.2.		9	2	2	2
	Gas boards		PC 5.5		9	2	3	
		Gas works, gas holder stations	PC 5.6.3		9	2	3	1
		Control stations and governors	PC 5.6.1		9	2	3	2
	Water undertakings		PC 5.7		9	2	4	
		Storage reservoirs			9	2	4	1
		Intakes and wells			9	2	4	2
		Waterworks and treatment plants			9	2	4	3
		Pumping stations			9	2	4	4
	Telecommunication services & other public utilities				9	2	5	
		Telephone link boxes, cabinets, and pillars	PC 5.4.2		9	2	5	1
		Public call boxes			9	2	5	2
		Letter boxes (isolated)	PC 5.4.3		9	2	5	3

Appendix 5.2 Standard flood damage data for residential properties (salt water flooding)

310 THE ECONOMICS OF COASTAL MANAGEMENT

ALL RESIDENTIAL PROPERTIES   LAND USE CODE   1   JANUARY 1990 PRICES   SALT WATER FLOOD DURATION LESS THAN 12 HOURS

					DEPTH ABOVE UPPER SURFACE OF GROUND FLOOR										
COMPONENTS OF DAMAGE	-0.3	0.0	0.05	0.1	0.2	0.3	0.6	0.9	1.2	1.5	1.8	2.1	2.4	2.7	3.0 (M)
PATHS & PAVED AREAS	0	0	0	0	0	5	9	13	17	22	28	35	45	66	73 (£)
COL PERCENT	0.0	0.0	0.0	0.0	0.0	0.1	0.2	0.2	0.2	0.3	0.3	0.3	0.3	0.4	0.5
ROW PERCENT	0.0	0.0	0.0	0.6	1.0	8.1	13.2	18.5	23.8	31.2	39.5	48.8	62.0	91.3	100.0
GARDENS/FENCES/SHEDS	0	0	0	0	49	104	168	294	521	751	1021	1377	1857	2125	2381 (£)
COL PERCENT	0.0	0.0	0.0	0.0	1.8	2.6	3.1	4.5	6.9	8.7	10.1	11.7	14.0	14.2	14.7
ROW PERCENT	0.0	0.0	0.0	0.0	2.1	4.4	7.1	12.4	21.9	31.5	42.9	57.8	78.0	89.3	100.0
EXT. MAIN BUILDING	227	295	312	313	350	401	427	450	500	606	822	1196	1524	1753	2074 (£)
COL PERCENT	89.3	86.7	34.1	21.2	12.3	10.0	7.8	6.8	6.7	7.0	8.1	10.2	11.5	11.7	12.8
ROW PERCENT	11.0	14.3	15.1	15.1	16.9	19.3	20.6	21.7	24.1	29.2	39.7	57.7	73.5	84.5	100.0
PLASTERWORK	0	0	0	5	129	280	437	586	798	1007	1391	1672	1922	2184	2392 (£)
COL PERCENT	0.0	0.0	0.0	0.4	4.6	7.0	8.0	8.9	10.6	11.6	13.7	14.2	14.5	14.6	14.7
ROW PERCENT	0.0	0.0	0.0	0.2	5.4	11.7	18.3	24.5	33.4	42.1	58.1	69.9	80.4	91.3	100.0
FLOORS	0	0	41	114	218	345	430	525	591	692	804	968	1201	1734	2150 (£)
COL PERCENT	0.0	0.0	4.6	7.7	7.7	8.6	7.9	8.0	7.9	8.0	7.9	8.2	9.1	11.6	13.3
ROW PERCENT	0.0	0.0	1.9	5.3	10.2	16.0	20.0	24.4	27.5	32.2	37.4	45.0	55.9	80.7	100.0
JOINERY	0	0	0	11	67	125	195	260	350	706	991	1366	1492	1817	1891 (£)
COL PERCENT	0.0	0.0	0.0	0.8	2.4	3.1	3.6	3.9	4.7	8.2	9.8	11.6	11.3	12.2	11.7
ROW PERCENT	0.0	0.0	0.0	0.6	3.6	6.6	10.3	13.8	18.5	37.4	52.4	72.3	78.9	96.1	100.0
INTERNAL DECORATIONS	15	15	16	122	203	240	316	347	370	392	408	420	426	437	445 (£)
COL PERCENT	6.2	4.6	1.8	8.3	7.2	6.0	5.8	5.3	4.9	4.5	4.0	3.6	3.2	2.9	2.7
ROW PERCENT	3.5	3.5	3.7	27.5	45.7	54.0	71.0	78.0	83.2	88.1	91.8	94.3	95.7	98.3	100.0
PLUMBING & ELECTRICAL	11	29	59	112	128	141	171	201	213	233	382	410	435	444	454 (£)
COL PERCENT	4.5	8.7	6.5	7.6	4.5	3.5	3.1	3.1	2.8	2.7	3.8	3.5	3.3	3.0	2.8
ROW PERCENT	2.5	6.5	13.2	24.8	28.3	31.1	37.7	44.5	47.0	51.4	84.2	90.3	95.9	97.9	100.0
DOMESTIC APPLIANCES	0	0	0	49	129	262	350	560	563	599	617	617	617	617	617 (£)
COL PERCENT	0.0	0.0	0.1	3.3	4.6	6.5	6.4	8.5	7.5	6.9	6.1	5.3	4.7	4.1	3.8
ROW PERCENT	0.0	0.0	0.1	8.0	20.9	42.5	56.8	90.8	91.3	97.1	100.0	100.0	100.0	100.0	100.0
HEATING EQUIPMENT	0	0	0	15	27	52	95	103	111	113	113	113	113	113	113 (£)
COL PERCENT	0.0	0.0	0.1	1.0	1.0	1.3	1.7	1.6	1.5	1.3	1.1	1.0	0.9	0.8	0.7
ROW PERCENT	0.0	0.0	0.7	13.6	24.2	46.5	83.6	90.5	98.1	100.0	100.0	100.0	100.0	100.0	100.0
AUDIO/VIDEO	0	0	29	108	166	206	334	463	482	482	482	482	482	482	482 (£)
COL PERCENT	0.0	0.0	3.2	7.3	5.9	5.1	6.1	7.0	6.4	5.6	4.8	4.1	3.6	3.2	3.0
ROW PERCENT	0.0	0.0	6.1	22.4	34.5	42.7	69.4	96.0	100.0	100.0	100.0	100.0	100.0	100.0	100.0
FURNITURE	0	0	38	117	360	640	939	999	1056	1061	1061	1061	1061	1061	1061 (£)
COL PERCENT	0.0	0.0	4.2	7.9	12.7	15.9	17.2	15.1	14.1	12.3	10.5	9.0	8.0	7.1	6.5
ROW PERCENT	0.0	0.0	3.6	11.1	34.0	60.3	88.5	94.1	99.5	100.0	100.0	100.0	100.0	100.0	100.0
PERSONAL EFFECTS	0	0	5	42	87	236	462	598	675	695	695	695	695	695	695 (£)
COL PERCENT	0.0	0.0	0.6	2.8	3.1	5.9	8.4	9.1	9.0	8.0	6.9	5.9	5.3	4.7	4.3
ROW PERCENT	0.0	0.0	0.8	6.0	12.5	33.9	66.4	86.0	97.0	100.0	100.0	100.0	100.0	100.0	100.0
FLOOR COVER/CURTAINS	0	0	246	267	534	547	669	675	718	718	718	718	718	718	718 (£)
COL PERCENT	0.0	0.0	26.9	18.1	18.8	13.6	12.2	10.2	9.6	8.3	7.1	6.1	5.4	4.8	4.4
ROW PERCENT	0.0	0.0	34.3	37.3	74.4	76.1	93.1	94.0	100.0	100.0	100.0	100.0	100.0	100.0	100.0
GARDEN/DIY/LEISURE	0	0	24	50	74	110	134	173	173	189	191	191	198	198	198 (£)
COL PERCENT	0.0	0.0	2.6	3.4	2.6	2.8	2.5	2.6	2.3	2.2	1.9	1.6	1.5	1.3	1.2
ROW PERCENT	0.0	0.0	12.2	25.7	37.8	55.9	68.0	87.8	87.8	95.4	96.6	96.6	100.0	100.0	100.0
DOMESTIC CLEAN-UP	0	0	140	147	308	319	331	344	358	379	403	427	451	477	477 (£)
COL PERCENT	0.0	0.0	15.3	10.0	10.9	8.0	6.1	5.2	4.8	4.4	4.0	3.6	3.4	3.2	2.9
ROW PERCENT	0.0	0.0	29.5	30.9	64.5	67.0	69.5	72.0	75.0	79.5	84.5	89.5	94.5	100.0	100.0

ALL RESIDENTIAL PROPERTIES   LAND USE CODE   1   JANUARY 1990 PRICES   SALT WATER FLOOD DURATION LESS THAN 12 HOURS

					DEPTH ABOVE UPPER SURFACE OF GROUND FLOOR										
COMPONENTS OF DAMAGE	-0.3	0.0	0.05	0.1	0.2	0.3	0.6	0.9	1.2	1.5	1.8	2.1	2.4	2.7	3.0 (M)
TOTAL BUILDING FABRIC	254	341	430	681	1150	1640	2160	2680	3360	4410	5850	7450	8910	10560	11860 (£)
COL PERCENT	100.0	100.0	47.0	46.0	40.5	40.9	39.4	40.6	44.8	51.0	57.7	63.4	67.2	70.8	73.1
ROW PERCENT	2.1	2.9	3.6	5.7	9.7	13.9	18.2	22.6	28.4	37.2	49.3	62.8	75.1	89.1	100.0
TOTAL INVENTORY	0	0	486	799	1690	2380	3320	3920	4140	4240	4280	4310	4340	4370	4370 (£)
COL PERCENT	0.0	0.0	53.0	54.0	59.5	59.1	60.6	59.4	55.2	49.0	42.3	36.6	32.8	29.2	26.9
ROW PERCENT	0.0	0.0	11.1	18.3	38.7	54.4	76.0	89.8	94.8	97.1	98.1	98.7	99.4	100.0	100.0
TOTAL DAMAGE	254	341	917	1480	2840	4020	5470	6600	7500	8650	10140	11760	13250	14930	16230 (£)
ROW PERCENT	1.6	2.1	5.7	9.1	17.5	24.8	33.7	40.7	46.2	53.3	62.5	72.4	81.6	92.0	100.0
U.C. LIMIT	289	367	966	1550	3020	4290	5870	7080	8070	9300	10850	12570	14140	15920	17310 (£)
L.C. LIMIT	219	315	872	1410	2660	3760	5080	6130	6950	8020	9430	10950	12360	13940	15150 (£)
TOTAL DAMAGE/SQ.M	5.2	7.1	17.7	28.5	53.2	75.4	102.2	123.4	139.8	161.3	189.4	220.2	247.9	278.9	302.2 (£)
ROW PERCENT	1.7	2.3	5.9	9.4	17.6	25.0	33.8	40.8	46.3	53.4	62.7	72.9	82.0	92.3	100.0
U.C. LIMIT	6.0	7.7	18.3	29.3	54.7	77.9	106.1	128.5	145.8	168.3	196.9	228.7	256.8	288.3	312.0 (£)
L.C. LIMIT	4.4	6.5	17.1	27.7	51.7	72.9	98.3	118.3	133.8	154.3	181.9	211.7	239.0	269.5	292.4 (£)

APPENDIX 5.2   311

ALL RESIDENTIAL PROPERTIES   LAND USE CODE 1   JANUARY 1990 PRICES   SALT WATER FLOOD DURATION MORE THAN 12 HOURS

DEPTH ABOVE UPPER SURFACE OF GROUND FLOOR

COMPONENTS OF DAMAGE	-0.3	0.0	0.05	0.1	0.2	0.3	0.6	0.9	1.2	1.5	1.8	2.1	2.4	2.7	3.0	(M)
PATHS & PAVED AREAS	0	3	8	13	32	57	73	83	88	94	100	105	112	120	127	(£)
COL PERCENT	0.0	0.7	0.4	0.3	0.5	0.7	0.7	0.7	0.6	0.6	0.6	0.6	0.6	0.6	0.6	
ROW PERCENT	0.0	3.0	6.6	10.4	25.4	45.5	57.8	66.0	69.8	74.4	78.9	82.9	88.7	94.9	100.0	
GARDENS/FENCES/SHEDS	0	3	34	67	232	450	654	727	1137	1556	1830	2189	2577	2730	2812	(£)
COL PERCENT	0.0	0.6	1.5	1.8	3.9	5.6	6.4	6.2	8.3	10.2	11.1	12.2	13.4	13.5	13.0	
ROW PERCENT	0.0	0.1	1.2	2.4	8.3	16.0	23.3	25.9	40.4	55.4	65.1	77.8	91.6	97.1	100.0	
EXT. MAIN BUILDING	242	338	353	407	509	608	742	894	1177	1628	2236	2862	3644	4475	5671	(£)
COL PERCENT	78.0	62.9	15.9	10.8	8.6	7.6	7.3	7.6	8.6	10.6	13.6	15.9	19.0	22.1	26.3	
ROW PERCENT	4.3	6.0	6.2	7.2	9.0	10.7	13.1	15.8	20.8	28.7	39.4	50.5	64.3	78.9	100.0	
PLASTERWORK	17	46	87	274	473	666	1045	1439	1826	2106	2205	2529	2614	2660	2658	(£)
COL PERCENT	5.7	8.6	3.9	7.3	8.0	8.4	10.3	12.2	13.3	13.8	13.4	14.1	13.6	13.1	12.3	
ROW PERCENT	0.7	1.7	3.3	10.3	17.8	25.1	39.3	54.1	68.7	79.2	82.9	95.1	98.4	100.1	100.0	
FLOORS	0	75	586	951	1360	1509	1636	1659	2108	2335	2282	2281	2281	2281	2281	(£)
COL PERCENT	0.0	14.0	26.3	25.2	23.0	18.9	16.1	14.1	15.3	15.3	13.9	12.7	11.9	11.2	10.6	
ROW PERCENT	0.0	3.3	25.7	41.7	59.6	66.2	71.7	72.7	92.4	102.3	100.0	100.0	100.0	100.0	100.0	
JOINERY	0	0	110	352	693	1081	1425	1805	2097	2171	2294	2445	2434	2434	2434	(£)
COL PERCENT	0.0	0.0	5.0	9.4	11.7	13.6	14.0	15.3	15.2	14.2	14.0	13.6	12.7	12.0	11.3	
ROW PERCENT	0.0	0.0	4.6	14.5	28.5	44.4	58.5	74.2	86.1	89.2	94.2	100.4	100.0	100.0	100.0	
INTERNAL DECORATIONS	14	19	44	243	358	387	399	419	421	427	433	435	443	461	461	(£)
COL PERCENT	4.7	3.7	2.0	6.5	6.0	4.9	3.9	3.6	3.1	2.8	2.6	2.4	2.3	2.3	2.1	
ROW PERCENT	3.2	4.3	9.7	52.7	77.7	83.9	86.5	90.9	91.4	92.6	94.0	94.4	96.2	100.1	100.0	
PLUMBING & ELECTRICAL	35	51	74	133	203	232	398	434	457	471	495	512	522	531	543	(£)
COL PERCENT	11.5	9.6	3.4	3.5	3.4	2.9	3.9	3.7	3.3	3.1	3.0	2.9	2.7	2.6	2.5	
ROW PERCENT	6.6	9.5	13.8	24.6	37.4	42.8	73.5	80.0	84.2	86.8	91.3	94.3	96.2	97.9	100.0	
DOMESTIC APPLIANCES	0	0	2	78	183	386	441	628	628	634	634	634	634	634	634	(£)
COL PERCENT	0.0	0.0	0.1	2.1	3.1	4.8	4.3	5.3	4.6	4.1	3.9	3.5	3.3	3.1	2.9	
ROW PERCENT	0.0	0.0	0.4	12.3	29.0	61.0	69.6	99.1	99.1	100.0	100.0	100.0	100.0	100.0	100.0	
HEATING EQUIPMENT	0	0	0	24	36	89	115	132	146	146	146	150	150	150	150	(£)
COL PERCENT	0.0	0.0	0.0	0.6	0.6	1.1	1.1	1.1	1.1	1.0	0.9	0.8	0.8	0.7	0.7	
ROW PERCENT	0.0	0.0	0.6	16.2	24.2	59.8	76.8	88.3	97.1	97.1	97.1	100.0	100.0	100.0	100.0	
AUDIO/VIDEO	0	0	36	122	181	222	372	470	483	483	483	483	483	483	483	(£)
COL PERCENT	0.0	0.0	1.6	3.3	3.1	2.8	3.7	4.0	3.5	3.2	2.9	2.7	2.5	2.4	2.2	
ROW PERCENT	0.0	0.0	7.5	25.4	37.6	46.0	77.0	97.2	100.0	100.0	100.0	100.0	100.0	100.0	100.0	
FURNITURE	0	0	54	156	434	746	1060	1126	1157	1161	1161	1161	1161	1161	1161	(£)
COL PERCENT	0.0	0.0	2.4	4.2	7.3	9.3	10.4	9.5	8.4	7.6	7.1	6.5	6.0	5.7	5.4	
ROW PERCENT	0.0	0.0	4.7	13.5	37.4	64.2	91.3	97.0	99.7	100.0	100.0	100.0	100.0	100.0	100.0	
PERSONAL EFFECTS	0	0	10	62	119	390	540	689	701	711	711	711	711	711	711	(£)
COL PERCENT	0.0	0.0	0.5	1.7	2.0	4.9	5.3	5.8	5.1	4.6	4.3	4.0	3.7	3.5	3.3	
ROW PERCENT	0.0	0.0	1.5	8.7	16.8	54.9	76.0	96.9	98.7	100.0	100.0	100.0	100.0	100.0	100.0	
FLOOR COVER/CURTAINS	0	0	640	664	690	690	736	747	756	756	756	756	756	756	756	(£)
COL PERCENT	0.0	0.0	28.8	17.6	11.6	8.6	7.3	6.3	5.5	4.9	4.6	4.2	3.9	3.7	3.5	
ROW PERCENT	0.0	0.0	84.6	87.8	91.2	91.2	97.3	98.8	100.0	100.0	100.0	100.0	100.0	100.0	100.0	
GARDEN/DIY/LEISURE	0	0	33	64	94	128	169	191	198	208	214	214	220	220	220	(£)
COL PERCENT	0.0	0.0	1.5	1.7	1.6	1.6	1.7	1.6	1.4	1.4	1.3	1.2	1.1	1.1	1.0	
ROW PERCENT	0.0	0.0	15.1	29.3	43.0	58.1	76.7	86.6	89.7	94.5	97.0	97.0	100.0	100.0	100.0	
DOMESTIC CLEAN-UP	0	0	147	152	319	331	344	358	379	403	451	477	477	477	477	(£)
COL PERCENT	0.0	0.0	6.6	4.0	5.4	4.2	3.4	3.0	2.8	2.6	2.7	2.7	2.5	2.4	2.2	
ROW PERCENT	0.0	0.0	30.9	31.9	67.0	69.5	72.0	75.0	79.5	84.5	94.5	100.0	100.0	100.0	100.0	

ALL RESIDENTIAL PROPERTIES   LAND USE CODE 1   JANUARY 1990 PRICES   SALT WATER FLOOD DURATION MORE THAN 12 HOURS

DEPTH ABOVE UPPER SURFACE OF GROUND FLOOR

COMPONENTS OF DAMAGE	-0.3	0.0	0.05	0.1	0.2	0.3	0.6	0.9	1.2	1.5	1.8	2.1	2.4	2.7	3.0	(M)
TOTAL BUILDING FABRIC	310	537	1300	2440	3860	4990	6380	7460	9320	10790	11880	13360	14630	15700	16990	(£)
COL PERCENT	100.0	100.0	58.4	64.8	65.2	62.6	62.8	63.2	67.7	70.5	72.3	74.4	76.1	77.4	78.7	
ROW PERCENT	1.8	3.2	7.7	14.4	22.7	29.4	37.5	43.9	54.8	63.5	69.9	78.6	86.1	92.4	100.0	
TOTAL INVENTORY	0	0	926	1330	2060	2990	3780	4340	4450	4510	4560	4590	4600	4600	4600	(£)
COL PERCENT	0.0	0.0	41.6	35.2	34.8	37.4	37.2	36.8	32.3	29.5	27.7	25.6	23.9	22.6	21.3	
ROW PERCENT	0.0	0.0	20.1	28.9	44.8	65.0	82.2	94.5	96.9	98.0	99.2	99.9	100.0	100.0	100.0	
TOTAL DAMAGE	310	537	2230	3770	5930	7980	10160	11810	13770	15300	16440	17950	19230	20290	21590	(£)
ROW PERCENT	1.4	2.5	10.3	17.5	27.5	37.0	47.0	54.7	63.8	70.9	76.1	83.2	89.1	94.0	100.0	
U.C. LIMIT	351	578	2430	4060	6380	8580	10900	12670	14760	16380	17580	19190	20580	21730	23100	(£)
L.C. LIMIT	269	496	2030	3480	5480	7390	9410	10960	12780	14220	15300	16720	17880	18860	20080	(£)
TOTAL DAMAGE/SQ.M	6.3	10.8	41.2	69.8	109.4	148.1	188.9	220.2	255.9	284.3	305.8	333.7	356.4	375.8	400.2	(£)
ROW PERCENT	1.6	2.7	10.3	17.4	27.3	37.0	47.2	55.0	63.9	71.0	76.4	83.4	89.1	93.9	100.0	
U.C. LIMIT	7.2	11.6	42.9	72.3	113.5	154.2	196.6	229.3	265.7	294.6	316.6	345.2	368.4	388.5	414.0	(£)
L.C. LIMIT	5.4	10.0	39.5	67.3	105.3	142.0	181.2	211.1	246.1	274.0	295.0	322.2	344.4	363.1	386.4	(£)

312 THE ECONOMICS OF COASTAL MANAGEMENT

```
DETACHED                    LAND USE CODE 11     JANUARY 1990 PRICES  SALT WATER FLOOD DURATION LESS THAN 12 HOURS
                                              DEPTH ABOVE UPPER SURFACE OF GROUND FLOOR
COMPONENTS OF DAMAGE   -0.3    0.0   0.05    0.1    0.2    0.3    0.6    0.9    1.2    1.5    1.8    2.1    2.4    2.7    3.0  (M)
------------------------------------------------------------------------------------------------------------------------------
 PATHS & PAVED AREAS      0      0      0      0      1      6     16     26     33     45     57     72     89    170    150  (£)
      COL PERCENT       0.0    0.0    0.0    0.0    0.0    0.1    0.2    0.3    0.3    0.4    0.4    0.4    0.5    0.8    0.6
      ROW PERCENT       0.0    0.0    0.0    0.4    0.7    4.4   11.0   17.9   22.4   30.1   38.3   48.0   59.5  113.1  100.0
 GARDENS/FENCES/SHEDS     0      0      0      0     66    144    237    534   1068   1521   2071   2719   3376   3874   4358 (£)
      COL PERCENT       0.0    0.0    0.0    0.1    1.7    2.6    3.2    5.8   10.0   12.1   14.0   15.6   17.3   17.4   17.8
      ROW PERCENT       0.0    0.0    0.0    0.0    1.5    3.3    5.4   12.3   24.5   34.9   47.5   62.4   77.5   88.9  100.0
 EXT. MAIN BUILDING     143    228    246    248    305    365    396    432    504    660   1004   1592   2093   2550   3073 (£)
      COL PERCENT      88.0   87.2   21.0   13.8    7.9    6.5    5.3    4.7    4.7    5.3    6.8    9.2   10.7   11.5   12.6
      ROW PERCENT       4.7    7.4    8.0    8.1    9.9   11.9   12.9   14.1   16.4   21.5   32.7   51.8   68.1   83.0  100.0
 PLASTERWORK              0      0      0      0    237    465    679    906   1160   1447   1856   2182   2529   2972   3477 (£)
      COL PERCENT       0.0    0.0    0.0    0.0    6.1    8.3    9.1    9.8   10.8   11.5   12.5   12.6   12.9   13.4   14.2
      ROW PERCENT       0.0    0.0    0.0    0.0    6.8   13.4   19.5   26.1   33.4   41.6   53.4   62.7   72.7   85.5  100.0
 FLOORS                   0      0     57    144    299    487    621    788    914   1093   1294   1579   1983   2771   3367 (£)
      COL PERCENT       0.0    0.0    4.9    8.0    7.7    8.7    8.3    8.5    8.5    8.7    8.7    9.1   10.2   12.5   13.8
      ROW PERCENT       0.0    0.0    1.7    4.3    8.9   14.5   18.5   23.4   27.2   32.5   38.4   46.9   58.9   82.3  100.0
 JOINERY                  0      0      0     32    120    199    288    372    504   1066   1531   2168   2328   2662   2780 (£)
      COL PERCENT       0.0    0.0    0.0    1.8    3.1    3.6    3.9    4.0    4.7    8.5   10.3   12.5   11.9   12.0   11.4
      ROW PERCENT       0.0    0.0    0.0    1.2    4.3    7.2   10.4   13.4   18.1   38.4   55.1   78.0   83.8   95.8  100.0
 INTERNAL DECORATIONS    14     14     15    105    276    409    527    557    602    639    662    679    688    708    717 (£)
      COL PERCENT       8.8    5.5    1.3    5.9    7.1    7.3    7.1    6.0    5.6    5.1    4.5    3.9    3.5    3.2    2.9
      ROW PERCENT       2.0    2.0    2.1   14.7   38.5   57.2   73.5   77.7   84.1   89.2   92.4   94.8   96.0   98.8  100.0
 PLUMBING & ELECTRICAL    5     19     51    110    133    152    191    231    246    271    457    472    488    500    514 (£)
      COL PERCENT       3.2    7.3    4.4    6.2    3.4    2.7    2.6    2.5    2.3    2.2    3.1    2.7    2.5    2.3    2.1
      ROW PERCENT       1.0    3.7   10.1   21.5   25.9   29.6   37.3   45.1   47.9   52.8   89.0   92.0   95.0   97.4  100.0

 DOMESTIC APPLIANCES      0      0      0     57    164    342    471    784    793    846    876    876    876    876    876 (£)
      COL PERCENT       0.0    0.0    0.1    3.2    4.2    6.1    6.3    8.5    7.4    6.7    5.9    5.0    4.5    3.9    3.6
      ROW PERCENT       0.0    0.0    0.1    6.5   18.8   39.1   53.8   89.5   90.5   96.6  100.0  100.0  100.0  100.0  100.0
 HEATING EQUIPMENT        0      0      0     11     33     67    134    144    156    159    159    159    159    159    159 (£)
      COL PERCENT       0.0    0.0    0.0    0.7    0.9    1.2    1.8    1.6    1.5    1.3    1.1    0.9    0.8    0.7    0.7
      ROW PERCENT       0.0    0.0    0.0    7.5   21.3   42.5   84.4   91.0   98.1  100.0  100.0  100.0  100.0  100.0  100.0
 AUDIO/VIDEO              0      0     45    162    247    308    483    707    742    742    742    742    742    742    742 (£)
      COL PERCENT       0.0    0.0    3.9    9.0    6.4    5.5    6.5    7.7    6.9    5.9    5.0    4.3    3.8    3.3    3.0
      ROW PERCENT       0.0    0.0    6.1   21.8   33.3   41.5   65.1   95.3  100.0  100.0  100.0  100.0  100.0  100.0  100.0
 FURNITURE                0      0     49    124    336    741   1046   1132   1164   1169   1169   1169   1169   1169   1169 (£)
      COL PERCENT       0.0    0.0    4.2    6.9    8.7   13.2   14.0   12.3   10.9    9.3    7.9    6.7    6.0    5.3    4.8
      ROW PERCENT       0.0    0.0    4.2   10.6   28.8   63.4   89.5   96.8   99.6  100.0  100.0  100.0  100.0  100.0  100.0
 PERSONAL EFFECTS         0      0      0      0     34    178    320    460    501    531    531    531    531    531    531 (£)
      COL PERCENT       0.0    0.0    0.0    0.0    0.9    3.2    4.3    5.0    4.7    4.2    3.6    3.1    2.7    2.4    2.2
      ROW PERCENT       0.0    0.0    0.0    0.0    6.4   33.5   60.3   86.6   94.3  100.0  100.0  100.0  100.0  100.0  100.0
 FLOOR COVER/CURTAINS     0      0    455    504   1051   1098   1369   1375   1515   1515   1515   1515   1515   1515   1515 (£)
      COL PERCENT       0.0    0.0   38.8   28.1   27.1   19.6   18.3   14.9   14.2   12.1   10.2    8.7    7.8    6.8    6.2
      ROW PERCENT       0.0    0.0   30.0   33.3   69.4   72.5   90.4   90.8  100.0  100.0  100.0  100.0  100.0  100.0  100.0
 GARDEN/DIY/LEISURE       0      0     28     61     89    135    168    225    225    245    248    248    258    258    258 (£)
      COL PERCENT       0.0    0.0    2.4    3.4    2.3    2.4    2.3    2.4    2.1    2.0    1.7    1.4    1.3    1.2    1.1
      ROW PERCENT       0.0    0.0   11.1   23.7   34.8   52.6   65.3   87.2   87.2   95.2   96.5   96.5  100.0  100.0  100.0
 DOMESTIC CLEAN-UP        0      0    222    233    486    505    523    543    565    599    637    674    712    754    754 (£)
      COL PERCENT       0.0    0.0   19.0   13.0   12.5    9.0    7.0    5.9    5.3    4.8    4.3    3.9    3.6    3.4    3.1
      ROW PERCENT       0.0    0.0   29.5   30.9   64.5   67.0   69.5   72.0   75.0   79.5   84.5   89.5   94.5  100.0  100.0


DETACHED                    LAND USE CODE 11     JANUARY 1990 PRICES  SALT WATER FLOOD DURATION LESS THAN 12 HOURS
                                              DEPTH ABOVE UPPER SURFACE OF GROUND FLOOR
COMPONENTS OF DAMAGE   -0.3    0.0   0.05    0.1    0.2    0.3    0.6    0.9    1.2    1.5    1.8    2.1    2.4    2.7    3.0  (M)
------------------------------------------------------------------------------------------------------------------------------

TOTAL BUILDING FABRIC  163    261    370    642   1440   2230   2960   3850   5030   6750   8940  11470  13580  16210  18440 (£)
      COL PERCENT    100.0  100.0   31.6   35.8   37.0   39.8   39.6   41.7   47.1   53.7   60.3   66.0   69.5   73.0   75.4
      ROW PERCENT       0.9    1.4    2.0    3.5    7.8   12.1   16.0   20.9   27.3   36.6   48.5   62.2   73.6   87.9  100.0
TOTAL INVENTORY          0      0    801   1150   2440   3380   4520   5370   5660   5810   5880   5920   5970   6010   6010 (£)
      COL PERCENT       0.0    0.0   68.4   64.2   63.0   60.2   60.4   58.3   52.9   46.3   39.7   34.0   30.5   27.0   24.6
      ROW PERCENT       0.0    0.0   13.3   19.2   40.7   56.2   75.2   89.5   94.3   96.7   97.9   98.5   99.3  100.0  100.0

TOTAL DAMAGE           163    261   1170   1800   3880   5610   7480   9220  10700  12560  14820  17390  19540  22220  24440 (£)
      ROW PERCENT       0.7    1.1    4.8    7.4   15.9   22.9   30.6   37.7   43.8   51.4   60.6   71.1   79.9   90.9  100.0
      U.C. LIMIT       239    322   1360   1990   4300   6250   8300  10220  11860  13800  16210  18960  21390  24530  27270 (£)
      L.C. LIMIT        87    200    986   1610   3470   4980   6660   8240   9540  11310  13430  15820  17710  19910  21630 (£)
TOTAL DAMAGE/SQ.M      1.7    3.0   14.0   22.7   49.1   71.1   95.3  118.2  137.2  162.4  191.3  224.7  250.4  280.7  305.1 (£)
      ROW PERCENT       0.6    1.0    4.6    7.4   16.1   23.3   31.2   38.7   45.0   53.2   62.7   73.6   82.1   92.0  100.0
      U.C. LIMIT       2.5    3.6   14.7   24.6   53.2   77.7  105.0  130.7  152.1  180.6  211.8  247.6  273.8  303.6  327.1 (£)
      L.C. LIMIT       0.9    2.4   13.3   20.8   45.0   64.5   85.6  105.7  122.3  144.2  171.4  201.8  227.0  257.8  283.1 (£)
```

APPENDIX 5.2   313

SEMI DETACHED   LAND USE CODE 12   JANUARY 1990 PRICES   SALT WATER FLOOD DURATION LESS THAN 12 HOURS

                                         DEPTH ABOVE UPPER SURFACE OF GROUND FLOOR
COMPONENTS OF DAMAGE -0.3   0.0   0.05   0.1   0.2   0.3   0.6   0.9   1.2   1.5   1.8   2.1   2.4   2.7   3.0 (M)

COMPONENTS OF DAMAGE	-0.3	0.0	0.05	0.1	0.2	0.3	0.6	0.9	1.2	1.5	1.8	2.1	2.4	2.7	3.0	
PATHS & PAVED AREAS	0	0	0	0	0	2	4	7	9	11	15	18	24	31	39	(£)
COL PERCENT	0.0	0.0	0.0	0.0	0.0	0.1	0.1	0.1	0.1	0.2	0.2	0.2	0.2	0.2	0.3	
ROW PERCENT	0.0	0.0	0.0	0.2	0.5	5.3	12.1	20.0	24.0	29.2	37.6	47.3	60.6	79.0	100.0	
GARDENS/FENCES/SHEDS	0	0	0	0	48	102	173	308	555	803	1082	1514	2148	2465	2769	(£)
COL PERCENT	0.0	0.0	0.0	0.1	1.9	2.8	3.6	5.3	8.3	10.5	11.9	14.3	17.6	17.6	18.0	
ROW PERCENT	0.0	0.0	0.0	0.0	1.8	3.7	6.3	11.2	20.1	29.0	39.1	54.7	77.6	89.0	100.0	
EXT. MAIN BUILDING	353	370	393	394	410	465	491	516	557	666	882	1259	1601	1811	2142	(£)
COL PERCENT	87.6	84.8	43.3	27.0	16.0	12.8	10.3	8.8	8.4	8.7	9.7	11.9	13.1	12.9	13.9	
ROW PERCENT	16.5	17.3	18.3	18.4	19.2	21.7	22.9	24.1	26.0	31.1	41.2	58.8	74.8	84.6	100.0	
PLASTERWORK	0	0	0	3	100	220	347	459	671	871	1310	1526	1752	1979	2131	(£)
COL PERCENT	0.0	0.0	0.0	0.2	3.9	6.1	7.3	7.9	10.1	11.4	14.4	14.4	14.3	14.1	13.8	
ROW PERCENT	0.0	0.0	0.0	0.2	4.7	10.3	16.3	21.5	31.5	40.9	61.5	71.6	82.2	92.8	100.0	
FLOORS	0	0	24	107	194	309	397	480	524	583	664	781	960	1557	2052	(£)
COL PERCENT	0.0	0.0	2.7	7.4	7.6	8.5	8.3	8.2	7.9	7.6	7.3	7.4	7.9	11.1	13.3	
ROW PERCENT	0.0	0.0	1.2	5.2	9.5	15.1	19.4	23.4	25.5	28.4	32.4	38.1	46.8	75.9	100.0	
JOINERY	0	0	0	10	101	202	303	406	545	790	1027	1291	1458	1859	1938	(£)
COL PERCENT	0.0	0.0	0.0	0.7	4.0	5.6	6.4	7.0	8.2	10.3	11.3	12.2	11.9	13.2	12.6	
ROW PERCENT	0.0	0.0	0.0	0.6	5.3	10.4	15.7	21.0	28.1	40.8	53.0	66.6	75.2	95.9	100.0	
INTERNAL DECORATIONS	29	29	30	188	231	247	278	297	312	329	344	350	359	371	382	(£)
COL PERCENT	7.4	6.8	3.3	12.9	9.1	6.8	5.8	5.1	4.7	4.3	3.8	3.3	2.9	2.6	2.5	
ROW PERCENT	7.8	7.8	7.9	49.2	60.6	64.8	72.9	77.7	81.8	86.3	90.0	91.8	94.0	97.3	100.0	
PLUMBING & ELECTRICAL	19	36	58	97	110	118	134	169	186	210	321	379	424	433	443	(£)
COL PERCENT	5.0	8.4	6.4	6.7	4.3	3.3	2.8	2.9	2.8	2.7	3.5	3.6	3.5	3.1	2.9	
ROW PERCENT	4.5	8.3	13.2	22.0	25.0	26.8	30.4	38.3	42.1	47.6	72.7	85.7	95.8	97.7	100.0	
DOMESTIC APPLIANCES	0	0	0	52	136	279	362	586	589	627	646	646	646	646	646	(£)
COL PERCENT	0.0	0.0	0.1	3.6	5.3	7.7	7.6	10.0	8.8	8.2	7.1	6.1	5.3	4.6	4.2	
ROW PERCENT	0.0	0.0	0.1	8.1	21.1	43.2	56.1	90.7	91.1	96.9	100.0	100.0	100.0	100.0	100.0	
HEATING EQUIPMENT	0	0	0	14	27	49	91	98	105	107	107	107	107	107	107	(£)
COL PERCENT	0.0	0.0	0.0	1.0	1.1	1.4	1.9	1.7	1.6	1.4	1.2	1.0	0.9	0.8	0.7	
ROW PERCENT	0.0	0.0	0.0	13.9	25.3	46.3	85.7	91.9	98.2	100.0	100.0	100.0	100.0	100.0	100.0	
AUDIO/VIDEO	0	0	26	107	167	205	337	459	479	479	479	479	479	479	479	(£)
COL PERCENT	0.0	0.0	2.9	7.4	6.5	5.7	7.1	7.9	7.2	6.2	5.3	4.5	3.9	3.4	3.1	
ROW PERCENT	0.0	0.0	5.4	22.5	34.9	42.9	70.4	95.9	100.0	100.0	100.0	100.0	100.0	100.0	100.0	
FURNITURE	0	0	32	91	229	499	709	775	796	802	802	802	802	802	802	(£)
COL PERCENT	0.0	0.0	3.6	6.3	9.0	13.8	14.9	13.3	11.9	10.5	8.8	7.6	6.6	5.7	5.2	
ROW PERCENT	0.0	0.0	4.1	11.4	28.6	62.3	88.4	96.7	99.2	100.0	100.0	100.0	100.0	100.0	100.0	
PERSONAL EFFECTS	0	0	0	0	25	99	174	256	277	300	300	300	300	300	300	(£)
COL PERCENT	0.0	0.0	0.0	0.0	1.0	2.7	3.7	4.4	4.2	3.9	3.3	2.8	2.5	2.1	1.9	
ROW PERCENT	0.0	0.0	0.0	0.0	8.4	33.0	58.1	85.5	92.4	100.0	100.0	100.0	100.0	100.0	100.0	
FLOOR COVER/CURTAINS	0	0	196	216	435	442	547	551	576	576	576	576	576	576	576	(£)
COL PERCENT	0.0	0.0	21.6	14.8	17.0	12.2	11.5	9.4	8.6	7.5	6.3	5.4	4.7	4.1	3.7	
ROW PERCENT	0.0	0.0	34.0	37.5	75.5	76.8	95.0	95.7	100.0	100.0	100.0	100.0	100.0	100.0	100.0	
GARDEN/DIY/LEISURE	0	0	21	46	68	103	127	167	167	183	185	185	192	192	192	(£)
COL PERCENT	0.0	0.0	2.4	3.2	2.7	2.8	2.7	2.9	2.5	2.4	2.0	1.8	1.6	1.4	1.2	
ROW PERCENT	0.0	0.0	11.1	24.0	35.6	53.6	66.1	87.2	87.2	95.3	96.4	96.4	100.0	100.0	100.0	
DOMESTIC CLEAN-UP	0	0	123	128	269	279	289	300	313	332	352	373	394	417	417	(£)
COL PERCENT	0.0	0.0	13.6	8.8	10.5	7.7	6.1	5.1	4.7	4.3	3.9	3.5	3.2	3.0	2.7	
ROW PERCENT	0.0	0.0	29.5	30.9	64.5	67.0	69.4	72.1	75.0	79.6	84.5	89.6	94.5	100.0	100.0	

SEMI DETACHED   LAND USE CODE 12   JANUARY 1990 PRICES   SALT WATER FLOOD DURATION LESS THAN 12 HOURS

                                         DEPTH ABOVE UPPER SURFACE OF GROUND FLOOR
COMPONENTS OF DAMAGE -0.3   0.0   0.05   0.1   0.2   0.3   0.6   0.9   1.2   1.5   1.8   2.1   2.4   2.7   3.0 (M)

COMPONENTS OF DAMAGE	-0.3	0.0	0.05	0.1	0.2	0.3	0.6	0.9	1.2	1.5	1.8	2.1	2.4	2.7	3.0	
TOTAL BUILDING FABRIC	403	437	506	802	1200	1670	2130	2650	3360	4270	5650	7120	8730	10510	11900	(£)
COL PERCENT	100.0	100.0	55.8	54.9	46.9	46.0	44.7	45.3	50.4	55.6	62.1	67.2	71.4	74.9	77.2	
ROW PERCENT	3.4	3.7	4.3	6.7	10.1	14.0	17.9	22.2	28.3	35.9	47.5	59.9	73.4	88.3	100.0	
TOTAL INVENTORY	0	0	400	658	1360	1960	2640	3200	3310	3410	3450	3470	3500	3520	3520	(£)
COL PERCENT	0.0	0.0	44.2	45.1	53.1	54.0	55.3	54.7	49.6	44.4	37.9	32.8	28.6	25.1	22.8	
ROW PERCENT	0.0	0.0	11.4	18.7	38.6	55.6	75.0	90.8	93.8	96.8	98.0	98.6	99.3	100.0	100.0	
TOTAL DAMAGE	403	437	906	1460	2560	3630	4770	5840	6670	7680	9100	10600	12230	14030	15420	(£)
ROW PERCENT	2.6	2.8	5.9	9.5	16.6	23.5	31.0	37.9	43.2	49.8	59.0	68.7	79.3	91.0	100.0	
U.C. LIMIT	474	497	969	1550	2710	3870	5140	6330	7250	8300	9770	11330	12990	14880	16310	(£)
L.C. LIMIT	332	377	847	1380	2410	3390	4410	5350	6090	7060	8440	9860	11470	13200	14540	(£)
TOTAL DAMAGE/SQ.M	8.3	9.2	19.3	31.0	54.6	77.7	102.4	125.4	143.1	165.0	195.6	228.1	262.8	300.7	330.0	(£)
ROW PERCENT	2.5	2.8	5.8	9.4	16.5	23.5	31.0	38.0	43.4	50.0	59.3	69.1	79.6	91.1	100.0	
U.C. LIMIT	9.9	10.5	20.5	32.4	57.7	83.2	111.1	137.1	156.9	180.4	212.4	247.7	282.9	321.5	351.0	(£)
L.C. LIMIT	6.7	7.9	18.1	29.6	51.5	72.2	93.7	113.7	129.3	149.6	178.8	208.5	242.7	279.9	309.0	(£)

314 THE ECONOMICS OF COASTAL MANAGEMENT

TERRACE   LAND USE CODE 13   JANUARY 1990 PRICES   SALT WATER FLOOD DURATION LESS THAN 12 HOURS
                                                        DEPTH ABOVE UPPER SURFACE OF GROUND FLOOR
COMPONENTS OF DAMAGE   -0.3   0.0   0.05   0.1   0.2   0.3   0.6   0.9   1.2   1.5   1.8   2.1   2.4   2.7   3.0 (M)

COMPONENTS OF DAMAGE	-0.3	0.0	0.05	0.1	0.2	0.3	0.6	0.9	1.2	1.5	1.8	2.1	2.4	2.7	3.0	
PATHS & PAVED AREAS	0	0	0	0	0	9	11	13	19	25	32	38	50	67	80	(£)
COL PERCENT	0.0	0.0	0.0	0.1	0.0	0.3	0.3	0.3	0.4	0.4	0.4	0.4	0.5	0.6	0.7	
ROW PERCENT	0.0	0.0	0.0	0.9	1.0	11.7	14.7	17.3	24.2	31.7	40.3	47.8	62.9	83.7	100.0	
GARDENS/FENCES/SHEDS	0	0	0	0	32	68	104	159	272	410	580	806	1202	1376	1540	(£)
COL PERCENT	0.0	0.0	0.0	0.1	1.6	2.3	2.6	3.4	5.1	6.6	7.8	9.2	12.2	12.4	12.8	
ROW PERCENT	0.0	0.0	0.0	0.0	2.1	4.4	6.8	10.4	17.7	26.6	37.7	52.4	78.0	89.4	100.0	
EXT. MAIN BUILDING	257	323	337	339	367	414	440	456	499	569	694	962	1169	1251	1445	(£)
COL PERCENT	92.6	88.3	42.8	27.6	17.4	13.9	11.1	9.6	9.4	9.2	9.4	11.0	11.9	11.3	12.0	
ROW PERCENT	17.8	22.4	23.4	23.5	25.4	28.7	30.5	31.6	34.6	39.4	48.1	66.6	80.9	86.6	100.0	
PLASTERWORK	0	0	0	13	111	242	396	543	747	941	1329	1726	1906	2093	2192	(£)
COL PERCENT	0.0	0.0	0.0	1.1	5.3	8.1	10.0	11.4	14.0	15.2	17.9	19.8	19.4	18.9	18.2	
ROW PERCENT	0.0	0.0	0.0	0.6	5.1	11.1	18.1	24.8	34.1	42.9	60.7	78.7	86.9	95.5	100.0	
FLOORS	0	0	35	89	167	256	310	366	402	456	520	617	759	1194	1557	(£)
COL PERCENT	0.0	0.0	4.5	7.3	7.9	8.6	7.8	7.7	7.6	7.4	7.0	7.1	7.7	10.8	13.0	
ROW PERCENT	0.0	0.0	2.3	5.8	10.8	16.4	19.9	23.5	25.8	29.3	33.4	39.6	48.7	76.6	100.0	
JOINERY	0	0	0	0	10	23	75	116	184	485	727	1001	1123	1456	1534	(£)
COL PERCENT	0.0	0.0	0.0	0.0	0.5	0.8	1.9	2.5	3.5	7.8	9.8	11.5	11.4	13.1	12.8	
ROW PERCENT	0.0	0.0	0.0	0.0	0.7	1.5	4.9	7.6	12.0	31.7	47.4	65.3	73.2	94.9	100.0	
INTERNAL DECORATIONS	14	14	15	91	131	163	234	263	279	297	308	315	321	334	342	(£)
COL PERCENT	5.2	4.0	2.0	7.5	6.3	5.5	5.9	5.5	5.2	4.8	4.2	3.6	3.3	3.0	2.9	
ROW PERCENT	4.2	4.2	4.6	26.7	38.5	47.7	68.3	76.8	81.5	86.6	90.0	92.1	93.8	97.6	100.0	
PLUMBING & ELECTRICAL	6	28	62	122	139	150	185	200	202	215	407	419	433	435	437	(£)
COL PERCENT	2.2	7.7	7.9	10.0	6.6	5.0	4.7	4.2	3.8	3.5	5.5	4.8	4.4	3.9	3.6	
ROW PERCENT	1.4	6.5	14.3	28.1	31.9	34.5	42.3	45.8	46.3	49.2	93.0	95.8	99.1	99.5	100.0	
DOMESTIC APPLIANCES	0	0	0	41	102	205	268	424	426	453	465	465	465	465	465	(£)
COL PERCENT	0.0	0.0	0.1	3.3	4.9	6.9	6.7	8.9	8.0	7.3	6.3	5.3	4.7	4.2	3.9	
ROW PERCENT	0.0	0.0	0.1	8.8	22.1	44.1	57.6	91.2	91.5	97.3	100.0	100.0	100.0	100.0	100.0	
HEATING EQUIPMENT	0	0	0	15	25	46	83	89	96	98	98	98	98	98	98	(£)
COL PERCENT	0.0	0.0	0.0	1.3	1.2	1.6	2.1	1.9	1.8	1.6	1.3	1.1	1.0	0.9	0.8	
ROW PERCENT	0.0	0.0	0.0	16.2	25.9	47.6	84.6	90.8	98.2	100.0	100.0	100.0	100.0	100.0	100.0	
AUDIO/VIDEO	0	0	23	92	143	175	291	393	409	409	409	409	409	409	409	(£)
COL PERCENT	0.0	0.0	3.0	7.6	6.8	5.9	7.3	8.3	7.7	6.6	5.5	4.7	4.2	3.7	3.4	
ROW PERCENT	0.0	0.0	5.7	22.7	35.2	43.0	71.3	96.0	100.0	100.0	100.0	100.0	100.0	100.0	100.0	
FURNITURE	0	0	31	94	228	491	665	711	729	733	733	733	733	733	733	(£)
COL PERCENT	0.0	0.0	4.0	7.7	10.8	16.4	16.8	15.0	13.7	11.8	9.9	8.4	7.5	6.6	6.1	
ROW PERCENT	0.0	0.0	4.4	12.9	31.1	67.1	90.8	97.0	99.4	100.0	100.0	100.0	100.0	100.0	100.0	
PERSONAL EFFECTS	0	0	0	0	17	71	126	185	201	216	216	216	216	216	216	(£)
COL PERCENT	0.0	0.0	0.0	0.0	0.8	2.4	3.2	3.9	3.8	3.5	2.9	2.5	2.2	2.0	1.8	
ROW PERCENT	0.0	0.0	0.0	0.0	8.2	33.0	58.4	85.6	92.7	100.0	100.0	100.0	100.0	100.0	100.0	
FLOOR COVER/CURTAINS	0	0	163	179	350	354	433	437	454	454	454	454	454	454	454	(£)
COL PERCENT	0.0	0.0	20.7	14.6	16.6	11.9	10.9	9.2	8.5	7.3	6.1	5.2	4.6	4.1	3.8	
ROW PERCENT	0.0	0.0	36.0	39.5	77.1	78.1	95.5	96.4	100.0	100.0	100.0	100.0	100.0	100.0	100.0	
GARDEN/DIY/LEISURE	0	0	18	40	61	91	112	146	146	160	162	162	168	168	168	(£)
COL PERCENT	0.0	0.0	2.4	3.3	2.9	3.1	2.8	3.1	2.8	2.6	2.2	1.9	1.7	1.5	1.4	
ROW PERCENT	0.0	0.0	11.2	24.3	36.2	54.2	66.6	87.1	87.1	95.2	96.3	96.3	100.0	100.0	100.0	
DOMESTIC CLEAN-UP	0	0	99	104	217	225	233	242	252	267	284	301	318	336	336	(£)
COL PERCENT	0.0	0.0	12.6	8.5	10.3	7.5	5.9	5.1	4.7	4.3	3.8	3.5	3.2	3.0	2.8	
ROW PERCENT	0.0	0.0	29.5	31.0	64.5	66.9	69.5	72.1	75.0	79.4	84.5	89.5	94.4	100.0	100.0	

TERRACE   LAND USE CODE 13   JANUARY 1990 PRICES   SALT WATER FLOOD DURATION LESS THAN 12 HOURS
                                                        DEPTH ABOVE UPPER SURFACE OF GROUND FLOOR
COMPONENTS OF DAMAGE   -0.3   0.0   0.05   0.1   0.2   0.3   0.6   0.9   1.2   1.5   1.8   2.1   2.4   2.7   3.0 (M)

COMPONENTS OF DAMAGE	-0.3	0.0	0.05	0.1	0.2	0.3	0.6	0.9	1.2	1.5	1.8	2.1	2.4	2.7	3.0	
TOTAL BUILDING FABRIC	277	366	451	658	961	1330	1760	2120	2610	3400	4600	5890	6970	8210	9130	(£)
COL PERCENT	100.0	100.0	57.2	53.6	45.6	44.4	44.3	44.6	49.0	54.9	62.0	67.4	70.9	74.0	76.0	
ROW PERCENT	3.0	4.0	4.9	7.2	10.5	14.6	19.3	23.2	28.6	37.2	50.4	64.5	76.3	89.9	100.0	
TOTAL INVENTORY	0	0	337	569	1150	1660	2220	2630	2720	2790	2820	2840	2860	2880	2880	(£)
COL PERCENT	0.0	0.0	42.8	46.4	54.4	55.6	55.7	55.4	51.0	45.1	38.0	32.6	29.1	26.0	24.0	
ROW PERCENT	0.0	0.0	11.7	19.7	39.8	57.7	76.8	91.3	94.2	96.9	98.0	98.6	99.3	100.0	100.0	
TOTAL DAMAGE	277	366	789	1230	2110	2990	3970	4750	5320	6190	7430	8730	9830	11090	12020	(£)
ROW PERCENT	2.3	3.1	6.6	10.2	17.5	24.9	33.1	39.6	44.3	51.6	61.8	72.7	81.8	92.3	100.0	
U.C. LIMIT	362	427	859	1340	2300	3280	4400	5300	5960	6960	8240	9640	10840	12250	13280	(£)
L.C. LIMIT	192	305	723	1120	1930	2710	3560	4210	4700	5440	6620	7830	8830	9950	10760	(£)
TOTAL DAMAGE/SQ.M	6.5	9.0	19.6	30.3	52.0	74.0	98.4	117.6	131.7	153.3	184.0	216.6	243.4	273.8	295.8	(£)
ROW PERCENT	2.2	3.0	6.6	10.2	17.6	25.0	33.3	39.8	44.5	51.8	62.2	73.2	82.3	92.6	100.0	
U.C. LIMIT	8.5	10.4	20.9	31.8	55.2	79.7	107.6	129.9	145.8	170.7	202.3	237.5	265.9	298.4	321.9	(£)
L.C. LIMIT	4.5	7.6	18.3	28.8	48.8	68.3	89.2	105.3	117.6	135.9	165.7	195.7	220.9	249.2	269.7	(£)

APPENDIX 5.2 315

```
BUNGALOW                       LAND USE CODE 14    JANUARY 1990 PRICES   SALT WATER FLOOD DURATION LESS THAN 12 HOURS
                                                DEPTH ABOVE UPPER SURFACE OF GROUND FLOOR
COMPONENTS OF DAMAGE    -0.3    0.0    0.05    0.1    0.2    0.3    0.6    0.9    1.2    1.5    1.8    2.1    2.4    2.7    3.0  (M)
-------------------------------------------------------------------------------------------------------------------------------------
PATHS & PAVED AREAS       0      0      0       1      3     11     17     24     31     41     52     68     84    110    127  (£)
  COL PERCENT            0.0    0.0    0.0     0.1    0.1    0.2    0.2    0.2    0.2    0.3    0.3    0.3    0.3    0.4    0.4
  ROW PERCENT            0.0    0.0    0.0     1.2    2.7    8.9   14.0   19.5   24.6   32.7   41.3   53.4   66.5   86.7  100.0
GARDENS/FENCES/SHEDS      0      0      0       1    129    267    411    665   1101   1556   2096   2684   3445   3880   4271  (£)
  COL PERCENT            0.0    0.0    0.0     0.0    2.5    3.7    3.9    5.3    7.6    9.4   11.0   12.2   14.1   14.4   14.9
  ROW PERCENT            0.0    0.0    0.0     0.0    3.0    6.3    9.6   15.6   25.8   36.4   49.1   62.8   80.7   90.8  100.0
EXT. MAIN BUILDING       50    188    201     203    319    372    402    427    525    703   1121   1775   2402   2983   3607  (£)
  COL PERCENT           69.4   81.4   15.7     9.0    6.2    5.2    3.8    3.4    3.6    4.2    5.9    8.1    9.8   11.1   12.6
  ROW PERCENT            1.4    5.2    5.6     5.6    8.9   10.3   11.2   11.9   14.6   19.5   31.1   49.2   66.6   82.7  100.0
PLASTERWORK               0      0      0       1    189    438    686    911   1187   1465   1874   2168   2632   3091   3447  (£)
  COL PERCENT            0.0    0.0    0.0     0.0    3.7    6.1    6.5    7.3    8.2    8.8    9.8    9.9   10.8   11.5   12.0
  ROW PERCENT            0.0    0.0    0.0     0.0    5.5   12.7   19.9   26.4   34.4   42.5   54.4   62.9   76.4   89.7  100.0
FLOORS                    0      0     89     193    389    618    761    952   1111   1390   1646   2023   2522   3097   3445  (£)
  COL PERCENT            0.0    0.0    7.0     8.5    7.6    8.6    7.2    7.6    7.7    8.4    8.6    9.2   10.3   11.5   12.0
  ROW PERCENT            0.0    0.0    2.6     5.6   11.3   17.9   22.1   27.6   32.3   40.3   47.8   58.7   73.2   89.9  100.0
JOINERY                   0      0      0      45    153    248    340    406    460   1197   1740   2607   2680   2936   3001  (£)
  COL PERCENT            0.0    0.0    0.0     2.0    3.0    3.4    3.2    3.2    3.2    7.2    9.1   11.9   11.0   10.9   10.5
  ROW PERCENT            0.0    0.0    0.0     1.5    5.1    8.3   11.4   13.5   15.4   39.9   58.0   86.9   89.3   97.8  100.0
INTERNAL DECORATIONS      3      3      3     113    307    333    499    580    621    656    689    721    724    729    732  (£)
  COL PERCENT            5.0    1.6    0.3     5.0    6.0    4.6    4.7    4.6    4.3    4.0    3.6    3.3    3.0    2.7    2.6
  ROW PERCENT            0.5    0.5    0.5    15.4   42.0   45.4   68.1   79.2   84.8   89.6   94.1   98.5   98.8   99.5  100.0
PLUMBING & ELECTRICAL    18     39     73     138    155    178    233    291    310    337    556    577    595    620    637  (£)
  COL PERCENT           25.6   17.0    5.7     6.1    3.0    2.5    2.2    2.3    2.1    2.0    2.9    2.6    2.4    2.3    2.2
  ROW PERCENT            2.9    6.2   11.5    21.7   24.4   28.0   36.7   45.7   48.7   52.9   87.3   90.5   93.3   97.3  100.0

DOMESTIC APPLIANCES       0      0      0      54    145    295    419    651    654    693    712    712    712    712    712  (£)
  COL PERCENT            0.0    0.0    0.1     2.4    2.8    4.1    4.0    5.2    4.5    4.2    3.7    3.2    2.9    2.7    2.5
  ROW PERCENT            0.0    0.0    0.1     7.6   20.4   41.5   58.8   91.3   91.8   97.3  100.0  100.0  100.0  100.0  100.0
HEATING EQUIPMENT         0      0      4      19     34     75    128    142    159    162    162    162    162    162    162  (£)
  COL PERCENT            0.0    0.0    0.3     0.9    0.7    1.0    1.2    1.1    1.1    1.0    0.9    0.7    0.7    0.6    0.6
  ROW PERCENT            0.0    0.0    2.5    12.1   21.3   46.5   78.6   87.5   97.7  100.0  100.0  100.0  100.0  100.0  100.0
AUDIO/VIDEO               0      0     44     136    206    261    415    585    605    605    605    605    605    605    605  (£)
  COL PERCENT            0.0    0.0    3.5     6.0    4.0    3.6    3.9    4.7    4.2    3.7    3.2    2.8    2.5    2.3    2.1
  ROW PERCENT            0.0    0.0    7.4    22.5   34.0   43.1   68.6   96.7  100.0  100.0  100.0  100.0  100.0  100.0  100.0
FURNITURE                 0      0     67     246   1041   1379   2238   2331   2598   2603   2603   2603   2603   2603   2603  (£)
  COL PERCENT            0.0    0.0    5.2    10.9   20.2   19.1   21.1   18.6   18.0   15.7   13.7   11.9   10.6    9.7    9.1
  ROW PERCENT            0.0    0.0    2.6     9.5   40.0   53.0   86.0   89.5   99.8  100.0  100.0  100.0  100.0  100.0  100.0
PERSONAL EFFECTS          0      0     39     272    429   1006   2058   2504   2889   2912   2912   2912   2912   2912   2912  (£)
  COL PERCENT            0.0    0.0    3.0    12.0    8.3   13.9   19.4   20.0   20.0   17.6   15.3   13.3   11.9   10.8   10.2
  ROW PERCENT            0.0    0.0    1.3     9.3   14.8   34.6   70.7   86.0   99.2  100.0  100.0  100.0  100.0  100.0  100.0
FLOOR COVER/CURTAINS      0      0    480     506   1008   1030   1238   1258   1345   1345   1345   1345   1345   1345   1345  (£)
  COL PERCENT            0.0    0.0   37.4    22.4   19.6   14.3   11.7   10.0    9.3    8.1    7.1    6.1    5.5    5.0    4.7
  ROW PERCENT            0.0    0.0   35.7    37.6   74.9   76.6   92.0   93.5  100.0  100.0  100.0  100.0  100.0  100.0  100.0
GARDEN/DIY/LEISURE        0      0     38      76    109    155    181    221    221    236    239    239    246    246    246  (£)
  COL PERCENT            0.0    0.0    3.0     3.4    2.1    2.1    1.7    1.8    1.5    1.4    1.3    1.1    1.0    0.9    0.9
  ROW PERCENT            0.0    0.0   15.7    31.2   44.3   63.0   73.6   89.8   89.8   96.1   97.2   97.2  100.0  100.0  100.0
DOMESTIC CLEAN-UP         0      0    242     254    530    550    571    591    616    653    694    735    777    822    822  (£)
  COL PERCENT            0.0    0.0   18.8    11.2   10.3    7.6    5.4    4.7    4.3    3.9    3.6    3.4    3.2    3.1    2.9
  ROW PERCENT            0.0    0.0   29.4    31.0   64.5   67.0   69.5   72.0   75.0   79.5   84.5   89.5   94.5  100.0  100.0
```

```
BUNGALOW                       LAND USE CODE 14    JANUARY 1990 PRICES   SALT WATER FLOOD DURATION LESS THAN 12 HOURS
                                                DEPTH ABOVE UPPER SURFACE OF GROUND FLOOR
COMPONENTS OF DAMAGE    -0.3    0.0    0.05    0.1    0.2    0.3    0.6    0.9    1.2    1.5    1.8    2.1    2.4    2.7    3.0  (M)
-------------------------------------------------------------------------------------------------------------------------------------
TOTAL BUILDING FABRIC    73    231    368     697   1650   2470   3350   4260   5350   7350   9780  12630  15090  17450  19270  (£)
  COL PERCENT          100.0  100.0   28.7    30.8   32.0   34.2   31.6   34.0   37.0   44.4   51.3   57.5   61.7   65.0   67.2
  ROW PERCENT            0.4    1.2    1.9     3.6    8.5   12.8   17.4   22.1   27.8   38.1   50.7   65.5   78.3   90.5  100.0
TOTAL INVENTORY           0      0    916    1570   3510   4750   7250   8290   9090   9210   9280   9320   9370   9410   9410  (£)
  COL PERCENT            0.0    0.0   71.3    69.2   68.0   65.8   68.4   66.0   63.0   55.6   48.7   42.5   38.3   35.0   32.8
  ROW PERCENT            0.0    0.0    9.7    16.6   37.2   50.5   77.0   88.0   96.6   97.9   98.6   99.0   99.5  100.0  100.0

TOTAL DAMAGE             73    231   1280    2260   5150   7220  10600  12550  14440  16560  19060  21950  24450  26860  28680  (£)
  ROW PERCENT            0.3    0.8    4.5     7.9   18.0   25.2   37.0   43.7   50.3   57.7   66.4   76.5   85.3   93.6  100.0
U.C. LIMIT              115    270   1370    2440   5650   7910  11710  13830  15960  18250  20810  23870  26400  28820  30690  (£)
L.C. LIMIT               31    192   1210    2090   4670   6540   9500  11270  12930  14880  17300  20030  22510  24900  26680  (£)
TOTAL DAMAGE/SQ.M       0.9    2.6   13.8    24.3   55.2   77.5  114.0  135.1  155.5  177.6  204.6  235.4  262.2  288.0  307.3  (£)
  ROW PERCENT            0.3    0.8    4.5     7.9   18.0   25.2   37.1   44.0   50.6   57.8   66.6   76.6   85.3   93.7  100.0
U.C. LIMIT              1.6    3.2   14.8    26.5   60.8   85.7  127.5  151.1  174.4  197.8  226.4  259.2  286.6  313.1  332.9  (£)
L.C. LIMIT              0.2    2.0   12.8    22.1   49.6   69.3  100.5  119.1  136.6  157.4  182.8  211.6  237.8  262.9  281.7  (£)
```

316 THE ECONOMICS OF COASTAL MANAGEMENT

FLAT   LAND USE CODE 15   JANUARY 1990 PRICES   SALT WATER FLOOD DURATION LESS THAN 12 HOURS

COMPONENTS OF DAMAGE	-0.3	0.0	0.05	0.1	DEPTH ABOVE UPPER SURFACE OF GROUND FLOOR										
					0.2	0.3	0.6	0.9	1.2	1.5	1.8	2.1	2.4	2.7	3.0 (M)
PATHS & PAVED AREAS	0	0	0	0	0	4	5	7	9	12	15	19	23	32	38 (£)
COL PERCENT	0.0	0.0	0.0	0.0	0.0	0.1	0.1	0.1	0.1	0.2	0.2	0.2	0.2	0.3	0.3
ROW PERCENT	0.0	0.0	0.0	0.0	0.5	11.0	15.4	19.8	24.8	33.5	41.3	50.1	61.4	84.8	100.0
GARDENS/FENCES/SHEDS	0	0	0	0	27	57	88	128	172	246	326	427	507	592	676 (£)
COL PERCENT	0.0	0.0	0.0	0.0	1.1	1.6	1.7	2.1	2.6	3.3	3.8	4.5	4.9	5.2	5.5
ROW PERCENT	0.0	0.0	0.0	0.0	4.1	8.5	13.1	18.9	25.6	36.4	48.2	63.3	75.1	87.6	100.0
EXT. MAIN BUILDING	107	224	235	236	268	309	330	351	392	473	645	888	1108	1264	1476 (£)
COL PERCENT	95.1	90.9	30.3	18.1	10.4	8.6	6.5	5.8	5.9	6.3	7.6	9.3	10.7	11.0	12.0
ROW PERCENT	7.3	15.2	15.9	16.0	18.2	20.9	22.4	23.8	26.6	32.1	43.7	60.2	75.1	85.7	100.0
PLASTERWORK	0	0	0	2	102	230	363	489	652	808	1055	1238	1468	1682	1873 (£)
COL PERCENT	0.0	0.0	0.0	0.2	4.0	6.4	7.1	8.1	9.7	10.7	12.5	13.0	14.1	14.7	15.3
ROW PERCENT	0.0	0.0	0.0	0.1	5.5	12.3	19.4	26.1	34.8	43.2	56.4	66.1	78.4	89.8	100.0
FLOORS	0	0	48	104	200	312	377	456	517	625	730	888	1104	1476	1739 (£)
COL PERCENT	0.0	0.0	6.2	8.0	7.7	8.6	7.4	7.6	7.7	8.3	8.6	9.3	10.6	12.9	14.2
ROW PERCENT	0.0	0.0	2.8	6.0	11.5	18.0	21.7	26.2	29.8	36.0	42.0	51.1	63.5	84.9	100.0
JOINERY	0	0	0	0	17	34	54	76	102	406	587	891	952	1160	1192 (£)
COL PERCENT	0.0	0.0	0.0	0.0	0.7	1.0	1.1	1.3	1.5	5.4	6.9	9.4	9.2	10.1	9.7
ROW PERCENT	0.0	0.0	0.0	0.0	1.5	2.9	4.6	6.4	8.6	34.0	49.3	74.7	79.9	97.3	100.0
INTERNAL DECORATIONS	0	0	0	72	165	181	269	304	325	344	358	370	373	377	381 (£)
COL PERCENT	0.0	0.0	0.1	5.5	6.4	5.0	5.3	5.1	4.9	4.6	4.2	3.9	3.6	3.3	3.1
ROW PERCENT	0.0	0.0	0.1	18.9	43.3	47.6	70.7	79.8	85.3	90.4	94.2	97.2	97.9	99.1	100.0
PLUMBING & ELECTRICAL	5	22	57	114	125	140	168	196	209	225	312	326	345	356	370 (£)
COL PERCENT	4.9	9.1	7.5	8.7	4.9	3.9	3.3	3.3	3.1	3.0	3.7	3.4	3.3	3.1	3.0
ROW PERCENT	1.5	6.1	15.6	30.8	34.0	38.0	45.6	53.1	56.5	61.0	84.5	88.2	93.4	96.4	100.0
DOMESTIC APPLIANCES	0	0	0	49	125	247	337	522	523	555	569	569	569	569	569 (£)
COL PERCENT	0.0	0.0	0.1	3.8	4.8	6.9	6.6	8.7	7.8	7.4	6.7	6.0	5.5	5.0	4.6
ROW PERCENT	0.0	0.0	0.1	8.7	22.0	43.5	59.2	91.7	92.0	97.6	100.0	100.0	100.0	100.0	100.0
HEATING EQUIPMENT	0	0	2	16	23	46	75	83	91	93	93	93	93	93	93 (£)
COL PERCENT	0.0	0.0	0.3	1.3	0.9	1.3	1.5	1.4	1.4	1.2	1.1	1.0	0.9	0.8	0.8
ROW PERCENT	0.0	0.0	2.3	17.5	25.6	49.7	81.1	89.2	97.9	100.0	100.0	100.0	100.0	100.0	100.0
AUDIO/VIDEO	0	0	25	80	123	152	249	342	352	352	352	352	352	352	352 (£)
COL PERCENT	0.0	0.0	3.3	6.2	4.8	4.2	4.9	5.7	5.3	4.7	4.2	3.7	3.4	3.1	2.9
ROW PERCENT	0.0	0.0	7.3	22.8	34.9	43.3	70.6	97.1	100.0	100.0	100.0	100.0	100.0	100.0	100.0
FURNITURE	0	0	38	130	482	687	1059	1094	1196	1200	1200	1200	1200	1200	1200 (£)
COL PERCENT	0.0	0.0	4.9	10.0	18.6	19.0	20.8	18.2	17.9	15.9	14.2	12.6	11.6	10.5	9.8
ROW PERCENT	0.0	0.0	3.2	10.9	40.2	57.2	88.2	91.2	99.6	100.0	100.0	100.0	100.0	100.0	100.0
PERSONAL EFFECTS	0	0	12	99	174	398	811	1009	1152	1170	1170	1170	1170	1170	1170 (£)
COL PERCENT	0.0	0.0	1.6	7.6	6.7	11.0	15.9	16.8	17.2	15.5	13.8	12.3	11.3	10.2	9.5
ROW PERCENT	0.0	0.0	1.1	8.5	14.9	34.0	69.3	86.2	98.4	100.0	100.0	100.0	100.0	100.0	100.0
FLOOR COVER/CURTAINS	0	0	199	209	394	399	470	479	503	503	503	503	503	503	503 (£)
COL PERCENT	0.0	0.0	25.7	16.1	15.2	11.1	9.2	8.0	7.5	6.7	5.9	5.3	4.9	4.4	4.1
ROW PERCENT	0.0	0.0	39.7	41.7	78.3	79.3	93.4	95.3	100.0	100.0	100.0	100.0	100.0	100.0	100.0
GARDEN/DIY/LEISURE	0	0	27	55	79	114	135	166	166	179	181	181	187	187	187 (£)
COL PERCENT	0.0	0.0	3.5	4.2	3.1	3.2	2.7	2.8	2.5	2.4	2.1	1.9	1.8	1.6	1.5
ROW PERCENT	0.0	0.0	14.5	29.5	42.5	61.2	72.2	88.9	88.9	95.8	96.9	96.9	100.0	100.0	100.0
DOMESTIC CLEAN-UP	0	0	128	135	281	292	303	314	327	347	369	391	412	437	437 (£)
COL PERCENT	0.0	0.0	16.6	10.4	10.9	8.1	6.0	5.2	4.9	4.6	4.4	4.1	4.0	3.8	3.6
ROW PERCENT	0.0	0.0	29.4	31.0	64.4	67.0	69.5	72.0	75.0	79.5	84.4	89.5	94.5	100.0	100.0

FLAT   LAND USE CODE 15   JANUARY 1990 PRICES   SALT WATER FLOOD DURATION LESS THAN 12 HOURS

COMPONENTS OF DAMAGE	-0.3	0.0	0.05	0.1	DEPTH ABOVE UPPER SURFACE OF GROUND FLOOR										
					0.2	0.3	0.6	0.9	1.2	1.5	1.8	2.1	2.4	2.7	3.0 (M)
TOTAL BUILDING FABRIC	113	247	342	529	907	1270	1660	2010	2380	3140	4030	5050	5890	6940	7750 (£)
COL PERCENT	100.0	100.0	44.1	40.5	35.0	35.2	32.5	33.4	35.6	41.6	47.6	53.1	56.7	60.6	63.2
ROW PERCENT	1.5	3.2	4.4	6.8	11.7	16.4	21.4	25.9	30.7	40.6	52.0	65.2	76.0	89.6	100.0
TOTAL INVENTORY	0	0	434	776	1680	2340	3440	4010	4310	4400	4440	4460	4490	4510	4510 (£)
COL PERCENT	0.0	0.0	55.9	59.5	65.0	64.8	67.5	66.6	64.4	58.4	52.4	46.9	43.3	39.4	36.8
ROW PERCENT	0.0	0.0	9.6	17.2	37.3	51.8	76.2	88.9	95.6	97.5	98.4	98.9	99.5	100.0	100.0
TOTAL DAMAGE	113	247	776	1310	2590	3610	5100	6020	6700	7550	8470	9510	10380	11460	12260 (£)
ROW PERCENT	0.9	2.0	6.3	10.7	21.1	29.4	41.6	49.1	54.6	61.5	69.1	77.6	84.6	93.4	100.0
U.C. LIMIT	176	292	852	1470	2950	4100	5900	6940	7740	8690	9700	10810	11720	12890	13790 (£)
L.C. LIMIT	50	202	706	1150	2240	3120	4300	5110	5660	6400	7250	8220	9030	10030	10740 (£)
TOTAL DAMAGE/SQ.M	2.4	5.8	16.8	27.6	54.6	76.1	106.8	126.8	140.6	158.6	178.4	201.4	219.9	242.8	259.1 (£)
ROW PERCENT	0.9	2.2	6.5	10.7	21.1	29.4	41.2	48.9	54.3	61.2	68.9	77.7	84.9	93.7	100.0
U.C. LIMIT	3.9	7.1	18.4	29.8	59.9	83.8	119.9	142.6	158.2	178.1	198.7	223.8	242.3	265.2	281.6 (£)
L.C. LIMIT	0.9	4.5	15.2	25.4	49.3	68.4	93.7	111.0	123.0	139.1	158.1	179.0	197.5	220.4	236.6 (£)

APPENDIX 5.2 317

```
PREFAB                  LAND USE CODE 17    JANUARY 1990 PRICES   SALT WATER FLOOD DURATION LESS THAN 12 HOURS
                                        DEPTH ABOVE UPPER SURFACE OF GROUND FLOOR
COMPONENTS OF DAMAGE  -0.3   0.0   0.05   0.1   0.2    0.3    0.6    0.9    1.2    1.5    1.8    2.1    2.4    2.7    3.0  (M)
-------------------------------------------------------------------------------------------------------------------------
PATHS & PAVED AREAS     0     0     0      0     0      2      3      4      5      7      8     10     12     17     21  (£)
    COL PERCENT        0.0   0.0   0.0    0.0   0.0    0.0    0.0    0.1    0.1    0.1    0.1    0.1    0.1    0.1    0.1
    ROW PERCENT        0.0   0.0   0.0    0.0   0.0    9.5   14.3   19.0   23.8   33.3   38.1   47.6   57.1   81.0  100.0
GARDENS/FENCES/SHEDS    0     0     0      0    82    168    257    347    438    612    786   1027   1210   1408   1573  (£)
    COL PERCENT        0.0   0.0   0.0    0.0   2.7    3.8    3.8    4.4    4.9    6.1    6.9    8.0    8.8    9.5   10.1
    ROW PERCENT        0.0   0.0   0.0    0.0   5.2   10.7   16.3   22.1   27.8   38.9   50.0   65.3   76.9   89.5  100.0
EXT. MAIN BUILDING      0    11    22     22    24     47     71     83    116    167    436    939   1164   1430   1682  (£)
    COL PERCENT        0.0  22.2   3.4    1.7   0.8    1.1    1.1    1.0    1.3    1.7    3.8    7.3    8.4    9.6   10.8
    ROW PERCENT        0.0   0.7   1.3    1.3   1.4    2.8    4.2    4.9    6.9    9.9   25.9   55.8   69.2   85.0  100.0
PLASTERWORK             0     0     0     69   302    462    603    757    898   1224   1600   1755   1733   1891   1941  (£)
    COL PERCENT        0.0   0.0   0.0    5.4  10.1   10.6    8.9    9.5   10.0   12.1   14.0   13.6   12.6   12.8   12.4
    ROW PERCENT        0.0   0.0   0.0    3.6  15.6   23.8   31.1   39.0   46.3   63.1   82.4   90.4   89.3   97.4  100.0
FLOORS                  0     0    39     42   150    228    298    421    566    832   1059   1422   1842   2183   2474  (£)
    COL PERCENT        0.0   0.0   5.9    3.3   5.0    5.2    4.4    5.3    6.3    8.2    9.3   11.0   13.4   14.7   15.9
    ROW PERCENT        0.0   0.0   1.6    1.7   6.1    9.2   12.0   17.0   22.9   33.6   42.8   57.5   74.5   88.2  100.0
JOINERY                 0     0     0      0     7    129    345    394    437    615    857   1002   1072   1093   1097  (£)
    COL PERCENT        0.0   0.0   0.0    0.0   0.2    3.0    5.1    5.0    4.9    6.1    7.5    7.8    7.8    7.4    7.0
    ROW PERCENT        0.0   0.0   0.0    0.0   0.6   11.8   31.4   35.9   39.8   56.1   78.1   91.3   97.7   99.6  100.0
INTERNAL DECORATIONS    0     0     0     62    81    108    302    358    475    502    502    516    523    523    523  (£)
    COL PERCENT        0.0   0.0   0.0    4.9   2.7    2.5    4.5    4.5    5.3    5.0    4.4    4.0    3.8    3.5    3.4
    ROW PERCENT        0.0   0.0   0.0   11.9  15.5   20.7   57.7   68.5   90.8   96.0   96.0   98.7  100.0  100.0  100.0
PLUMBING & ELECTRICAL   0    38    39     77    82     85     89    104    107    108    116    117    123    129    136  (£)
    COL PERCENT      100.0  77.8   6.0    6.0   2.8    2.0    1.3    1.3    1.2    1.1    1.0    0.9    0.9    0.9    0.9
    ROW PERCENT        0.1  28.3  29.1   56.6  60.5   63.0   65.8   76.7   79.1   79.4   85.6   85.9   90.7   95.3  100.0

DOMESTIC APPLIANCES     0     0     0     53   134    272    374    572    573    607    622    622    622    622    622  (£)
    COL PERCENT        0.0   0.0   0.1    4.2   4.5    6.2    5.6    7.2    6.4    6.0    5.4    4.8    4.5    4.2    4.0
    ROW PERCENT        0.0   0.0   0.1    8.6  21.6   43.7   60.2   92.0   92.1   97.5  100.0  100.0  100.0  100.0  100.0
HEATING EQUIPMENT       0     0     3     20    31     61     99    109    120    122    122    122    122    122    122  (£)
    COL PERCENT        0.0   0.0   0.6    1.6   1.1    1.4    1.5    1.4    1.3    1.2    1.1    1.0    0.9    0.8    0.8
    ROW PERCENT        0.0   0.0   3.1   17.0  25.7   50.1   81.2   89.0   97.8  100.0  100.0  100.0  100.0  100.0  100.0
AUDIO/VIDEO             0     0    35    111   169    208    339    460    472    472    472    472    472    472    472  (£)
    COL PERCENT        0.0   0.0   5.5    8.7   5.7    4.8    5.0    5.8    5.3    4.7    4.1    3.7    3.4    3.2    3.0
    ROW PERCENT        0.0   0.0   7.6   23.7  35.8   44.2   71.9   97.5  100.0  100.0  100.0  100.0  100.0  100.0  100.0
FURNITURE               0     0    42    161   688    949   1477   1497   1655   1661   1661   1661   1661   1661   1661  (£)
    COL PERCENT        0.0   0.0   6.5   12.6  23.0   21.7   21.9   18.9   18.5   16.5   14.5   12.9   12.0   11.2   10.7
    ROW PERCENT        0.0   0.0   2.5    9.7  41.4   57.1   88.9   90.1   99.6  100.0  100.0  100.0  100.0  100.0  100.0
PERSONAL EFFECTS        0     0    17    153   264    617   1320   1597   1833   1852   1852   1852   1852   1852   1852  (£)
    COL PERCENT        0.0   0.0   2.6   12.0   8.9   14.1   19.6   20.2   20.5   18.4   16.2   14.4   13.4   12.5   11.9
    ROW PERCENT        0.0   0.0   0.9    8.3  14.3   33.3   71.3   86.2   99.0  100.0  100.0  100.0  100.0  100.0  100.0
FLOOR COVER/CURTAINS    0     0   272    280   547    550    648    662    690    690    690    690    690    690    690  (£)
    COL PERCENT        0.0   0.0  41.6   21.9  18.3   12.6    9.6    8.4    7.7    6.8    6.0    5.4    5.0    4.7    4.4
    ROW PERCENT        0.0   0.0  39.5   40.6  79.3   79.7   94.0   95.9  100.0  100.0  100.0  100.0  100.0  100.0  100.0
GARDEN/DIY/LEISURE      0     0    33     67    96    137    160    194    194    207    210    210    216    216    216  (£)
    COL PERCENT        0.0   0.0   5.2    5.3   3.2    3.1    2.4    2.5    2.2    2.1    1.8    1.6    1.6    1.5    1.4
    ROW PERCENT        0.0   0.0  15.7   31.2  44.6   63.4   73.9   89.7   89.7   96.0   97.1   97.1  100.0  100.0  100.0
DOMESTIC CLEAN-UP       0     0   148    156   326    339    351    364    380    402    428    452    478    506    506  (£)
    COL PERCENT        0.0   0.0  22.7   12.3  10.9    7.8    5.2    4.6    4.2    4.0    3.7    3.5    3.5    3.4    3.2
    ROW PERCENT        0.0   0.0  29.4   31.0  64.6   67.0   69.4   72.1   75.1   79.6   84.6   89.5   94.5  100.0  100.0
```

```
PREFAB                  LAND USE CODE 17    JANUARY 1990 PRICES   SALT WATER FLOOD DURATION LESS THAN 12 HOURS
                                        DEPTH ABOVE UPPER SURFACE OF GROUND FLOOR
COMPONENTS OF DAMAGE  -0.3   0.0   0.05   0.1   0.2    0.3    0.6    0.9    1.2    1.5    1.8    2.1    2.4    2.7    3.0  (M)
-------------------------------------------------------------------------------------------------------------------------
TOTAL BUILDING FABRIC   0    49   100    272   728   1230   1970   2470   3040   4070   5360   6790   7680   8670   9450  (£)
    COL PERCENT      100.0 100.0  15.3   21.3  24.4   28.2   29.2   31.1   33.9   40.3   47.0   52.7   55.7   58.5   60.6
    ROW PERCENT        0.0   0.5   1.1    2.9   7.7   13.0   20.8   26.1   32.2   43.1   56.8   71.9   81.3   91.8  100.0
TOTAL INVENTORY         0     0   554   1010  2260   3140   4770   5460   5920   6020   6060   6090   6120   6150   6150  (£)
    COL PERCENT        0.0   0.0  84.7   78.7  75.6   71.8   70.8   68.9   66.1   59.7   53.0   47.3   44.3   41.5   39.4
    ROW PERCENT        0.0   0.0   9.0   16.4  36.8   51.0   77.6   88.8   96.3   97.9   98.6   99.0   99.5  100.0  100.0

TOTAL DAMAGE            0    49   655   1280  2990   4370   6740   7930   8960  10090  11430  12870  13800  14820  15590  (£)
    ROW PERCENT        0.0   0.3   4.2    8.2  19.2   28.0   43.2   50.8   57.5   64.7   73.3   82.6   88.5   95.0  100.0
    U.C. LIMIT         0    49   737   1440  3490   5110   8100   9460  10680  11820  13170  14620  15540  16560  17340  (£)
    L.C. LIMIT         0    49   553   1080  2400   3490   5190   6170   7010   8100   9430  10870  11790  12810  13590  (£)
TOTAL DAMAGE/SQ.M     0.0   0.9  11.5   22.5  52.5   76.8  118.6  139.5  157.9  177.9  201.7  227.6  244.1  262.3  276.1  (£)
    ROW PERCENT        0.0   0.3   4.2    8.1  19.0   27.8   43.0   50.5   57.2   64.4   73.1   82.4   88.4   95.0  100.0
    U.C. LIMIT        0.0   0.9  13.1   25.7  62.2   91.3  144.6  168.9  190.7  211.1  235.2  261.0  277.5  295.8  309.5  (£)
    L.C. LIMIT        0.0   0.9   9.9   19.3  42.8   62.3   92.6  110.1  125.1  144.7  168.2  194.2  210.7  228.8  242.7  (£)
```

318 THE ECONOMICS OF COASTAL MANAGEMENT

DETACHED — LAND USE CODE 11 — JANUARY 1990 PRICES — SALT WATER FLOOD DURATION MORE THAN 12 HOURS

COMPONENTS OF DAMAGE	-0.3	0.0	0.05	0.1	0.2	0.3	0.6	0.9	1.2	1.5	1.8	2.1	2.4	2.7	3.0 (M)
PATHS & PAVED AREAS	0	6	12	21	72	140	176	204	219	238	257	270	300	329	352 (£)
COL PERCENT	0.0	1.0	0.3	0.4	0.8	1.1	1.1	1.1	1.0	1.0	1.0	1.0	1.0	1.0	1.0
ROW PERCENT	0.0	1.7	3.5	6.0	20.5	39.8	50.0	58.0	62.4	67.7	72.9	76.8	85.1	93.3	100.0
GARDENS/FENCES/SHEDS	0	3	44	93	482	1000	1426	1568	2177	3035	3614	4238	5008	5252	5382 (£)
COL PERCENT	0.0	0.5	1.2	1.6	5.1	7.9	9.1	8.7	10.4	13.0	14.3	15.4	16.7	16.4	15.9
ROW PERCENT	0.0	0.1	0.8	1.7	9.0	18.6	26.5	29.1	40.5	56.4	67.2	78.7	93.0	97.6	100.0
EXT. MAIN BUILDING	164	295	324	401	554	691	889	1102	1631	2356	3254	4272	5524	6877	8714 (£)
COL PERCENT	59.9	48.1	8.6	6.7	5.9	5.5	5.7	6.1	7.8	10.1	12.9	15.5	18.4	21.5	25.7
ROW PERCENT	1.9	3.4	3.7	4.6	6.4	7.9	10.2	12.6	18.7	27.0	37.3	49.0	63.4	78.9	100.0
PLASTERWORK	65	154	275	499	833	1147	1704	2247	2806	3165	3285	3763	4124	4431	4389 (£)
COL PERCENT	23.8	25.2	7.3	8.3	8.8	9.1	10.9	12.5	13.4	13.5	13.0	13.6	13.7	13.9	12.9
ROW PERCENT	1.5	3.5	6.3	11.4	19.0	26.1	38.8	51.2	63.9	72.1	74.8	85.7	94.0	101.0	100.0
FLOORS	0	91	968	1597	2356	2600	2781	2824	3482	3817	3733	3733	3733	3733	3733 (£)
COL PERCENT	0.0	14.8	25.8	26.5	24.9	20.6	17.8	15.6	16.6	16.3	14.8	13.5	12.4	11.7	11.0
ROW PERCENT	0.0	2.4	25.9	42.8	63.1	69.7	74.5	75.6	93.3	102.2	100.0	100.0	100.0	100.0	100.0
JOINERY	0	0	195	712	1228	1826	2249	2839	3217	3278	3457	3549	3543	3543	3543 (£)
COL PERCENT	0.0	0.0	5.2	11.8	13.0	14.5	14.4	15.7	15.3	14.0	13.7	12.9	11.8	11.1	10.4
ROW PERCENT	0.0	0.0	5.5	20.1	34.7	51.5	63.5	80.1	90.8	92.5	97.6	100.2	100.0	100.0	100.0
INTERNAL DECORATIONS	14	22	55	320	576	612	629	662	669	676	684	687	698	724	723 (£)
COL PERCENT	5.3	3.7	1.5	5.3	6.1	4.9	4.0	3.7	3.2	2.9	2.7	2.5	2.3	2.3	2.1
ROW PERCENT	2.0	3.2	7.6	44.3	79.7	84.6	87.1	91.6	92.5	93.5	94.6	95.0	96.6	100.2	100.0
PLUMBING & ELECTRICAL	30	40	71	150	236	283	501	557	585	598	642	654	665	679	699 (£)
COL PERCENT	11.1	6.6	1.9	2.5	2.5	2.2	3.2	3.1	2.8	2.6	2.5	2.4	2.2	2.1	2.1
ROW PERCENT	4.4	5.8	10.3	21.6	33.9	40.5	71.8	79.8	83.7	85.6	91.9	93.6	95.2	97.2	100.0
DOMESTIC APPLIANCES	0	0	3	95	246	541	619	900	900	918	918	918	918	918	918 (£)
COL PERCENT	0.0	0.0	0.1	1.6	2.6	4.3	4.0	5.0	4.3	3.9	3.6	3.3	3.1	2.9	2.7
ROW PERCENT	0.0	0.0	0.4	10.4	26.8	58.9	67.5	98.1	98.1	100.0	100.0	100.0	100.0	100.0	100.0
HEATING EQUIPMENT	0	0	0	18	49	118	161	184	202	202	202	208	208	208	208 (£)
COL PERCENT	0.0	0.0	0.0	0.3	0.5	0.9	1.0	1.0	1.0	0.9	0.8	0.8	0.7	0.7	0.6
ROW PERCENT	0.0	0.0	0.0	8.9	23.9	57.0	77.5	88.6	97.1	97.1	97.1	100.0	100.0	100.0	100.0
AUDIO/VIDEO	0	0	56	187	276	340	545	726	745	745	745	745	745	745	745 (£)
COL PERCENT	0.0	0.0	1.5	3.1	2.9	2.7	3.5	4.0	3.6	3.2	2.9	2.7	2.5	2.3	2.2
ROW PERCENT	0.0	0.0	7.6	25.1	37.1	45.7	73.2	97.4	100.0	100.0	100.0	100.0	100.0	100.0	100.0
FURNITURE	0	0	70	175	409	868	1231	1322	1362	1370	1370	1370	1370	1370	1370 (£)
COL PERCENT	0.0	0.0	1.9	2.9	4.3	6.9	7.9	7.3	6.5	5.8	5.4	5.0	4.6	4.3	4.0
ROW PERCENT	0.0	0.0	5.1	12.8	29.9	63.4	89.9	96.5	99.4	100.0	100.0	100.0	100.0	100.0	100.0
PERSONAL EFFECTS	0	0	0	0	43	270	382	518	518	532	532	532	532	532	532 (£)
COL PERCENT	0.0	0.0	0.0	0.0	0.5	2.1	2.4	2.9	2.5	2.3	2.1	1.9	1.8	1.7	1.6
ROW PERCENT	0.0	0.0	0.0	0.0	8.2	50.8	71.8	97.4	97.4	100.0	100.0	100.0	100.0	100.0	100.0
FLOOR COVER/CURTAINS	0	0	1399	1436	1479	1479	1562	1576	1591	1591	1591	1591	1591	1591	1591 (£)
COL PERCENT	0.0	0.0	37.3	23.8	15.6	11.7	10.0	8.7	7.6	6.8	6.3	5.8	5.3	5.0	4.7
ROW PERCENT	0.0	0.0	88.0	90.3	93.0	93.0	98.2	99.1	100.0	100.0	100.0	100.0	100.0	100.0	100.0
GARDEN/DIY/LEISURE	0	0	39	77	113	159	215	249	258	272	280	280	289	289	289 (£)
COL PERCENT	0.0	0.0	1.1	1.3	1.2	1.3	1.4	1.4	1.2	1.2	1.1	1.0	1.0	0.9	0.9
ROW PERCENT	0.0	0.0	13.7	26.8	39.2	55.0	74.6	86.1	89.3	94.2	96.9	96.9	100.0	100.0	100.0
DOMESTIC CLEAN-UP	0	0	233	241	505	523	543	565	599	637	712	754	754	754	754 (£)
COL PERCENT	0.0	0.0	6.2	4.0	5.3	4.2	3.5	3.1	2.9	2.7	2.8	2.7	2.5	2.4	2.2
ROW PERCENT	0.0	0.0	30.9	32.0	67.0	69.5	72.0	75.0	79.5	84.5	94.5	100.0	100.0	100.0	100.0

DETACHED — LAND USE CODE 11 — JANUARY 1990 PRICES — SALT WATER FLOOD DURATION MORE THAN 12 HOURS

COMPONENTS OF DAMAGE	-0.3	0.0	0.05	0.1	0.2	0.3	0.6	0.9	1.2	1.5	1.8	2.1	2.4	2.7	3.0 (M)
TOTAL BUILDING FABRIC	274	613	1950	3800	6340	8300	10360	12010	14790	17170	18930	21170	23600	25570	27540 (£)
COL PERCENT	100.0	100.0	51.9	63.0	67.0	65.9	66.3	66.5	70.5	73.2	74.9	76.8	78.6	80.0	81.1
ROW PERCENT	1.0	2.2	7.1	13.8	23.0	30.1	37.6	43.6	53.7	62.3	68.7	76.9	85.7	92.9	100.0
TOTAL INVENTORY	0	0	1800	2230	3120	4300	5260	6040	6180	6270	6350	6400	6410	6410	6410 (£)
COL PERCENT	0.0	0.0	48.1	37.0	33.0	34.1	33.7	33.5	29.5	26.8	25.1	23.2	21.4	20.0	18.9
ROW PERCENT	0.0	0.0	28.1	34.8	48.7	67.1	82.1	94.3	96.4	97.8	99.1	99.9	100.0	100.0	100.0
TOTAL DAMAGE	274	613	3750	6030	9460	12610	15620	18050	20970	23440	25280	27570	30010	31980	33950 (£)
ROW PERCENT	0.8	1.8	11.0	17.8	27.9	37.1	46.0	53.2	61.8	69.0	74.5	81.2	88.4	94.2	100.0
U.C. LIMIT	405	806	4640	7150	11040	14390	17630	20260	23380	26230	28270	30860	33610	36010	38390 (£)
L.C. LIMIT	143	420	2870	4910	7890	10830	13610	15850	18550	20650	22300	24280	26410	27950	29510 (£)
TOTAL DAMAGE/SQ.M	2.7	6.5	42.7	71.1	114.2	156.0	195.4	226.7	263.8	292.9	315.8	343.8	373.4	395.6	418.6 (£)
ROW PERCENT	0.6	1.6	10.2	17.0	27.3	37.3	46.7	54.2	63.0	70.0	75.4	82.1	89.2	94.5	100.0
U.C. LIMIT	3.8	7.3	46.6	76.1	123.4	170.2	213.8	248.0	287.1	316.6	340.4	370.0	400.2	422.4	447.3 (£)
L.C. LIMIT	1.6	5.7	38.8	66.1	105.0	141.8	177.0	205.4	240.5	269.2	291.2	317.6	346.6	368.8	389.9 (£)

APPENDIX 5.2 319

```
SEMI DETACHED            LAND USE CODE 12     JANUARY 1990 PRICES   SALT WATER FLOOD DURATION MORE THAN 12 HOURS
                                          DEPTH ABOVE UPPER SURFACE OF GROUND FLOOR
COMPONENTS OF DAMAGE  -0.3   0.0   0.05   0.1   0.2   0.3   0.6   0.9   1.2   1.5   1.8   2.1   2.4   2.7   3.0  (M)
---------------------------------------------------------------------------------------------------------------
PATHS & PAVED AREAS      0     1     3     5    17    34    42    47    51    56    60    64    69    76    81  (£)
     COL PERCENT       0.0   0.3   0.2   0.2   0.3   0.5   0.5   0.5   0.4   0.4   0.4   0.4   0.4   0.4   0.4
     ROW PERCENT       0.0   1.9   4.2   7.0  21.3  41.9  52.4  58.1  63.0  68.8  73.9  78.5  85.0  93.1 100.0
GARDENS/FENCES/SHEDS     0     5    39    76   215   408   625   710  1362  1791  2067  2475  2901  3131  3255  (£)
     COL PERCENT       0.0   0.9   1.9   2.2   4.0   5.7   6.9   6.8  11.0  12.9  13.6  14.6  15.9  16.1  15.4
     ROW PERCENT       0.0   0.2   1.2   2.3   6.6  12.5  19.2  21.8  41.8  55.0  63.5  76.0  89.1  96.2 100.0
EXT. MAIN BUILDING     389   418   436   491   576   672   785   937  1082  1537  2224  2876  3695  4649  6206  (£)
     COL PERCENT      85.4  71.3  20.6  14.1  10.7   9.4   8.7   9.0   8.8  11.0  14.6  17.0  20.3  23.9  29.4
     ROW PERCENT       6.3   6.7   7.0   7.9   9.3  10.8  12.7  15.1  17.4  24.8  35.8  46.3  59.5  74.9 100.0
PLASTERWORK              5    14    29   211   356   502   762  1045  1311  1561  1771  2157  2162  2161  2160  (£)
     COL PERCENT       1.2   2.5   1.4   6.1   6.6   7.0   8.4  10.0  10.6  11.2  11.6  12.7  11.9  11.1  10.2
     ROW PERCENT       0.3   0.7   1.4   9.8  16.5  23.2  35.3  48.4  60.7  72.3  82.0  99.9 100.1 100.1 100.0
FLOORS                   0    62   619   969  1361  1449  1557  1561  2019  2250  2195  2195  2195  2195  2195  (£)
     COL PERCENT       0.0  10.6  29.3  27.8  25.3  20.2  17.2  15.0  16.3  16.2  14.4  12.9  12.0  11.3  10.4
     ROW PERCENT       0.0   2.8  28.2  44.2  62.0  66.0  70.9  71.1  92.0 102.5 100.0 100.0 100.0 100.0 100.0
JOINERY                  0     0   116   317   741  1179  1635  1912  2174  2296  2378  2626  2603  2603  2603  (£)
     COL PERCENT       0.0   0.0   5.5   9.1  13.8  16.4  18.0  18.3  17.6  16.5  15.6  15.5  14.3  13.4  12.3
     ROW PERCENT       0.0   0.0   4.5  12.2  28.5  45.3  62.8  73.5  83.5  88.2  91.4 100.9 100.0 100.0 100.0
INTERNAL DECORATIONS    26    28    47   227   303   325   334   347   349   356   363   364   374   391   391  (£)
     COL PERCENT       5.7   4.9   2.2   6.5   5.6   4.5   3.7   3.3   2.8   2.6   2.4   2.2   2.1   2.0   1.9
     ROW PERCENT       6.7   7.4  12.1  58.1  77.6  83.3  85.6  88.9  89.2  91.0  92.8  93.2  95.8 100.1 100.0
PLUMBING & ELECTRICAL   34    55    77   113   170   197   303   357   409   428   454   471   487   504   521  (£)
     COL PERCENT       7.6   9.5   3.7   3.3   3.2   2.7   3.4   3.4   3.3   3.1   3.0   2.8   2.7   2.6   2.5
     ROW PERCENT       6.7  10.7  14.8  21.8  32.6  37.9  58.3  68.6  78.5  82.3  87.2  90.5  93.4  96.8 100.0

DOMESTIC APPLIANCES      0     0     2    83   194   409   458   658   658   663   663   663   663   663   663  (£)
     COL PERCENT       0.0   0.0   0.1   2.4   3.6   5.7   5.1   6.3   5.3   4.8   4.4   3.9   3.6   3.4   3.1
     ROW PERCENT       0.0   0.0   0.4  12.5  29.3  61.7  69.0  99.2  99.2 100.0 100.0 100.0 100.0 100.0 100.0
HEATING EQUIPMENT        0     0     0    24    35    85   110   124   135   135   135   139   139   139   139  (£)
     COL PERCENT       0.0   0.0   0.0   0.7   0.7   1.2   1.2   1.2   1.1   1.0   0.9   0.8   0.8   0.7   0.7
     ROW PERCENT       0.0   0.0   0.0  17.3  25.4  61.1  79.4  89.4  97.3  97.3  97.3 100.0 100.0 100.0 100.0
AUDIO/VIDEO              0     0    31   121   181   221   375   466   481   481   481   481   481   481   481  (£)
     COL PERCENT       0.0   0.0   1.5   3.5   3.4   3.1   4.1   4.5   3.9   3.5   3.2   2.8   2.6   2.5   2.3
     ROW PERCENT       0.0   0.0   6.6  25.2  37.6  45.9  77.9  97.0 100.0 100.0 100.0 100.0 100.0 100.0 100.0
FURNITURE                0     0    47   123   273   581   804   878   896   899   899   899   899   899   899  (£)
     COL PERCENT       0.0   0.0   2.3   3.5   5.1   8.1   8.9   8.4   7.3   6.5   5.9   5.3   4.9   4.6   4.3
     ROW PERCENT       0.0   0.0   5.3  13.7  30.4  64.6  89.4  97.7  99.6 100.0 100.0 100.0 100.0 100.0 100.0
PERSONAL EFFECTS         0     0     0     0    32   150   207   289   289   300   300   300   300   300   300  (£)
     COL PERCENT       0.0   0.0   0.0   0.0   0.6   2.1   2.3   2.8   2.3   2.2   2.0   1.8   1.6   1.5   1.4
     ROW PERCENT       0.0   0.0   0.0   0.0  10.8  50.1  69.2  96.5  96.5 100.0 100.0 100.0 100.0 100.0 100.0
FLOOR COVER/CURTAINS     0     0   502   528   556   556   599   606   611   611   611   611   611   611   611  (£)
     COL PERCENT       0.0   0.0  23.8  15.2  10.3   7.7   6.6   5.8   4.9   4.4   4.0   3.6   3.4   3.1   2.9
     ROW PERCENT       0.0   0.0  82.3  86.5  91.0  91.0  98.1  99.2 100.0 100.0 100.0 100.0 100.0 100.0 100.0
GARDEN/DIY/LEISURE       0     0    29    58    86   120   162   185   192   203   208   208   215   215   215  (£)
     COL PERCENT       0.0   0.0   1.4   1.7   1.6   1.7   1.8   1.8   1.6   1.5   1.4   1.2   1.2   1.1   1.0
     ROW PERCENT       0.0   0.0  13.6  27.2  40.2  55.9  75.5  86.0  89.2  94.2  96.8  96.8 100.0 100.0 100.0
DOMESTIC CLEAN-UP        0     0   128   133   279   289   300   313   332   352   394   417   417   417   417  (£)
     COL PERCENT       0.0   0.0   6.1   3.8   5.2   4.0   3.3   3.0   2.7   2.5   2.6   2.5   2.3   2.1   2.0
     ROW PERCENT       0.0   0.0  30.9  31.9  67.0  69.4  72.1  75.0  79.6  84.5  94.5 100.0 100.0 100.0 100.0
```

```
SEMI DETACHED            LAND USE CODE 12     JANUARY 1990 PRICES   SALT WATER FLOOD DURATION MORE THAN 12 HOURS
                                          DEPTH ABOVE UPPER SURFACE OF GROUND FLOOR
COMPONENTS OF DAMAGE  -0.3   0.0   0.05   0.1   0.2   0.3   0.6   0.9   1.2   1.5   1.8   2.1   2.4   2.7   3.0  (M)
---------------------------------------------------------------------------------------------------------------
TOTAL BUILDING FABRIC  456   586  1370  2410  3740  4770  6050  6920  8760 10280 11520 13230 14490 15710 17420  (£)
     COL PERCENT     100.0 100.0  64.8  69.2  69.5  66.4  66.7  66.3  70.9  73.8  75.7  78.0  79.5  80.8  82.4
     ROW PERCENT       2.6   3.4   7.9  13.9  21.5  27.4  34.7  39.7  50.3  59.0  66.1  76.0  83.2  90.2 100.0
TOTAL INVENTORY          0     0   742  1070  1640  2420  3020  3520  3600  3650  3700  3720  3730  3730  3730  (£)
     COL PERCENT       0.0   0.0  35.2  30.8  30.5  33.6  33.3  33.7  29.1  26.2  24.3  22.0  20.5  19.2  17.6
     ROW PERCENT       0.0   0.0  19.9  28.8  44.0  64.8  81.0  94.5  96.5  97.8  99.1  99.8 100.0 100.0 100.0

TOTAL DAMAGE           456   586  2110  3490  5380  7180  9070 10440 12360 13930 15210 16950 18220 19440 21150  (£)
     ROW PERCENT       2.2   2.8  10.0  16.5  25.4  34.0  42.9  49.4  58.4  65.9  71.9  80.2  86.2  91.9 100.0
     U.C. LIMIT       534   649  2290  3720  5780  7770  9820 11330 13350 14940 16230 18090 19340 20600 22370  (£)
     L.C. LIMIT       378   523  1940  3250  4980  6600  8320  9570 11370 12920 14190 15830 17100 18290 19930  (£)
TOTAL DAMAGE/SQ.M      9.5  12.5  45.1  74.4 115.1 154.2 194.8 224.7 265.0 298.9 326.4 363.1 390.5 416.6 452.8  (£)
     ROW PERCENT       2.1   2.8  10.0  16.4  25.4  34.1  43.0  49.6  58.5  66.0  72.1  80.2  86.2  92.0 100.0
     U.C. LIMIT      11.3  14.0  49.1  79.8 124.1 168.5 213.2 246.7 288.3 323.7 351.9 390.1 418.2 445.4 483.4  (£)
     L.C. LIMIT       7.7  11.0  41.1  69.0 105.8 139.9 176.4 202.7 241.7 274.1 300.9 336.1 362.8 387.8 422.2  (£)
```

TERRACE   LAND USE CODE 13   JANUARY 1990 PRICES   SALT WATER FLOOD DURATION MORE THAN 12 HOURS
                                           DEPTH ABOVE UPPER SURFACE OF GROUND FLOOR
COMPONENTS OF DAMAGE -0.3   0.0   0.05   0.1   0.2   0.3   0.6   0.9   1.2   1.5   1.8   2.1   2.4   2.7   3.0 (M)

COMPONENTS OF DAMAGE	-0.3	0.0	0.05	0.1	0.2	0.3	0.6	0.9	1.2	1.5	1.8	2.1	2.4	2.7	3.0	
PATHS & PAVED AREAS	0	4	9	15	27	46	56	61	64	68	71	75	76	77	78	(£)
COL PERCENT	0.0	0.7	0.6	0.5	0.6	0.8	0.7	0.7	0.6	0.6	0.6	0.6	0.5	0.5	0.5	
ROW PERCENT	0.0	5.2	12.4	19.4	34.8	59.2	71.0	77.9	82.0	86.3	90.6	96.3	96.8	98.5	100.0	
GARDENS/FENCES/SHEDS	0	1	22	44	134	296	431	487	690	995	1129	1330	1543	1677	1749	(£)
COL PERCENT	0.1	0.3	1.3	1.5	3.0	4.9	5.7	5.4	6.7	8.7	9.2	10.1	11.1	11.5	11.3	
ROW PERCENT	0.0	0.1	1.3	2.5	7.7	16.9	24.7	27.8	39.5	56.9	64.6	76.1	88.2	95.9	100.0	
EXT. MAIN BUILDING	275	366	373	408	484	555	663	784	933	1193	1618	1898	2380	2921	3753	(£)
COL PERCENT	78.2	64.8	21.3	13.9	10.8	9.3	8.7	8.8	9.1	10.5	13.2	14.4	17.1	20.0	24.3	
ROW PERCENT	7.3	9.8	10.0	10.9	12.9	14.8	17.7	20.9	24.9	31.8	43.1	50.6	63.4	77.8	100.0	
PLASTERWORK	22	59	97	234	395	562	884	1228	1590	1881	2062	2307	2296	2298	2292	(£)
COL PERCENT	6.4	10.5	5.6	7.9	8.8	9.4	11.6	13.7	15.5	16.5	16.8	17.5	16.5	15.8	14.8	
ROW PERCENT	1.0	2.6	4.2	10.2	17.3	24.5	38.6	53.6	69.4	82.1	90.0	100.7	100.2	100.3	100.0	
FLOORS	0	61	467	699	986	1075	1167	1175	1523	1699	1663	1663	1663	1663	1663	(£)
COL PERCENT	0.0	10.9	26.7	23.7	22.0	17.9	15.3	13.1	14.8	14.9	13.6	12.6	12.0	11.4	10.7	
ROW PERCENT	0.0	3.7	28.1	42.1	59.3	64.7	70.2	70.7	91.6	102.2	100.0	100.0	100.0	100.0	100.0	
JOINERY	0	0	76	310	581	900	1166	1544	1733	1789	1887	2067	2061	2061	2061	(£)
COL PERCENT	0.0	0.0	4.4	10.5	13.0	15.0	15.3	17.3	16.9	15.7	15.4	15.7	14.8	14.1	13.3	
ROW PERCENT	0.0	0.0	3.7	15.1	28.2	43.7	56.6	74.9	84.1	86.8	91.6	100.3	100.0	100.0	100.0	
INTERNAL DECORATIONS	14	18	39	206	300	321	330	346	349	353	360	358	363	375	374	(£)
COL PERCENT	4.1	3.3	2.2	7.0	6.7	5.4	4.3	3.9	3.4	3.1	2.9	2.7	2.6	2.6	2.4	
ROW PERCENT	3.9	5.0	10.5	55.1	80.2	85.9	88.3	92.4	93.3	94.4	96.1	95.8	97.1	100.1	100.0	
PLUMBING & ELECTRICAL	39	53	72	143	208	230	417	433	435	439	448	453	458	459	461	(£)
COL PERCENT	11.2	9.4	4.1	4.9	4.7	3.8	5.5	4.8	4.2	3.9	3.7	3.4	3.3	3.2	3.0	
ROW PERCENT	8.5	11.5	15.7	31.1	45.1	49.9	90.5	93.8	94.3	95.2	97.1	98.1	99.2	99.5	100.0	
DOMESTIC APPLIANCES	0	0	2	63	143	295	333	472	472	474	474	474	474	474	474	(£)
COL PERCENT	0.0	0.0	0.1	2.2	3.2	4.9	4.4	5.3	4.6	4.2	3.9	3.6	3.4	3.3	3.1	
ROW PERCENT	0.0	0.0	0.5	13.5	30.3	62.3	70.2	99.5	99.5	100.0	100.0	100.0	100.0	100.0	100.0	
HEATING EQUIPMENT	0	0	0	25	32	80	101	115	125	125	125	129	129	129	129	(£)
COL PERCENT	0.0	0.0	0.0	0.9	0.7	1.3	1.3	1.3	1.2	1.1	1.0	1.0	0.9	0.9	0.8	
ROW PERCENT	0.0	0.0	0.0	19.7	25.0	62.0	78.1	88.9	97.3	97.3	97.3	100.0	100.0	100.0	100.0	
AUDIO/VIDEO	0	0	28	104	155	188	323	398	410	410	410	410	410	410	410	(£)
COL PERCENT	0.0	0.0	1.6	3.5	3.5	3.1	4.2	4.5	4.0	3.6	3.3	3.1	3.0	2.8	2.7	
ROW PERCENT	0.0	0.0	7.0	25.4	37.9	45.9	78.8	97.0	100.0	100.0	100.0	100.0	100.0	100.0	100.0	
FURNITURE	0	0	46	124	271	565	738	790	800	802	802	802	802	802	802	(£)
COL PERCENT	0.0	0.0	2.7	4.2	6.1	9.4	9.7	8.8	7.8	7.0	6.5	6.1	5.8	5.5	5.2	
ROW PERCENT	0.0	0.0	5.8	15.5	33.9	70.5	92.0	98.6	99.7	100.0	100.0	100.0	100.0	100.0	100.0	
PERSONAL EFFECTS	0	0	0	0	22	108	150	209	209	216	216	216	216	216	216	(£)
COL PERCENT	0.0	0.0	0.0	0.0	0.5	1.8	2.0	2.3	2.0	1.9	1.8	1.6	1.6	1.5	1.4	
ROW PERCENT	0.0	0.0	0.0	0.0	10.5	50.2	69.5	96.6	96.6	100.0	100.0	100.0	100.0	100.0	100.0	
FLOOR COVER/CURTAINS	0	0	386	409	432	432	469	476	481	481	481	481	481	481	481	(£)
COL PERCENT	0.0	0.0	22.0	13.9	9.7	7.2	6.2	5.3	4.7	4.2	3.9	3.7	3.5	3.3	3.1	
ROW PERCENT	0.0	0.0	80.3	85.0	89.8	89.8	97.5	98.9	100.0	100.0	100.0	100.0	100.0	100.0	100.0	
GARDEN/DIY/LEISURE	0	0	25	52	77	106	143	162	168	178	183	183	189	189	189	(£)
COL PERCENT	0.0	0.0	1.5	1.8	1.7	1.8	1.9	1.8	1.6	1.6	1.5	1.4	1.4	1.3	1.2	
ROW PERCENT	0.0	0.0	13.7	27.6	40.9	56.5	76.0	85.9	89.1	94.2	96.7	96.7	100.0	100.0	100.0	
DOMESTIC CLEAN-UP	0	0	104	107	225	233	242	252	267	284	318	336	336	336	336	(£)
COL PERCENT	0.0	0.0	5.9	3.6	5.0	3.9	3.2	2.8	2.6	2.5	2.6	2.6	2.4	2.3	2.2	
ROW PERCENT	0.0	0.0	31.0	31.9	66.9	69.5	72.1	75.0	79.4	84.5	94.4	100.0	100.0	100.0	100.0	

TERRACE   LAND USE CODE 13   JANUARY 1990 PRICES   SALT WATER FLOOD DURATION MORE THAN 12 HOURS
                                           DEPTH ABOVE UPPER SURFACE OF GROUND FLOOR

COMPONENTS OF DAMAGE	-0.3	0.0	0.05	0.1	0.2	0.3	0.6	0.9	1.2	1.5	1.8	2.1	2.4	2.7	3.0	
TOTAL BUILDING FABRIC	351	565	1160	2060	3120	3990	5120	6060	7320	8420	9240	10150	10840	11530	12430	(£)
COL PERCENT	100.0	100.0	66.1	69.9	69.6	66.5	67.2	67.8	71.4	73.9	75.4	77.0	78.1	79.1	80.3	
ROW PERCENT	2.8	4.5	9.3	16.6	25.1	32.1	41.2	48.7	58.9	67.7	74.3	81.7	87.2	92.8	100.0	
TOTAL INVENTORY	0	0	594	887	1360	2010	2500	2880	2940	2980	3010	3040	3040	3040	3040	(£)
COL PERCENT	0.0	0.0	33.9	30.1	30.4	33.5	32.8	32.2	28.6	26.1	24.6	23.0	21.9	20.9	19.7	
ROW PERCENT	0.0	0.0	19.5	29.2	44.8	66.2	82.3	94.6	96.5	97.8	99.1	99.8	100.0	100.0	100.0	
TOTAL DAMAGE	351	565	1750	2950	4480	6000	7620	8940	10260	11400	12260	13190	13890	14580	15480	(£)
ROW PERCENT	2.3	3.7	11.3	19.1	28.9	38.8	49.2	57.8	66.3	73.6	79.2	85.2	89.7	94.2	100.0	
U.C. LIMIT	451	656	1950	3290	5000	6710	8510	9980	11420	12660	13590	14620	15380	16130	17160	(£)
L.C. LIMIT	251	474	1560	2610	3970	5300	6730	7910	9100	10140	10930	11760	12400	13020	13800	(£)
TOTAL DAMAGE/SQ.M	8.4	13.9	43.3	72.4	110.2	148.2	188.5	221.4	253.4	281.1	302.0	324.7	341.6	358.6	380.4	(£)
ROW PERCENT	2.2	3.7	11.4	19.0	29.0	39.0	49.6	58.2	66.6	73.9	79.4	85.4	89.8	94.3	100.0	
U.C. LIMIT	10.8	16.2	47.6	79.4	120.8	164.0	208.5	245.0	278.8	308.1	330.0	354.2	372.0	390.6	415.1	(£)
L.C. LIMIT	6.0	11.6	39.0	65.4	99.6	132.4	168.5	197.8	228.0	254.1	274.0	295.2	311.2	326.6	345.7	(£)

APPENDIX 5.2   321

```
BUNGALOW                     LAND USE CODE 14      JANUARY 1990 PRICES   SALT WATER FLOOD DURATION MORE THAN 12 HOURS
                                                DEPTH ABOVE UPPER SURFACE OF GROUND FLOOR
COMPONENTS OF DAMAGE  -0.3    0.0   0.05    0.1    0.2    0.3    0.6    0.9    1.2    1.5    1.8    2.1    2.4    2.7    3.0  (M)
--------------------  ----   ----  -----  -----  -----  -----  -----  -----  -----  -----  -----  -----  -----  -----  -----
PATHS & PAVED AREAS      0      7     17     27     63    106    146    172    180    191    201    211    228    244    258  (£)
    COL PERCENT        0.0    1.5    0.6    0.5    0.6    0.8    0.8    0.8    0.7    0.7    0.7    0.7    0.7    0.7    0.7
    ROW PERCENT        0.0    3.0    6.9   10.7   24.7   41.2   56.7   66.5   69.9   74.1   78.1   81.6   88.2   94.7  100.0
GARDENS/FENCES/SHEDS     0      5     71    148    559    979   1313   1411   2043   2748   3342   4091   4850   4905   4927  (£)
    COL PERCENT        0.0    1.2    2.2    2.5    5.6    7.1    7.3    6.7    8.2   10.0   11.5   13.0   14.2   13.8   13.4
    ROW PERCENT        0.0    0.1    1.5    3.0   11.3   19.9   26.7   28.6   41.5   55.8   67.8   83.0   98.4   99.5  100.0
EXT. MAIN BUILDING      54    281    297    379    580    739    943   1151   2040   2959   3978   5401   6877   8165   9548  (£)
    COL PERCENT       62.3   56.2    9.2    6.4    5.8    5.4    5.3    5.5    8.2   10.8   13.7   17.1   20.2   23.0   25.9
    ROW PERCENT        0.6    2.9    3.1    4.0    6.1    7.7    9.9   12.1   21.4   31.0   41.7   56.6   72.0   85.5  100.0
PLASTERWORK              1      4     55    395    760   1094   1892   2696   3427   3819   3541   3821   4011   4012   4012  (£)
    COL PERCENT        1.8    0.8    1.7    6.7    7.7    8.0   10.5   12.8   13.8   13.9   12.2   12.1   11.8   11.3   10.9
    ROW PERCENT        0.0    0.1    1.4    9.9   19.0   27.3   47.2   67.2   85.4   95.2   88.2   95.2  100.0  100.0  100.0
FLOORS                   0    132    680   1317   1915   2291   2525   2609   3202   3507   3438   3438   3438   3438   3438  (£)
    COL PERCENT        0.0   26.5   21.1   22.2   19.3   16.7   14.1   12.4   12.9   12.8   11.9   10.9   10.1    9.7    9.3
    ROW PERCENT        0.0    3.8   19.8   38.3   55.7   66.6   73.4   75.9   93.1  102.0  100.0  100.0  100.0  100.0  100.0
JOINERY                  0      0    172    509    898   1372   1770   2448   3145   3202   3389   3427   3423   3423   3423  (£)
    COL PERCENT        0.0    0.0    5.3    8.6    9.0   10.0    9.9   11.7   12.6   11.7   11.7   10.9   10.1    9.7    9.3
    ROW PERCENT        0.0    0.0    5.0   14.9   26.2   40.1   51.7   71.5   91.9   93.6   99.0  100.1  100.0  100.0  100.0
INTERNAL DECORATIONS     3     14     77    386    557    611    626    658    665    671    678    686    701    738    739  (£)
    COL PERCENT        4.2    2.9    2.4    6.5    5.6    4.5    3.5    3.1    2.7    2.5    2.3    2.2    2.1    2.1    2.0
    ROW PERCENT        0.5    2.0   10.5   52.3   75.4   82.7   84.8   89.1   90.0   90.8   91.7   92.9   94.8   99.9  100.0
PLUMBING & ELECTRICAL   27     54     89    161    275    314    543    591    611    626    666    680    694    711    732  (£)
    COL PERCENT       31.6   10.9    2.8    2.7    2.8    2.3    3.0    2.8    2.5    2.3    2.3    2.2    2.0    2.0    2.0
    ROW PERCENT        3.8    7.4   12.3   22.1   37.6   42.9   74.2   80.7   83.6   85.6   91.0   92.9   94.8   97.2  100.0

DOMESTIC APPLIANCES      0      0      2     85    203    431    521    724    724    731    731    731    731    731    731  (£)
    COL PERCENT        0.0    0.0    0.1    1.4    2.1    3.1    2.9    3.4    2.9    2.7    2.5    2.3    2.1    2.1    2.0
    ROW PERCENT        0.0    0.0    0.4   11.7   27.9   59.0   71.3   99.1   99.1  100.0  100.0  100.0  100.0  100.0  100.0
HEATING EQUIPMENT        0      0      5     29     46    124    158    191    216    216    216    223    223    223    223  (£)
    COL PERCENT        0.0    0.0    0.2    0.5    0.5    0.9    0.9    0.9    0.8    0.7    0.7    0.7    0.7    0.6    0.6
    ROW PERCENT        0.0    0.0    2.3   13.2   20.7   55.6   70.9   85.6   96.7   96.7   96.7  100.0  100.0  100.0  100.0
AUDIO/VIDEO              0      0     56    156    227    282    461    592    607    607    607    607    607    607    607  (£)
    COL PERCENT        0.0    0.0    1.8    2.6    2.3    2.1    2.6    2.8    2.4    2.2    2.1    1.9    1.8    1.7    1.6
    ROW PERCENT        0.0    0.0    9.3   25.8   37.5   46.6   76.0   97.7  100.0  100.0  100.0  100.0  100.0  100.0  100.0
FURNITURE                0      0     84    326   1266   1629   2525   2616   2744   2748   2748   2748   2748   2748   2748  (£)
    COL PERCENT        0.0    0.0    2.6    5.5   12.8   11.9   14.1   12.5   11.0   10.0    9.5    8.7    8.1    7.8    7.4
    ROW PERCENT        0.0    0.0    3.1   11.9   46.1   59.3   91.9   95.2   99.9  100.0  100.0  100.0  100.0  100.0  100.0
PERSONAL EFFECTS         0      0     70    403    604   1729   2369   2911   2996   3006   3006   3006   3006   3006   3006  (£)
    COL PERCENT        0.0    0.0    2.2    6.8    6.1   12.6   13.2   13.9   12.0   11.0   10.4    9.5    8.8    8.5    8.1
    ROW PERCENT        0.0    0.0    2.3   13.4   20.1   57.5   78.8   96.8   99.7  100.0  100.0  100.0  100.0  100.0  100.0
FLOOR COVER/CURTAINS     0      0   1231   1256   1284   1284   1350   1381   1407   1407   1407   1407   1407   1407   1407  (£)
    COL PERCENT        0.0    0.0   38.2   21.1   12.9    9.3    7.5    6.6    5.6    5.1    4.9    4.5    4.1    4.0    3.8
    ROW PERCENT        0.0    0.0   87.5   89.3   91.2   91.2   96.0   98.2  100.0  100.0  100.0  100.0  100.0  100.0  100.0
GARDEN/DIY/LEISURE       0      0     53     97    139    174    216    239    246    257    262    262    269    269    269  (£)
    COL PERCENT        0.0    0.0    1.7    1.6    1.4    1.3    1.2    1.1    1.0    0.9    0.9    0.8    0.8    0.8    0.7
    ROW PERCENT        0.0    0.0   19.9   36.1   51.7   64.8   80.4   88.9   91.5   95.4   97.5  100.0  100.0  100.0  100.0
DOMESTIC CLEAN-UP        0      0    254    262    550    571    591    616    653    694    777    822    822    822    822  (£)
    COL PERCENT        0.0    0.0    7.9    4.4    5.5    4.2    3.3    2.9    2.6    2.5    2.7    2.6    2.4    2.3    2.2
    ROW PERCENT        0.0    0.0   31.0   32.0   67.0   69.5   72.0   75.0   79.5   84.5   94.5  100.0  100.0  100.0  100.0
```

```
BUNGALOW                     LAND USE CODE 14      JANUARY 1990 PRICES   SALT WATER FLOOD DURATION MORE THAN 12 HOURS
                                                DEPTH ABOVE UPPER SURFACE OF GROUND FLOOR
COMPONENTS OF DAMAGE  -0.3    0.0   0.05    0.1    0.2    0.3    0.6    0.9    1.2    1.5    1.8    2.1    2.4    2.7    3.0  (M)
--------------------  ----  -----  -----  -----  -----  -----  -----  -----  -----  -----  -----  -----  -----  -----  -----
TOTAL BUILDING FABRIC   86    500   1460   3330   5610   7510   9760  11740  15320  17730  19240  21760  24230  25640  27080  (£)
    COL PERCENT      100.0  100.0   45.4   56.0   56.5   54.4   55.9   61.5   64.7   66.3   68.9   71.2   72.3   73.4
    ROW PERCENT        0.3    1.8    5.4   12.3   20.7   27.7   36.1   43.4   56.6   65.5   71.0   80.3   89.5   94.7  100.0
TOTAL INVENTORY          0      0   1760   2620   4320   6230   8200   9270   9600   9670   9760   9810   9820   9820   9820  (£)
    COL PERCENT        0.0    0.0   54.6   44.0   43.5   45.6   44.1   38.5   35.3   33.7   31.1   28.8   27.7   26.6
    ROW PERCENT        0.0    0.0   17.9   26.7   44.0   63.4   83.5   94.5   97.8   98.5   99.4   99.9  100.0  100.0  100.0

TOTAL DAMAGE            86    500   3220   5950   9930  13740  17960  21010  24920  27400  28990  31570  34040  35460  36900  (£)
    ROW PERCENT        0.2    1.4    8.7   16.1   26.9   37.2   48.7   57.0   67.5   74.3   78.6   85.6   92.3   96.1  100.0
    U.C. LIMIT         143    536   3590   6580  10920  15150  19790  23100  27200  29690  31310  33930  36480  37940  39410  (£)
    L.C. LIMIT          29    464   2860   5320   8960  12330  16130  18930  22640  25110  26680  29210  31610  32970  34380  (£)
TOTAL DAMAGE/SQ.M      1.1    5.5   34.4   63.3  106.4  147.2  192.4  225.0  266.2  293.0  310.2  337.8  363.8  378.9  394.4  (£)
    ROW PERCENT        0.3    1.4    8.7   16.0   27.0   37.3   48.8   57.0   67.5   74.3   78.7   85.6   92.2   96.1  100.0
    U.C. LIMIT         1.9    6.2   38.5   70.1  118.0  163.9  213.9  249.5  292.4  320.0  338.0  366.5  393.1  408.7  424.9  (£)
    L.C. LIMIT         0.3    4.8   30.3   56.5   94.8  130.5  170.9  200.5  240.0  266.0  282.4  309.1  334.5  349.1  363.9  (£)
```

322 THE ECONOMICS OF COASTAL MANAGEMENT

FLAT   LAND USE CODE 15   JANUARY 1990 PRICES   SALT WATER FLOOD DURATION MORE THAN 12 HOURS

DEPTH ABOVE UPPER SURFACE OF GROUND FLOOR

COMPONENTS OF DAMAGE	-0.3	0.0	0.05	0.1	0.2	0.3	0.6	0.9	1.2	1.5	1.8	2.1	2.4	2.7	3.0	(M)
PATHS & PAVED AREAS	0	4	7	10	21	34	45	53	52	53	53	54	55	56	56	(£)
COL PERCENT	0.0	1.1	0.5	0.4	0.5	0.5	0.5	0.6	0.5	0.4	0.4	0.4	0.4	0.4	0.4	
ROW PERCENT	0.0	7.2	12.8	17.9	38.5	60.2	79.9	94.5	92.9	93.7	94.8	95.4	97.4	98.9	100.0	
GARDENS/FENCES/SHEDS	0	0	15	32	79	115	177	196	257	385	493	632	773	806	822	(£)
COL PERCENT	0.0	0.0	0.9	1.1	1.7	1.8	2.1	2.0	2.3	3.2	3.9	4.6	5.4	5.4	5.3	
ROW PERCENT	0.0	0.0	1.8	3.9	9.7	14.0	21.6	23.9	31.3	46.8	60.0	76.8	94.0	98.0	100.0	
EXT. MAIN BUILDING	79	213	227	277	368	457	591	718	981	1300	1650	2103	2609	3085	3665	(£)
COL PERCENT	64.3	57.0	14.3	9.7	7.9	7.2	7.1	7.4	8.8	10.8	13.0	15.5	18.1	20.6	23.5	
ROW PERCENT	2.2	5.8	6.2	7.6	10.0	12.5	16.1	19.6	26.8	35.5	45.0	57.4	71.2	84.2	100.0	
PLASTERWORK	5	22	53	222	396	555	904	1267	1615	1820	1764	2016	2146	2174	2208	(£)
COL PERCENT	4.5	6.0	3.4	7.7	8.5	8.8	10.9	13.0	14.5	15.1	13.9	14.8	14.9	14.5	14.2	
ROW PERCENT	0.3	1.0	2.4	10.1	17.9	25.2	41.0	57.4	73.2	82.4	79.9	91.3	97.2	98.5	100.0	
FLOORS	0	80	397	677	966	1131	1252	1286	1658	1838	1793	1787	1787	1787	1787	(£)
COL PERCENT	0.0	21.5	25.0	23.5	20.8	17.8	15.1	13.2	14.9	15.2	14.2	13.1	12.4	12.0	11.5	
ROW PERCENT	0.0	4.5	22.3	37.9	54.0	63.3	70.1	72.0	92.8	102.8	100.3	100.0	100.0	100.0	100.0	
JOINERY	0	0	63	146	305	519	706	978	1224	1261	1421	1458	1454	1454	1454	(£)
COL PERCENT	0.0	0.0	4.0	5.1	6.6	8.2	8.5	10.1	11.0	10.4	11.2	10.7	10.1	9.7	9.3	
ROW PERCENT	0.0	0.0	4.4	10.1	21.0	35.7	48.6	67.2	84.2	86.7	97.7	100.3	100.0	100.0	100.0	
INTERNAL DECORATIONS	0	5	25	205	293	327	341	367	365	368	372	376	381	396	395	(£)
COL PERCENT	0.0	1.4	1.6	7.1	6.3	5.2	4.1	3.8	3.3	3.1	2.9	2.8	2.6	2.6	2.5	
ROW PERCENT	0.0	1.4	6.5	51.9	74.1	82.8	86.3	92.9	92.3	93.0	94.2	95.0	96.3	100.1	100.0	
PLUMBING & ELECTRICAL	38	48	68	127	195	220	394	409	413	430	457	495	500	502	507	(£)
COL PERCENT	31.2	13.0	4.3	4.4	4.2	3.5	4.8	4.2	3.7	3.6	3.6	3.6	3.5	3.4	3.3	
ROW PERCENT	7.7	9.6	13.6	25.0	38.4	43.5	77.7	80.6	81.5	84.8	90.2	97.6	98.6	99.0	100.0	
DOMESTIC APPLIANCES	0	0	1	75	172	354	413	576	576	578	578	578	578	578	578	(£)
COL PERCENT	0.0	0.0	0.1	2.6	3.7	5.6	5.0	5.9	5.2	4.8	4.6	4.3	4.0	3.9	3.7	
ROW PERCENT	0.0	0.0	0.3	13.1	29.8	61.3	71.5	99.6	99.6	100.0	100.0	100.0	100.0	100.0	100.0	
HEATING EQUIPMENT	0	0	2	24	29	76	92	109	121	121	121	125	125	125	125	(£)
COL PERCENT	0.0	0.0	0.2	0.9	0.6	1.2	1.1	1.1	1.1	1.0	1.0	0.9	0.9	0.8	0.8	
ROW PERCENT	0.0	0.0	2.1	19.7	23.8	60.6	73.9	87.0	96.9	96.9	96.9	100.0	100.0	100.0	100.0	
AUDIO/VIDEO	0	0	32	91	134	164	276	344	353	353	353	353	353	353	353	(£)
COL PERCENT	0.0	0.0	2.0	3.2	2.9	2.6	3.3	3.5	3.2	2.9	2.8	2.6	2.5	2.4	2.3	
ROW PERCENT	0.0	0.0	9.2	26.0	38.1	46.4	78.2	97.5	100.0	100.0	100.0	100.0	100.0	100.0	100.0	
FURNITURE	0	0	51	169	580	796	1174	1215	1253	1255	1255	1255	1255	1255	1255	(£)
COL PERCENT	0.0	0.0	3.3	5.9	12.5	12.6	14.1	12.5	11.3	10.4	9.9	9.2	8.7	8.4	8.1	
ROW PERCENT	0.0	0.0	4.1	13.5	46.2	63.4	93.6	96.8	99.8	100.0	100.0	100.0	100.0	100.0	100.0	
PERSONAL EFFECTS	0	0	22	146	242	683	949	1174	1201	1209	1209	1209	1209	1209	1209	(£)
COL PERCENT	0.0	0.0	1.4	5.1	5.2	10.8	11.4	12.1	10.8	10.0	9.5	8.9	8.4	8.1	7.8	
ROW PERCENT	0.0	0.0	1.9	12.1	20.0	56.5	78.5	97.1	99.3	100.0	100.0	100.0	100.0	100.0	100.0	
FLOOR COVER/CURTAINS	0	0	445	458	472	472	501	517	526	526	526	526	526	526	526	(£)
COL PERCENT	0.0	0.0	28.0	15.9	10.2	7.4	6.0	5.3	4.7	4.4	4.2	3.9	3.7	3.5	3.4	
ROW PERCENT	0.0	0.0	84.5	87.0	89.6	89.6	95.3	98.1	100.0	100.0	100.0	100.0	100.0	100.0	100.0	
GARDEN/DIY/LEISURE	0	0	37	70	101	130	164	181	187	196	201	201	206	206	206	(£)
COL PERCENT	0.0	0.0	2.4	2.4	2.2	2.1	2.0	1.9	1.7	1.6	1.5	1.4	1.4	1.3		
ROW PERCENT	0.0	0.0	18.1	34.0	49.2	63.1	79.7	87.9	90.8	95.1	97.3	97.3	100.0	100.0	100.0	
DOMESTIC CLEAN-UP	0	0	135	139	292	303	314	327	347	369	412	437	437	437	437	(£)
COL PERCENT	0.0	0.0	8.5	4.9	6.3	4.8	3.8	3.4	3.1	3.1	3.3	3.2	3.0	2.9	2.8	
ROW PERCENT	0.0	0.0	31.0	31.9	67.0	69.5	72.0	75.0	79.5	84.4	94.5	100.0	100.0	100.0	100.0	

FLAT   LAND USE CODE 15   JANUARY 1990 PRICES   SALT WATER FLOOD DURATION MORE THAN 12 HOURS

DEPTH ABOVE UPPER SURFACE OF GROUND FLOOR

COMPONENTS OF DAMAGE	-0.3	0.0	0.05	0.1	0.2	0.3	0.6	0.9	1.2	1.5	1.8	2.1	2.4	2.7	3.0	(M)
TOTAL BUILDING FABRIC	124	375	859	1700	2630	3360	4410	5280	6570	7460	8010	8920	9710	10260	10900	(£)
COL PERCENT	100.0	100.0	54.1	59.1	56.4	53.0	53.2	54.3	59.0	61.8	63.2	65.6	67.4	68.6	69.9	
ROW PERCENT	1.1	3.4	7.9	15.6	24.1	30.8	40.5	48.4	60.3	68.4	73.5	81.9	89.1	94.2	100.0	
TOTAL INVENTORY	0	0	729	1180	2030	2980	3890	4450	4570	4610	4660	4690	4690	4690	4690	(£)
COL PERCENT	0.0	0.0	45.9	40.9	43.6	47.0	46.8	45.7	41.0	38.2	36.8	34.4	32.6	31.4	30.1	
ROW PERCENT	0.0	0.0	15.5	25.1	43.2	63.5	82.8	94.7	97.3	98.3	99.3	99.9	100.0	100.0	100.0	
TOTAL DAMAGE	124	375	1590	2880	4650	6340	8300	9720	11140	12070	12670	13610	14400	14960	15590	(£)
ROW PERCENT	0.8	2.4	10.2	18.4	29.8	40.7	53.2	62.4	71.4	77.4	81.2	87.3	92.4	95.9	100.0	
U.C. LIMIT	176	421	1840	3310	5380	7340	9590	11170	12710	13710	14310	15320	16140	16740	17460	(£)
L.C. LIMIT	72	329	1340	2450	3930	5350	7020	8280	9570	10440	11020	11910	12670	13180	13730	(£)
TOTAL DAMAGE/SQ.M	2.5	8.5	33.1	59.7	96.5	132.1	173.1	203.6	233.5	253.4	266.9	287.3	304.4	316.0	329.0	(£)
ROW PERCENT	0.8	2.6	10.1	18.1	29.3	40.2	52.6	61.9	71.0	77.0	81.1	87.3	92.5	96.0	100.0	
U.C. LIMIT	3.6	10.0	36.7	65.0	106.1	146.3	191.7	225.3	256.4	276.3	290.3	311.4	328.8	340.7	354.2	(£)
L.C. LIMIT	1.4	7.0	29.5	54.4	86.9	117.9	154.5	181.9	210.6	230.5	243.5	263.2	280.0	291.3	303.8	(£)

PREFAB   LAND USE CODE 17   JANUARY 1990 PRICES   SALT WATER FLOOD DURATION MORE THAN 12 HOURS
                                             DEPTH ABOVE UPPER SURFACE OF GROUND FLOOR
COMPONENTS OF DAMAGE -0.3   0.0   0.05   0.1   0.2   0.3   0.6   0.9   1.2   1.5   1.8   2.1   2.4   2.7   3.0 (M)

PATHS & PAVED AREAS   0   2   4   6   12   19   25   30   29   29   29   29   29   29   29 (£)
  COL PERCENT   0.0   2.1   0.2   0.2   0.2   0.3   0.3   0.3   0.2   0.2   0.2   0.2   0.2   0.2   0.2
  ROW PERCENT   0.0   6.9   13.8 20.7   41.4   65.5   86.2 103.4 100.0 100.0 100.0 100.0 100.0 100.0 100.0
GARDENS/FENCES/SHEDS   0   0   41   92   176   211   345   380   497   813   1078   1408   1737   1737   1737 (£)
  COL PERCENT   0.0   0.0   2.5   3.0   3.4   2.8   3.5   3.3   3.8   5.7   7.1   8.8   10.3   9.9   9.7
  ROW PERCENT   0.0   0.0   2.4   5.3   10.1   12.1   19.9   21.9   28.6   46.8   62.1   81.1 100.0 100.0 100.0
EXT. MAIN BUILDING   0   46   48   66   143   212   369   518   906   1199   1575   1962   2433   3107   3581 (£)
  COL PERCENT   0.0 48.8   3.0   2.2   2.7   2.8   3.8   4.5   7.0   8.5   10.4   12.3   14.5   17.8   19.9
  ROW PERCENT   0.0   1.3   1.3   1.8   4.0   5.9   10.3   14.5   25.3   33.5   44.0   54.8   67.9   86.8 100.0
PLASTERWORK   0   0   370   737   1096   1393   1501   1619   1636   1723   1844   1941   1941   1941   1941 (£)
  COL PERCENT   0.0   0.0   22.8 24.4   20.9   18.4   15.3   14.0   12.6   12.2   12.2   12.2   11.6   11.1   10.8
  ROW PERCENT   0.0   0.0   19.1 38.0   56.5   71.8   77.3   83.4   84.3   88.8   95.0 100.0 100.0 100.0 100.0
FLOORS   0   0   28   248   504   846   1161   1482   1924   2379   2458   2458   2458   2458   2458 (£)
  COL PERCENT   0.0   0.0   1.7   8.2   9.6   11.2   11.8   12.8   14.9   16.8   16.3   15.4   14.6   14.1   13.7
  ROW PERCENT   0.0   0.0   1.1 10.1   20.5   34.4   47.2   60.3   78.3   96.8 100.0 100.0 100.0 100.0 100.0
JOINERY   0   0   94   162   279   440   538   789   1046   1079   1115   1115   1115   1115   1115 (£)
  COL PERCENT   0.0   0.0   5.8   5.4   5.3   5.8   5.5   6.8   8.1   7.6   7.4   7.0   6.6   6.4   6.2
  ROW PERCENT   0.0   0.0   8.4 14.5   25.0   39.5   48.3   70.8   93.8   96.8 100.0 100.0 100.0 100.0 100.0
INTERNAL DECORATIONS   0   5   25   77   187   284   360   508   489   490   492   498   504   523   523 (£)
  COL PERCENT   0.0   5.3   1.5   2.5   3.6   3.7   3.7   4.4   3.8   3.5   3.3   3.1   3.0   3.0   2.9
  ROW PERCENT   0.0   1.0   4.8 14.7   35.8   54.3   68.8   97.1   93.5   93.7   94.1   95.2   96.4 100.0 100.0
PLUMBING & ELECTRICAL   39   41   44   77   127   141   145   146   159   162   165   172   172   183   184 (£)
  COL PERCENT   100.0 43.8   2.8   2.6   2.4   1.9   1.5   1.3   1.2   1.1   1.1   1.1   1.0   1.1   1.0
  ROW PERCENT   21.2 22.5   24.4 42.1   69.2   76.7   79.3   79.5   86.6   88.5   90.0   93.8   93.9   99.9 100.0

DOMESTIC APPLIANCES   0   0   2   82   184   381   456   629   629   630   630   630   630   630   630 (£)
  COL PERCENT   0.0   0.0   0.1   2.7   3.5   5.0   4.6   5.5   4.9   4.4   4.2   3.9   3.8   3.6   3.5
  ROW PERCENT   0.0   0.0   0.4 13.0   29.3   60.5   72.3   99.8   99.8 100.0 100.0 100.0 100.0 100.0 100.0
HEATING EQUIPMENT   0   0   5   31   39   98   122   143   159   159   159   164   164   164   164 (£)
  COL PERCENT   0.0   0.0   0.3   1.0   0.7   1.3   1.2   1.2   1.2   1.1   1.1   1.0   1.0   0.9   0.9
  ROW PERCENT   0.0   0.0   3.1 19.4   24.1   59.6   74.1   86.9   96.8   96.8   96.8 100.0 100.0 100.0 100.0
AUDIO/VIDEO   0   0   46   127   184   222   375   461   473   473   473   473   473   473   473 (£)
  COL PERCENT   0.0   0.0   2.8   4.2   3.5   2.9   3.8   4.0   3.7   3.3   3.1   3.0   2.8   2.7   2.6
  ROW PERCENT   0.0   0.0   9.8 27.0   38.9   47.1   79.4   97.5 100.0 100.0 100.0 100.0 100.0 100.0 100.0
FURNITURE   0   0   53   210   822   1106   1648   1669   1726   1729   1729   1729   1729   1729   1729 (£)
  COL PERCENT   0.0   0.0   3.3   6.9   15.7   14.6   16.8   14.5   13.3   12.2   11.4   10.8   10.3   9.9   9.6
  ROW PERCENT   0.0   0.0   3.1 12.2   47.5   64.0   95.3   96.6   99.9 100.0 100.0 100.0 100.0 100.0 100.0
PERSONAL EFFECTS   0   0   31   225   365   1066   1540   1876   1925   1934   1934   1934   1934   1934   1934 (£)
  COL PERCENT   0.0   0.0   1.9   7.4   7.0   14.1   15.7   16.3   14.9   13.6   12.8   12.1   11.5   11.1   10.8
  ROW PERCENT   0.0   0.0   1.6 11.6   18.9   55.1   79.6   97.0   99.6 100.0 100.0 100.0 100.0 100.0 100.0
FLOOR COVER/CURTAINS   0   0   625   636   648   648   684   705   720   720   720   720   720   720   720 (£)
  COL PERCENT   0.0   0.0   38.5 21.0   12.4   8.6   7.0   6.1   5.6   5.1   4.8   4.5   4.3   4.1   4.0
  ROW PERCENT   0.0   0.0   86.7 88.3   90.0   90.0   94.9   97.9 100.0 100.0 100.0 100.0 100.0 100.0 100.0
GARDEN/DIY/LEISURE   0   0   46   85   123   154   191   210   216   225   231   231   236   236   236 (£)
  COL PERCENT   0.0   0.0   2.9   2.8   2.4   2.0   1.9   1.8   1.7   1.6   1.5   1.4   1.4   1.4   1.3
  ROW PERCENT   0.0   0.0   19.7 36.2   52.1   65.3   80.7   88.8   91.4   95.4   97.5   97.5 100.0 100.0 100.0
DOMESTIC CLEAN-UP   0   0   156   161   339   351   364   380   402   428   478   506   506   506   506 (£)
  COL PERCENT   0.0   0.0   9.7   5.3   6.5   4.6   3.7   3.3   3.1   3.0   3.2   3.2   3.0   2.9   2.8
  ROW PERCENT   0.0   0.0   31.0 32.0   67.0   69.4   72.1   75.1   79.6   84.6   94.5 100.0 100.0 100.0 100.0

PREFAB   LAND USE CODE 17   JANUARY 1990 PRICES   SALT WATER FLOOD DURATION MORE THAN 12 HOURS
                                             DEPTH ABOVE UPPER SURFACE OF GROUND FLOOR
COMPONENTS OF DAMAGE -0.3   0.0   0.05   0.1   0.2   0.3   0.6   0.9   1.2   1.5   1.8   2.1   2.4   2.7   3.0 (M)

TOTAL BUILDING FABRIC   39   94   654 1470   2520   3550   4440   5470   6690   7870   8760   9580 10390 11090 11570 (£)
  COL PERCENT   100.0 100.0   40.4 48.4   48.3   46.8   45.2   47.4   51.7   55.5   57.9   60.0   61.9   63.4   64.4
  ROW PERCENT   0.3   0.8   5.7 12.7   21.8   30.7   38.4   47.3   57.8   68.1   75.7   82.8   89.8   95.9 100.0
TOTAL INVENTORY   0   0   966 1560   2710   4030   5380   6080   6250   6300   6360   6390   6400   6400   6400 (£)
  COL PERCENT   0.0   0.0   59.6 51.6   51.7   53.2   54.8   52.6   48.3   44.5   42.1   40.0   38.1   36.6   35.6
  ROW PERCENT   0.0   0.0   15.1 24.4   42.3   63.0   84.1   95.0   97.8   98.5   99.4   99.9 100.0 100.0 100.0

TOTAL DAMAGE   39   94   1620 3030   5230   7580   9830 11550 12940 14180 15110 15970 16790 17490 17960 (£)
  ROW PERCENT   0.2   0.5   9.0 16.8   29.1   42.2   54.7   64.3   72.0   78.9   84.1   88.9   93.4   97.4 100.0
  U.C. LIMIT   38   93   1930 3440   5900   8630 11380 13330 14810 16060 16990 17850 18670 19370 19840 (£)
  L.C. LIMIT   38   93   1280 2540   4440   6360   8050   9510 10800 12020 12960 13810 14620 15330 15800 (£)
TOTAL DAMAGE/SQ.M   0.7   1.7   28.6 53.4   92.4 133.8 173.4 203.9 228.7 250.7 267.4 282.8 297.2 309.8 318.3 (£)
  ROW PERCENT   0.2   0.5   9.0 16.8   29.0   42.0   54.5   64.1   71.9   78.8   84.0   88.8   93.4   97.3 100.0
  U.C. LIMIT   0.7   1.7   34.4 61.5 105.5 154.0 203.1 238.0 264.5 286.7 303.4 318.3 333.3 345.9 354.4 (£)
  L.C. LIMIT   0.7   1.7   22.8 45.3   79.3 113.6 143.7 169.8 192.9 214.7 231.4 246.8 261.1 273.7 282.2 (£)

APPENDIX 5.2

PRE 1918 DETACHED — LAND USE CODE 111 — JANUARY 1990 PRICES — SALT WATER FLOOD DURATION LESS THAN 12 HOURS

DEPTH ABOVE UPPER SURFACE OF GROUND FLOOR

COMPONENTS OF DAMAGE	-0.3	0.0	0.05	0.1	0.2	0.3	0.6	0.9	1.2	1.5	1.8	2.1	2.4	2.7	3.0 (M)
PATHS & PAVED AREAS	0	0	0	1	1	10	13	15	21	29	38	49	71	101	143 (£)
GARDENS/FENCES/SHEDS	0	0	0	1	11	44	106	430	959	1462	2015	2903	3974	4755	5525 (£)
EXT. MAIN BUILDING	431	431	466	467	478	550	580	672	752	1043	1484	2226	3005	3700	4522 (£)
PLASTERWORK	0	0	0	0	546	971	1352	1832	2287	2893	3629	4328	5018	6070	7684 (£)
FLOORS	0	0	58	118	325	554	748	1015	1185	1362	1673	2096	2752	4871	6528 (£)
JOINERY	0	0	0	0	0	0	59	140	412	536	964	1226	1696	2320	2589 (£)
INTERNAL DECORATIONS	57	57	60	98	197	775	993	1059	1124	1201	1246	1265	1293	1366	1394 (£)
PLUMBING & ELECTRICAL	0	38	67	134	148	169	208	245	263	294	549	574	574	574	574 (£)
DOMESTIC APPLIANCES	0	0	0	56	158	327	462	758	766	816	843	843	843	843	843 (£)
HEATING EQUIPMENT	0	0	0	12	32	70	135	148	162	166	166	166	166	166	166 (£)
AUDIO/VIDEO	0	0	47	157	238	296	465	687	722	722	722	722	722	722	722 (£)
FURNITURE	0	0	64	162	435	923	1277	1357	1397	1401	1401	1401	1401	1401	1401 (£)
PERSONAL EFFECTS	0	0	0	0	31	159	286	412	449	477	477	477	477	477	477 (£)
FLOOR COVER/CURTAINS	0	0	848	919	1900	1981	2421	2428	2652	2652	2652	2652	2652	2652	2652 (£)
GARDEN/DIY/LEISURE	0	0	26	57	84	127	158	210	210	229	232	232	241	241	241 (£)
DOMESTIC CLEAN-UP	0	0	409	430	895	930	964	999	1041	1103	1173	1242	1311	1388	1388 (£)

FLOOD DURATION LESS THAN 12 HOURS

	-0.3	0.0	0.05	0.1	0.2	0.3	0.6	0.9	1.2	1.5	1.8	2.1	2.4	2.7	3.0
TOTAL BUILDING FABRIC	488	526	651	820	1710	3080	4060	5410	7000	8820	11600	14670	18390	23760	28960 (£)
COL PERCENT	100.0	100.0	31.8	31.3	31.1	39.0	39.7	43.6	48.6	53.8	60.2	65.5	70.2	75.1	78.6
ROW PERCENT	1.7	1.8	2.2	2.8	5.9	10.6	14.0	18.7	24.2	30.5	40.1	50.7	63.5	82.0	100.0
TOTAL INVENTORY	0	0	1400	1800	3780	4820	6170	7000	7400	7570	7670	7740	7820	7890	7890 (£)
COL PERCENT	0.0	0.0	68.2	68.7	68.9	61.0	60.3	56.4	51.4	46.2	39.8	34.5	29.8	24.9	21.4
ROW PERCENT	0.0	0.0	17.7	22.8	47.9	61.0	78.2	88.7	93.8	95.9	97.2	98.0	99.0	100.0	100.0
TOTAL DAMAGE	488	526	2050	2620	5490	7890	10230	12420	14410	16390	19270	22410	26200	31650	36860 (£)
ROW PERCENT	1.3	1.4	5.6	7.1	14.9	21.4	27.8	33.7	39.1	44.5	52.3	60.8	71.1	85.9	100.0
U.C. LIMIT	715	648	2380	2900	6080	8780	11350	13750	15970	18010	21080	24440	28660	34940	41110 (£)
L.C. LIMIT	260	403	1720	2340	4900	7000	9110	11090	12850	14770	17460	20380	23740	28360	32610 (£)
TOTAL DAMAGE/SQ.METRE	3.1	3.4	13.0	16.7	34.9	50.3	65.2	79.1	91.8	104.4	122.8	142.8	167.0	201.6	234.8 (£)
U.C. LIMIT	4.6	4.1	13.6	18.1	37.8	55.0	71.8	87.5	101.8	116.1	135.6	157.4	182.6	218.0	251.7 (£)
L.C. LIMIT	1.6	2.7	12.4	15.3	32.0	45.6	58.6	70.7	81.8	92.7	110.0	128.2	151.4	185.2	217.9 (£)

INTER WAR DETACHED — LAND USE CODE 112 — JANUARY 1990 PRICES — SALT WATER FLOOD DURATION LESS THAN 12 HOURS

DEPTH ABOVE UPPER SURFACE OF GROUND FLOOR

COMPONENTS OF DAMAGE	-0.3	0.0	0.05	0.1	0.2	0.3	0.6	0.9	1.2	1.5	1.8	2.1	2.4	2.7	3.0 (M)
PATHS & PAVED AREAS	0	0	0	0	1	4	27	51	62	74	92	110	132	177	231 (£)
GARDENS/FENCES/SHEDS	0	0	0	1	42	94	145	371	809	1127	1517	2073	2372	2751	3108 (£)
EXT. MAIN BUILDING	251	251	286	287	292	353	373	404	440	611	854	1305	1765	2121	2600 (£)
PLASTERWORK	0	0	0	0	105	235	364	472	731	968	1532	1720	1987	2266	2453 (£)
FLOORS	0	0	0	59	107	226	341	456	519	589	702	866	1110	1924	2607 (£)
JOINERY	0	0	0	0	25	102	175	298	532	794	1189	1393	1523	1644	1751 (£)
INTERNAL DECORATIONS	11	11	11	290	339	345	358	375	391	412	427	436	450	473	488 (£)
PLUMBING & ELECTRICAL	29	46	61	86	100	106	112	170	208	243	169	192	217	227	239 (£)
DOMESTIC APPLIANCES	0	0	0	57	167	347	465	781	790	844	875	875	875	875	875 (£)
HEATING EQUIPMENT	0	0	0	11	34	64	129	138	147	149	149	149	149	149	149 (£)
AUDIO/VIDEO	0	0	35	152	236	295	466	662	691	691	691	691	691	691	691 (£)
FURNITURE	0	0	37	97	258	606	878	965	992	996	996	996	996	996	996 (£)
PERSONAL EFFECTS	0	0	0	0	34	184	331	475	517	548	548	548	548	548	548 (£)
FLOOR COVER/CURTAINS	0	0	276	316	661	691	882	886	977	977	977	977	977	977	977 (£)
GARDEN/DIY/LEISURE	0	0	29	62	91	137	171	228	228	249	253	253	262	262	262 (£)
DOMESTIC CLEAN-UP	0	0	141	148	308	320	332	344	358	380	403	427	451	477	477 (£)

FLOOD DURATION LESS THAN 12 HOURS

	-0.3	0.0	0.05	0.1	0.2	0.3	0.6	0.9	1.2	1.5	1.8	2.1	2.4	2.7	3.0
TOTAL BUILDING FABRIC	291	308	358	724	1010	1470	1900	2600	3700	4820	6490	8100	9560	11590	13480 (£)
COL PERCENT	100.0	100.0	40.7	46.1	36.1	35.7	34.2	36.7	44.0	49.9	57.0	62.2	65.9	69.9	73.0
ROW PERCENT	2.2	2.3	2.7	5.4	7.5	10.9	14.1	19.3	27.4	35.8	48.1	60.1	70.9	86.0	100.0
TOTAL INVENTORY	0	0	520	845	1790	2650	3660	4480	4700	4840	4900	4920	4950	4980	4980 (£)
COL PERCENT	0.0	0.0	59.3	53.9	63.9	64.3	65.8	63.3	56.0	50.1	43.0	37.8	34.1	30.1	27.0
ROW PERCENT	0.0	0.0	10.5	17.0	36.0	53.2	73.4	90.0	94.5	97.2	98.3	98.8	99.5	100.0	100.0
TOTAL DAMAGE	291	308	878	1570	2810	4110	5550	7080	8400	9660	11380	13020	14510	16560	18460 (£)
ROW PERCENT	1.6	1.7	4.8	8.5	15.2	22.3	30.1	38.4	45.5	52.3	61.7	70.5	78.6	89.7	100.0
U.C. LIMIT	426	379	1020	1740	3110	4570	6160	7840	9310	10620	12450	14200	15870	18280	20590 (£)
L.C. LIMIT	155	236	737	1400	2510	3650	4940	6320	7490	8700	10310	11840	13150	14840	16330 (£)
TOTAL DAMAGE/SQ.METRE	5.4	5.7	16.3	29.1	52.0	76.3	102.9	131.2	155.6	178.9	210.8	241.1	268.8	306.8	341.9 (£)
U.C. LIMIT	7.9	6.8	17.1	31.5	56.3	83.4	113.4	145.1	172.5	198.9	232.7	265.7	293.9	331.8	366.6 (£)
L.C. LIMIT	2.9	4.6	15.5	26.7	47.7	69.2	92.4	117.3	138.7	158.9	188.9	216.5	243.7	281.8	317.2 (£)

APPENDIX 5.2 325

POST WAR DETACHED   LAND USE CODE 113   JANUARY 1990 PRICES   SALT WATER FLOOD DURATION LESS THAN 12 HOURS

COMPONENTS OF DAMAGE	-0.3	0.0	0.05	0.1	0.2	0.3	0.6	0.9	1.2	1.5	1.8	2.1	2.4	2.7	3.0	(M)
PATHS & PAVED AREAS	0	0	0	0	0	4	5	8	10	15	21	29	36	50	59	(£)
GARDENS/FENCES/SHEDS	0	0	0	1	35	75	150	422	872	1187	1600	1986	2264	2562	2905	(£)
EXT. MAIN BUILDING	22	251	251	251	311	336	361	376	442	562	775	1243	1788	2199	2731	(£)
PLASTERWORK	0	0	0	0	151	336	520	675	859	1044	1259	1444	1883	2333	2733	(£)
FLOORS	0	0	105	229	481	771	951	1219	1450	1873	2255	2841	3574	4196	4464	(£)
JOINERY	0	0	0	0	98	160	215	277	341	481	1811	2833	2902	3645	3725	(£)
INTERNAL DECORATIONS	0	0	0	71	353	364	534	552	629	668	694	722	722	722	722	(£)
PLUMBING & ELECTRICAL	0	1	25	86	96	104	148	181	183	199	431	433	441	451	460	(£)
DOMESTIC APPLIANCES	0	0	0	57	168	350	487	812	821	877	908	908	908	908	908	(£)
HEATING EQUIPMENT	0	0	0	11	35	70	141	152	164	167	167	167	167	167	167	(£)
AUDIO/VIDEO	0	0	52	175	265	330	514	767	808	808	808	808	808	808	808	(£)
FURNITURE	0	0	51	127	351	766	1077	1165	1198	1203	1203	1203	1203	1203	1203	(£)
PERSONAL EFFECTS	0	0	0	0	35	188	340	487	531	562	562	562	562	562	562	(£)
FLOOR COVER/CURTAINS	0	0	453	503	1055	1108	1383	1391	1548	1548	1548	1548	1548	1548	1548	(£)
GARDEN/DIY/LEISURE	0	0	29	62	92	139	173	231	231	252	256	256	265	265	265	(£)
DOMESTIC CLEAN-UP	0	0	216	227	474	492	510	529	551	584	621	657	694	735	735	(£)

FLOOD DURATION LESS THAN 12 HOURS

	-0.3	0.0	0.05	0.1	0.2	0.3	0.6	0.9	1.2	1.5	1.8	2.1	2.4	2.7	3.0	
TOTAL BUILDING FABRIC	22	252	382	638	1530	2150	2890	3710	4790	6030	8850	11530	13610	16160	17800	(£)
COL PERCENT	100.0	100.0	32.2	35.4	38.2	38.4	38.4	40.1	45.0	50.1	59.3	65.4	68.9	72.3	74.2	
ROW PERCENT	0.1	1.4	2.1	3.6	8.6	12.1	16.2	20.9	26.9	33.9	49.7	64.8	76.5	90.8	100.0	
TOTAL INVENTORY	0	0	803	1170	2480	3450	4630	5540	5860	6000	6080	6110	6160	6200	6200	(£)
COL PERCENT	0.0	0.0	67.8	64.6	61.8	61.6	61.6	59.9	55.0	49.9	40.7	34.6	31.1	27.7	25.8	
ROW PERCENT	0.0	0.0	13.0	18.8	40.0	55.6	74.6	89.3	94.5	96.9	98.0	98.6	99.3	100.0	100.0	
TOTAL DAMAGE	22	252	1190	1800	4010	5600	7510	9250	10640	12040	14930	17650	19770	22360	24000	(£)
ROW PERCENT	0.1	1.1	4.9	7.5	16.7	23.3	31.3	38.5	44.3	50.1	62.2	73.5	82.4	93.2	100.0	
U.C. LIMIT	32	310	1380	1990	4440	6230	8330	10240	11790	13230	16330	19250	21630	24680	26770	(£)
L.C. LIMIT	11	193	999	1610	3580	4970	6690	8260	9490	10850	13530	16050	17910	20040	21230	(£)
TOTAL DAMAGE/SQ.METRE	0.3	3.0	14.3	21.8	48.3	67.5	90.5	111.5	128.3	145.0	179.9	212.7	238.3	269.4	289.2	(£)
U.C. LIMIT	0.4	3.6	15.0	23.6	52.3	73.8	99.7	123.3	142.2	161.2	198.6	234.4	260.6	291.4	310.1	(£)
L.C. LIMIT	0.2	2.4	13.6	20.0	44.3	61.2	81.3	99.7	114.4	128.8	161.2	191.0	216.0	247.4	268.3	(£)

MODERN DETACHED   LAND USE CODE 114   JANUARY 1990 PRICES   SALT WATER FLOOD DURATION LESS THAN 12 HOURS

COMPONENTS OF DAMAGE	-0.3	0.0	0.05	0.1	0.2	0.3	0.6	0.9	1.2	1.5	1.8	2.1	2.4	2.7	3.0	(M)
PATHS & PAVED AREAS	0	0	0	0	1	1	26	51	62	81	100	122	140	481	211	(£)
GARDENS/FENCES/SHEDS	0	0	0	1	188	385	586	937	1607	2314	3127	4003	5004	5724	6374	(£)
EXT. MAIN BUILDING	0	98	109	112	215	285	321	336	418	472	834	1439	1769	2137	2490	(£)
PLASTERWORK	0	0	0	0	92	206	319	413	526	640	771	884	1147	1421	1666	(£)
FLOORS	0	0	68	170	308	460	550	638	717	830	929	1064	1247	1395	1525	(£)
JOINERY	0	0	0	0	66	110	151	196	241	1347	1449	2272	2324	2400	2460	(£)
INTERNAL DECORATIONS	0	0	0	43	266	270	356	369	405	429	450	470	470	470	470	(£)
PLUMBING & ELECTRICAL	0	14	44	118	158	187	240	277	285	308	557	575	599	614	637	(£)
DOMESTIC APPLIANCES	0	0	0	57	164	344	458	767	775	827	858	858	858	858	858	(£)
HEATING EQUIPMENT	0	0	0	12	33	62	126	134	143	145	145	145	145	145	145	(£)
AUDIO/VIDEO	0	0	35	149	232	289	459	648	677	677	677	677	677	677	677	(£)
FURNITURE	0	0	37	97	258	598	868	955	982	987	987	987	987	987	987	(£)
PERSONAL EFFECTS	0	0	0	0	34	174	313	451	491	521	521	521	521	521	521	(£)
FLOOR COVER/CURTAINS	0	0	329	369	776	809	1017	1021	1117	1117	1117	1117	1117	1117	1117	(£)
GARDEN/DIY/LEISURE	0	0	28	60	89	135	167	224	224	244	248	248	257	257	257	(£)
DOMESTIC CLEAN-UP	0	0	161	169	353	366	380	394	410	435	462	490	517	547	547	(£)

FLOOD DURATION LESS THAN 12 HOURS

	-0.3	0.0	0.05	0.1	0.2	0.3	0.6	0.9	1.2	1.5	1.8	2.1	2.4	2.7	3.0	
TOTAL BUILDING FABRIC	0	112	221	445	1300	1910	2550	3220	4260	6420	8220	10830	12700	14640	15840	(£)
COL PERCENT	0.0	100.0	27.2	32.7	40.0	40.7	40.2	41.2	46.9	56.4	62.1	68.2	71.4	74.1	75.6	
ROW PERCENT	0.0	0.7	1.4	2.8	8.2	12.0	16.1	20.3	26.9	40.6	51.9	68.4	80.2	92.5	100.0	
TOTAL INVENTORY	0	0	592	915	1940	2780	3790	4600	4820	4960	5020	5050	5080	5110	5110	(£)
COL PERCENT	0.0	0.0	72.8	67.3	60.0	59.3	59.8	58.8	53.1	43.6	37.9	31.8	28.6	25.9	24.4	
ROW PERCENT	0.0	0.0	11.6	17.9	38.0	54.4	74.2	89.9	94.3	97.0	98.2	98.7	99.4	100.0	100.0	
TOTAL DAMAGE	0	112	814	1360	3240	4690	6340	7820	9090	11380	13240	15880	17780	19760	20950	(£)
ROW PERCENT	0.0	0.5	3.9	6.5	15.5	22.4	30.3	37.3	43.4	54.3	63.2	75.8	84.9	94.3	100.0	
U.C. LIMIT	0	138	944	1510	3590	5220	7030	8660	10080	12510	14480	17320	19450	21810	23360	(£)
L.C. LIMIT	0	85	683	1210	2890	4160	5650	6980	8100	10250	12000	14440	16110	17710	18540	(£)
TOTAL DAMAGE/SQ.METRE	0.0	1.8	13.2	22.0	52.3	75.7	102.4	126.1	146.6	183.6	213.6	256.1	286.9	318.7	337.9	(£)
U.C. LIMIT	0.0	2.2	13.9	23.8	56.7	82.7	112.8	139.4	162.5	204.2	235.8	282.2	313.7	344.7	362.3	(£)
L.C. LIMIT	0.0	1.4	12.5	20.2	47.9	68.7	92.0	112.8	130.7	163.0	191.4	230.0	260.1	292.7	313.5	(£)

326 THE ECONOMICS OF COASTAL MANAGEMENT

POST 1975   DETACHED   LAND USE CODE 115   JANUARY 1990 PRICES   SALT WATER FLOOD DURATION LESS THAN 12 HOURS

```
                                        DEPTH ABOVE UPPER SURFACE OF GROUND FLOOR
COMPONENTS OF DAMAGE  -0.3   0.0   0.05   0.1   0.2   0.3   0.6   0.9   1.2   1.5   1.8   2.1   2.4   2.7   3.0  (M)
----------------------------------------------------------------------------------------------------------------------
  PATHS & PAVED AREAS    0     0     0     0     0     9    13    17    22    37    48    64    80   108   128  (£)
  GARDENS/FENCES/SHEDS   0     0     0     0    73   160   253   562  1146  1592  2185  2740  3356  3690  4007  (£)
  EXT. MAIN BUILDING     0    98   109   112   215   285   326   348   438   560   989  1628  1990  2406  2801  (£)
  PLASTERWORK            0     0     0     0   228   469   688   928  1150  1389  1737  2094  2170  2294  2294  (£)
  FLOORS                 0     0    52   145   269   411   500   587   668   780   869   979  1161  1297  1450  (£)
  JOINERY                0     0     0   137   360   554   745   848   893  2052  2153  3016  3067  3143  3202  (£)
  INTERNAL DECORATIONS   0     0     0    46   245   260   344   372   403   423   431   440   440   440   440  (£)
  PLUMBING & ELECTRICAL  0     0    59   124   158   187   240   277   285   308   548   557   581   605   629  (£)

  DOMESTIC APPLIANCES    0     0     0    57   167   349   484   806   815   870   900   900   900   900   900  (£)
  HEATING EQUIPMENT      0     0     0    11    34    69   138   149   161   165   165   165   165   165   165  (£)
  AUDIO/VIDEO            0     0    52   173   262   327   509   759   799   799   799   799   799   799   799  (£)
  FURNITURE              0     0    50   127   350   763  1073  1160  1193  1198  1198  1198  1198  1198  1198  (£)
  PERSONAL EFFECTS       0     0     0     0    35   186   335   481   524   555   555   555   555   555   555  (£)
  FLOOR COVER/CURTAINS   0     0   320   362   761   796  1012  1019  1135  1135  1135  1135  1135  1135  1135  (£)
  GARDEN/DIY/LEISURE     0     0    29    62    92   138   172   230   230   251   255   255   264   264   264  (£)
  DOMESTIC CLEAN-UP      0     0   161   169   353   366   380   394   410   435   462   490   517   547   547  (£)

                                                                       FLOOD DURATION LESS THAN 12 HOURS
TOTAL BUILDING FABRIC    0    98   220   565  1550  2340  3110  3940  5010  7140  8960 11520 12850 13990 14950  (£)
  COL PERCENT          0.0 100.0  26.4  37.0  43.0  43.8  43.1  44.1  48.7  56.9  62.1  67.7  69.9  71.5  72.9
  ROW PERCENT          0.0   0.7   1.5   3.8  10.4  15.6  20.8  26.4  33.5  47.8  59.9  77.0  85.9  93.5 100.0
TOTAL INVENTORY          0     0   614   964  2060  3000  4110  5000  5270  5410  5470  5500  5540  5570  5570  (£)
  COL PERCENT          0.0   0.0  73.6  63.0  57.0  56.2  56.9  55.9  51.3  43.1  37.9  32.3  30.1  28.5  27.1
  ROW PERCENT          0.0   0.0  11.0  17.3  36.9  53.9  73.8  89.9  94.7  97.2  98.3  98.8  99.5 100.0 100.0

TOTAL DAMAGE             0    98   835  1530  3610  5340  7220  8940 10280 12550 14430 17020 18390 19550 20520  (£)
  ROW PERCENT          0.0   0.5   4.1   7.5  17.6  26.0  35.2  43.6  50.1  61.2  70.3  83.0  89.6  95.3 100.0
  U.C. LIMIT             0   120   968  1690  4000  5940  8010  9900 11390 13790 15790 18560 20120 21580 22890  (£)
  L.C. LIMIT             0    75   701  1370  3220  4740  6430  7980  9170 11310 13070 15480 16660 17520 18150  (£)
TOTAL DAMAGE/SQ.METRE  0.0   1.6  13.6  24.8  58.2  86.1 116.5 144.3 165.8 202.5 232.9 274.6 296.6 315.4 331.0  (£)
  U.C. LIMIT           0.0   1.9  14.3  26.9  63.1  94.1 128.4 159.6 183.8 225.2 257.1 302.6 324.3 341.1 354.9  (£)
  L.C. LIMIT           0.0   1.3  12.9  22.7  53.3  78.1 104.6 129.0 147.8 179.8 208.7 246.6 268.9 289.7 307.1  (£)
```

POST 1975 UTILITY DETACHED   LAND USE CODE 117   JANUARY 1990 PRICES   SALT WATER FLOOD DURATION LESS THAN 12 HOURS

```
                                        DEPTH ABOVE UPPER SURFACE OF GROUND FLOOR
COMPONENTS OF DAMAGE  -0.3   0.0   0.05   0.1   0.2   0.3   0.6   0.9   1.2   1.5   1.8   2.1   2.4   2.7   3.0  (M)
----------------------------------------------------------------------------------------------------------------------
  PATHS & PAVED AREAS    0     0     0     0     1     8    11    13    17    25    30    40    50    67    77  (£)
  GARDENS/FENCES/SHEDS   0     0     0     0    84   179   280   495   938  1305  1823  2285  2785  3105  3400  (£)
  EXT. MAIN BUILDING     0    98   109   112   195   265   303   321   397   464   706  1165  1344  1569  1776  (£)
  PLASTERWORK            0     0     0     0    53   144   230   325   412   505   643   779   850   897   897  (£)
  FLOORS                 0     0    11    35    71   115   144   180   211   259   294   335   422   483   542  (£)
  JOINERY                0     0     0     0     0     0     0     0     0   516   516   907   907   907   907  (£)
  INTERNAL DECORATIONS   0     0     0    17    88    94   132   149   157   167   170   174   174   174   174  (£)
  PLUMBING & ELECTRICAL  0     0     0     0     0     0     0     0    14    14    14    17    17    20    23    26  (£)
  DOMESTIC APPLIANCES    0     0     0     0     1    68   186   324   355   356   358   358   358   358   358  (£)
  HEATING EQUIPMENT      0     0     0     0     0     0     0     0     0     0     0     0     0     0     0  (£)
  AUDIO/VIDEO            0     0     2     2     2     3     6    23    23    23    23    23    23    23    23  (£)
  FURNITURE              0     0     3     9    18    19    25    25    26    26    26    26    26    26    26  (£)
  PERSONAL EFFECTS       0     0     0     0     0     0     0     0     0     0     0     0     0     0     0  (£)
  FLOOR COVER/CURTAINS   0     0   208   214   394   409   468   475   522   524   524   524   524   524   524  (£)
  GARDEN/DIY/LEISURE     0     0    27    58    86   129   160   214   214   233   237   237   245   245   245  (£)
  DOMESTIC CLEAN-UP      0     0    60    63   131   136   141   147   153   162   172   182   192   204   204  (£)

                                                                       FLOOD DURATION LESS THAN 12 HOURS
TOTAL BUILDING FABRIC    0    98   120   165   494   806  1100  1500  2150  3260  4200  5700  6550  7230  7800  (£)
  COL PERCENT          0.0 100.0  28.5  32.2  43.8  51.3  52.7  55.3  62.4  71.1  75.8  80.9  82.7  84.0  85.0
  ROW PERCENT          0.0   1.3   1.5   2.1   6.3  10.3  14.1  19.2  27.5  41.8  53.9  73.1  84.0  92.6 100.0
TOTAL INVENTORY          0     0   301   348   633   766   989  1210  1290  1330  1340  1350  1370  1380  1380  (£)
  COL PERCENT          0.0   0.0  71.5  67.8  56.2  48.7  47.3  44.7  37.6  28.9  24.2  19.1  17.3  16.0  15.0
  ROW PERCENT          0.0   0.0  21.8  25.2  45.9  55.5  71.6  87.6  93.6  95.9  97.1  97.8  99.1 100.0 100.0

TOTAL DAMAGE             0    98   421   513  1130  1570  2090  2710  3440  4580  5540  7050  7920  8610  9180  (£)
  ROW PERCENT          0.0   1.1   4.6   5.6  12.3  17.1  22.8  29.5  37.5  49.9  60.4  76.8  86.3  93.8 100.0
  U.C. LIMIT             0   120   488   568  1250  1750  2320  3000  3810  5030  6060  7690  8660  9500 10240  (£)
  L.C. LIMIT             0    75   353   457  1010  1390  1860  2420  3070  4130  5020  6410  7180  7720  8120  (£)
TOTAL DAMAGE/SQ.METRE  0.0   4.3  18.4  22.4  49.2  68.5  91.1 117.9 149.7 199.3 241.1 306.8 344.6 374.3 399.3  (£)
  U.C. LIMIT           0.0   5.2  19.3  24.3  53.3  74.9 100.4 130.4 166.0 221.6 266.2 338.1 376.8 404.8 428.1  (£)
  L.C. LIMIT           0.0   3.4  17.5  20.5  45.1  62.1  81.8 105.4 133.4 177.0 216.0 275.5 312.4 343.8 370.5  (£)
```

APPENDIX 5.2 327

PRE 1918 SEMI DETACHED   LAND USE CODE 121   JANUARY 1990 PRICES   SALT WATER FLOOD DURATION LESS THAN 12 HOURS

```
                                    DEPTH ABOVE UPPER SURFACE OF GROUND FLOOR
COMPONENTS OF DAMAGE  -0.3   0.0   0.05  0.1   0.2   0.3   0.6   0.9   1.2   1.5   1.8   2.1   2.4   2.7   3.0   (M)
----------------------------------------------------------------------------------------------------------------
PATHS & PAVED AREAS      0     0     0     0     0     1     2     3     4     5     7     8    13    17    24  (£)
GARDENS/FENCES/SHEDS     0     0     0     1     9    22    50   192   487   703   945  1352  1983  2274  2577  (£)
EXT. MAIN BUILDING     431   431   448   449   451   489   505   520   542   645   808  1180  1570  1870  2280  (£)
PLASTERWORK              0     0     0    27   112   231   413   582   785   987  1381  1886  2019  2141  2121  (£)
FLOORS                   0     0    22    45   111   169   212   271   308   353   423   526   679  1205  1635  (£)
JOINERY                  0     0     0     0     0     0    33    74   188   247   511   633   895  1338  1491  (£)
INTERNAL DECORATIONS    34    34    36    40    57   151   197   211   225   239   252   261   271   301   314  (£)
PLUMBING & ELECTRICAL    0    38    67   134   148   169   208   222   222   236   464   471   471   471   471  (£)

DOMESTIC APPLIANCES      0     0     0    53   145   299   392   643   647   690   713   713   713   713   713  (£)
HEATING EQUIPMENT        0     0     0    14    29    54   103   110   117   119   119   119   119   119   119  (£)
AUDIO/VIDEO              0     0    28   120   187   231   375   519   541   541   541   541   541   541   541  (£)
FURNITURE                0     0    34    93   238   531   759   830   853   859   859   859   859   859   859  (£)
PERSONAL EFFECTS         0     0     0     0    27   123   220   320   347   372   372   372   372   372   372  (£)
FLOOR COVER/CURTAINS     0     0   184   209   426   437   553   556   594   594   594   594   594   594   594  (£)
GARDEN/DIY/LEISURE       0     0    23    50    75   113   139   185   185   202   204   204   212   212   212  (£)
DOMESTIC CLEAN-UP        0     0    96   101   211   219   228   236   246   260   277   293   309   327   327  (£)
                                                                                FLOOD DURATION LESS THAN 12 HOURS
TOTAL BUILDING FABRIC  465   503   573   696   889  1230  1620  2080  2760  3420  4790  6320  7910  9620 10920  (£)
    COL PERCENT      100.0 100.0  60.9  52.0  39.9  38.1  36.9  37.9  43.9  48.4  56.5  63.1  68.0  72.0  74.5
    ROW PERCENT        4.3   4.6   5.3   6.4   8.2  11.3  14.9  19.0  25.3  31.3  43.9  57.9  72.4  88.1 100.0
TOTAL INVENTORY          0     0   368   643  1340  2010  2770  3400  3530  3640  3680  3700  3720  3740  3740  (£)
    COL PERCENT        0.0   0.0  39.1  48.0  60.1  61.9  63.1  62.1  56.1  51.6  43.5  36.9  32.0  28.0  25.5
    ROW PERCENT        0.0   0.0   9.9  17.2  35.8  53.7  74.1  90.9  94.5  97.3  98.5  98.9  99.5 100.0 100.0

TOTAL DAMAGE           465   503   941  1340  2230  3240  4400  5480  6300  7060  8480 10020 11630 13360 14660  (£)
    ROW PERCENT        3.2   3.4   6.4   9.1  15.2  22.1  30.0  37.4  43.0  48.2  57.8  68.4  79.3  91.2 100.0
    U.C. LIMIT         546   572  1000  1420  2360  3450  4730  5940  6850  7630  9100 10720 12350 14160 15500  (£)
    L.C. LIMIT         383   433   877  1260  2100  3030  4070  5020  5750  6490  7860  9320 10910 12560 13820  (£)
TOTAL DAMAGE/SQ.METRE 12.6  13.6  25.5  36.2  60.3  87.6 118.6 147.9 170.0 190.5 228.8 270.6 314.0 360.7 395.7  (£)
    U.C. LIMIT        15.0  15.5  27.1  37.8  63.7  93.8 128.7 161.7 186.4 208.3 248.5 293.9 338.0 385.7 420.9  (£)
    L.C. LIMIT        10.2  11.7  23.9  34.6  56.9  81.4 108.5 134.1 153.6 172.7 209.1 247.3 290.0 335.7 370.5  (£)
```

INTER WAR SEMI DETACHED   LAND USE CODE 122   JANUARY 1990 PRICES   SALT WATER FLOOD DURATION LESS THAN 12 HOURS

```
                                    DEPTH ABOVE UPPER SURFACE OF GROUND FLOOR
COMPONENTS OF DAMAGE  -0.3   0.0   0.05  0.1   0.2   0.3   0.6   0.9   1.2   1.5   1.8   2.1   2.4   2.7   3.0   (M)
----------------------------------------------------------------------------------------------------------------
PATHS & PAVED AREAS      0     0     0     0     0     2     5     8    10    12    15    19    25    33    42  (£)
GARDENS/FENCES/SHEDS     0     0     0     1    48   100   177   313   547   786  1038  1490  2143  2468  2776  (£)
EXT. MAIN BUILDING     431   431   457   458   462   517   543   572   611   728   924  1259  1611  1771  2093  (£)
PLASTERWORK              0     0     0     0    99   219   338   439   678   898  1419  1598  1851  2110  2287  (£)
FLOORS                   0     0    16   114   201   330   435   531   575   633   720   849  1047  1788  2404  (£)
JOINERY                  0     0     0     0   106   231   351   481   647   832  1105  1329  1518  2022  2099  (£)
INTERNAL DECORATIONS    37    37    37   254   286   289   302   317   330   348   363   367   377   390   403  (£)
PLUMBING & ELECTRICAL   29    46    61    84    96   100   103   146   171   198   258   345   403   413   425  (£)

DOMESTIC APPLIANCES      0     0     0    51   134   274   357   575   578   616   635   635   635   635   635  (£)
HEATING EQUIPMENT        0     0     0    15    27    49    91    97   104   106   106   106   106   106   106  (£)
AUDIO/VIDEO              0     0    25   105   163   201   329   448   468   468   468   468   468   468   468  (£)
FURNITURE                0     0    32    91   226   492   696   760   780   786   786   786   786   786   786  (£)
PERSONAL EFFECTS         0     0     0     0    24    96   170   250   271   293   293   293   293   293   293  (£)
FLOOR COVER/CURTAINS     0     0   198   218   433   442   544   548   574   574   574   574   574   574   574  (£)
GARDEN/DIY/LEISURE       0     0    20    45    67   100   124   163   163   178   181   181   187   187   187  (£)
DOMESTIC CLEAN-UP        0     0   133   139   291   302   313   325   338   359   381   404   426   451   451  (£)
                                                                                FLOOD DURATION LESS THAN 12 HOURS
TOTAL BUILDING FABRIC  497   514   571   911  1300  1790  2260  2810  3570  4440  5850  7260  8980 11000 12530  (£)
    COL PERCENT      100.0 100.0  58.2  57.8  48.7  47.8  46.2  47.0  52.1  56.7  63.0  67.8  72.1  75.8  78.2
    ROW PERCENT        4.0   4.1   4.6   7.3  10.4  14.3  18.0  22.4  28.5  35.4  46.6  57.9  71.6  87.8 100.0
TOTAL INVENTORY          0     0   411   666  1370  1960  2630  3170  3280  3380  3430  3450  3480  3500  3500  (£)
    COL PERCENT        0.0   0.0  41.8  42.2  51.3  52.2  53.8  53.0  47.9  43.3  37.0  32.2  27.9  24.2  21.8
    ROW PERCENT        0.0   0.0  11.7  19.0  39.1  55.9  75.0  90.5  93.6  96.6  97.8  98.5  99.3 100.0 100.0

TOTAL DAMAGE           497   514   982  1580  2670  3750  4880  5980  6850  7820  9270 10710 12460 14500 16040  (£)
    ROW PERCENT        3.1   3.2   6.1   9.8  16.7  23.4  30.5  37.3  42.7  48.8  57.8  66.8  77.7  90.4 100.0
    U.C. LIMIT         584   584  1050  1670  2820  4000  5250  6480  7450  8450  9950 11450 13230 15370 16960  (£)
    L.C. LIMIT         409   443   916  1490  2520  3500  4510  5480  6250  7190  8590  9970 11690 13630 15120  (£)
TOTAL DAMAGE/SQ.METRE  9.7  10.1  19.3  31.0  52.4  73.6  95.8 117.3 134.4 153.3 181.8 210.0 244.2 284.3 314.4  (£)
    U.C. LIMIT        11.6  11.5  20.5  32.4  55.4  78.8 103.9 128.2 147.4 167.6 197.4 228.0 262.9 304.0 334.4  (£)
    L.C. LIMIT         7.8   8.7  18.1  29.6  49.4  68.4  87.7 106.4 121.4 139.0 166.2 192.0 225.5 264.6 294.4  (£)
```

328 THE ECONOMICS OF COASTAL MANAGEMENT

```
POST WAR SEMI DETACHED      LAND USE CODE 123      JANUARY 1990 PRICES   SALT WATER FLOOD DURATION LESS THAN 12 HOURS
                                            DEPTH ABOVE UPPER SURFACE OF GROUND FLOOR
COMPONENTS OF DAMAGE  -0.3   0.0   0.05   0.1   0.2   0.3   0.6   0.9   1.2   1.5   1.8   2.1   2.4   2.7   3.0  (M)
----------------------------------------------------------------------------------------------------------------------
  PATHS & PAVED AREAS    0     0     0      0     0     1     4     7     8    10    13    16    21    28    36  (£)
  GARDENS/FENCES/SHEDS   0     0     0      1    54   112   191   317   525   764  1017  1456  2111  2438  2746  (£)
  EXT. MAIN BUILDING   431   431   457    458   462   517   542   571   610   730   923  1258  1609  1769  2091  (£)
  PLASTERWORK            0     0     0      0    99   219   338   439   678   898  1419  1598  1851  2110  2287  (£)
  FLOORS                 0     0    18    118   210   338   440   532   575   631   715   839  1032  1771  2384  (£)
  JOINERY                0     0     0      0   100   219   332   456   615   793  1053  1264  1443  1911  1984  (£)
  INTERNAL DECORATIONS  37    37    37    258   288   292   304   319   332   350   364   369   379   391   404  (£)
  PLUMBING & ELECTRICAL 29    46    61     83    95    98   101   140   163   188   242   319   384   394   406  (£)

  DOMESTIC APPLIANCES    0     0     0     51   133   272   352   566   568   604   623   623   623   623   623  (£)
  HEATING EQUIPMENT      0     0     0     15    26    47    87    94   100   102   102   102   102   102   102  (£)
  AUDIO/VIDEO            0     0    25    103   159   196   323   438   457   457   457   457   457   457   457  (£)
  FURNITURE              0     0    32     91   227   490   693   757   777   783   783   783   783   783   783  (£)
  PERSONAL EFFECTS       0     0     0      0    24    91   159   236   255   277   277   277   277   277   277  (£)
  FLOOR COVER/CURTAINS   0     0   198    217   435   441   541   545   566   566   566   566   566   566   566  (£)
  GARDEN/DIY/LEISURE     0     0    20     44    66    99   122   161   161   176   179   179   185   185   185  (£)
  DOMESTIC CLEAN-UP      0     0   133    139   291   302   313   325   338   359   381   404   426   451   451  (£)

                                                                              FLOOD DURATION LESS THAN 12 HOURS

TOTAL BUILDING FABRIC  497   514   573    920  1310  1800  2260  2780  3510  4370  5750  7120  8830 10820 12340  (£)
  COL PERCENT        100.0 100.0  58.3   58.1  49.0  48.1  46.5  47.1  52.1  56.7  63.0  67.7  72.1  75.8  78.2
  ROW PERCENT          4.0   4.2   4.7    7.5  10.6  14.6  18.3  22.6  28.4  35.4  46.6  57.7  71.6  87.6 100.0
TOTAL INVENTORY          0     0   410    662  1360  1940  2600  3130  3230  3330  3370  3390  3420  3450  3450  (£)
  COL PERCENT          0.0   0.0  41.7   41.9  51.0  51.9  53.5  52.9  47.9  43.3  37.0  32.3  27.9  24.2  21.8
  ROW PERCENT          0.0   0.0  11.9   19.2  39.5  56.3  75.3  90.7  93.6  96.5  97.8  98.4  99.3 100.0 100.0

TOTAL DAMAGE           497   514   984   1580  2670  3740  4850  5910  6730  7690  9120 10520 12260 14260 15790  (£)
  ROW PERCENT          3.1   3.3   6.2   10.0  16.9  23.7  30.7  37.4  42.7  48.7  57.8  66.6  77.6  90.3 100.0
  U.C. LIMIT           584   584  1050   1670  2820  3990  5220  6410  7320  8310  9790 11250 13020 15110 16690  (£)
  L.C. LIMIT           409   443   917   1490  2520  3490  4480  5410  6140  7070  8450  9790 11500 13410 14890  (£)
TOTAL DAMAGE/SQ.METRE  9.7  10.1  19.3   31.1  52.5  73.4  95.1 115.8 132.0 150.8 178.8 206.2 240.3 279.6 309.5  (£)
  U.C. LIMIT          11.6  11.5  20.5   32.5  55.5  78.6 103.2 126.6 144.7 164.9 194.2 223.9 258.7 298.9 329.2  (£)
  L.C. LIMIT           7.8   8.7  18.1   29.7  49.5  68.2  87.0 105.0 119.3 136.7 163.4 188.5 221.9 260.3 289.8  (£)


MODERN SEMI DETACHED        LAND USE CODE 124      JANUARY 1990 PRICES   SALT WATER FLOOD DURATION LESS THAN 12 HOURS
                                            DEPTH ABOVE UPPER SURFACE OF GROUND FLOOR
COMPONENTS OF DAMAGE  -0.3   0.0   0.05   0.1   0.2   0.3   0.6   0.9   1.2   1.5   1.8   2.1   2.4   2.7   3.0  (M)
----------------------------------------------------------------------------------------------------------------------
  PATHS & PAVED AREAS    0     0     0      0     1     1     7    14    17    20    26    33    37    48    56  (£)
  GARDENS/FENCES/SHEDS   0     0     0      1    65   137   211   355   666   983  1407  1814  2448  2772  3096  (£)
  EXT. MAIN BUILDING     0    98   109    112   176   245   279   293   350   442   830  1419  1743  2140  2498  (£)
  PLASTERWORK            0     0     0      0    78   173   268   349   444   539   650   746   985  1205  1481  (£)
  FLOORS                 0     0    61    135   240   358   425   484   539   626   701   803   948  1066  1157  (£)
  JOINERY                0     0     0     88   240   373   504   579   677  1268  1409  1970  2012  2082  2131  (£)
  INTERNAL DECORATIONS   0     0     0     36   148   154   252   286   317   337   354   370   370   370   370  (£)
  PLUMBING & ELECTRICAL  0     0    35    107   132   155   206   235   237   267   504   513   525   537   549  (£)

  DOMESTIC APPLIANCES    0     0     0     53   141   297   384   624   627   667   690   690   690   690   690  (£)
  HEATING EQUIPMENT      0     0     0     14    27    50    95   102   110   112   112   112   112   112   112  (£)
  AUDIO/VIDEO            0     0    28    117   181   223   367   500   522   522   522   522   522   522   522  (£)
  FURNITURE              0     0    33     93   240   523   751   824   847   854   854   854   854   854   854  (£)
  PERSONAL EFFECTS       0     0     0      0    27   108   191   282   304   329   329   329   329   329   329  (£)
  FLOOR COVER/CURTAINS   0     0   212    233   485   492   613   617   643   643   643   643   643   643   643  (£)
  GARDEN/DIY/LEISURE     0     0    22     49    73   110   136   180   180   197   199   199   207   207   207  (£)
  DOMESTIC CLEAN-UP      0     0   111    117   244   254   263   273   284   301   320   339   358   379   379  (£)

                                                                              FLOOD DURATION LESS THAN 12 HOURS

TOTAL BUILDING FABRIC    0    98   205    481  1080  1600  2150  2600  3250  4490  5880  7670  9070 10220 11340  (£)
  COL PERCENT          0.0 100.0  33.4   41.5  43.2  43.7  43.5  43.3  48.0  55.3  61.6  67.5  70.9  73.2  75.2
  ROW PERCENT          0.0   0.9   1.8    4.2   9.5  14.1  19.0  22.9  28.6  39.5  51.9  67.6  80.0  90.1 100.0
TOTAL INVENTORY          0     0   409    679  1420  2060  2800  3410  3520  3630  3670  3690  3720  3740  3740  (£)
  COL PERCENT          0.0   0.0  66.6   58.5  56.8  56.3  56.5  56.7  52.0  44.7  38.4  32.5  29.1  26.8  24.8
  ROW PERCENT          0.0   0.0  11.0   18.2  38.0  55.1  75.0  91.1  94.2  97.0  98.2  98.7  99.4 100.0 100.0

TOTAL DAMAGE             0    98   614   1160  2500  3660  4960  6010  6770  8110  9560 11360 12790 13960 15080  (£)
  ROW PERCENT          0.0   0.6   4.1    7.7  16.6  24.3  32.9  39.8  44.9  53.8  63.4  75.3  84.8  92.6 100.0
  U.C. LIMIT             0   111   655   1230  2640  3900  5340  6510  7360  8770 10260 12150 13580 14800 15940  (£)
  L.C. LIMIT             0    84   572   1090  2360  3420  4580  5510  6180  7450  8860 10570 12000 13120 14220  (£)
TOTAL DAMAGE/SQ.METRE  0.0   2.3  14.3   27.0  58.3  85.1 115.3 139.5 157.4 188.6 222.1 264.1 297.3 324.5 350.5  (£)
  U.C. LIMIT           0.0   2.6  15.2   28.2  61.6  91.1 125.1 152.5 172.6 206.2 241.2 286.8 320.0 346.9 372.8  (£)
  L.C. LIMIT           0.0   2.0  13.4   25.8  55.0  79.1 105.5 126.5 142.2 171.0 203.0 241.4 274.6 302.1 328.2  (£)
```

APPENDIX 5.2   329

POST 1975 SEMI DETACHED   LAND USE CODE 125   JANUARY 1990 PRICES   SALT WATER FLOOD DURATION LESS THAN 12 HOURS

COMPONENTS OF DAMAGE	-0.3	0.0	0.05	0.1	DEPTH ABOVE UPPER SURFACE OF GROUND FLOOR											
					0.2	0.3	0.6	0.9	1.2	1.5	1.8	2.1	2.4	2.7	3.0	(M)
PATHS & PAVED AREAS	0	0	0	0	1	4	6	8	10	15	18	23	30	39	45	(£)
GARDENS/FENCES/SHEDS	0	0	0	0	66	147	237	393	713	996	1364	1753	2164	2412	2637	(£)
EXT. MAIN BUILDING	0	98	109	112	195	265	301	317	391	450	673	1113	1276	1482	1676	(£)
PLASTERWORK	0	0	0	0	133	310	474	654	819	1000	1263	1504	1606	1693	1693	(£)
FLOORS	0	0	33	80	150	233	284	338	387	459	518	601	716	812	896	(£)
JOINERY	0	0	0	0	26	43	60	77	95	800	844	1369	1389	1420	1443	(£)
INTERNAL DECORATIONS	0	0	0	34	139	153	227	269	276	293	303	315	316	316	316	(£)
PLUMBING & ELECTRICAL	0	0	59	152	158	164	208	236	236	250	482	482	488	494	500	(£)
DOMESTIC APPLIANCES	0	0	0	52	134	278	358	576	578	615	634	634	634	634	634	(£)
HEATING EQUIPMENT	0	0	0	15	26	48	89	95	102	104	104	104	104	104	104	(£)
AUDIO/VIDEO	0	0	25	106	164	201	333	450	470	470	470	470	470	470	470	(£)
FURNITURE	0	0	32	92	230	494	703	770	790	797	797	797	797	797	797	(£)
PERSONAL EFFECTS	0	0	0	0	24	90	159	236	254	276	276	276	276	276	276	(£)
FLOOR COVER/CURTAINS	0	0	164	182	370	374	466	469	486	486	486	486	486	486	486	(£)
GARDEN/DIY/LEISURE	0	0	20	45	67	101	124	164	164	179	181	181	188	188	188	(£)
DOMESTIC CLEAN-UP	0	0	91	96	199	207	215	224	233	247	262	278	293	310	310	(£)

FLOOD DURATION LESS THAN 12 HOURS

TOTAL BUILDING FABRIC	0	98	201	379	869	1320	1800	2300	2930	4270	5470	7160	7990	8670	9210	(£)
COL PERCENT	0.0	100.0	37.5	39.2	41.7	42.4	42.3	43.5	48.7	57.3	63.0	68.9	71.1	72.6	73.8	
ROW PERCENT	0.0	1.1	2.2	4.1	9.4	14.4	19.5	24.9	31.8	46.3	59.4	77.8	86.7	94.2	100.0	
TOTAL INVENTORY	0	0	336	590	1220	1800	2450	2990	3080	3180	3210	3230	3250	3270	3270	(£)
COL PERCENT	0.0	0.0	62.5	60.8	58.3	57.6	57.7	56.5	51.3	42.7	37.0	31.1	28.9	27.4	26.2	
ROW PERCENT	0.0	0.0	10.3	18.1	37.3	55.0	75.0	91.4	94.2	97.2	98.3	98.8	99.5	100.0	100.0	
TOTAL DAMAGE	0	98	537	970	2090	3120	4250	5280	6010	7440	8680	10390	11240	11940	12480	(£)
ROW PERCENT	0.0	0.8	4.3	7.8	16.7	25.0	34.1	42.3	48.2	59.7	69.6	83.3	90.1	95.7	100.0	
U.C. LIMIT	0	111	573	1030	2210	3330	4570	5720	6540	8040	9310	11110	11940	12650	13190	(£)
L.C. LIMIT	0	84	500	913	1970	2910	3930	4840	5480	6840	8050	9670	10540	11230	11770	(£)
TOTAL DAMAGE/SQ.METRE	0.0	2.8	15.4	27.7	59.6	89.0	121.1	150.5	171.3	212.1	247.5	296.3	320.6	340.5	355.9	(£)
U.C. LIMIT	0.0	3.2	16.4	29.0	63.0	95.3	131.4	164.5	187.8	231.9	268.8	321.8	345.1	364.1	378.5	(£)
L.C. LIMIT	0.0	2.4	14.4	26.4	56.2	82.7	110.8	136.5	154.8	192.3	226.2	270.8	296.1	316.9	333.3	(£)

POST 1975 UTILITY SEMI DET LAND USE CODE 127   JANUARY 1990 PRICES   SALT WATER FLOOD DURATION LESS THAN 12 HOURS

COMPONENTS OF DAMAGE	-0.3	0.0	0.05	0.1	DEPTH ABOVE UPPER SURFACE OF GROUND FLOOR											
					0.2	0.3	0.6	0.9	1.2	1.5	1.8	2.1	2.4	2.7	3.0	(M)
PATHS & PAVED AREAS	0	0	0	0	0	5	7	9	11	17	21	27	34	45	52	(£)
GARDENS/FENCES/SHEDS	0	0	0	0	52	114	183	331	693	977	1352	1719	2166	2406	2629	(£)
EXT. MAIN BUILDING	0	98	109	112	195	265	302	319	393	456	683	1129	1296	1506	1702	(£)
PLASTERWORK	0	0	0	0	92	207	313	430	538	654	818	977	1036	1083	1083	(£)
FLOORS	0	0	13	41	85	138	173	216	255	313	356	405	511	585	657	(£)
JOINERY	0	0	0	0	0	0	0	0	0	631	631	1108	1108	1108	1108	(£)
INTERNAL DECORATIONS	0	0	0	20	109	115	161	180	189	201	205	210	210	210	210	(£)
PLUMBING & ELECTRICAL	0	0	0	0	0	0	0	16	16	16	20	20	24	27	31	(£)
DOMESTIC APPLIANCES	0	0	0	0	0	83	235	408	439	440	441	441	441	441	441	(£)
HEATING EQUIPMENT	0	0	0	0	0	0	0	0	0	0	0	0	0	0	0	(£)
AUDIO/VIDEO	0	0	1	1	1	3	5	20	20	20	20	20	20	20	20	(£)
FURNITURE	0	0	4	11	20	22	31	31	32	32	32	32	32	32	32	(£)
PERSONAL EFFECTS	0	0	0	0	0	0	0	0	0	0	0	0	0	0	0	(£)
FLOOR COVER/CURTAINS	0	0	271	276	487	496	557	567	604	606	606	606	606	606	606	(£)
GARDEN/DIY/LEISURE	0	0	28	61	90	136	168	223	223	243	246	246	255	255	255	(£)
DOMESTIC CLEAN-UP	0	0	72	75	157	163	169	176	183	194	206	218	230	244	244	(£)

FLOOD DURATION LESS THAN 12 HOURS

TOTAL BUILDING FABRIC	0	98	122	174	535	845	1140	1500	2100	3270	4090	5600	6390	6970	7470	(£)
COL PERCENT	0.0	100.0	24.4	29.1	41.4	48.3	49.4	51.3	58.3	68.0	72.5	78.1	80.1	81.3	82.4	
ROW PERCENT	0.0	1.3	1.6	2.3	7.2	11.3	15.3	20.1	28.1	43.7	54.7	74.9	85.4	93.3	100.0	
TOTAL INVENTORY	0	0	378	426	758	905	1170	1430	1500	1540	1550	1570	1590	1600	1600	(£)
COL PERCENT	0.0	0.0	75.6	70.9	58.6	51.7	50.6	48.7	41.7	32.0	27.5	21.9	19.9	18.7	17.6	
ROW PERCENT	0.0	0.0	23.6	26.6	47.4	56.5	72.9	89.2	93.9	96.0	97.0	97.8	99.1	100.0	100.0	
TOTAL DAMAGE	0	98	500	601	1290	1750	2310	2930	3600	4800	5640	7160	7970	8570	9080	(£)
ROW PERCENT	0.0	1.1	5.5	6.6	14.3	19.3	25.4	32.3	39.7	52.9	62.2	78.9	87.9	94.5	100.0	
U.C. LIMIT	0	111	533	636	1360	1870	2490	3180	3910	5190	6050	7660	8470	9080	9600	(£)
L.C. LIMIT	0	84	466	565	1220	1630	2130	2680	3290	4410	5230	6660	7470	8060	8560	(£)
TOTAL DAMAGE/SQ.METRE	0.0	4.3	19.1	22.7	49.4	67.7	89.5	114.1	142.1	189.1	225.1	287.4	321.7	347.0	368.3	(£)
U.C. LIMIT	0.0	4.9	20.3	23.7	52.2	72.5	97.1	124.7	155.8	206.7	244.4	312.1	346.3	371.0	391.7	(£)
L.C. LIMIT	0.0	3.7	17.9	21.7	46.6	62.9	81.9	103.5	128.4	171.5	205.8	262.7	297.1	323.0	344.9	(£)

330 THE ECONOMICS OF COASTAL MANAGEMENT

PRE 1918 TERRACE   LAND USE CODE 131   JANUARY 1990 PRICES   SALT WATER FLOOD DURATION LESS THAN 12 HOURS

					DEPTH	ABOVE	UPPER	SURFACE	OF	GROUND	FLOOR					
COMPONENTS OF DAMAGE	-0.3	0.0	0.05	0.1	0.2	0.3	0.6	0.9	1.2	1.5	1.8	2.1	2.4	2.7	3.0	(M)
PATHS & PAVED AREAS	0	0	0	1	1	19	23	26	38	49	63	73	97	129	154	(£)
GARDENS/FENCES/SHEDS	0	0	0	1	3	6	8	10	34	130	248	454	993	1192	1389	(£)
EXT. MAIN BUILDING	431	431	448	449	452	496	520	533	559	639	744	984	1236	1325	1559	(£)
PLASTERWORK	0	0	0	33	148	298	502	711	966	1206	1710	2460	2561	2688	2692	(£)
FLOORS	0	0	31	66	153	217	254	309	344	390	457	562	716	1328	1856	(£)
JOINERY	0	0	0	0	0	0	103	169	289	411	679	772	1012	1511	1642	(£)
INTERNAL DECORATIONS	22	22	25	30	43	108	184	201	214	230	243	250	260	286	300	(£)
PLUMBING & ELECTRICAL	0	38	67	134	148	169	208	222	222	236	464	471	471	471	471	(£)
DOMESTIC APPLIANCES	0	0	0	50	125	250	324	515	516	550	565	565	565	565	565	(£)
HEATING EQUIPMENT	0	0	0	16	25	49	87	95	103	105	105	105	105	105	105	(£)
AUDIO/VIDEO	0	0	22	90	141	173	287	385	401	401	401	401	401	401	401	(£)
FURNITURE	0	0	31	90	215	465	647	702	719	725	725	725	725	725	725	(£)
PERSONAL EFFECTS	0	0	0	0	21	74	130	193	208	228	228	228	228	228	228	(£)
FLOOR COVER/CURTAINS	0	0	156	172	330	335	411	418	439	439	439	439	439	439	439	(£)
GARDEN/DIY/LEISURE	0	0	18	40	60	89	110	144	144	157	159	159	165	165	165	(£)
DOMESTIC CLEAN-UP	0	0	102	107	223	231	240	249	259	274	292	309	326	345	345	(£)

FLOOD DURATION LESS THAN 12 HOURS

TOTAL BUILDING FABRIC	453	491	571	715	949	1320	1800	2180	2670	3290	4610	6030	7350	8930	10070	(£)
COL PERCENT	100.0	100.0	63.3	55.8	45.3	44.1	44.6	44.7	48.9	53.3	61.3	67.3	71.3	75.0	77.2	
ROW PERCENT	4.5	4.9	5.7	7.1	9.4	13.1	17.9	21.7	26.5	32.7	45.8	59.9	73.0	88.7	100.0	
TOTAL INVENTORY	0	0	330	566	1140	1670	2240	2700	2790	2880	2920	2930	2960	2980	2980	(£)
COL PERCENT	0.0	0.0	36.7	44.2	54.7	55.9	55.4	55.3	51.1	46.7	38.7	32.7	28.7	25.0	22.8	
ROW PERCENT	0.0	0.0	11.1	19.0	38.5	56.1	75.3	90.9	93.9	96.8	98.0	98.6	99.4	100.0	100.0	
TOTAL DAMAGE	453	491	901	1280	2090	2980	4040	4890	5460	6180	7530	8960	10310	11910	13040	(£)
ROW PERCENT	3.5	3.8	6.9	9.8	16.1	22.9	31.0	37.5	41.9	47.4	57.7	68.7	79.0	91.3	100.0	
U.C. LIMIT	592	572	978	1390	2270	3260	4470	5450	6100	6940	8350	9890	11370	13150	14410	(£)
L.C. LIMIT	313	409	823	1170	1910	2700	3610	4330	4820	5420	6710	8030	9250	10670	11670	(£)
TOTAL DAMAGE/SQ.METRE	11.6	12.6	23.2	33.0	53.7	76.6	103.7	125.4	140.1	158.4	193.0	229.9	264.3	305.4	334.4	(£)
U.C. LIMIT	15.2	14.6	24.7	34.6	57.0	82.5	113.4	138.5	155.1	176.9	212.2	252.1	288.7	332.8	363.9	(£)
L.C. LIMIT	8.0	10.6	21.7	31.4	50.4	70.7	94.0	112.3	125.1	140.4	173.8	207.7	239.9	278.0	304.9	(£)

INTER WAR TERRACE   LAND USE CODE 132   JANUARY 1990 PRICES   SALT WATER FLOOD DURATION LESS THAN 12 HOURS

					DEPTH	ABOVE	UPPER	SURFACE	OF	GROUND	FLOOR					
COMPONENTS OF DAMAGE	-0.3	0.0	0.05	0.1	0.2	0.3	0.6	0.9	1.2	1.5	1.8	2.1	2.4	2.7	3.0	(M)
PATHS & PAVED AREAS	0	0	0	0	0	1	2	3	4	6	7	9	12	17	23	(£)
GARDENS/FENCES/SHEDS	0	0	0	1	39	83	127	197	337	474	616	866	1357	1501	1632	(£)
EXT. MAIN BUILDING	431	431	457	458	462	510	528	548	576	679	808	1046	1285	1409	1641	(£)
PLASTERWORK	0	0	0	0	137	304	471	610	944	1249	1973	2218	2553	2910	3149	(£)
FLOORS	0	0	0	79	124	253	359	460	505	572	670	811	1029	1878	2617	(£)
JOINERY	0	0	0	0	11	41	56	91	153	243	398	484	563	919	989	(£)
INTERNAL DECORATIONS	31	31	31	366	422	428	441	460	476	496	513	517	529	544	558	(£)
PLUMBING & ELECTRICAL	32	49	64	84	95	97	99	112	124	131	163	213	285	286	289	(£)
DOMESTIC APPLIANCES	0	0	0	50	125	252	333	526	528	561	576	576	576	576	576	(£)
HEATING EQUIPMENT	0	0	0	16	25	49	86	94	102	104	104	104	104	104	104	(£)
AUDIO/VIDEO	0	0	22	91	142	174	289	387	404	404	404	404	404	404	404	(£)
FURNITURE	0	0	44	124	304	622	859	916	938	944	944	944	944	944	944	(£)
PERSONAL EFFECTS	0	0	0	0	21	74	129	193	208	227	227	227	227	227	227	(£)
FLOOR COVER/CURTAINS	0	0	207	230	452	457	563	566	582	582	582	582	582	582	582	(£)
GARDEN/DIY/LEISURE	0	0	18	40	60	90	111	145	145	159	161	161	167	167	167	(£)
DOMESTIC CLEAN-UP	0	0	135	142	295	307	318	330	344	364	387	410	433	458	458	(£)

FLOOD DURATION LESS THAN 12 HOURS

TOTAL BUILDING FABRIC	494	511	552	988	1290	1720	2090	2480	3120	3850	5150	6170	7620	9470	10900	(£)
COL PERCENT	100.0	100.0	56.3	58.7	47.5	45.9	43.7	44.0	49.0	53.5	60.3	64.4	68.9	73.2	75.9	
ROW PERCENT	4.5	4.7	5.1	9.1	11.9	15.8	19.1	22.8	28.6	35.3	47.3	56.6	69.9	86.9	100.0	
TOTAL INVENTORY	0	0	429	695	1430	2030	2690	3160	3250	3350	3390	3410	3440	3470	3470	(£)
COL PERCENT	0.0	0.0	43.7	41.3	52.5	54.1	56.3	56.0	51.0	46.5	39.7	35.6	31.1	26.8	24.1	
ROW PERCENT	0.0	0.0	12.4	20.1	41.2	58.5	77.7	91.2	93.9	96.6	97.8	98.4	99.3	100.0	100.0	
TOTAL DAMAGE	494	511	981	1680	2720	3750	4780	5650	6380	7200	8540	9580	11060	12930	14370	(£)
ROW PERCENT	3.4	3.6	6.8	11.7	18.9	26.1	33.3	39.3	44.4	50.1	59.4	66.7	77.0	90.0	100.0	
U.C. LIMIT	645	596	1070	1830	2960	4110	5280	6300	7130	8090	9470	10580	12190	14270	15880	(£)
L.C. LIMIT	342	425	896	1530	2480	3390	4280	5000	5630	6310	7610	8580	9930	11590	12860	(£)
TOTAL DAMAGE/SQ.METRE	9.5	9.9	18.9	32.4	52.3	72.1	91.9	108.5	122.6	138.5	164.2	184.2	212.6	248.6	276.2	(£)
U.C. LIMIT	12.4	11.4	20.2	34.0	55.5	77.7	100.5	119.8	135.7	154.2	180.5	202.0	232.3	270.9	300.6	(£)
L.C. LIMIT	6.6	8.4	17.6	30.8	49.1	66.5	83.3	97.2	109.5	122.8	147.9	166.4	192.9	226.3	251.8	(£)

APPENDIX 5.2 331

POST WAR TERRACE   LAND USE CODE 133   JANUARY 1990 PRICES   SALT WATER FLOOD DURATION LESS THAN 12 HOURS

                                                DEPTH ABOVE UPPER SURFACE OF GROUND FLOOR
COMPONENTS OF DAMAGE   -0.3   0.0   0.05   0.1   0.2   0.3   0.6   0.9   1.2   1.5   1.8   2.1   2.4   2.7   3.0 (M)

PATHS & PAVED AREAS   0   0   0   0   0   0   1   1   2   3   4   5   6   9   12 (£)
GARDENS/FENCES/SHEDS   0   0   0   0   49   103   159   252   411   544   740   921 1026 1128 1214 (£)
EXT. MAIN BUILDING   22   251   251   251   293   319   341   357   412   451   523   650   809   727   824 (£)
PLASTERWORK   0   0   0   0   53   117   182   236   301   365   441   505   662   824   964 (£)
FLOORS   0   0   72   147   243   339   385   406   423   460   491   542   611   669   656 (£)
JOINERY   0   0   0   0   0   8   18   27   36   123   557   861   878 1153 1170 (£)
INTERNAL DECORATIONS   0   0   0   24   61   69   141   185   203   212   217   221   221   221   221 (£)
PLUMBING & ELECTRICAL   0   0   52   110   144   150   194   208   208   222   450   450   453   457   460 (£)

DOMESTIC APPLIANCES   0   0   0   3   7   10   11   12   12   12   12   12   12   12   12 (£)
HEATING EQUIPMENT   0   0   0   15   24   36   65   67   69   71   71   71   71   71   71 (£)
AUDIO/VIDEO   0   0   20   87   133   160   269   357   368   368   368   368   368   368   368 (£)
FURNITURE   0   0   16   70   159   390   459   459   469   469   469   469   469   469   469 (£)
PERSONAL EFFECTS   0   0   0   0   0   48   92   124   137   137   137   137   137   137   137 (£)
FLOOR COVER/CURTAINS   0   0   127   135   265   267   316   316   319   319   319   319   319   319   319 (£)
GARDEN/DIY/LEISURE   0   0   18   40   59   89   110   143   143   157   158   158   165   165   165 (£)
DOMESTIC CLEAN-UP   0   0   60   63   131   136   141   147   153   162   172   182   192   204   204 (£)

                                                                                FLOOD DURATION LESS THAN 12 HOURS

TOTAL BUILDING FABRIC   22   251   375   533   845 1110 1420 1680 2000 2380 3430 4160 4670 5190 5520 (£)
    COL PERCENT   100.0 100.0 60.6 56.2 51.9 49.3 49.3 50.7 54.4 58.4 66.7 70.7 72.9 74.8 76.0
    ROW PERCENT   0.4   4.5   6.8   9.7 15.3 20.0 25.8 30.3 36.2 43.1 62.0 75.3 84.5 94.0 100.0
TOTAL INVENTORY   0   0   244   415   782 1140 1470 1630 1680 1700 1710 1720 1740 1750 1750 (£)
    COL PERCENT   0.0   0.0 39.4 43.8 48.1 50.7 50.7 49.3 45.6 41.6 33.3 29.3 27.1 25.2 24.0
    ROW PERCENT   0.0   0.0 14.0 23.8 44.8 65.2 83.8 93.1 95.8 97.1 97.8 98.4 99.3 100.0 100.0

TOTAL DAMAGE   22   251   619   949 1630 2250 2890 3300 3670 4080 5140 5880 6410 6940 7270 (£)
    ROW PERCENT   0.3   3.5   8.5 13.1 22.4 30.9 39.7 45.4 50.5 56.1 70.6 80.8 88.1 95.4 100.0
    U.C. LIMIT   28   292   672 1030 1770 2460 3200 3680 4100 4580 5700 6490 7070 7660 8030 (£)
    L.C. LIMIT   15   209   565   866 1490 2040 2580 2920 3240 3580 4580 5270 5750 6220 6510 (£)
TOTAL DAMAGE/SQ.METRE   0.6   7.2 17.8 27.2 46.6 64.2 82.6 94.3 104.9 116.5 146.7 168.0 183.0 198.2 207.7 (£)
    U.C. LIMIT   0.8   8.3 19.0 28.5 49.5 69.1 90.3 104.2 116.1 129.7 161.3 184.2 199.9 216.0 226.0 (£)
    L.C. LIMIT   0.4   6.1 16.6 25.9 43.7 59.3 74.9 84.4 93.7 103.3 132.1 151.8 166.1 180.4 189.4 (£)

MODERN TERRACE   LAND USE CODE 134   JANUARY 1990 PRICES   SALT WATER FLOOD DURATION LESS THAN 12 HOURS

                                                DEPTH ABOVE UPPER SURFACE OF GROUND FLOOR
COMPONENTS OF DAMAGE   -0.3   0.0   0.05   0.1   0.2   0.3   0.6   0.9   1.2   1.5   1.8   2.1   2.4   2.7   3.0 (M)

PATHS & PAVED AREAS   0   0   0   1   2   6   9   11   14   19   24   33   41   53   59 (£)
GARDENS/FENCES/SHEDS   0   0   0   0   111   228   348   524   817 1136 1529 1939 2330 2648 2928 (£)
EXT. MAIN BUILDING   0   98   109   112   216   297   346   368   465   530   791 1377 1547 1783 2001 (£)
PLASTERWORK   0   0   0   0   97   216   335   435   554   673   813   931 1226 1524 1783 (£)
FLOORS   0   0   51   114   201   322   389   453   514   609   697   809   975 1112 1225 (£)
JOINERY   0   0   0   0   38   67   95   125   155 1439 1529 2476 2513 2567 2608 (£)
INTERNAL DECORATIONS   0   0   0   44   148   161   306   364   402   438   452   466   466   466   466 (£)
PLUMBING & ELECTRICAL   0   38   67   142   158   164   208   236   236   250   481   481   487   493   499 (£)

DOMESTIC APPLIANCES   0   0   0   51   132   272   358   572   574   610   628   628   628   628   628 (£)
HEATING EQUIPMENT   0   0   0   14   25   50   92   100   109   111   111   111   111   111   111 (£)
AUDIO/VIDEO   0   0   34   115   174   215   349   492   516   516   516   516   516   516   516 (£)
FURNITURE   0   0   41   112   291   593   813   877   901   908   908   908   908   908   908 (£)
PERSONAL EFFECTS   0   0   0   0   24   87   153   227   244   266   266   266   266   266   266 (£)
FLOOR COVER/CURTAINS   0   0   215   233   469   474   582   589   613   613   613   613   613   613   613 (£)
GARDEN/DIY/LEISURE   0   0   20   44   66   99   122   161   161   176   178   178   185   185   185 (£)
DOMESTIC CLEAN-UP   0   0   117   123   257   267   277   287   299   316   336   356   376   398   398 (£)

                                                                                FLOOD DURATION LESS THAN 12 HOURS

TOTAL BUILDING FABRIC   0   136   227   414   974 1460 2040 2520 3160 5100 6320 8520 9590 10650 11570 (£)
    COL PERCENT   0.0 100.0 34.6 37.3 40.3 41.5 42.6 43.2 48.0 59.1 64.0 70.4 72.7 74.6 76.1
    ROW PERCENT   0.0   1.2   2.0   3.6   8.4 12.7 17.6 21.8 27.3 44.0 54.6 73.6 82.9 92.0 100.0
TOTAL INVENTORY   0   0   430   695 1440 2060 2750 3310 3420 3520 3560 3580 3610 3630 3630 (£)
    COL PERCENT   0.0   0.0 65.4 62.7 59.7 58.5 57.4 56.8 52.0 40.9 36.0 29.6 27.3 25.4 23.9
    ROW PERCENT   0.0   0.0 11.9 19.2 39.7 56.8 75.8 91.2 94.3 97.0 98.1 98.7 99.4 100.0 100.0

TOTAL DAMAGE   0   136   658 1110 2420 3520 4790 5830 6580 8620 9880 12100 13200 14280 15200 (£)
    ROW PERCENT   0.0   0.9   4.3   7.3 15.9 23.2 31.5 38.3 43.3 56.7 65.0 79.6 86.8 93.9 100.0
    U.C. LIMIT   0   158   714 1210 2630 3860 5300 6500 7360 9680 10960 13360 14550 15760 16800 (£)
    L.C. LIMIT   0   113   601 1010 2210 3180 4280 5160 5800 7560 8800 10840 11850 12800 13600 (£)
TOTAL DAMAGE/SQ.METRE   0.0   3.0 14.7 24.7 53.7 78.3 106.3 129.4 146.1 191.4 219.4 268.6 293.0 317.1 337.6 (£)
    U.C. LIMIT   0.0   3.5 15.7 25.9 57.0 84.3 116.2 142.9 161.7 213.1 241.2 294.5 320.1 345.6 367.4 (£)
    L.C. LIMIT   0.0   2.5 13.7 23.5 50.4 72.3 96.4 115.9 130.5 169.7 197.6 242.7 265.9 288.6 307.8 (£)

332 THE ECONOMICS OF COASTAL MANAGEMENT

POST 1975 TERRACE   LAND USE CODE 135   JANUARY 1990 PRICES   SALT WATER FLOOD DURATION LESS THAN 12 HOURS

DEPTH ABOVE UPPER SURFACE OF GROUND FLOOR

COMPONENTS OF DAMAGE	-0.3	0.0	0.05	0.1	0.2	0.3	0.6	0.9	1.2	1.5	1.8	2.1	2.4	2.7	3.0	(M)
PATHS & PAVED AREAS	0	0	0	0	0	6	8	9	12	16	20	24	29	41	49	(£)
GARDENS/FENCES/SHEDS	0	0	0	0	27	58	91	138	281	412	568	745	990	1130	1258	(£)
EXT. MAIN BUILDING	0	98	109	112	176	245	279	293	350	391	555	938	1043	1190	1326	(£)
PLASTERWORK	0	0	0	0	54	193	329	476	612	760	967	1186	1333	1405	1405	(£)
FLOORS	0	0	30	69	124	192	232	268	309	361	405	465	549	618	680	(£)
JOINERY	0	0	0	0	39	69	96	127	158	949	1039	1615	1653	1708	1750	(£)
INTERNAL DECORATIONS	0	0	0	28	86	100	177	215	219	240	246	255	255	255	255	(£)
PLUMBING & ELECTRICAL	0	0	59	152	158	164	208	222	222	236	464	464	467	470	473	(£)
DOMESTIC APPLIANCES	0	0	0	50	129	259	338	541	543	578	595	595	595	595	595	(£)
HEATING EQUIPMENT	0	0	0	15	26	47	86	92	98	100	100	100	100	100	100	(£)
AUDIO/VIDEO	0	0	23	96	150	184	304	412	429	429	429	429	429	429	429	(£)
FURNITURE	0	0	32	90	218	472	663	721	740	745	745	745	745	745	745	(£)
PERSONAL EFFECTS	0	0	0	0	22	84	149	220	238	258	258	258	258	258	258	(£)
FLOOR COVER/CURTAINS	0	0	135	153	298	303	378	381	399	399	399	399	399	399	399	(£)
GARDEN/DIY/LEISURE	0	0	19	42	63	94	116	152	152	166	168	168	174	174	174	(£)
DOMESTIC CLEAN-UP	0	0	86	90	189	196	203	210	219	232	247	262	276	293	293	(£)

FLOOD DURATION LESS THAN 12 HOURS

TOTAL BUILDING FABRIC	0	98	198	362	666	1030	1420	1750	2170	3370	4270	5700	6320	6820	7200	(£)
COL PERCENT	0.0	100.0	40.0	40.2	37.7	38.5	38.8	39.0	43.4	53.6	59.2	65.8	68.0	69.5	70.6	
ROW PERCENT	0.0	1.4	2.8	5.0	9.3	14.3	19.8	24.3	30.1	46.8	59.3	79.1	87.8	94.7	100.0	
TOTAL INVENTORY	0	0	297	539	1100	1650	2240	2730	2820	2910	2940	2960	2980	3000	3000	(£)
COL PERCENT	0.0	0.0	60.0	59.8	62.3	61.5	61.2	61.0	56.6	46.4	40.8	34.2	32.0	30.5	29.4	
ROW PERCENT	0.0	0.0	9.9	18.0	36.7	54.9	74.7	91.2	94.1	97.1	98.2	98.8	99.4	100.0	100.0	
TOTAL DAMAGE	0	98	495	901	1770	2670	3660	4480	4990	6280	7210	8660	9300	9820	10200	(£)
ROW PERCENT	0.0	1.0	4.9	8.8	17.3	26.2	35.9	44.0	48.9	61.6	70.7	84.9	91.2	96.3	100.0	
U.C. LIMIT	0	114	537	979	1930	2920	4050	4990	5580	7050	8000	9560	10250	10840	11270	(£)
L.C. LIMIT	0	81	452	822	1610	2420	3270	3970	4400	5510	6420	7760	8350	8800	9130	(£)
TOTAL DAMAGE/SQ.METRE	0.0	3.0	15.1	27.4	53.5	81.0	111.0	135.8	151.1	190.1	218.4	262.1	281.7	297.3	308.8	(£)
U.C. LIMIT	0.0	3.5	16.1	28.8	56.8	87.2	121.4	150.0	167.3	211.7	240.1	287.4	307.7	324.0	336.0	(£)
L.C. LIMIT	0.0	2.5	14.1	26.0	50.2	74.8	100.6	121.6	134.9	168.5	196.7	236.8	255.7	270.6	281.6	(£)

POST 1975 UTILITY TERRACE   LAND USE CODE 137   JANUARY 1990 PRICES   SALT WATER FLOOD DURATION LESS THAN 12 HOURS

DEPTH ABOVE UPPER SURFACE OF GROUND FLOOR

COMPONENTS OF DAMAGE	-0.3	0.0	0.05	0.1	0.2	0.3	0.6	0.9	1.2	1.5	1.8	2.1	2.4	2.7	3.0	(M)
PATHS & PAVED AREAS	0	0	0	0	1	4	6	8	10	14	17	22	28	37	42	(£)
GARDENS/FENCES/SHEDS	0	0	0	0	72	155	243	393	683	954	1306	1665	2037	2288	2514	(£)
EXT. MAIN BUILDING	0	98	109	112	195	265	301	317	390	448	669	1107	1269	1473	1665	(£)
PLASTERWORK	0	0	0	0	113	240	358	487	605	733	912	1083	1135	1182	1182	(£)
FLOORS	0	0	17	44	86	135	166	202	233	282	319	365	453	518	575	(£)
JOINERY	0	0	0	0	0	0	0	0	0	517	517	908	908	908	908	(£)
INTERNAL DECORATIONS	0	0	0	19	82	90	138	164	172	183	187	192	192	192	192	(£)
PLUMBING & ELECTRICAL	0	0	0	0	0	0	0	16	16	16	19	19	23	26	29	(£)
DOMESTIC APPLIANCES	0	0	0	0	0	106	303	527	560	560	561	561	561	561	561	(£)
HEATING EQUIPMENT	0	0	0	0	0	0	0	0	0	0	0	0	0	0	0	(£)
AUDIO/VIDEO	0	0	1	1	1	2	3	14	14	14	14	14	14	14	14	(£)
FURNITURE	0	0	3	11	17	20	28	28	28	28	28	28	28	28	28	(£)
PERSONAL EFFECTS	0	0	0	0	0	0	0	0	0	0	0	0	0	0	0	(£)
FLOOR COVER/CURTAINS	0	0	346	349	566	572	620	633	663	665	665	665	665	665	665	(£)
GARDEN/DIY/LEISURE	0	0	23	50	75	112	139	182	182	199	201	201	209	209	209	(£)
DOMESTIC CLEAN-UP	0	0	69	72	150	156	162	169	175	186	197	209	220	234	234	(£)

FLOOD DURATION LESS THAN 12 HOURS

TOTAL BUILDING FABRIC	0	98	126	176	550	891	1220	1590	2110	3150	3950	5360	6050	6630	7110	(£)
COL PERCENT	0.0	100.0	22.2	26.7	40.4	47.9	49.2	50.5	56.5	65.5	70.3	76.1	78.0	79.4	80.6	
ROW PERCENT	0.0	1.4	1.8	2.5	7.7	12.5	17.1	22.4	29.7	44.3	55.6	75.4	85.0	93.2	100.0	
TOTAL INVENTORY	0	0	444	484	812	971	1260	1560	1630	1660	1670	1680	1700	1720	1720	(£)
COL PERCENT	0.0	0.0	77.8	73.3	59.6	52.1	50.8	49.5	43.5	34.5	29.7	23.9	22.0	20.6	19.4	
ROW PERCENT	0.0	0.0	25.9	28.3	47.4	56.6	73.3	90.8	94.8	96.5	97.4	98.1	99.2	100.0	100.0	
TOTAL DAMAGE	0	98	570	661	1360	1860	2470	3150	3740	4810	5620	7050	7750	8340	8830	(£)
ROW PERCENT	0.0	1.1	6.5	7.5	15.4	21.1	28.0	35.7	42.3	54.5	63.7	79.8	87.8	94.5	100.0	
U.C. LIMIT	0	114	619	718	1480	2040	2730	3510	4180	5400	6230	7780	8540	9210	9760	(£)
L.C. LIMIT	0	81	520	603	1240	1680	2210	2790	3300	4220	5010	6320	6960	7470	7900	(£)
TOTAL DAMAGE/SQ.METRE	0.0	4.3	22.8	26.2	54.2	74.8	99.5	126.9	151.9	194.4	229.5	288.5	318.3	343.5	364.2	(£)
U.C. LIMIT	0.0	5.0	24.3	27.5	57.5	80.6	108.8	140.2	168.2	216.5	252.3	316.3	347.7	374.4	396.3	(£)
L.C. LIMIT	0.0	3.6	21.3	24.9	50.9	69.0	90.2	113.6	135.6	172.3	206.7	260.7	288.9	312.6	332.1	(£)

APPENDIX 5.2 333

```
PRE 1918 BUNGALOW          LAND USE CODE 141      JANUARY 1990 PRICES   SALT WATER FLOOD DURATION LESS THAN 12 HOURS
                                          DEPTH ABOVE UPPER SURFACE OF GROUND FLOOR
COMPONENTS OF DAMAGE   -0.3   0.0   0.05   0.1   0.2   0.3   0.6   0.9   1.2   1.5   1.8   2.1   2.4   2.7   3.0  (M)
---------------------  ----  ----  ----  ----  ----  ----  ----  ----  ----  ----  ----  ----  ----  ----  ----
  PATHS & PAVED AREAS     0     0     0     0     0     1     1     2     8    11    15    17    34    32    46  (£)
  GARDENS/FENCES/SHEDS    0     0     0     1    11    28    63   291   680   945  1243  1710  2374  2666  2978  (£)
  EXT. MAIN BUILDING    431   431   466   467   473   541   569   606   650   793   961  1237  1522  1640  1903  (£)
  PLASTERWORK             0     0     0    37   149   300   552   815  1086  1444  2037  2214  2394  2623  2659  (£)
  FLOORS                  0     0    35    73   177   262   321   404   463   529   641   780  1023  1833  2428  (£)
  JOINERY                 0     0     0     0     0     0    22    45    82   111   364   459   712  1422  1663  (£)
  INTERNAL DECORATIONS    0     0     1    12   176   200   213   227   247   264   280   289   297   315   325  (£)
  PLUMBING & ELECTRICAL   0    14    39   101   110   122   165   179   179   193   296   303   303   303   303  (£)

  DOMESTIC APPLIANCES     0     0     0    59   165   345   494   779   785   832   858   858   858   858   858  (£)
  HEATING EQUIPMENT       0     0     3    18    38    71   128   138   150   152   152   152   152   152   152  (£)
  AUDIO/VIDEO             0     0    55   168   253   319   499   709   727   727   727   727   727   727   727  (£)
  FURNITURE               0     0    51   176   873  1200  1964  2002  2230  2237  2237  2237  2237  2237  2237  (£)
  PERSONAL EFFECTS        0     0    23   194   313   799  1729  2095  2417  2444  2444  2444  2444  2444  2444  (£)
  FLOOR COVER/CURTAINS    0     0   260   277   590   605   748   762   831   831   831   831   831   831   831  (£)
  GARDEN/DIY/LEISURE      0     0    46    92   130   185   217   267   267   285   289   289   297   297   297  (£)
  DOMESTIC CLEAN-UP       0     0   131   139   289   300   311   322   335   356   378   400   423   447   447  (£)
                                                                                   FLOOD DURATION LESS THAN 12 HOURS

TOTAL BUILDING FABRIC   431   446   541   693  1100  1460  1910  2570  3400  4290  5840  7010  8660 10840 12310  (£)
    COL PERCENT       100.0 100.0  48.5  38.1  29.3  27.5  23.9  26.6  30.5  35.3  42.4  46.9  52.1  57.5  60.6
    ROW PERCENT         3.5   3.6   4.4   5.6   8.9  11.8  15.5  20.9  27.6  34.9  47.4  57.0  70.4  88.1 100.0
TOTAL INVENTORY           0     0   574  1130  2650  3830  6100  7080  7750  7870  7920  7940  7970  8000  8000  (£)
    COL PERCENT         0.0   0.0  51.5  61.9  70.7  72.5  76.1  73.4  69.5  64.7  57.6  53.1  47.9  42.5  39.4
    ROW PERCENT         0.0   0.0   7.2  14.1  33.2  47.9  76.2  88.5  96.9  98.4  99.0  99.3  99.7 100.0 100.0

TOTAL DAMAGE            431   446  1120  1820  3750  5290  8010  9650 11140 12160 13760 14960 16630 18840 20310  (£)
    ROW PERCENT         2.1   2.2   5.5   9.0  18.5  26.0  39.4  47.5  54.9  59.9  67.8  73.6  81.9  92.8 100.0
    U.C. LIMIT          678   521  1190  1960  4110  5790  8850 10640 12310 13390 15030 16270 17950 20220 21730  (£)
    L.C. LIMIT          183   370  1050  1680  3390  4790  7170  8660  9970 10930 12490 13650 15310 17460 18890  (£)
TOTAL DAMAGE/SQ.METRE   8.5   8.8  22.0  35.7  73.6 103.5 156.9 189.1 218.4 238.4 269.7 293.2 326.2 369.2 398.0  (£)
    U.C. LIMIT         15.1  10.8  23.6  38.9  81.1 114.5 175.5 211.5 244.5 265.5 298.4 322.8 356.6 401.4 431.2  (£)
    L.C. LIMIT          1.9   6.8  20.4  32.5  66.1  92.5 138.3 166.7 191.9 211.3 241.0 263.6 295.8 337.0 364.8  (£)


INTER WAR BUNGALOW         LAND USE CODE 142      JANUARY 1990 PRICES   SALT WATER FLOOD DURATION LESS THAN 12 HOURS
                                          DEPTH ABOVE UPPER SURFACE OF GROUND FLOOR
COMPONENTS OF DAMAGE   -0.3   0.0   0.05   0.1   0.2   0.3   0.6   0.9   1.2   1.5   1.8   2.1   2.4   2.7   3.0  (M)
---------------------  ----  ----  ----  ----  ----  ----  ----  ----  ----  ----  ----  ----  ----  ----  ----
  PATHS & PAVED AREAS     0     0     0     0     1    15    22    30    38    49    62    76    95   129   159  (£)
  GARDENS/FENCES/SHEDS    0     0     0     1   126   262   415   671  1024  1441  1926  2579  3385  3893  4349  (£)
  EXT. MAIN BUILDING    251   251   308   311   319   417   442   516   584   806  1121  1642  2174  2466  2970  (£)
  PLASTERWORK             0     0     0     0   182   406   629   815  1262  1671  2637  2972  3436  3916  4241  (£)
  FLOORS                  0     0    60   188   390   576   713   868   949  1057  1218  1466  1846  3336  4545  (£)
  JOINERY                 0     0     0     0   100   219   332   456   615   794  1118  1365  1579  1948  2058  (£)
  INTERNAL DECORATIONS   31    31    31   369   421   432   461   498   532   567   596   609   630   666   695  (£)
  PLUMBING & ELECTRICAL  29    46    61    84    96    98   101   140   163   188   243   320   385   395   407  (£)

  DOMESTIC APPLIANCES     0     0     0    52   136   274   383   588   591   626   642   642   642   642   642  (£)
  HEATING EQUIPMENT       0     0     4    20    32    71   117   130   146   149   149   149   149   149   149  (£)
  AUDIO/VIDEO             0     0    38   119   181   228   367   512   529   529   529   529   529   529   529  (£)
  FURNITURE               0     0    65   243   973  1286  2053  2136  2374  2379  2379  2379  2379  2379  2379  (£)
  PERSONAL EFFECTS        0     0    35   253   409   935  1901  2312  2664  2684  2684  2684  2684  2684  2684  (£)
  FLOOR COVER/CURTAINS    0     0   425   446   865   877  1042  1060  1117  1117  1117  1117  1117  1117  1117  (£)
  GARDEN/DIY/LEISURE      0     0    34    69    98   140   163   198   198   212   214   214   221   221   221  (£)
  DOMESTIC CLEAN-UP       0     0   265   279   581   604   627   649   676   717   762   807   852   902   902  (£)
                                                                                   FLOOD DURATION LESS THAN 12 HOURS

TOTAL BUILDING FABRIC   311   328   460   955  1640  2430  3120  4000  5170  6580  8920 11030 13530 16750 19430  (£)
    COL PERCENT       100.0 100.0  34.6  39.2  33.3  35.5  31.9  34.5  38.4  43.9  51.3  56.4  61.2  66.0  69.3
    ROW PERCENT         1.6   1.7   2.4   4.9   8.4  12.5  16.1  20.6  26.6  33.9  45.9  56.8  69.7  86.2 100.0
TOTAL INVENTORY           0     0   871  1480  3280  4420  6660  7590  8300  8420  8480  8530  8580  8630  8630  (£)
    COL PERCENT         0.0   0.0  65.4  60.8  66.7  64.5  68.1  65.5  61.6  56.1  48.7  43.6  38.8  34.0  30.7
    ROW PERCENT         0.0   0.0  10.1  17.2  38.0  51.2  77.2  88.0  96.2  97.6  98.3  98.8  99.4 100.0 100.0

TOTAL DAMAGE            311   328  1330  2440  4920  6850  9780 11590 13470 14990 17400 19560 22110 25380 28050  (£)
    ROW PERCENT         1.1   1.2   4.7   8.7  17.5  24.4  34.8  41.3  48.0  53.4  62.0  69.7  78.8  90.5 100.0
    U.C. LIMIT          489   383  1410  2630  5390  7500 10800 12780 14880 16510 19000 21270 23870 27230 30010  (£)
    L.C. LIMIT          132   272  1250  2250  4450  6200  8760 10400 12060 13470 15800 17850 20350 23530 26090  (£)
TOTAL DAMAGE/SQ.METRE   3.4   3.6  14.5  26.5  53.4  74.3 106.1 125.8 146.2 162.8 188.9 212.3 240.1 275.5 304.6  (£)
    U.C. LIMIT          6.0   4.4  15.6  28.9  58.8  82.2 118.7 140.7 164.0 181.3 209.0 233.8 262.4 299.5 330.0  (£)
    L.C. LIMIT          0.8   2.8  13.4  24.1  48.0  66.4  93.5 110.9 128.4 144.3 168.8 190.8 217.8 251.5 279.2  (£)
```

334 THE ECONOMICS OF COASTAL MANAGEMENT

POST WAR BUNGALOW   LAND USE CODE 143   JANUARY 1990 PRICES   SALT WATER FLOOD DURATION LESS THAN 12 HOURS

DEPTH ABOVE UPPER SURFACE OF GROUND FLOOR

COMPONENTS OF DAMAGE	-0.3	0.0	0.05	0.1	0.2	0.3	0.6	0.9	1.2	1.5	1.8	2.1	2.4	2.7	3.0	(M)
PATHS & PAVED AREAS	0	0	0	1	2	8	16	25	31	40	50	64	78	101	117	(£)
GARDENS/FENCES/SHEDS	0	0	0	1	133	275	419	668	1111	1581	2126	2703	3462	3900	4291	(£)
EXT. MAIN BUILDING	22	251	251	251	387	408	430	445	553	748	1079	1764	2551	3289	4091	(£)
PLASTERWORK	0	0	0	0	151	336	525	678	867	1052	1269	1454	1896	2352	2756	(£)
FLOORS	0	0	89	164	368	618	772	1027	1248	1669	2053	2638	3374	4002	4232	(£)
JOINERY	0	0	0	79	212	330	446	512	544	653	1747	2564	2603	2980	3023	(£)
INTERNAL DECORATIONS	0	0	0	69	252	265	432	514	548	576	621	668	668	668	668	(£)
PLUMBING & ELECTRICAL	37	57	90	152	173	196	266	330	339	362	602	611	621	636	646	(£)
DOMESTIC APPLIANCES	0	0	0	54	144	293	417	647	651	690	709	709	709	709	709	(£)
HEATING EQUIPMENT	0	0	4	19	34	76	128	143	160	163	163	163	163	163	163	(£)
AUDIO/VIDEO	0	0	44	135	205	260	413	582	602	602	602	602	602	602	602	(£)
FURNITURE	0	0	67	248	1043	1380	2238	2331	2599	2603	2603	2603	2603	2603	2603	(£)
PERSONAL EFFECTS	0	0	39	273	431	1011	2060	2514	2899	2922	2922	2922	2922	2922	2922	(£)
FLOOR COVER/CURTAINS	0	0	449	476	943	965	1162	1183	1270	1270	1270	1270	1270	1270	1270	(£)
GARDEN/DIY/LEISURE	0	0	38	76	108	154	180	219	219	234	237	237	244	244	244	(£)
DOMESTIC CLEAN-UP	0	0	231	243	506	526	545	565	589	624	663	702	742	785	785	(£)

FLOOD DURATION LESS THAN 12 HOURS

TOTAL BUILDING FABRIC	59	308	430	718	1680	2440	3310	4200	5240	6680	9550	12470	15260	17930	19830	(£)
COL PERCENT	100.0	100.0	33.0	32.0	33.0	34.3	31.7	33.9	36.8	42.3	51.0	57.5	62.2	65.8	68.1	
ROW PERCENT	0.3	1.6	2.2	3.6	8.5	12.3	16.7	21.2	26.5	33.7	48.2	62.9	76.9	90.4	100.0	
TOTAL INVENTORY	0	0	875	1530	3420	4670	7150	8190	8990	9110	9170	9210	9260	9300	9300	(£)
COL PERCENT	0.0	0.0	67.0	68.0	67.0	65.7	68.3	66.1	63.2	57.7	49.0	42.5	37.8	34.2	31.9	
ROW PERCENT	0.0	0.0	9.4	16.4	36.7	50.2	76.8	88.0	96.7	98.0	98.6	99.0	99.5	100.0	100.0	
TOTAL DAMAGE	59	308	1310	2250	5100	7100	10450	12390	14240	15800	18720	21680	24520	27230	29130	(£)
ROW PERCENT	0.2	1.1	4.5	7.7	17.5	24.4	35.9	42.5	48.9	54.2	64.3	74.4	84.2	93.5	100.0	
U.C. LIMIT	92	359	1390	2420	5580	7770	11540	13660	15730	17400	20440	23580	26470	29220	31170	(£)
L.C. LIMIT	25	256	1230	2080	4620	6430	9360	11120	12750	14200	17000	19780	22570	25240	27090	(£)
TOTAL DAMAGE/SQ.METRE	0.7	3.5	14.7	25.2	57.3	79.8	117.5	139.2	160.0	177.5	210.4	243.6	275.4	306.0	327.3	(£)
U.C. LIMIT	1.2	4.3	15.8	27.5	63.1	88.2	131.4	155.7	179.4	197.7	232.8	268.2	301.0	332.7	354.6	(£)
L.C. LIMIT	0.2	2.7	13.6	22.9	51.5	71.4	103.6	122.7	140.6	157.3	188.0	219.0	249.8	279.3	300.0	(£)

MODERN BUNGALOW   LAND USE CODE 144   JANUARY 1990 PRICES   SALT WATER FLOOD DURATION LESS THAN 12 HOURS

DEPTH ABOVE UPPER SURFACE OF GROUND FLOOR

COMPONENTS OF DAMAGE	-0.3	0.0	0.05	0.1	0.2	0.3	0.6	0.9	1.2	1.5	1.8	2.1	2.4	2.7	3.0	(M)
PATHS & PAVED AREAS	0	0	0	3	7	19	26	32	41	58	74	100	126	164	184	(£)
GARDENS/FENCES/SHEDS	0	0	0	1	124	255	390	631	1069	1524	2059	2625	3379	3814	4207	(£)
EXT. MAIN BUILDING	0	98	109	112	254	316	342	359	450	623	1260	1908	2575	3231	3859	(£)
PLASTERWORK	0	0	0	0	217	483	748	969	1234	1500	1810	2076	2706	3361	3934	(£)
FLOORS	0	0	119	267	492	754	915	1084	1235	1474	1676	1953	2336	2650	2888	(£)
JOINERY	0	0	0	0	62	104	142	185	228	1944	2061	3334	3401	3496	3570	(£)
INTERNAL DECORATIONS	0	0	0	101	372	419	653	755	803	853	881	909	909	909	909	(£)
PLUMBING & ELECTRICAL	0	38	67	142	158	187	240	299	335	367	619	636	641	691	718	(£)
DOMESTIC APPLIANCES	0	0	0	54	146	298	423	657	661	700	720	720	720	720	720	(£)
HEATING EQUIPMENT	0	0	4	19	34	76	129	144	161	165	165	165	165	165	165	(£)
AUDIO/VIDEO	0	0	45	137	209	265	421	594	615	615	615	615	615	615	615	(£)
FURNITURE	0	0	68	249	1060	1401	2281	2378	2651	2656	2656	2656	2656	2656	2656	(£)
PERSONAL EFFECTS	0	0	40	277	437	1024	2096	2545	2939	2961	2961	2961	2961	2961	2961	(£)
FLOOR COVER/CURTAINS	0	0	630	660	1316	1345	1605	1625	1731	1731	1731	1731	1731	1731	1731	(£)
GARDEN/DIY/LEISURE	0	0	39	77	110	156	183	223	223	239	241	241	248	248	248	(£)
DOMESTIC CLEAN-UP	0	0	286	301	627	651	676	700	729	773	822	870	919	972	972	(£)

FLOOD DURATION LESS THAN 12 HOURS

TOTAL BUILDING FABRIC	0	136	295	627	1690	2540	3460	4320	5400	8350	10440	13550	16080	18320	20270	(£)
COL PERCENT	0.0	100.0	21.0	26.1	30.0	32.7	30.7	32.7	35.7	45.9	51.3	57.6	61.6	64.5	66.8	
ROW PERCENT	0.0	0.7	1.5	3.1	8.3	12.5	17.1	21.3	26.6	41.2	51.5	66.8	79.3	90.4	100.0	
TOTAL INVENTORY	0	0	1110	1780	3940	5220	7820	8870	9710	9840	9920	9960	10020	10070	10070	(£)
COL PERCENT	0.0	0.0	79.0	73.9	70.0	67.3	69.3	67.3	64.3	54.1	48.7	42.4	38.4	35.5	33.2	
ROW PERCENT	0.0	0.0	11.1	17.7	39.1	51.8	77.6	88.1	96.4	97.7	98.4	98.9	99.5	100.0	100.0	
TOTAL DAMAGE	0	136	1410	2410	5630	7760	11280	13190	15110	18190	20360	23510	26090	28390	30340	(£)
ROW PERCENT	0.0	0.4	4.6	7.9	18.6	25.6	37.2	43.5	49.8	60.0	67.1	77.5	86.0	93.6	100.0	
U.C. LIMIT	0	158	1500	2600	6160	8490	12460	14540	16700	20040	22230	25570	28160	30460	32460	(£)
L.C. LIMIT	0	113	1320	2220	5100	7030	10100	11840	13520	16340	18490	21450	24020	26320	28220	(£)
TOTAL DAMAGE/SQ.METRE	0.0	1.1	11.3	19.3	45.1	62.0	90.2	105.5	120.8	145.4	162.8	188.0	208.7	227.0	242.6	(£)
U.C. LIMIT	0.0	1.4	12.1	21.0	49.7	68.6	100.9	118.0	135.5	161.9	180.1	207.0	228.1	246.8	262.8	(£)
L.C. LIMIT	0.0	0.8	10.5	17.6	40.5	55.4	79.5	93.0	106.1	128.9	145.5	169.0	189.3	207.2	222.4	(£)

APPENDIX 5.2 335

POST 1975 BUNGALOW   LAND USE CODE 145   JANUARY 1990 PRICES   SALT WATER FLOOD DURATION LESS THAN 12 HOURS

                                        DEPTH ABOVE UPPER SURFACE OF GROUND FLOOR
COMPONENTS OF DAMAGE -0.3   0.0   0.05   0.1   0.2   0.3   0.6   0.9   1.2   1.5   1.8   2.1   2.4   2.7   3.0 (M)

PATHS & PAVED AREAS   0   0   0   0   1   6   8   10   14   19   24   31   39   51   58 (£)
GARDENS/FENCES/SHEDS   0   0   0   1   145   306   474   762 1237 1718 2327 2945 3705 4109 4467 (£)
EXT. MAIN BUILDING   0   98   109   112   235   317   371   398   513   633 1042 1783 2093 2464 2811 (£)
PLASTERWORK   0   0   0   0   250   654 1029 1446 1827 2241 2838 3417 3736 3941 3941 (£)
FLOORS   0   0   74   177   321   507   617   731   831   982 1109 1272 1511 1702 1883 (£)
JOINERY   0   0   0   71   209   324   437   505   561 1819 1900 2825 2870 2934 2985 (£)
INTERNAL DECORATIONS   0   0   0   80   288   323   503   594   647   682   700   721   721   721   721 (£)
PLUMBING & ELECTRICAL   0   0   59   142   158   187   249   308   326   358   611   629   655   682   709 (£)

DOMESTIC APPLIANCES   0   0   0   55   148   305   433   675   679   719   740   740   740   740   740 (£)
HEATING EQUIPMENT   0   0   4   19   34   77   131   147   164   168   168   168   168   168   168 (£)
AUDIO/VIDEO   0   0   46   142   215   273   434   614   636   636   636   636   636   636   636 (£)
FURNITURE   0   0   68   251 1085 1437 2350 2451 2733 2739 2739 2739 2739 2739 2739 (£)
PERSONAL EFFECTS   0   0   41   284   446 1050 2158 2612 3020 3043 3043 3043 3043 3043 3043 (£)
FLOOR COVER/CURTAINS   0   0   406   432   878   897 1094 1114 1199 1199 1199 1199 1199 1199 1199 (£)
GARDEN/DIY/LEISURE   0   0   40   80   113   161   189   231   231   247   250   250   257   257   257 (£)
DOMESTIC CLEAN-UP   0   0   206   216   450   468   485   503   524   590   625   660   698   698   698 (£)
                                                                              FLOOD DURATION LESS THAN 12 HOURS

TOTAL BUILDING FABRIC   0   98   242   585 1610 2630 3690 4760 5960 8460 10550 13630 15330 16610 17580 (£)
  COL PERCENT   0.0 100.0 22.9 28.3 32.3 36.0 33.6 36.3 39.3 47.6 53.0 59.2 61.9 63.7 65.0
  ROW PERCENT   0.0   0.6   1.4   3.3   9.2 14.9 21.0 27.1 33.9 48.1 60.0 77.5 87.2 94.5 100.0
TOTAL INVENTORY   0   0   814 1480 3370 4670 7280 8350 9190 9310 9370 9400 9450 9480 9480 (£)
  COL PERCENT   0.0   0.0 77.1 71.7 67.7 64.0 66.4 63.7 60.7 52.4 47.0 40.8 38.1 36.3 35.0
  ROW PERCENT   0.0   0.0   8.6 15.6 35.6 49.3 76.8 88.1 96.9 98.2 98.8 99.2 99.6 100.0 100.0

TOTAL DAMAGE   0   98 1060 2070 4980 7300 10970 13110 15150 17770 19920 23030 24780 26090 27060 (£)
  ROW PERCENT   0.0   0.4   3.9   7.6 18.4 27.0 40.5 48.4 56.0 65.6 73.6 85.1 91.6 96.4 100.0
  U.C. LIMIT   0   114 1130 2230 5450 7990 12120 14450 16740 19570 21750 25040 26750 28000 28950 (£)
  L.C. LIMIT   0   81   992 1910 4510 6610 9820 11770 13560 15970 18090 21020 22810 24180 25170 (£)
TOTAL DAMAGE/SQ.METRE 0.0   1.2 13.4 26.2 63.1 92.4 138.8 165.9 191.6 224.7 252.0 291.3 313.5 330.1 342.4 (£)
  U.C. LIMIT   0.0   1.5 14.4 28.6 69.5 102.2 155.2 185.7 214.9 250.3 278.9 320.8 342.7 358.9 370.9 (£)
  L.C. LIMIT   0.0   0.9 12.4 23.8 56.7 82.6 122.4 146.3 168.3 199.1 225.1 261.8 284.3 301.3 313.9 (£)

PRE 1918 FLAT   LAND USE CODE 151   JANUARY 1990 PRICES   SALT WATER FLOOD DURATION LESS THAN 12 HOURS

                                        DEPTH ABOVE UPPER SURFACE OF GROUND FLOOR
COMPONENTS OF DAMAGE -0.3   0.0   0.05   0.1   0.2   0.3   0.6   0.9   1.2   1.5   1.8   2.1   2.4   2.7   3.0 (M)

PATHS & PAVED AREAS   0   0   0   0   0   0   0   0   0   0   0   0   0   0   0 (£)
GARDENS/FENCES/SHEDS   0   0   0   0   0   0   0   0   0   0   0   0   0   0   0 (£)
EXT. MAIN BUILDING   431   431   448   449   452   486   502   531   560   649   765   927 1098 1244 1434 (£)
PLASTERWORK   0   0   0   14   90   191   325   487   665   811   996 1264 1360 1450 1546 (£)
FLOORS   0   0   51   106   233   298   319   359   380   415   465   551   676 1270 1746 (£)
JOINERY   0   0   0   0   0   0   18   37   60   81   219   246   435   658   709 (£)
INTERNAL DECORATIONS   0   0   3   6   135   142   237   247   255   269   278   282   289   304   315 (£)
PLUMBING & ELECTRICAL   0   38   58   122   130   151   190   204   204   218   91   98   98   98   98 (£)

DOMESTIC APPLIANCES   0   0   0   50   139   289   417   657   661   701   721   721   721   721   721 (£)
HEATING EQUIPMENT   0   0   4   4   6   21   28   35   42   43   43   43   43   43   43 (£)
AUDIO/VIDEO   0   0   21   32   45   64   92   153   153   153   153   153   153   153   153 (£)
FURNITURE   0   0   33   107   623   649 1211 1239 1419 1423 1423 1423 1423 1423 1423 (£)
PERSONAL EFFECTS   0   0   20   172   284   621 1342 1630 1879 1902 1902 1902 1902 1902 1902 (£)
FLOOR COVER/CURTAINS   0   0   138   141   284   291   347   359   399   399   399   399   399   399   399 (£)
GARDEN/DIY/LEISURE   0   0   39   78   111   159   185   226   226   242   245   245   252   252   252 (£)
DOMESTIC CLEAN-UP   0   0   104   109   227   236   245   254   264   280   298   316   333   353   353 (£)
                                                                              FLOOD DURATION LESS THAN 12 HOURS

TOTAL BUILDING FABRIC 431   469   560   697 1040 1270 1590 1870 2130 2440 2820 3370 3960 5030 5850 (£)
  COL PERCENT   100.0 100.0 60.7 50.0 37.7 35.2 29.1 29.1 29.6 32.2 35.2 39.3 43.1 48.9 52.7
  ROW PERCENT   7.4   8.0   9.6 11.9 17.8 21.7 27.2 31.9 36.3 41.8 48.1 57.6 67.6 85.9 100.0
TOTAL INVENTORY   0   0   362   697 1720 2330 3870 4560 5050 5150 5190 5210 5230 5250 5250 (£)
  COL PERCENT   0.0   0.0 39.3 50.0 62.3 64.8 70.9 70.9 70.4 67.8 64.8 60.7 56.9 51.1 47.3
  ROW PERCENT   0.0   0.0   6.9 13.3 32.8 44.4 73.7 86.8 96.1 98.0 98.8 99.2 99.6 100.0 100.0

TOTAL DAMAGE   431   469   922 1390 2760 3600 5460 6420 7170 7590 8000 8580 9190 10280 11100 (£)
  ROW PERCENT   3.9   4.2   8.3 12.6 24.9 32.4 49.2 57.9 64.6 68.4 72.1 77.3 82.8 92.6 100.0
  U.C. LIMIT   671   554 1010 1560 3130 4090 6310 7400 8280 8740 9150 9750 10380 11560 12480 (£)
  L.C. LIMIT   190   383   835 1220 2390 3110 4610 5440 6060 6440 6850 7410 8000 9000 9720 (£)
TOTAL DAMAGE/SQ.METRE 10.8 11.7 23.1 34.9 69.0 89.9 136.4 160.4 179.0 189.5 199.8 214.1 229.4 256.5 277.1 (£)
  U.C. LIMIT   17.5 14.3 25.3 37.7 75.7 99.0 153.1 180.4 201.4 212.8 222.5 237.9 252.8 280.2 301.2 (£)
  L.C. LIMIT   4.0   9.1 20.9 32.1 62.3 80.8 119.7 140.4 156.6 166.2 177.1 190.3 206.0 232.8 253.0 (£)

336 THE ECONOMICS OF COASTAL MANAGEMENT

INTER WAR FLAT   LAND USE CODE 152   JANUARY 1990 PRICES   SALT WATER FLOOD DURATION LESS THAN 12 HOURS

                                              DEPTH ABOVE UPPER SURFACE OF GROUND FLOOR
COMPONENTS OF DAMAGE	-0.3	0.0	0.05	0.1	0.2	0.3	0.6	0.9	1.2	1.5	1.8	2.1	2.4	2.7	3.0	(M)
PATHS & PAVED AREAS	0	0	0	0	0	0	0	0	0	0	0	0	0	0	0	(£)
GARDENS/FENCES/SHEDS	0	0	0	0	0	0	0	0	0	0	0	0	0	0	0	(£)
EXT. MAIN BUILDING	251	251	286	287	294	351	366	427	477	642	849	1123	1394	1602	1875	(£)
PLASTERWORK	0	0	0	0	169	375	582	753	1166	1544	2437	2751	3190	3638	3942	(£)
FLOORS	0	0	0	69	120	276	402	537	607	708	854	1062	1389	2606	3633	(£)
JOINERY	0	0	0	0	61	140	211	296	411	551	817	1001	1165	1389	1488	(£)
INTERNAL DECORATIONS	0	0	0	376	430	438	460	486	506	536	557	561	576	592	607	(£)
PLUMBING & ELECTRICAL	32	49	64	84	94	96	98	110	121	127	157	203	278	279	282	(£)
DOMESTIC APPLIANCES	0	0	0	50	123	241	334	504	505	535	547	547	547	547	547	(£)
HEATING EQUIPMENT	0	0	4	22	30	61	97	108	119	122	122	122	122	122	122	(£)
AUDIO/VIDEO	0	0	31	95	145	179	292	400	409	409	409	409	409	409	409	(£)
FURNITURE	0	0	54	222	816	1084	1617	1637	1828	1832	1832	1832	1832	1832	1832	(£)
PERSONAL EFFECTS	0	0	30	223	373	828	1651	2037	2333	2350	2350	2350	2350	2350	2350	(£)
FLOOR COVER/CURTAINS	0	0	339	346	635	639	732	748	779	779	779	779	779	779	779	(£)
GARDEN/DIY/LEISURE	0	0	29	58	84	119	138	166	166	178	180	180	186	186	186	(£)
DOMESTIC CLEAN-UP	0	0	213	224	468	487	505	523	545	577	613	650	686	727	727	(£)

                                                                                     FLOOD DURATION LESS THAN 12 HOURS

	-0.3	0.0	0.05	0.1	0.2	0.3	0.6	0.9	1.2	1.5	1.8	2.1	2.4	2.7	3.0	
TOTAL BUILDING FABRIC	283	300	350	817	1170	1680	2120	2610	3290	4110	5670	6700	7990	10110	11830	(£)
COL PERCENT	100.0	100.0	33.3	39.6	30.4	31.6	28.3	29.9	33.0	37.7	45.4	49.4	53.6	59.2	63.0	
ROW PERCENT	2.4	2.5	3.0	6.9	9.9	14.2	17.9	22.1	27.8	34.7	48.0	56.7	67.6	85.4	100.0	
TOTAL INVENTORY	0	0	702	1240	2680	3640	5370	6130	6690	6790	6840	6870	6910	6950	6950	(£)
COL PERCENT	0.0	0.0	66.7	60.4	69.6	68.4	71.7	70.1	67.0	62.3	54.6	50.6	46.4	40.8	37.0	
ROW PERCENT	0.0	0.0	10.1	17.9	38.5	52.4	77.2	88.1	96.2	97.6	98.3	98.8	99.4	100.0	100.0	
TOTAL DAMAGE	283	300	1050	2060	3850	5320	7490	8740	9980	10900	12510	13580	14910	17060	18780	(£)
ROW PERCENT	1.5	1.6	5.6	11.0	20.5	28.3	39.9	46.5	53.1	58.0	66.6	72.3	79.4	90.8	100.0	
U.C. LIMIT	440	354	1150	2310	4370	6040	8660	10070	11530	12550	14310	15430	16840	19180	21120	(£)
L.C. LIMIT	125	245	951	1810	3330	4600	6320	7410	8430	9250	10710	11730	12980	14940	16440	(£)
TOTAL DAMAGE/SQ.METRE	3.4	3.6	12.9	25.1	46.9	64.7	91.2	106.3	121.5	132.6	152.3	165.3	181.5	207.7	228.7	(£)
U.C. LIMIT	5.5	4.4	14.1	27.1	51.5	71.2	102.4	119.5	136.7	148.9	169.6	183.7	200.0	226.9	248.6	(£)
L.C. LIMIT	1.3	2.8	11.7	23.1	42.3	58.2	80.0	93.1	106.3	116.3	135.0	146.9	163.0	188.5	208.8	(£)

POST WAR FLAT   LAND USE CODE 153   JANUARY 1990 PRICES   SALT WATER FLOOD DURATION LESS THAN 12 HOURS

                                              DEPTH ABOVE UPPER SURFACE OF GROUND FLOOR
COMPONENTS OF DAMAGE	-0.3	0.0	0.05	0.1	0.2	0.3	0.6	0.9	1.2	1.5	1.8	2.1	2.4	2.7	3.0	(M)
PATHS & PAVED AREAS	0	0	0	0	0	1	3	5	7	9	11	13	15	21	26	(£)
GARDENS/FENCES/SHEDS	0	0	0	0	60	124	192	273	363	518	676	893	1067	1247	1431	(£)
EXT. MAIN BUILDING	22	251	251	251	292	316	344	363	422	486	591	779	996	1041	1213	(£)
PLASTERWORK	0	0	0	0	56	123	193	249	319	386	467	534	705	838	1027	(£)
FLOORS	0	0	42	71	153	252	310	404	483	645	792	1024	1313	1561	1637	(£)
JOINERY	0	0	0	0	0	0	0	0	0	0	278	486	486	851	851	(£)
INTERNAL DECORATIONS	0	0	0	25	132	137	194	196	224	232	246	257	257	257	257	(£)
PLUMBING & ELECTRICAL	5	5	49	91	101	102	108	133	139	148	190	196	205	215	230	(£)
DOMESTIC APPLIANCES	0	0	0	48	118	231	300	469	470	499	511	511	511	511	511	(£)
HEATING EQUIPMENT	0	0	0	16	24	43	73	78	84	85	85	85	85	85	85	(£)
AUDIO/VIDEO	0	0	19	79	123	151	253	335	349	349	349	349	349	349	349	(£)
FURNITURE	0	0	30	87	206	434	598	647	664	668	668	668	668	668	668	(£)
PERSONAL EFFECTS	0	0	0	0	19	59	102	154	165	182	182	182	182	182	182	(£)
FLOOR COVER/CURTAINS	0	0	125	138	260	262	318	321	328	328	328	328	328	328	328	(£)
GARDEN/DIY/LEISURE	0	0	16	36	55	82	100	130	130	142	143	143	149	149	149	(£)
DOMESTIC CLEAN-UP	0	0	91	96	199	207	215	223	232	246	261	277	292	309	309	(£)

                                                                                     FLOOD DURATION LESS THAN 12 HOURS

	-0.3	0.0	0.05	0.1	0.2	0.3	0.6	0.9	1.2	1.5	1.8	2.1	2.4	2.7	3.0	
TOTAL BUILDING FABRIC	27	256	343	439	795	1060	1350	1630	1960	2430	3250	4190	5050	6030	6680	(£)
COL PERCENT	100.0	100.0	54.7	46.5	44.1	41.8	40.7	40.8	44.7	49.2	56.2	62.1	66.3	70.0	72.1	
ROW PERCENT	0.4	3.8	5.2	6.6	11.9	15.8	20.2	24.4	29.3	36.4	48.7	62.7	75.6	90.4	100.0	
TOTAL INVENTORY	0	0	284	504	1010	1470	1960	2360	2420	2500	2530	2550	2570	2590	2590	(£)
COL PERCENT	0.0	0.0	45.3	53.5	55.9	58.2	59.3	59.2	55.3	50.8	43.8	37.9	33.7	30.0	27.9	
ROW PERCENT	0.0	0.0	11.0	19.5	38.9	56.9	75.9	91.3	93.7	96.8	97.9	98.5	99.3	100.0	100.0	
TOTAL DAMAGE	27	256	628	944	1800	2530	3310	3990	4380	4930	5790	6730	7620	8620	9260	(£)
ROW PERCENT	0.3	2.8	6.8	10.2	19.5	27.3	35.7	43.0	47.3	53.2	62.5	72.7	82.2	93.1	100.0	
U.C. LIMIT	42	302	686	1060	2040	2870	3830	4600	5060	5680	6620	7650	8610	9690	10410	(£)
L.C. LIMIT	11	209	569	831	1560	2190	2790	3380	3700	4180	4960	5810	6630	7550	8110	(£)
TOTAL DAMAGE/SQ.METRE	0.8	7.4	18.0	27.0	51.5	72.2	94.5	113.8	125.1	140.8	165.2	192.2	217.4	246.1	264.4	(£)
U.C. LIMIT	1.3	9.1	19.7	29.2	56.5	79.5	106.1	128.0	140.8	158.1	184.0	213.6	239.5	268.8	287.4	(£)
L.C. LIMIT	0.3	5.7	16.3	24.8	46.5	64.9	82.9	99.6	109.4	123.5	146.4	170.8	195.3	223.4	241.4	(£)

APPENDIX 5.2   337

MODERN FLAT   LAND USE CODE 154   JANUARY 1990 PRICES   SALT WATER FLOOD DURATION LESS THAN 12 HOURS

```
                              DEPTH ABOVE UPPER SURFACE OF GROUND FLOOR
COMPONENTS OF DAMAGE   -0.3   0.0   0.05   0.1   0.2   0.3   0.6   0.9   1.2   1.5   1.8   2.1   2.4   2.7   3.0  (M)
-------------------------------------------------------------------------------------------------------------------
PATHS & PAVED AREAS      0     0     0      0     0    10    13    16    20    27    35    43    54    74    85  (£)
GARDENS/FENCES/SHEDS     0     0     0      0    22    46    69   102   144   205   282   357   416   490   555  (£)
EXT. MAIN BUILDING       0    98   109    112   156   208   225   231   256   341   655  1021  1322  1640  1946  (£)
PLASTERWORK              0     0     0      0   123   272   419   546   693   842  1017  1166  1535  1912  2236  (£)
FLOORS                   0     0    81    177   308   464   551   624   694   796   884  1003  1178  1322  1447  (£)
JOINERY                  0     0     0      0    31    56    81   107   133   921  1007  1576  1610  1657  1694  (£)
INTERNAL DECORATIONS     0     0     0     58   148   190   338   426   455   486   504   525   525   525   525  (£)
PLUMBING & ELECTRICAL    0    38    67    142   158   187   240   285   321   353   602   619   643   667   691  (£)

DOMESTIC APPLIANCES      0     0     0     51   128   254   355   541   543   575   589   589   589   589   589  (£)
HEATING EQUIPMENT        0     0     4     21    31    60    97   106   116   119   119   119   119   119   119  (£)
AUDIO/VIDEO              0     0    34    105   159   198   320   441   452   452   452   452   452   452   452  (£)
FURNITURE                0     0    44    165   663   914  1402  1419  1566  1571  1571  1571  1571  1571  1571  (£)
PERSONAL EFFECTS         0     0    16    146   252   589  1243  1531  1750  1767  1767  1767  1767  1767  1767  (£)
FLOOR COVER/CURTAINS     0     0   319    329   619   625   726   740   777   777   777   777   777   777   777  (£)
GARDEN/DIY/LEISURE       0     0    31     63    90   128   149   181   181   194   196   196   202   202   202  (£)
DOMESTIC CLEAN-UP        0     0   187    197   410   426   442   458   477   506   537   569   602   637   637  (£)

                                                                                  FLOOD DURATION LESS THAN 12 HOURS

TOTAL BUILDING FABRIC    0   136   257    490   949  1440  1940  2340  2720  3970  4990  6310  7290  8290  9180  (£)
    COL PERCENT        0.0 100.0  28.7   31.2  28.7  31.0  29.0  30.2  31.7  40.0  45.4  51.1  54.5  57.5  60.0
    ROW PERCENT        0.0   1.5   2.8    5.3  10.3  15.6  21.1  25.5  29.6  43.3  54.3  68.7  79.3  90.3 100.0
TOTAL INVENTORY          0     0   638   1080  2360  3200  4740  5420  5860  5960  6010  6040  6080  6120  6120  (£)
    COL PERCENT        0.0   0.0  71.3   68.8  71.3  69.0  71.0  69.8  68.3  60.0  54.6  48.9  45.5  42.5  40.0
    ROW PERCENT        0.0   0.0  10.4   17.7  38.5  52.3  77.5  88.6  95.9  97.5  98.3  98.8  99.4 100.0 100.0

TOTAL DAMAGE             0   136   895   1570  3310  4640  6680  7760  8580  9940 11000 12360 13370 14410 15300  (£)
    ROW PERCENT        0.0   0.9   5.9   10.3  21.6  30.3  43.6  50.7  56.1  65.0  71.9  80.8  87.4  94.2 100.0
    U.C. LIMIT           0   160   978   1760  3760  5270  7720  8940  9910 11450 12580 14040 15100 16200 17210  (£)
    L.C. LIMIT           0   111   811   1380  2860  4010  5640  6580  7250  8430  9420 10680 11640 12620 13390  (£)
TOTAL DAMAGE/SQ.METRE  0.0   1.9  12.5   21.8  45.9  64.3  92.6 107.7 119.1 137.8 152.6 171.4 185.4 199.9 212.3  (£)
    U.C. LIMIT         0.0   2.3  13.7   23.5  50.4  70.8 104.0 121.1 134.0 154.7 170.0 190.5 204.3 218.3 230.7  (£)
    L.C. LIMIT         0.0   1.5  11.3   20.1  41.4  57.8  81.2  94.3 104.2 120.9 135.2 152.3 166.5 181.5 193.9  (£)
```

POST 1975 FLAT   LAND USE CODE 155   JANUARY 1990 PRICES   SALT WATER FLOOD DURATION LESS THAN 12 HOURS

```
                              DEPTH ABOVE UPPER SURFACE OF GROUND FLOOR
COMPONENTS OF DAMAGE   -0.3   0.0   0.05   0.1   0.2   0.3   0.6   0.9   1.2   1.5   1.8   2.1   2.4   2.7   3.0  (M)
-------------------------------------------------------------------------------------------------------------------
PATHS & PAVED AREAS      0     0     0      0     0     9    11    13    17    23    29    34    42    59    70  (£)
GARDENS/FENCES/SHEDS     0     0     0      0     7    20    34    55    81   112   157   210   248   277   304  (£)
EXT. MAIN BUILDING       0    98   109    112   156   212   233   241   271   309   452   705   821   963  1096  (£)
PLASTERWORK              0     0     0      0   147   364   566   788   991  1212  1534  1855  1996  2106  2106  (£)
FLOORS                   0     0    45     96   173   268   321   372   417   489   551   633   754   856   936  (£)
JOINERY                  0     0     0      0    23    43    64    85   106   869   950  1496  1525  1566  1597  (£)
INTERNAL DECORATIONS     0     0     0     41    94   118   233   299   307   334   345   358   358   358   358  (£)
PLUMBING & ELECTRICAL    0     0    59    142   158   187   240   277   285   308   549   558   573   588   603  (£)

DOMESTIC APPLIANCES      0     0     0     49   122   237   312   481   482   513   524   524   524   524   524  (£)
HEATING EQUIPMENT        0     0     0     17    25    50    85    94   103   106   106   106   106   106   106  (£)
AUDIO/VIDEO              0     0    27     89   139   168   279   385   399   399   399   399   399   399   399  (£)
FURNITURE                0     0    40    133   451   690  1050  1103  1181  1185  1185  1185  1185  1185  1185  (£)
PERSONAL EFFECTS         0     0    15     85   151   335   596   762   869   886   886   886   886   886   886  (£)
FLOOR COVER/CURTAINS     0     0   150    164   303   307   371   380   401   401   401   401   401   401   401  (£)
GARDEN/DIY/LEISURE       0     0    29     59    84   120   139   167   167   179   181   181   187   187   187  (£)
DOMESTIC CLEAN-UP        0     0    88     93   193   201   208   216   225   239   254   269   284   301   301  (£)

                                                                                  FLOOD DURATION LESS THAN 12 HOURS

TOTAL BUILDING FABRIC    0    98   213    391   760  1220  1700  2130  2480  3660  4570  5850  6320  6780  7070  (£)
    COL PERCENT        0.0 100.0  37.7   36.1  34.1  36.7  35.9  37.2  39.3  48.3  53.7  59.7  61.4  62.9  63.9
    ROW PERCENT        0.0   1.4   3.0    5.5  10.7  17.3  24.1  30.1  35.0  51.7  64.6  82.8  89.4  95.8 100.0
TOTAL INVENTORY          0     0   352    693  1470  2110  3040  3590  3830  3910  3940  3950  3970  3990  3990  (£)
    COL PERCENT        0.0   0.0  62.3   63.9  65.9  63.3  64.1  62.8  60.7  51.7  46.3  40.3  38.6  37.1  36.1
    ROW PERCENT        0.0   0.0   8.8   17.4  36.9  52.9  76.2  90.0  96.0  98.0  98.7  99.1  99.6 100.0 100.0

TOTAL DAMAGE             0    98   565   1090  2230  3340  4750  5730  6310  7570  8510  9810 10300 10770 11060  (£)
    ROW PERCENT        0.0   0.9   5.1    9.8  20.2  30.1  42.9  51.7  57.0  68.4  76.9  88.6  93.1  97.3 100.0
    U.C. LIMIT           0   115   617   1220  2530  3800  5490  6600  7290  8720  9740 11150 11630 12110 12440  (£)
    L.C. LIMIT           0    80   512    960  1930  2880  4010  4860  5330  6420  7280  8470  8970  9430  9680  (£)
TOTAL DAMAGE/SQ.METRE  0.0   2.9  16.7   31.9  65.5  97.8 139.2 167.9 185.0 222.0 249.7 287.7 302.1 315.9 324.6  (£)
    U.C. LIMIT         0.0   3.5  18.3   34.4  71.9 107.7 156.3 188.8 208.2 249.3 278.1 319.7 332.9 345.0 352.8  (£)
    L.C. LIMIT         0.0   2.3  15.1   29.4  59.1  87.9 122.1 147.0 161.8 194.7 221.3 255.7 271.3 286.8 296.4  (£)
```

338 THE ECONOMICS OF COASTAL MANAGEMENT

PRE 1918 DETACHED   LAND USE CODE 111   JANUARY 1990 PRICES   SALT WATER FLOOD DURATION MORE THAN 12 HOURS

DEPTH ABOVE UPPER SURFACE OF GROUND FLOOR

COMPONENTS OF DAMAGE	-0.3	0.0	0.05	0.1	0.2	0.3	0.6	0.9	1.2	1.5	1.8	2.1	2.4	2.7	3.0	(M)
PATHS & PAVED AREAS	0	10	17	30	74	179	202	210	222	245	261	271	286	313	336	(£)
GARDENS/FENCES/SHEDS	0	0	38	94	381	626	1002	1262	1754	3414	4071	4770	5497	5876	6088	(£)
EXT. MAIN BUILDING	491	591	678	819	1028	1225	1456	1763	2210	3362	4817	6263	8330	10868	15050	(£)
PLASTERWORK	298	708	987	1218	1764	2384	3280	4088	5269	6120	7022	8241	9605	11009	10817	(£)
FLOORS	0	144	2400	3369	4916	5129	5441	5442	6980	7751	7533	7533	7533	7534	7535	(£)
JOINERY	0	0	297	1290	1927	2717	3283	3993	3801	3857	3918	4314	4315	4315	4315	(£)
INTERNAL DECORATIONS	57	71	122	430	1221	1239	1261	1298	1299	1317	1338	1337	1355	1404	1400	(£)
PLUMBING & ELECTRICAL	38	50	78	172	238	317	582	582	582	582	589	589	589	589	589	(£)
DOMESTIC APPLIANCES	0	0	3	92	234	510	602	865	865	880	880	880	880	880	880	(£)
HEATING EQUIPMENT	0	0	0	20	46	122	164	192	214	214	214	222	222	222	222	(£)
AUDIO/VIDEO	0	0	60	182	268	329	527	707	725	725	725	725	725	725	725	(£)
FURNITURE	0	0	92	229	528	1080	1492	1578	1613	1620	1620	1620	1620	1620	1620	(£)
PERSONAL EFFECTS	0	0	0	0	41	243	343	467	467	480	480	480	480	480	480	(£)
FLOOR COVER/CURTAINS	0	0	2523	2570	2625	2625	2727	2742	2758	2758	2758	2758	2758	2758	2758	(£)
GARDEN/DIY/LEISURE	0	0	37	72	106	149	202	233	241	255	262	262	270	270	270	(£)
DOMESTIC CLEAN-UP	0	0	430	444	930	964	999	1041	1103	1173	1311	1388	1388	1388	1388	(£)

FLOOD DURATION MORE THAN 12 HOURS

TOTAL BUILDING FABRIC	884	1580	4620	7430	11550	13820	16510	18640	22120	26650	29550	33320	37510	41910	46130	(£)
COL PERCENT	100.0	100.0	59.5	67.3	70.7	69.6	70.1	70.4	73.5	76.7	78.2	80.0	81.8	83.4	84.7	
ROW PERCENT	1.9	3.4	10.0	16.1	25.0	30.0	35.8	40.4	47.9	57.8	64.1	72.2	81.3	90.8	100.0	
TOTAL INVENTORY	0	0	3150	3610	4780	6030	7060	7830	7990	8110	8250	8340	8350	8350	8350	(£)
COL PERCENT	0.0	0.0	40.5	32.7	29.3	30.4	29.9	29.6	26.5	23.3	21.8	20.0	18.2	16.6	15.3	
ROW PERCENT	0.0	0.0	37.7	43.3	57.3	72.2	84.6	93.8	95.7	97.1	98.9	99.9	100.0	100.0	100.0	
TOTAL DAMAGE	884	1580	7770	11040	16340	19850	23570	26470	30110	34760	37810	41660	45860	50260	54480	(£)
ROW PERCENT	1.6	2.9	14.3	20.3	30.0	36.4	43.3	48.6	55.3	63.8	69.4	76.5	84.2	92.2	100.0	
U.C. LIMIT	1310	2080	9600	13090	19050	22660	26600	29700	33580	38890	42280	46630	51360	56600	61600	(£)
L.C. LIMIT	461	1080	5940	8990	13630	17040	20540	23240	26640	30630	33340	36690	40360	43920	47360	(£)
TOTAL DAMAGE/SQ.METRE	5.6	10.0	49.5	70.3	104.0	126.4	150.1	168.6	191.8	221.4	240.8	265.3	292.1	320.0	346.9	(£)
U.C. LIMIT	7.9	11.2	54.0	75.2	112.4	137.9	164.2	184.4	208.7	239.3	259.6	285.5	313.1	341.7	370.7	(£)
L.C. LIMIT	3.3	8.8	45.0	65.4	95.6	114.9	136.0	152.8	174.9	203.5	222.0	245.1	271.1	298.3	323.1	(£)

INTER WAR DETACHED   LAND USE CODE 112   JANUARY 1990 PRICES   SALT WATER FLOOD DURATION MORE THAN 12 HOURS

DEPTH ABOVE UPPER SURFACE OF GROUND FLOOR

COMPONENTS OF DAMAGE	-0.3	0.0	0.05	0.1	0.2	0.3	0.6	0.9	1.2	1.5	1.8	2.1	2.4	2.7	3.0	(M)
PATHS & PAVED AREAS	0	3	6	15	91	197	225	248	270	288	313	326	364	398	422	(£)
GARDENS/FENCES/SHEDS	0	0	11	44	288	498	1223	1317	1957	2443	2909	3304	3713	3928	4045	(£)
EXT. MAIN BUILDING	318	352	385	463	562	641	731	870	1081	1718	2681	3549	4821	6277	8785	(£)
PLASTERWORK	0	0	0	235	385	536	795	1076	1334	1613	1911	2455	2436	2436	2436	(£)
FLOORS	0	0	847	1304	1884	1988	2107	2108	2719	3022	2949	2949	2950	2952	2954	(£)
JOINERY	0	0	22	68	880	1583	2070	2443	2531	2695	2801	2834	2802	2802	2802	(£)
INTERNAL DECORATIONS	11	14	24	231	385	415	430	455	454	463	475	475	488	505	505	(£)
PLUMBING & ELECTRICAL	38	60	80	87	125	136	170	373	493	509	524	533	540	560	579	(£)
DOMESTIC APPLIANCES	0	0	3	96	250	550	614	900	900	919	919	919	919	919	919	(£)
HEATING EQUIPMENT	0	0	0	18	50	113	154	172	186	186	186	191	191	191	191	(£)
AUDIO/VIDEO	0	0	43	173	259	323	523	675	695	695	695	695	695	695	695	(£)
FURNITURE	0	0	55	139	317	715	1044	1136	1178	1186	1186	1186	1186	1186	1186	(£)
PERSONAL EFFECTS	0	0	0	0	45	280	397	538	539	553	553	553	553	553	553	(£)
FLOOR COVER/CURTAINS	0	0	872	907	947	947	1021	1030	1039	1039	1039	1039	1039	1039	1039	(£)
GARDEN/DIY/LEISURE	0	0	40	78	115	161	219	253	262	277	284	284	294	294	294	(£)
DOMESTIC CLEAN-UP	0	0	148	152	320	332	345	359	381	404	452	478	478	478	478	(£)

FLOOD DURATION MORE THAN 12 HOURS

TOTAL BUILDING FABRIC	367	429	1380	2450	4600	6000	7750	8890	10840	12750	14570	16430	18120	19860	22530	(£)
COL PERCENT	100.0	100.0	54.2	61.0	66.6	63.6	64.2	63.7	67.7	70.8	73.3	75.4	77.2	78.8	80.8	
ROW PERCENT	1.6	1.9	6.1	10.9	20.4	26.6	34.4	39.5	48.1	56.6	64.6	72.9	80.4	88.1	100.0	
TOTAL INVENTORY	0	0	1160	1570	2310	3430	4320	5070	5180	5260	5320	5350	5360	5360	5360	(£)
COL PERCENT	0.0	0.0	45.8	39.0	33.4	36.4	35.8	36.3	32.3	29.2	26.7	24.6	22.8	21.2	19.2	
ROW PERCENT	0.0	0.0	21.7	29.2	43.1	63.9	80.6	94.5	96.7	98.2	99.2	99.8	100.0	100.0	100.0	
TOTAL DAMAGE	367	429	2540	4020	6910	9420	12080	13960	16030	18020	19880	21780	23470	25220	27890	(£)
ROW PERCENT	1.3	1.5	9.1	14.4	24.8	33.8	43.3	50.0	57.5	64.6	71.3	78.1	84.2	90.4	100.0	
U.C. LIMIT	542	564	3140	4770	8060	10750	13630	15670	17880	20160	22230	24380	26280	28400	31540	(£)
L.C. LIMIT	191	293	1940	3270	5760	8090	10530	12250	14180	15880	17530	19180	20660	22040	24240	(£)
TOTAL DAMAGE/SQ.METRE	6.8	8.0	47.1	74.4	127.9	174.4	223.5	258.3	296.5	333.4	368.0	403.0	434.4	466.6	516.1	(£)
U.C. LIMIT	9.6	9.0	51.4	79.6	138.2	190.3	244.5	282.6	322.7	360.4	396.7	433.7	465.6	498.2	551.5	(£)
L.C. LIMIT	4.0	7.0	42.8	69.2	117.6	158.5	202.5	234.0	270.3	306.4	339.3	372.3	403.2	435.0	480.7	(£)

APPENDIX 5.2   339

POST WAR DETACHED   LAND USE CODE 113   JANUARY 1990 PRICES   SALT WATER FLOOD DURATION MORE THAN 12 HOURS

DEPTH ABOVE UPPER SURFACE OF GROUND FLOOR

COMPONENTS OF DAMAGE	-0.3	0.0	0.05	0.1	0.2	0.3	0.6	0.9	1.2	1.5	1.8	2.1	2.4	2.7	3.0	(M)
PATHS & PAVED AREAS	0	4	8	13	33	54	78	98	104	112	120	127	139	152	164	(£)
GARDENS/FENCES/SHEDS	0	6	28	47	353	1029	1290	1385	1994	2379	2711	3202	3716	4003	4158	(£)
EXT. MAIN BUILDING	0	303	305	347	459	581	754	946	1756	2531	3376	4594	5822	6827	7676	(£)
PLASTERWORK	0	0	0	336	552	767	1537	2306	3076	3444	3076	3341	3503	3503	3503	(£)
FLOORS	0	0	559	1475	2248	2805	2986	3148	4064	4522	4407	4407	4408	4408	4408	(£)
JOINERY	0	0	177	604	907	1249	1647	2659	3398	3431	4107	4107	4108	4108	4108	(£)
INTERNAL DECORATIONS	0	11	40	363	480	552	590	664	663	668	671	681	692	724	722	(£)
PLUMBING & ELECTRICAL	0	10	69	148	262	304	532	541	550	569	593	603	613	613	613	(£)
DOMESTIC APPLIANCES	0	0	3	97	253	558	642	934	934	953	953	953	953	953	953	(£)
HEATING EQUIPMENT	0	0	0	17	51	124	169	193	212	212	212	219	219	219	219	(£)
AUDIO/VIDEO	0	0	66	203	299	368	583	791	811	811	811	811	811	811	811	(£)
FURNITURE	0	0	72	179	427	896	1269	1362	1405	1413	1413	1413	1413	1413	1413	(£)
PERSONAL EFFECTS	0	0	0	0	45	286	405	548	548	562	562	562	562	562	562	(£)
FLOOR COVER/CURTAINS	0	0	1432	1468	1508	1508	1591	1608	1627	1627	1627	1627	1627	1627	1627	(£)
GARDEN/DIY/LEISURE	0	0	40	79	116	163	222	256	266	280	288	288	297	297	297	(£)
DOMESTIC CLEAN-UP	0	0	227	235	492	510	529	551	584	621	694	735	735	735	735	(£)

FLOOD DURATION MORE THAN 12 HOURS

	-0.3	0.0	0.05	0.1	0.2	0.3	0.6	0.9	1.2	1.5	1.8	2.1	2.4	2.7	3.0	
TOTAL BUILDING FABRIC	0	334	1190	3340	5300	7340	9420	11750	15610	17660	19060	21070	23000	24340	25360	(£)
COL PERCENT	100.0	100.0	39.2	59.4	62.4	62.4	63.5	65.3	71.0	73.1	74.4	76.1	77.6	78.6	79.3	
ROW PERCENT	0.0	1.3	4.7	13.2	20.9	29.0	37.1	46.3	61.6	69.6	75.2	83.1	90.7	96.0	100.0	
TOTAL INVENTORY	0	0	1840	2280	3190	4420	5410	6250	6390	6480	6560	6610	6620	6620	6620	(£)
COL PERCENT	0.0	0.0	60.8	40.6	37.6	37.6	36.5	34.7	29.0	26.9	25.6	23.9	22.4	21.4	20.7	
ROW PERCENT	0.0	0.0	27.8	34.5	48.3	66.7	81.8	94.4	96.5	97.9	99.1	99.9	100.0	100.0	100.0	
TOTAL DAMAGE	0	334	3030	5620	8490	11760	14830	18000	22000	24140	25630	27680	29620	30960	31980	(£)
ROW PERCENT	0.0	1.0	9.5	17.6	26.6	36.8	46.4	56.3	68.8	75.5	80.1	86.5	92.6	96.8	100.0	
U.C. LIMIT	0	439	3740	6660	9900	13420	16740	20200	24530	27010	28660	30980	33170	34860	36160	(£)
L.C. LIMIT	0	228	2320	4580	7080	10100	12920	15800	19470	21270	22600	24380	26070	27060	27800	(£)
TOTAL DAMAGE/SQ.METRE	0.0	4.0	36.5	67.7	102.3	141.7	178.7	216.8	265.0	290.8	308.7	333.4	356.8	373.0	385.2	(£)
U.C. LIMIT	0.0	4.5	39.8	72.5	110.5	154.6	195.5	237.2	288.4	314.3	332.7	358.8	382.4	398.3	411.6	(£)
L.C. LIMIT	0.0	3.5	33.2	62.9	94.1	128.8	161.9	196.4	241.6	267.3	284.7	308.0	331.2	347.7	358.8	(£)

MODERN DETACHED   LAND USE CODE 114   JANUARY 1990 PRICES   SALT WATER FLOOD DURATION MORE THAN 12 HOURS

DEPTH ABOVE UPPER SURFACE OF GROUND FLOOR

COMPONENTS OF DAMAGE	-0.3	0.0	0.05	0.1	0.2	0.3	0.6	0.9	1.2	1.5	1.8	2.1	2.4	2.7	3.0	(M)
PATHS & PAVED AREAS	0	1	6	12	87	147	200	248	272	297	326	348	407	449	482	(£)
GARDENS/FENCES/SHEDS	0	6	96	208	770	1458	2001	2166	2977	4065	5043	6297	7584	7772	7852	(£)
EXT. MAIN BUILDING	0	97	101	141	275	384	584	770	1341	1776	2313	2853	3537	4497	5240	(£)
PLASTERWORK	0	0	0	206	338	469	941	1412	1884	2110	1884	1946	2138	2138	2138	(£)
FLOORS	0	166	473	844	1234	1407	1521	1541	1679	1750	1750	1750	1751	1751	1751	(£)
JOINERY	0	0	148	524	782	1343	1667	2081	2700	2735	2771	2772	2773	2773	2773	(£)
INTERNAL DECORATIONS	0	6	34	281	358	391	405	428	443	446	450	456	464	488	487	(£)
PLUMBING & ELECTRICAL	38	43	68	190	293	368	634	681	692	707	829	853	875	903	951	(£)
DOMESTIC APPLIANCES	0	0	3	95	245	539	602	882	882	899	899	899	899	899	899	(£)
HEATING EQUIPMENT	0	0	0	19	48	110	150	168	182	182	182	186	186	186	186	(£)
AUDIO/VIDEO	0	0	42	170	254	315	514	661	681	681	681	681	681	681	681	(£)
FURNITURE	0	0	54	138	315	707	1029	1121	1161	1168	1168	1168	1168	1168	1168	(£)
PERSONAL EFFECTS	0	0	0	0	44	266	376	511	511	525	525	525	525	525	525	(£)
FLOOR COVER/CURTAINS	0	0	1018	1052	1091	1091	1163	1171	1180	1180	1180	1180	1180	1180	1180	(£)
GARDEN/DIY/LEISURE	0	0	39	77	113	158	215	248	257	271	279	279	288	288	288	(£)
DOMESTIC CLEAN-UP	0	0	169	175	366	380	394	410	436	463	518	548	548	548	548	(£)

FLOOD DURATION MORE THAN 12 HOURS

	-0.3	0.0	0.05	0.1	0.2	0.3	0.6	0.9	1.2	1.5	1.8	2.1	2.4	2.7	3.0	
TOTAL BUILDING FABRIC	38	320	928	2410	4140	5970	7960	9330	11990	13890	15370	17280	19530	20770	21680	(£)
COL PERCENT	100.0	100.0	41.1	58.2	62.5	62.6	64.1	64.3	69.4	72.1	73.9	76.0	78.1	79.1	79.8	
ROW PERCENT	0.2	1.5	4.3	11.1	19.1	27.5	36.7	43.0	55.3	64.1	70.9	79.7	90.1	95.8	100.0	
TOTAL INVENTORY	0	0	1330	1730	2480	3570	4450	5180	5290	5370	5430	5470	5480	5480	5480	(£)
COL PERCENT	0.0	0.0	58.9	41.8	37.5	37.4	35.9	35.7	30.6	27.9	26.1	24.0	21.9	20.9	20.2	
ROW PERCENT	0.0	0.0	24.3	31.6	45.3	65.2	81.2	94.5	96.6	98.1	99.2	99.8	100.0	100.0	100.0	
TOTAL DAMAGE	38	320	2260	4140	6620	9540	12400	14500	17280	19260	20800	22750	25010	26250	27160	(£)
ROW PERCENT	0.1	1.2	8.3	15.2	24.4	35.1	45.7	53.4	63.6	70.9	76.6	83.8	92.1	96.7	100.0	
U.C. LIMIT	56	420	2790	4910	7720	10890	13990	16270	19270	21550	23260	25470	28010	29560	30710	(£)
L.C. LIMIT	19	219	1730	3370	5520	8190	10810	12730	15290	16970	18340	20030	22010	22940	23610	(£)
TOTAL DAMAGE/SQ.METRE	0.6	5.2	36.4	66.7	106.7	153.8	199.9	233.8	278.6	310.4	335.3	366.6	403.1	423.1	437.7	(£)
U.C. LIMIT	0.8	5.8	39.7	71.4	115.3	167.8	218.7	255.8	303.2	335.5	361.4	394.5	432.0	451.8	467.7	(£)
L.C. LIMIT	0.4	4.6	33.1	62.0	98.1	139.8	181.1	211.8	254.0	285.3	309.2	338.7	374.2	394.4	407.7	(£)

340 THE ECONOMICS OF COASTAL MANAGEMENT

POST 1975 DETACHED   LAND USE CODE 115   JANUARY 1990 PRICES   SALT WATER FLOOD DURATION MORE THAN 12 HOURS

DEPTH ABOVE UPPER SURFACE OF GROUND FLOOR

COMPONENTS OF DAMAGE	-0.3	0.0	0.05	0.1	0.2	0.3	0.6	0.9	1.2	1.5	1.8	2.1	2.4	2.7	3.0	(M)
PATHS & PAVED AREAS	0	9	19	29	76	127	178	223	238	257	274	292	322	353	380	(£)
GARDENS/FENCES/SHEDS	0	4	53	88	625	1374	1674	1769	2320	2951	3468	3858	4790	4937	5016	(£)
EXT. MAIN BUILDING	0	114	128	200	388	550	815	1034	1622	2190	2828	3745	4659	5413	6178	(£)
PLASTERWORK	0	0	251	384	902	1270	1627	1993	2100	2155	2066	2283	2294	2294	2294	(£)
FLOORS	0	134	432	831	1254	1428	1581	1616	1659	1705	1705	1705	1705	1705	1705	(£)
JOINERY	0	0	284	895	1440	2021	2357	2781	3439	3473	3509	3510	3510	3510	3510	(£)
INTERNAL DECORATIONS	0	7	43	283	364	392	396	406	421	422	422	425	430	440	440	(£)
PLUMBING & ELECTRICAL	38	40	65	154	261	288	566	608	617	635	696	715	735	759	792	(£)
DOMESTIC APPLIANCES	0	0	3	96	251	553	636	926	926	945	945	945	945	945	945	(£)
HEATING EQUIPMENT	0	0	0	17	50	122	166	191	209	209	209	216	216	216	216	(£)
AUDIO/VIDEO	0	0	66	201	295	364	578	782	802	802	802	802	802	802	802	(£)
FURNITURE	0	0	71	178	426	893	1264	1356	1398	1407	1407	1407	1407	1407	1407	(£)
PERSONAL EFFECTS	0	0	0	0	45	283	400	542	542	557	557	557	557	557	557	(£)
FLOOR COVER/CURTAINS	0	0	1012	1047	1088	1088	1170	1187	1206	1206	1206	1206	1206	1206	1206	(£)
GARDEN/DIY/LEISURE	0	0	40	79	116	163	221	255	264	279	287	287	296	296	296	(£)
DOMESTIC CLEAN-UP	0	0	169	175	366	380	394	410	435	462	517	547	547	547	547	(£)

FLOOD DURATION MORE THAN 12 HOURS

	-0.3	0.0	0.05	0.1	0.2	0.3	0.6	0.9	1.2	1.5	1.8	2.1	2.4	2.7	3.0	
TOTAL BUILDING FABRIC	38	309	1280	2870	5310	7450	9200	10430	12420	13790	14970	16540	18450	19410	20320	(£)
COL PERCENT	100.0	100.0	48.4	61.5	66.8	65.9	65.6	64.9	68.2	70.1	71.6	73.5	75.5	76.5	77.3	
ROW PERCENT	0.2	1.5	6.3	14.1	26.2	36.7	45.3	51.3	61.1	67.9	73.7	81.4	90.8	95.6	100.0	
TOTAL INVENTORY	0	0	1360	1800	2640	3850	4830	5650	5790	5870	5930	5970	5980	5980	5980	(£)
COL PERCENT	0.0	0.0	51.6	38.5	33.2	34.1	34.4	35.1	31.8	29.9	28.4	26.5	24.5	23.5	22.7	
ROW PERCENT	0.0	0.0	22.8	30.1	44.2	64.4	80.8	94.6	96.8	98.2	99.2	99.8	100.0	100.0	100.0	
TOTAL DAMAGE	38	309	2640	4670	7950	11300	14030	16090	18210	19660	20900	22510	24430	25390	26300	(£)
ROW PERCENT	0.1	1.2	10.0	17.7	30.3	43.0	53.4	61.2	69.2	74.8	79.5	85.6	92.9	96.6	100.0	
U.C. LIMIT	56	406	3260	5540	9270	12900	15830	18060	20310	22000	23370	25200	27360	28590	29740	(£)
L.C. LIMIT	19	211	2020	3800	6630	9700	12230	14120	16110	17320	18430	19820	21500	22190	22860	(£)
TOTAL DAMAGE/SQ.METRE	0.6	5.0	42.6	75.3	128.3	182.3	226.3	259.4	293.6	317.0	337.1	362.9	393.8	409.4	424.0	(£)
U.C. LIMIT	0.8	5.6	46.5	80.6	138.6	198.9	247.6	283.8	319.5	342.7	363.4	390.6	422.1	437.1	453.1	(£)
L.C. LIMIT	0.4	4.4	38.7	70.0	118.0	165.7	205.0	235.0	267.7	291.3	310.8	335.2	365.5	381.7	394.9	(£)

POST 1975 UTILITY DETACHED   LAND USE CODE 117   JANUARY 1990 PRICES   SALT WATER FLOOD DURATION MORE THAN 12 HOURS

DEPTH ABOVE UPPER SURFACE OF GROUND FLOOR

COMPONENTS OF DAMAGE	-0.3	0.0	0.05	0.1	0.2	0.3	0.6	0.9	1.2	1.5	1.8	2.1	2.4	2.7	3.0	(M)
PATHS & PAVED AREAS	0	7	14	20	46	74	99	117	120	124	127	131	139	146	152	(£)
GARDENS/FENCES/SHEDS	0	3	48	93	552	1222	1479	1565	2018	2507	2912	3452	4003	4181	4279	(£)
EXT. MAIN BUILDING	0	90	99	158	302	437	673	869	1158	1462	1844	2274	2752	3145	3566	(£)
PLASTERWORK	0	0	59	90	290	455	618	783	846	878	811	897	897	897	897	(£)
FLOORS	0	25	114	270	442	520	589	612	639	671	675	680	689	696	701	(£)
JOINERY	0	0	0	31	99	317	452	561	907	957	957	996	996	996	996	(£)
INTERNAL DECORATIONS	0	3	18	112	154	167	174	179	187	188	188	190	192	196	196	(£)
PLUMBING & ELECTRICAL	0	0	0	0	24	37	46	64	64	64	82	82	77	89	96	(£)
DOMESTIC APPLIANCES	0	0	0	0	3	159	392	441	445	445	445	445	445	445	445	(£)
HEATING EQUIPMENT	0	0	0	0	0	0	0	0	0	0	0	0	0	0	0	(£)
AUDIO/VIDEO	0	0	2	2	2	4	6	24	24	24	24	24	24	24	24	(£)
FURNITURE	0	0	8	17	30	31	41	41	41	41	41	41	41	41	41	(£)
PERSONAL EFFECTS	0	0	0	0	0	0	0	0	0	0	0	0	0	0	0	(£)
FLOOR COVER/CURTAINS	0	0	557	557	585	585	601	617	628	628	628	628	628	628	628	(£)
GARDEN/DIY/LEISURE	0	0	40	79	116	163	219	255	263	279	286	286	296	296	296	(£)
DOMESTIC CLEAN-UP	0	0	71	74	154	160	167	174	183	195	216	230	231	233	233	(£)

FLOOD DURATION MORE THAN 12 HOURS

	-0.3	0.0	0.05	0.1	0.2	0.3	0.6	0.9	1.2	1.5	1.8	2.1	2.4	2.7	3.0	
TOTAL BUILDING FABRIC	0	129	355	777	1910	3230	4130	4750	5940	6850	7600	8700	9750	10350	10890	(£)
COL PERCENT	0.0	100.0	34.3	51.6	68.1	74.5	74.3	75.4	78.9	80.9	82.2	84.0	85.4	86.1	86.7	
ROW PERCENT	0.0	1.2	3.3	7.1	17.6	29.7	38.0	43.7	54.6	62.9	69.8	80.0	89.5	95.1	100.0	
TOTAL INVENTORY	0	0	680	730	893	1100	1430	1550	1590	1610	1640	1660	1670	1670	1670	(£)
COL PERCENT	0.0	0.0	65.7	48.4	31.9	25.5	25.7	24.6	21.1	19.1	17.8	16.0	14.6	13.9	13.3	
ROW PERCENT	0.0	0.0	40.8	43.8	53.6	66.2	85.6	93.1	95.1	96.7	98.4	99.2	99.9	100.0	100.0	
TOTAL DAMAGE	0	129	1040	1510	2810	4340	5560	6310	7530	8470	9240	10360	11410	12020	12550	(£)
ROW PERCENT	0.0	1.0	8.2	12.0	22.3	34.5	44.3	50.2	60.0	67.4	73.6	82.5	90.9	95.7	100.0	
U.C. LIMIT	0	169	1280	1790	3280	4950	6270	7080	8400	9480	10330	11600	12780	13540	14190	(£)
L.C. LIMIT	0	88	795	1230	2340	3730	4850	5540	6660	7460	8150	9120	10040	10500	10910	(£)
TOTAL DAMAGE/SQ.METRE	0.0	5.7	42.2	62.3	116.5	181.8	233.0	263.1	315.7	353.9	387.3	433.9	479.3	505.1	528.2	(£)
U.C. LIMIT	0.0	6.4	46.1	66.7	125.9	198.3	254.9	287.8	343.6	382.5	417.5	467.0	513.7	539.3	564.4	(£)
L.C. LIMIT	0.0	5.0	38.3	57.9	107.1	165.3	211.1	238.4	287.8	325.3	357.1	400.8	444.9	470.9	492.0	(£)

APPENDIX 5.2

PRE 1918 SEMI DETACHED   LAND USE CODE 121   JANUARY 1990 PRICES   SALT WATER FLOOD DURATION MORE THAN 12 HOURS

DEPTH ABOVE UPPER SURFACE OF GROUND FLOOR

COMPONENTS OF DAMAGE	-0.3	0.0	0.05	0.1	0.2	0.3	0.6	0.9	1.2	1.5	1.8	2.1	2.4	2.7	3.0	(M)
PATHS & PAVED AREAS	0	1	3	5	12	28	33	34	36	40	43	44	46	50	54	(£)
GARDENS/FENCES/SHEDS	0	0	28	60	180	270	427	521	710	1735	2007	2315	2631	2742	2805	(£)
EXT. MAIN BUILDING	480	493	506	534	589	634	684	776	912	1405	2267	2921	3973	5273	7489	(£)
PLASTERWORK	43	112	163	210	320	453	692	977	1338	1673	2027	2188	2137	2131	2121	(£)
FLOORS	0	56	544	778	1136	1203	1279	1279	1671	1865	1814	1814	1814	1814	1814	(£)
JOINERY	0	0	166	542	922	1321	1651	2086	2027	2060	2096	2444	2444	2444	2444	(£)
INTERNAL DECORATIONS	34	36	46	110	238	251	266	292	292	300	309	310	319	344	343	(£)
PLUMBING & ELECTRICAL	38	54	78	172	238	253	471	471	471	471	471	471	471	471	471	(£)
DOMESTIC APPLIANCES	0	0	2	86	210	450	503	728	728	738	738	738	738	738	738	(£)
HEATING EQUIPMENT	0	0	0	22	40	93	123	139	151	151	151	155	155	155	155	(£)
AUDIO/VIDEO	0	0	35	136	204	250	418	527	543	543	543	543	543	543	543	(£)
FURNITURE	0	0	50	128	286	620	873	952	977	982	982	982	982	982	982	(£)
PERSONAL EFFECTS	0	0	0	0	35	187	262	360	360	372	372	372	372	372	372	(£)
FLOOR COVER/CURTAINS	0	0	510	538	569	569	622	629	635	635	635	635	635	635	635	(£)
GARDEN/DIY/LEISURE	0	0	32	64	94	132	179	204	212	224	230	230	238	238	238	(£)
DOMESTIC CLEAN-UP	0	0	101	104	219	228	236	246	260	277	309	327	327	327	327	(£)

FLOOD DURATION MORE THAN 12 HOURS

	-0.3	0.0	0.05	0.1	0.2	0.3	0.6	0.9	1.2	1.5	1.8	2.1	2.4	2.7	3.0	
TOTAL BUILDING FABRIC	595	754	1540	2410	3640	4420	5510	6440	7460	9550	11040	12510	13840	15270	17540	(£)
COL PERCENT	100.0	100.0	67.7	69.0	68.7	63.5	63.1	63.0	65.8	70.9	73.6	75.8	77.6	79.3	81.5	
ROW PERCENT	3.4	4.3	8.8	13.8	20.7	25.2	31.4	36.7	42.5	54.5	62.9	71.3	78.9	87.1	100.0	
TOTAL INVENTORY	0	0	732	1080	1660	2530	3220	3790	3870	3920	3960	3990	3990	3990	3990	(£)
COL PERCENT	0.0	0.0	32.3	31.0	31.3	36.5	36.9	37.0	34.2	29.1	26.4	24.2	22.4	20.7	18.5	
ROW PERCENT	0.0	0.0	18.3	27.1	41.6	63.4	80.6	94.9	96.9	98.3	99.3	99.8	100.0	100.0	100.0	
TOTAL DAMAGE	595	754	2270	3500	5300	6950	8720	10230	11330	13480	15000	16490	17830	19260	21540	(£)
ROW PERCENT	2.8	3.5	10.5	16.2	24.6	32.3	40.5	47.5	52.6	62.6	69.7	76.6	82.8	89.5	100.0	
U.C. LIMIT	696	835	2460	3740	5690	7520	9440	11090	12230	14450	16010	17590	18930	20410	22780	(£)
L.C. LIMIT	493	672	2080	3260	4910	6380	8000	9370	10430	12510	13990	15390	16730	18110	20300	(£)
TOTAL DAMAGE/SQ.METRE	16.1	20.3	61.2	94.3	142.9	187.4	235.2	276.0	305.7	363.6	404.8	445.2	481.3	520.1	581.4	(£)
U.C. LIMIT	19.2	22.7	66.6	101.1	154.4	204.8	257.5	303.0	332.2	393.8	436.4	478.3	515.4	556.1	620.7	(£)
L.C. LIMIT	13.0	17.9	55.8	87.5	131.4	170.0	213.1	249.0	278.8	333.4	373.2	412.1	447.2	484.1	542.1	(£)

INTER WAR SEMI DETACHED   LAND USE CODE 122   JANUARY 1990 PRICES   SALT WATER FLOOD DURATION MORE THAN 12 HOURS

DEPTH ABOVE UPPER SURFACE OF GROUND FLOOR

COMPONENTS OF DAMAGE	-0.3	0.0	0.05	0.1	0.2	0.3	0.6	0.9	1.2	1.5	1.8	2.1	2.4	2.7	3.0	(M)
PATHS & PAVED AREAS	0	1	3	5	18	37	45	50	54	59	63	68	73	80	86	(£)
GARDENS/FENCES/SHEDS	0	6	38	73	182	312	530	611	1394	1714	1972	2377	2802	3042	3172	(£)
EXT. MAIN BUILDING	474	495	517	582	666	767	869	1024	1095	1546	2242	2889	3682	4670	6362	(£)
PLASTERWORK	0	0	0	219	358	499	738	999	1238	1499	1777	2285	2269	2269	2269	(£)
FLOORS	0	42	726	1141	1595	1677	1789	1789	2352	2634	2564	2564	2564	2564	2564	(£)
JOINERY	0	0	96	231	711	1191	1755	2003	2273	2437	2542	2868	2835	2835	2835	(£)
INTERNAL DECORATIONS	30	32	49	251	325	348	356	366	367	375	383	384	395	410	410	(£)
PLUMBING & ELECTRICAL	38	60	80	92	137	166	217	293	371	397	423	442	459	479	498	(£)
DOMESTIC APPLIANCES	0	0	2	82	191	403	450	646	646	651	651	651	651	651	651	(£)
HEATING EQUIPMENT	0	0	0	24	35	85	110	124	134	134	134	138	138	138	138	(£)
AUDIO/VIDEO	0	0	30	118	177	216	366	455	469	469	469	469	469	469	469	(£)
FURNITURE	0	0	47	122	270	571	787	859	876	880	880	880	880	880	880	(£)
PERSONAL EFFECTS	0	0	0	0	31	147	203	283	283	293	293	293	293	293	293	(£)
FLOOR COVER/CURTAINS	0	0	502	527	554	554	597	603	608	608	608	608	608	608	608	(£)
GARDEN/DIY/LEISURE	0	0	28	57	85	118	159	181	187	198	203	203	210	210	210	(£)
DOMESTIC CLEAN-UP	0	0	139	144	302	313	325	338	359	381	426	451	451	451	451	(£)

FLOOD DURATION MORE THAN 12 HOURS

	-0.3	0.0	0.05	0.1	0.2	0.3	0.6	0.9	1.2	1.5	1.8	2.1	2.4	2.7	3.0	
TOTAL BUILDING FABRIC	543	638	1510	2600	4000	5000	6300	7140	9150	10660	11970	13880	15080	16350	18200	(£)
COL PERCENT	100.0	100.0	66.8	70.7	70.8	67.5	67.8	67.1	71.9	74.7	76.5	79.0	80.3	81.5	83.1	
ROW PERCENT	3.0	3.5	8.3	14.3	22.0	27.5	34.6	39.2	50.3	58.6	65.8	76.3	82.9	89.9	100.0	
TOTAL INVENTORY	0	0	751	1080	1650	2410	3000	3490	3570	3620	3670	3700	3700	3700	3700	(£)
COL PERCENT	0.0	0.0	33.2	29.3	29.2	32.5	32.2	32.9	28.1	25.3	23.5	21.0	19.7	18.5	16.9	
ROW PERCENT	0.0	0.0	20.3	29.1	44.5	65.0	81.0	94.3	96.3	97.7	99.0	99.8	100.0	100.0	100.0	
TOTAL DAMAGE	543	638	2270	3670	5640	7410	9300	10630	12710	14280	15640	17580	18790	20060	21900	(£)
ROW PERCENT	2.5	2.9	10.3	16.8	25.8	33.8	42.5	48.5	58.0	65.2	71.4	80.3	85.8	91.6	100.0	
U.C. LIMIT	635	706	2460	3920	6060	8010	10070	11530	13720	15310	16690	18750	19950	21250	23160	(£)
L.C. LIMIT	450	569	2080	3420	5220	6810	8530	9730	11700	13250	14590	16410	17630	18870	20640	(£)
TOTAL DAMAGE/SQ.METRE	10.7	12.5	44.5	72.1	110.7	145.3	182.3	208.4	249.2	280.0	306.6	344.6	368.3	393.2	429.4	(£)
U.C. LIMIT	12.7	14.0	48.4	77.3	119.6	158.8	199.5	228.8	271.1	303.2	330.6	370.2	394.4	420.4	458.4	(£)
L.C. LIMIT	8.7	11.0	40.6	66.9	101.8	131.8	165.1	188.0	227.3	256.8	282.6	319.0	342.2	366.0	400.4	(£)

POST WAR SEMI DETACHED   LAND USE CODE 123   JANUARY 1990 PRICES   SALT WATER FLOOD DURATION MORE THAN 12 HOURS

DEPTH ABOVE UPPER SURFACE OF GROUND FLOOR

COMPONENTS OF DAMAGE	-0.3	0.0	0.05	0.1	0.2	0.3	0.6	0.9	1.2	1.5	1.8	2.1	2.4	2.7	3.0	(M)
PATHS & PAVED AREAS	0	1	2	4	15	31	39	43	47	52	56	61	65	72	78	(£)
GARDENS/FENCES/SHEDS	0	6	41	79	183	310	526	606	1352	1667	1925	2329	2752	2984	3109	(£)
EXT. MAIN BUILDING	475	496	518	581	666	770	873	1028	1084	1534	2224	2872	3679	4667	6359	(£)
PLASTERWORK	0	0	0	219	358	499	738	999	1238	1499	1777	2285	2269	2269	2269	(£)
FLOORS	0	48	704	1110	1547	1626	1738	1738	2301	2584	2514	2514	2514	2514	2514	(£)
JOINERY	0	0	91	219	669	1119	1647	1879	2173	2337	2442	2710	2677	2677	2677	(£)
INTERNAL DECORATIONS	32	34	51	268	329	351	359	368	369	377	385	386	397	412	412	(£)
PLUMBING & ELECTRICAL	38	60	80	91	134	160	207	276	342	371	400	421	440	460	479	(£)
DOMESTIC APPLIANCES	0	0	2	81	188	395	441	633	633	637	637	637	637	637	637	(£)
HEATING EQUIPMENT	0	0	0	24	33	82	106	119	130	130	130	134	134	134	134	(£)
AUDIO/VIDEO	0	0	30	115	172	210	359	445	459	459	459	459	459	459	459	(£)
FURNITURE	0	0	46	122	269	569	781	854	869	872	872	872	872	872	872	(£)
PERSONAL EFFECTS	0	0	0	0	31	138	190	266	266	277	277	277	277	277	277	(£)
FLOOR COVER/CURTAINS	0	0	495	521	547	547	588	594	598	598	598	598	598	598	598	(£)
GARDEN/DIY/LEISURE	0	0	28	56	84	116	157	179	185	196	201	201	208	208	208	(£)
DOMESTIC CLEAN-UP	0	0	139	144	302	313	325	338	359	381	426	451	451	451	451	(£)

FLOOD DURATION MORE THAN 12 HOURS

	-0.3	0.0	0.05	0.1	0.2	0.3	0.6	0.9	1.2	1.5	1.8	2.1	2.4	2.7	3.0	
TOTAL BUILDING FABRIC	545	646	1490	2580	3900	4870	6130	6940	8910	10420	11730	13580	14800	16060	17900	(£)
COL PERCENT	100.0	100.0	66.7	70.7	70.6	67.2	67.5	66.9	71.8	74.6	76.5	78.9	80.3	81.5	83.1	
ROW PERCENT	3.0	3.6	8.3	14.4	21.8	27.2	34.2	38.8	49.8	58.2	65.5	75.9	82.7	89.7	100.0	
TOTAL INVENTORY	0	0	743	1070	1630	2370	2950	3430	3500	3550	3600	3630	3640	3640	3640	(£)
COL PERCENT	0.0	0.0	33.3	29.3	29.4	32.8	32.5	33.1	28.2	25.4	23.5	21.1	19.7	18.5	16.9	
ROW PERCENT	0.0	0.0	20.4	29.3	44.8	65.2	81.1	94.3	96.3	97.6	99.0	99.8	100.0	100.0	100.0	
TOTAL DAMAGE	545	646	2230	3640	5530	7240	9080	10370	12410	13980	15330	17210	18430	19700	21540	(£)
ROW PERCENT	2.5	3.0	10.4	16.9	25.7	33.6	42.2	48.7	57.6	64.9	71.2	79.9	85.6	91.4	100.0	
U.C. LIMIT	638	715	2420	3890	5940	7830	9830	11250	13400	14990	16360	18360	19570	20870	22780	(£)
L.C. LIMIT	451	576	2040	3390	5120	6650	8330	9490	11420	12970	14300	16060	17290	18530	20300	(£)
TOTAL DAMAGE/SQ.METRE	10.7	12.7	43.8	71.4	108.5	142.0	178.0	203.3	243.3	274.0	300.6	337.4	361.4	386.1	422.3	(£)
U.C. LIMIT	12.7	14.2	47.7	76.6	117.3	155.2	194.8	223.2	264.7	296.7	324.1	362.5	387.0	412.8	450.8	(£)
L.C. LIMIT	8.7	11.2	39.9	66.2	99.7	128.8	161.2	183.4	221.9	251.3	277.1	312.3	335.8	359.4	393.8	(£)

MODERN SEMI DETACHED   LAND USE CODE 124   JANUARY 1990 PRICES   SALT WATER FLOOD DURATION MORE THAN 12 HOURS

DEPTH ABOVE UPPER SURFACE OF GROUND FLOOR

COMPONENTS OF DAMAGE	-0.3	0.0	0.05	0.1	0.2	0.3	0.6	0.9	1.2	1.5	1.8	2.1	2.4	2.7	3.0	(M)
PATHS & PAVED AREAS	0	1	2	5	23	38	50	60	65	70	76	81	93	102	108	(£)
GARDENS/FENCES/SHEDS	0	6	44	82	357	909	1207	1316	1976	2390	2740	3271	3831	4213	4418	(£)
EXT. MAIN BUILDING	0	62	64	102	203	307	500	680	1239	1765	2376	3186	4038	4710	5366	(£)
PLASTERWORK	0	0	0	173	285	396	793	1192	1589	1779	1589	1649	1834	1834	1834	(£)
FLOORS	0	150	357	568	789	937	1048	1067	1206	1276	1276	1276	1276	1276	1276	(£)
JOINERY	0	0	222	653	1040	1431	1655	1995	2420	2453	2489	2489	2489	2489	2489	(£)
INTERNAL DECORATIONS	0	6	34	184	261	293	305	328	334	337	340	344	352	371	370	(£)
PLUMBING & ELECTRICAL	19	40	65	160	258	296	539	573	592	593	646	665	687	710	736	(£)
DOMESTIC APPLIANCES	0	0	2	86	204	435	486	703	703	708	708	708	708	708	708	(£)
HEATING EQUIPMENT	0	0	0	23	35	87	116	131	142	142	142	146	146	146	146	(£)
AUDIO/VIDEO	0	0	34	132	196	240	407	509	524	524	524	524	524	524	524	(£)
FURNITURE	0	0	48	127	284	614	859	942	961	965	965	965	965	965	965	(£)
PERSONAL EFFECTS	0	0	0	0	35	165	228	318	318	329	329	329	329	329	329	(£)
FLOOR COVER/CURTAINS	0	0	565	593	623	623	670	677	682	682	682	682	682	682	682	(£)
GARDEN/DIY/LEISURE	0	0	31	62	92	129	174	199	207	218	224	224	232	232	232	(£)
DOMESTIC CLEAN-UP	0	0	117	121	254	263	273	284	301	320	358	379	379	379	379	(£)

FLOOD DURATION MORE THAN 12 HOURS

	-0.3	0.0	0.05	0.1	0.2	0.3	0.6	0.9	1.2	1.5	1.8	2.1	2.4	2.7	3.0	
TOTAL BUILDING FABRIC	19	265	791	1930	3220	4610	6100	7210	9420	10660	11530	12960	14600	15710	16600	(£)
COL PERCENT	100.0	100.0	49.7	62.7	65.1	64.3	65.5	65.7	71.0	73.3	74.6	76.6	78.6	79.8	80.7	
ROW PERCENT	0.1	1.6	4.8	11.6	19.4	27.8	36.8	43.5	56.8	64.2	69.5	78.1	88.0	94.6	100.0	
TOTAL INVENTORY	0	0	799	1150	1730	2560	3220	3770	3840	3890	3940	3960	3970	3970	3970	(£)
COL PERCENT	0.0	0.0	50.3	37.3	34.9	35.7	34.5	34.3	29.0	26.7	25.4	23.4	21.4	20.2	19.3	
ROW PERCENT	0.0	0.0	20.1	28.9	43.5	64.4	81.1	94.9	96.8	98.1	99.2	99.8	100.0	100.0	100.0	
TOTAL DAMAGE	19	265	1590	3080	4950	7170	9320	10980	13270	14560	15470	16930	18570	19680	20570	(£)
ROW PERCENT	0.1	1.3	7.7	15.0	24.0	34.9	45.3	53.4	64.5	70.8	75.2	82.3	90.3	95.7	100.0	
U.C. LIMIT	22	293	1720	3290	5320	7750	10090	11910	14330	15610	16510	18060	19710	20850	21760	(£)
L.C. LIMIT	15	236	1460	2870	4580	6590	8550	10050	12210	13510	14430	15800	17430	18510	19380	(£)
TOTAL DAMAGE/SQ.METRE	0.4	6.2	37.0	71.6	114.9	166.6	216.5	255.2	308.3	338.3	359.5	393.4	431.7	457.4	478.1	(£)
U.C. LIMIT	0.5	6.9	40.3	76.8	124.2	182.0	236.9	280.2	335.4	366.4	387.6	422.7	462.3	489.0	510.4	(£)
L.C. LIMIT	0.3	5.5	33.7	66.4	105.6	151.2	196.1	230.2	281.2	310.2	331.4	364.1	401.1	425.8	445.8	(£)

APPENDIX 5.2 343

POST 1975 SEMI DETACHED   LAND USE CODE 125   JANUARY 1990 PRICES   SALT WATER FLOOD DURATION MORE THAN 12 HOURS

COMPONENTS OF DAMAGE	-0.3	0.0	0.05	0.1	0.2	0.3	0.6	0.9	1.2	1.5	1.8	2.1	2.4	2.7	3.0	(M)
PATHS & PAVED AREAS	0	4	8	12	27	43	56	66	66	69	71	73	77	81	84	(£)
GARDENS/FENCES/SHEDS	0	2	53	95	371	808	1005	1071	1441	1876	2221	2638	3063	3177	3239	(£)
EXT. MAIN BUILDING	0	85	91	144	269	385	596	781	1053	1336	1716	2098	2546	2907	3281	(£)
PLASTERWORK	0	0	133	202	574	868	1153	1443	1546	1606	1518	1654	1654	1654	1654	(£)
FLOORS	0	87	236	438	644	767	887	922	964	1010	1010	1010	1010	1010	1010	(£)
JOINERY	0	0	61	219	361	692	888	1144	1540	1555	1571	1571	1571	1571	1571	(£)
INTERNAL DECORATIONS	0	5	33	186	265	285	285	286	292	293	294	298	303	316	316	(£)
PLUMBING & ELECTRICAL	21	40	69	144	239	278	528	533	538	543	551	559	568	582	592	(£)
DOMESTIC APPLIANCES	0	0	2	82	191	402	449	644	644	648	648	648	648	648	648	(£)
HEATING EQUIPMENT	0	0	0	25	34	83	108	122	133	133	133	136	136	136	136	(£)
AUDIO/VIDEO	0	0	31	119	177	216	369	457	472	472	472	472	472	472	472	(£)
FURNITURE	0	0	47	123	272	576	793	869	884	887	887	887	887	887	887	(£)
PERSONAL EFFECTS	0	0	0	0	31	138	189	266	266	276	276	276	276	276	276	(£)
FLOOR COVER/CURTAINS	0	0	414	440	466	466	507	513	518	518	518	518	518	518	518	(£)
GARDEN/DIY/LEISURE	0	0	28	57	85	118	159	181	188	198	204	204	211	211	211	(£)
DOMESTIC CLEAN-UP	0	0	96	99	207	215	224	233	247	262	293	310	310	310	310	(£)

FLOOD DURATION MORE THAN 12 HOURS

	-0.3	0.0	0.05	0.1	0.2	0.3	0.6	0.9	1.2	1.5	1.8	2.1	2.4	2.7	3.0	
TOTAL BUILDING FABRIC	21	224	688	1440	2750	4130	5400	6250	7440	8290	8960	9910	10800	11300	11750	(£)
COL PERCENT	100.0	100.0	52.6	60.4	65.2	65.1	65.8	65.5	68.9	70.9	72.3	74.1	75.7	76.6	77.2	
ROW PERCENT	0.2	1.9	5.9	12.3	23.4	35.1	46.0	53.2	63.3	70.6	76.2	84.3	91.9	96.2	100.0	
TOTAL INVENTORY	0	0	621	947	1470	2220	2800	3290	3360	3400	3430	3460	3460	3460	3460	(£)
COL PERCENT	0.0	0.0	47.4	39.6	34.8	34.9	34.2	34.5	31.1	29.1	27.7	25.9	24.3	23.4	22.8	
ROW PERCENT	0.0	0.0	17.9	27.4	42.4	64.0	81.0	95.0	96.9	98.1	99.2	99.8	100.0	100.0	100.0	
TOTAL DAMAGE	21	224	1310	2390	4220	6350	8200	9540	10800	11690	12390	13360	14260	14760	15210	(£)
ROW PERCENT	0.1	1.5	8.6	15.7	27.7	41.7	53.9	62.7	71.0	76.8	81.4	87.8	93.7	97.0	100.0	
U.C. LIMIT	24	248	1420	2550	4530	6870	8880	10350	11660	12540	13220	14250	15140	15640	16090	(£)
L.C. LIMIT	17	199	1200	2230	3910	5830	7520	8730	9940	10840	11560	12470	13380	13880	14330	(£)
TOTAL DAMAGE/SQ.METRE	0.6	6.4	37.3	68.1	120.3	180.6	233.8	271.8	307.8	333.2	353.2	380.9	406.6	421.0	433.8	(£)
U.C. LIMIT	0.7	7.2	40.6	73.0	130.0	197.6	255.9	298.4	334.9	360.8	380.8	409.2	435.4	450.4	463.1	(£)
L.C. LIMIT	0.5	5.6	34.0	63.2	110.6	164.0	211.7	245.2	280.7	305.6	325.6	352.6	377.8	391.9	404.5	(£)

POST 1975 UTILITY SEMI DET   LAND USE CODE 127   JANUARY 1990 PRICES   SALT WATER FLOOD DURATION MORE THAN 12 HOURS

COMPONENTS OF DAMAGE	-0.3	0.0	0.05	0.1	0.2	0.3	0.6	0.9	1.2	1.5	1.8	2.1	2.4	2.7	3.0	(M)
PATHS & PAVED AREAS	0	5	9	14	32	50	68	80	81	85	87	90	96	101	105	(£)
GARDENS/FENCES/SHEDS	0	3	40	73	362	874	1069	1135	1542	1937	2261	2675	3096	3211	3275	(£)
EXT. MAIN BUILDING	0	87	94	149	280	402	619	806	1082	1369	1737	2146	2602	2972	3356	(£)
PLASTERWORK	0	0	101	155	405	595	780	969	1032	1064	997	1083	1083	1083	1083	(£)
FLOORS	0	28	134	323	529	618	697	721	751	783	784	785	787	788	789	(£)
JOINERY	0	0	0	37	121	387	552	686	1108	1118	1118	1126	1126	1126	1126	(£)
INTERNAL DECORATIONS	0	3	22	134	180	194	195	197	205	205	205	207	209	214	214	(£)
PLUMBING & ELECTRICAL	0	0	0	0	28	44	55	74	74	74	96	96	90	103	112	(£)
DOMESTIC APPLIANCES	0	0	0	0	2	182	438	463	464	464	464	464	464	464	464	(£)
HEATING EQUIPMENT	0	0	0	0	0	0	0	0	0	0	0	0	0	0	0	(£)
AUDIO/VIDEO	0	0	1	1	2	3	5	20	20	20	20	20	20	20	20	(£)
FURNITURE	0	0	9	18	32	32	40	40	40	40	40	40	40	40	40	(£)
PERSONAL EFFECTS	0	0	0	0	0	0	0	0	0	0	0	0	0	0	0	(£)
FLOOR COVER/CURTAINS	0	0	591	591	596	596	612	629	638	638	638	638	638	638	638	(£)
GARDEN/DIY/LEISURE	0	0	39	78	115	161	218	250	259	274	281	281	291	291	291	(£)
DOMESTIC CLEAN-UP	0	0	77	79	166	173	180	187	198	211	235	250	250	250	250	(£)

FLOOD DURATION MORE THAN 12 HOURS

	-0.3	0.0	0.05	0.1	0.2	0.3	0.6	0.9	1.2	1.5	1.8	2.1	2.4	2.7	3.0	
TOTAL BUILDING FABRIC	0	128	402	888	1940	3170	4040	4670	5880	6640	7290	8210	9090	9600	10060	(£)
COL PERCENT	0.0	100.0	35.9	53.6	67.9	73.4	73.0	74.6	78.4	80.1	81.3	82.9	84.2	84.9	85.5	
ROW PERCENT	0.0	1.3	4.0	8.8	19.3	31.5	40.1	46.4	58.4	66.0	72.4	81.6	90.4	95.4	100.0	
TOTAL INVENTORY	0	0	719	769	916	1150	1500	1590	1620	1650	1680	1700	1710	1710	1710	(£)
COL PERCENT	0.0	0.0	64.1	46.4	32.1	26.6	27.0	25.4	21.6	19.9	18.7	17.1	15.8	15.1	14.5	
ROW PERCENT	0.0	0.0	42.2	45.1	53.7	67.4	87.7	93.3	95.1	96.7	98.6	99.4	100.0	100.0	100.0	
TOTAL DAMAGE	0	128	1120	1660	2860	4320	5530	6260	7500	8290	8970	9910	10800	11310	11770	(£)
ROW PERCENT	0.0	1.1	9.5	14.1	24.3	36.7	47.0	53.2	63.7	70.4	76.2	84.2	91.8	96.1	100.0	
U.C. LIMIT	0	141	1210	1770	3070	4670	5990	6790	8100	8890	9570	10570	11470	11980	12450	(£)
L.C. LIMIT	0	114	1030	1550	2650	3970	5070	5730	6900	7690	8370	9250	10130	10640	11090	(£)
TOTAL DAMAGE/SQ.METRE	0.0	5.3	41.7	62.2	110.2	169.1	217.3	246.6	296.8	330.1	359.3	399.5	438.0	460.0	479.9	(£)
U.C. LIMIT	0.0	5.9	45.4	66.7	119.1	184.8	237.8	270.7	322.9	357.5	387.4	429.2	469.1	491.8	512.3	(£)
L.C. LIMIT	0.0	4.7	38.0	57.7	101.3	153.4	196.8	222.5	270.7	302.7	331.2	369.8	406.9	428.2	447.5	(£)

344 THE ECONOMICS OF COASTAL MANAGEMENT

PRE 1918 TERRACE   LAND USE CODE 131   JANUARY 1990 PRICES   SALT WATER FLOOD DURATION MORE THAN 12 HOURS

DEPTH ABOVE UPPER SURFACE OF GROUND FLOOR

COMPONENTS OF DAMAGE	-0.3	0.0	0.05	0.1	0.2	0.3	0.6	0.9	1.2	1.5	1.8	2.1	2.4	2.7	3.0	(M)
PATHS & PAVED AREAS	0	6	16	27	46	77	90	97	104	110	116	126	124	124	124	(£)
GARDENS/FENCES/SHEDS	0	0	8	16	23	127	233	280	412	823	864	983	1122	1380	1519	(£)
EXT. MAIN BUILDING	464	495	502	523	575	619	666	767	883	1175	1764	1984	2575	3332	4590	(£)
PLASTERWORK	56	147	225	294	444	621	934	1294	1753	2182	2607	2800	2702	2707	2692	(£)
FLOORS	0	79	524	763	1117	1191	1283	1283	1759	1997	1943	1943	1943	1943	1943	(£)
JOINERY	0	0	77	508	878	1262	1568	1981	1963	1996	2032	2464	2464	2464	2464	(£)
INTERNAL DECORATIONS	22	27	38	141	244	254	266	285	286	293	304	300	304	319	318	(£)
PLUMBING & ELECTRICAL	38	52	78	172	238	253	471	471	471	471	471	471	471	471	471	(£)
DOMESTIC APPLIANCES	0	0	2	77	175	361	403	573	573	576	576	576	576	576	576	(£)
HEATING EQUIPMENT	0	0	0	25	32	85	107	123	136	136	136	140	140	140	140	(£)
AUDIO/VIDEO	0	0	26	101	152	185	318	390	402	402	402	402	402	402	402	(£)
FURNITURE	0	0	45	118	257	533	715	780	791	794	794	794	794	794	794	(£)
PERSONAL EFFECTS	0	0	0	0	27	113	154	219	219	228	228	228	228	228	228	(£)
FLOOR COVER/CURTAINS	0	0	363	385	409	409	449	459	467	467	467	467	467	467	467	(£)
GARDEN/DIY/LEISURE	0	0	25	51	76	105	141	159	165	174	179	179	185	185	185	(£)
DOMESTIC CLEAN-UP	0	0	107	110	231	240	249	259	274	292	326	345	345	345	345	(£)

FLOOD DURATION MORE THAN 12 HOURS

	-0.3	0.0	0.05	0.1	0.2	0.3	0.6	0.9	1.2	1.5	1.8	2.1	2.4	2.7	3.0	
TOTAL BUILDING FABRIC	580	809	1470	2450	3570	4410	5510	6460	7630	9050	10100	11070	11710	12740	14120	(£)
COL PERCENT	100.0	100.0	72.0	73.8	72.4	68.4	68.5	68.5	71.6	74.7	76.5	77.9	78.9	80.2	81.8	
ROW PERCENT	4.1	5.7	10.4	17.3	25.3	31.2	39.0	45.7	54.0	64.1	71.5	78.4	82.9	90.2	100.0	
TOTAL INVENTORY	0	0	571	871	1360	2030	2540	2960	3030	3070	3110	3130	3140	3140	3140	(£)
COL PERCENT	0.0	0.0	28.0	26.2	27.6	31.6	31.5	31.5	28.4	25.3	23.5	22.1	21.1	19.8	18.2	
ROW PERCENT	0.0	0.0	18.2	27.7	43.4	64.8	80.9	94.4	96.5	97.8	99.1	99.8	100.0	100.0	100.0	
TOTAL DAMAGE	580	809	2040	3320	4930	6440	8050	9420	10660	12120	13220	14210	14850	15880	17260	(£)
ROW PERCENT	3.4	4.7	11.8	19.2	28.6	37.3	46.6	54.6	61.8	70.2	76.6	82.3	86.0	92.0	100.0	
U.C. LIMIT	745	939	2270	3700	5490	7200	8990	10510	11870	13460	14660	15750	16440	17570	19130	(£)
L.C. LIMIT	414	678	1810	2940	4370	5680	7110	8330	9450	10780	11780	12670	13260	14190	15390	(£)
TOTAL DAMAGE/SQ.METRE	14.9	20.8	52.4	85.1	126.4	165.2	206.5	241.7	273.5	310.8	338.9	364.3	380.7	407.2	442.6	(£)
U.C. LIMIT	19.2	24.2	57.6	93.3	138.6	182.8	228.4	267.5	300.9	340.7	370.3	397.4	414.6	443.5	483.0	(£)
L.C. LIMIT	10.6	17.4	47.2	76.9	114.2	147.6	184.6	215.9	246.1	280.9	307.5	331.2	346.8	370.9	402.2	(£)

INTER WAR TERRACE   LAND USE CODE 132   JANUARY 1990 PRICES   SALT WATER FLOOD DURATION MORE THAN 12 HOURS

DEPTH ABOVE UPPER SURFACE OF GROUND FLOOR

COMPONENTS OF DAMAGE	-0.3	0.0	0.05	0.1	0.2	0.3	0.6	0.9	1.2	1.5	1.8	2.1	2.4	2.7	3.0	(M)
PATHS & PAVED AREAS	0	1	3	5	11	24	30	31	34	37	39	41	43	47	51	(£)
GARDENS/FENCES/SHEDS	0	6	35	64	133	176	403	445	937	1118	1272	1487	1706	1748	1770	(£)
EXT. MAIN BUILDING	481	499	516	570	638	688	745	843	970	1293	1810	2152	2765	3467	4669	(£)
PLASTERWORK	0	0	0	304	499	693	1028	1389	1723	2084	2468	3161	3128	3128	3128	(£)
FLOORS	0	40	734	1185	1680	1784	1910	1910	2579	2913	2845	2845	2845	2845	2845	(£)
JOINERY	0	0	9	41	338	649	1005	1292	1380	1544	1650	1683	1650	1650	1650	(£)
INTERNAL DECORATIONS	31	33	46	495	604	628	640	657	658	665	674	671	680	691	691	(£)
PLUMBING & ELECTRICAL	46	74	75	90	105	118	136	200	209	231	255	278	299	299	299	(£)
DOMESTIC APPLIANCES	0	0	2	77	175	362	412	584	584	586	586	586	586	586	586	(£)
HEATING EQUIPMENT	0	0	0	25	32	84	106	123	136	136	136	140	140	140	140	(£)
AUDIO/VIDEO	0	0	27	102	153	186	320	393	405	405	405	405	405	405	405	(£)
FURNITURE	0	0	64	165	363	717	951	1016	1027	1030	1030	1030	1030	1030	1030	(£)
PERSONAL EFFECTS	0	0	0	0	28	113	154	218	218	227	227	227	227	227	227	(£)
FLOOR COVER/CURTAINS	0	0	490	524	559	559	609	614	618	618	618	618	618	618	618	(£)
GARDEN/DIY/LEISURE	0	0	25	51	76	106	142	161	167	176	181	181	187	187	187	(£)
DOMESTIC CLEAN-UP	0	0	142	146	307	318	330	344	364	387	433	458	458	458	458	(£)

FLOOD DURATION MORE THAN 12 HOURS

	-0.3	0.0	0.05	0.1	0.2	0.3	0.6	0.9	1.2	1.5	1.8	2.1	2.4	2.7	3.0	
TOTAL BUILDING FABRIC	558	654	1420	2760	4010	4760	5900	6770	8490	9890	11010	12320	13120	13880	15100	(£)
COL PERCENT	100.0	100.0	65.4	71.6	70.3	66.1	66.1	66.2	70.7	73.5	75.3	77.2	78.2	79.2	80.5	
ROW PERCENT	3.7	4.3	9.4	18.3	26.6	31.5	39.1	44.8	56.2	65.5	72.9	81.6	86.8	91.9	100.0	
TOTAL INVENTORY	0	0	752	1090	1700	2450	3030	3460	3520	3570	3620	3650	3660	3660	3660	(£)
COL PERCENT	0.0	0.0	34.6	28.4	29.7	33.9	33.9	33.8	29.3	26.5	24.7	22.8	21.8	20.8	19.5	
ROW PERCENT	0.0	0.0	20.6	29.9	46.4	67.0	82.8	94.6	96.4	97.6	99.0	99.8	100.0	100.0	100.0	
TOTAL DAMAGE	558	654	2170	3850	5710	7210	8930	10230	12020	13460	14630	15970	16770	17530	18760	(£)
ROW PERCENT	3.0	3.5	11.6	20.5	30.4	38.4	47.6	54.5	64.0	71.7	78.0	85.1	89.4	93.5	100.0	
U.C. LIMIT	716	759	2410	4290	6360	8060	9970	11420	13380	14950	16220	17700	18570	19400	20800	(£)
L.C. LIMIT	399	548	1930	3410	5060	6360	7890	9040	10660	11970	13040	14240	14970	15660	16720	(£)
TOTAL DAMAGE/SQ.METRE	10.7	12.6	41.8	74.1	109.8	138.7	171.6	196.6	231.0	258.7	281.3	307.0	322.4	337.0	360.7	(£)
U.C. LIMIT	13.8	14.7	46.0	81.3	120.4	153.5	189.8	217.6	254.2	283.5	307.4	334.9	351.1	367.1	393.6	(£)
L.C. LIMIT	7.6	10.5	37.6	66.9	99.2	123.9	153.4	175.6	207.8	233.9	255.2	279.1	293.7	306.9	327.8	(£)

APPENDIX 5.2 345

POST WAR TERRACE   LAND USE CODE 133   JANUARY 1990 PRICES   SALT WATER FLOOD DURATION MORE THAN 12 HOURS

DEPTH ABOVE UPPER SURFACE OF GROUND FLOOR

COMPONENTS OF DAMAGE	-0.3	0.0	0.05	0.1	0.2	0.3	0.6	0.9	1.2	1.5	1.8	2.1	2.4	2.7	3.0	(M)
PATHS & PAVED AREAS	0	0	0	0	0	0	0	0	0	0	0	0	0	0	0	(£)
GARDENS/FENCES/SHEDS	0	0	22	46	237	482	553	573	643	826	975	1150	1324	1324	1324	(£)
EXT. MAIN BUILDING	0	292	297	334	420	517	670	800	901	1042	1184	1409	1633	1822	1998	(£)
PLASTERWORK	0	0	0	117	193	268	537	807	1076	1206	1076	1173	1233	1233	1233	(£)
FLOORS	0	0	341	414	481	544	587	600	669	703	694	694	694	694	694	(£)
JOINERY	0	0	88	96	200	239	329	704	974	1001	1284	1285	1285	1285	1285	(£)
INTERNAL DECORATIONS	0	4	35	108	169	191	195	204	212	213	214	216	221	221	221	(£)
PLUMBING & ELECTRICAL	38	50	78	148	238	253	471	471	471	471	471	471	471	471	471	(£)
DOMESTIC APPLIANCES	0	0	2	6	9	11	11	12	12	12	12	12	12	12	12	(£)
HEATING EQUIPMENT	0	0	0	25	30	63	77	82	85	85	85	87	87	87	87	(£)
AUDIO/VIDEO	0	0	25	96	141	170	296	357	369	369	369	369	369	369	369	(£)
FURNITURE	0	0	25	90	189	448	506	506	506	506	506	506	506	506	506	(£)
PERSONAL EFFECTS	0	0	0	0	0	73	109	137	137	137	137	137	137	137	137	(£)
FLOOR COVER/CURTAINS	0	0	296	307	318	318	332	332	332	332	332	332	332	332	332	(£)
GARDEN/DIY/LEISURE	0	0	25	51	75	104	140	158	165	174	179	179	185	185	185	(£)
DOMESTIC CLEAN-UP	0	0	63	65	136	141	147	153	162	172	192	204	204	204	204	(£)

FLOOD DURATION MORE THAN 12 HOURS

	-0.3	0.0	0.05	0.1	0.2	0.3	0.6	0.9	1.2	1.5	1.8	2.1	2.4	2.7	3.0	
TOTAL BUILDING FABRIC	38	346	863	1260	1940	2500	3340	4160	4950	5460	5900	6400	6860	7050	7230	(£)
COL PERCENT	100.0	100.0	66.3	66.3	68.3	65.2	67.3	70.5	73.6	75.3	76.5	77.8	78.9	79.3	79.7	
ROW PERCENT	0.5	4.8	11.9	17.5	26.8	34.5	46.3	57.6	68.5	75.6	81.6	88.5	94.9	97.6	100.0	
TOTAL INVENTORY	0	0	438	643	901	1330	1620	1740	1770	1790	1820	1830	1840	1840	1840	(£)
COL PERCENT	0.0	0.0	33.7	33.7	31.7	34.8	32.7	29.5	26.4	24.7	23.5	22.2	21.1	20.7	20.3	
ROW PERCENT	0.0	0.0	23.9	35.0	49.1	72.5	88.4	94.8	96.5	97.6	98.9	99.7	100.0	100.0	100.0	
TOTAL DAMAGE	38	346	1300	1910	2840	3830	4970	5900	6720	7260	7720	8230	8690	8890	9060	(£)
ROW PERCENT	0.4	3.8	14.4	21.1	31.4	42.2	54.8	65.1	74.1	80.1	85.1	90.8	95.9	98.1	100.0	
U.C. LIMIT	48	401	1450	2130	3160	4280	5550	6580	7480	8060	8560	9120	9620	9840	10040	(£)
L.C. LIMIT	27	290	1150	1690	2520	3380	4390	5220	5960	6460	6880	7340	7760	7940	8080	(£)
TOTAL DAMAGE/SQ.METRE	1.1	9.9	37.3	54.5	81.2	109.3	141.9	168.6	191.9	207.2	220.3	235.0	248.3	253.8	258.8	(£)
U.C. LIMIT	1.4	11.5	41.0	59.8	89.0	121.0	157.0	186.6	211.1	227.1	240.7	256.4	270.4	276.4	282.4	(£)
L.C. LIMIT	0.8	8.3	33.6	49.2	73.4	97.6	126.8	150.6	172.7	187.3	199.9	213.6	226.2	231.2	235.2	(£)

MODERN TERRACE   LAND USE CODE 134   JANUARY 1990 PRICES   SALT WATER FLOOD DURATION MORE THAN 12 HOURS

DEPTH ABOVE UPPER SURFACE OF GROUND FLOOR

COMPONENTS OF DAMAGE	-0.3	0.0	0.05	0.1	0.2	0.3	0.6	0.9	1.2	1.5	1.8	2.1	2.4	2.7	3.0	(M)
PATHS & PAVED AREAS	0	4	10	16	34	56	72	83	86	89	93	96	103	109	114	(£)
GARDENS/FENCES/SHEDS	0	1	51	113	409	773	1006	1076	1382	1859	2279	2816	3367	3447	3482	(£)
EXT. MAIN BUILDING	0	97	101	149	296	427	696	945	1240	1558	1927	2408	2918	3323	3724	(£)
PLASTERWORK	0	0	0	216	356	494	991	1488	1985	2223	1985	2056	2280	2280	2280	(£)
FLOORS	0	128	312	533	770	943	1075	1096	1260	1343	1343	1343	1343	1343	1343	(£)
JOINERY	0	0	122	394	638	1219	1565	2006	2785	2818	2854	2854	2854	2854	2854	(£)
INTERNAL DECORATIONS	0	6	52	215	336	390	402	427	437	438	440	443	450	466	466	(£)
PLUMBING & ELECTRICAL	38	40	56	132	227	268	539	571	571	571	608	608	614	620	634	(£)
DOMESTIC APPLIANCES	0	0	2	81	187	392	447	638	638	641	641	641	641	641	641	(£)
HEATING EQUIPMENT	0	0	0	24	32	87	113	131	145	145	145	149	149	149	149	(£)
AUDIO/VIDEO	0	0	44	131	193	234	390	504	517	517	517	517	517	517	517	(£)
FURNITURE	0	0	58	148	343	684	911	984	998	1001	1001	1001	1001	1001	1001	(£)
PERSONAL EFFECTS	0	0	0	0	31	133	182	256	256	266	266	266	266	266	266	(£)
FLOOR COVER/CURTAINS	0	0	531	556	582	582	628	640	649	649	649	649	649	649	649	(£)
GARDEN/DIY/LEISURE	0	0	28	56	83	116	157	178	185	195	200	200	207	207	207	(£)
DOMESTIC CLEAN-UP	0	0	123	127	267	277	287	299	316	336	376	398	398	398	398	(£)

FLOOD DURATION MORE THAN 12 HOURS

	-0.3	0.0	0.05	0.1	0.2	0.3	0.6	0.9	1.2	1.5	1.8	2.1	2.4	2.7	3.0	
TOTAL BUILDING FABRIC	38	278	707	1770	3070	4570	6350	7690	9750	10900	11530	12630	13930	14440	14900	(£)
COL PERCENT	100.0	100.0	47.3	61.1	64.1	64.6	67.1	67.9	72.4	74.4	75.2	76.7	78.4	79.0	79.5	
ROW PERCENT	0.3	1.9	4.7	11.9	20.6	30.7	42.6	51.6	65.4	73.2	77.4	84.7	93.5	96.9	100.0	
TOTAL INVENTORY	0	0	788	1130	1720	2510	3120	3630	3710	3750	3800	3830	3830	3830	3830	(£)
COL PERCENT	0.0	0.0	52.7	38.9	35.9	35.4	32.9	32.1	27.6	25.6	24.8	23.3	21.6	21.0	20.5	
ROW PERCENT	0.0	0.0	20.6	29.4	44.9	65.4	81.3	94.8	96.7	97.9	99.1	99.8	100.0	100.0	100.0	
TOTAL DAMAGE	38	278	1500	2900	4790	7080	9470	11330	13460	14660	15330	16450	17760	18280	18730	(£)
ROW PERCENT	0.2	1.5	8.0	15.5	25.6	37.8	50.5	60.5	71.8	78.2	81.8	87.8	94.8	97.6	100.0	
U.C. LIMIT	48	322	1670	3230	5340	7910	10580	12640	14980	16280	17000	18230	19660	20230	20760	(£)
L.C. LIMIT	27	233	1330	2570	4240	6250	8360	10020	11940	13040	13660	14670	15860	16330	16700	(£)
TOTAL DAMAGE/SQ.METRE	0.8	6.2	33.3	64.4	106.4	157.2	210.2	251.6	298.8	325.5	340.4	365.4	394.5	405.9	416.1	(£)
U.C. LIMIT	1.0	7.2	36.6	70.6	116.6	174.0	232.5	278.4	328.8	356.8	372.0	398.6	429.6	442.1	454.1	(£)
L.C. LIMIT	0.6	5.2	30.0	58.2	96.2	140.4	187.9	224.8	268.8	294.2	308.8	332.2	359.4	369.7	378.1	(£)

346 THE ECONOMICS OF COASTAL MANAGEMENT

POST 1975 TERRACE   LAND USE CODE 135   JANUARY 1990 PRICES SALT WATER FLOOD DURATION MORE THAN 12 HOURS

COMPONENTS OF DAMAGE	-0.3	0.0	0.05	0.1	0.2	0.3	0.6	0.9	1.2	1.5	1.8	2.1	2.4	2.7	3.0	(M)
PATHS & PAVED AREAS	0	6	11	14	29	45	60	71	69	69	69	69	69	69	69	(£)
GARDENS/FENCES/SHEDS	2	4	25	44	99	334	454	612	738	927	1082	1299	1527	1660	1731	(£)
EXT. MAIN BUILDING	0	62	64	102	198	294	479	582	837	1042	1283	1590	1908	2160	2416	(£)
PLASTERWORK	0	0	59	131	447	725	1000	1253	1359	1382	1286	1405	1405	1405	1405	(£)
FLOORS	0	79	199	345	489	581	672	697	727	761	761	761	761	761	761	(£)
JOINERY	0	0	123	357	550	925	1142	1501	1928	1961	1997	1997	1997	1997	1997	(£)
INTERNAL DECORATIONS	0	4	28	132	207	234	236	241	240	240	240	243	246	255	255	(£)
PLUMBING & ELECTRICAL	38	40	56	132	203	259	491	494	497	501	507	511	517	524	530	(£)
DOMESTIC APPLIANCES	0	0	2	79	182	379	423	604	604	609	609	609	609	609	609	(£)
HEATING EQUIPMENT	0	0	0	25	34	82	103	117	127	127	127	130	130	130	130	(£)
AUDIO/VIDEO	0	0	28	108	162	198	338	417	430	430	430	430	430	430	430	(£)
FURNITURE	0	0	46	119	261	544	740	808	822	825	825	825	825	825	825	(£)
PERSONAL EFFECTS	0	0	0	0	29	129	177	249	249	258	258	258	258	258	258	(£)
FLOOR COVER/CURTAINS	0	0	331	354	378	378	417	423	427	427	427	427	427	427	427	(£)
GARDEN/DIY/LEISURE	0	0	26	53	79	110	148	168	174	184	189	189	196	196	196	(£)
DOMESTIC CLEAN-UP	0	0	90	93	196	203	210	219	232	247	276	293	293	293	293	(£)

FLOOD DURATION MORE THAN 12 HOURS

	-0.3	0.0	0.05	0.1	0.2	0.3	0.6	0.9	1.2	1.5	1.8	2.1	2.4	2.7	3.0	
TOTAL BUILDING FABRIC	40	195	567	1260	2230	3400	4540	5450	6400	6890	7230	7880	8430	8830	9170	(£)
COL PERCENT	100.0	100.0	51.9	60.2	62.7	62.7	63.9	64.5	67.6	68.9	69.7	71.3	72.7	73.6	74.3	
ROW PERCENT	0.4	2.1	6.2	13.7	24.3	37.1	49.5	59.5	69.8	75.1	78.9	85.9	92.0	96.4	100.0	
TOTAL INVENTORY	0	0	525	834	1330	2030	2560	3010	3070	3110	3150	3170	3170	3170	3170	(£)
COL PERCENT	0.0	0.0	48.1	39.8	37.3	37.3	36.1	35.5	32.4	31.1	30.3	28.7	27.3	26.4	25.7	
ROW PERCENT	0.0	0.0	16.6	26.3	41.8	63.9	80.7	94.8	96.8	98.1	99.2	99.8	100.0	100.0	100.0	
TOTAL DAMAGE	40	195	1090	2090	3550	5430	7100	8460	9470	10000	10370	11040	11600	12010	12340	(£)
ROW PERCENT	0.3	1.6	8.9	17.0	28.8	44.0	57.5	68.6	76.7	81.0	84.1	89.5	94.1	97.3	100.0	
U.C. LIMIT	51	226	1210	2330	3960	6070	7930	9440	10540	11110	11500	12240	12840	13290	13680	(£)
L.C. LIMIT	28	163	967	1850	3140	4790	6270	7480	8400	8890	9240	9840	10360	10730	11000	(£)
TOTAL DAMAGE/SQ.METRE	1.2	5.9	33.2	63.5	107.6	164.4	215.0	256.3	286.7	302.7	314.1	334.4	351.4	363.6	373.7	(£)
U.C. LIMIT	1.5	6.9	36.5	69.6	117.9	181.9	237.8	283.6	315.4	331.8	343.2	364.8	382.7	396.0	407.8	(£)
L.C. LIMIT	0.9	4.9	29.9	57.4	97.3	146.9	192.2	229.0	258.0	273.6	285.0	304.0	320.1	331.2	339.6	(£)

POST 1975 UTILITY TERRACE   LAND USE CODE 137   JANUARY 1990 PRICES SALT WATER FLOOD DURATION MORE THAN 12 HOURS

COMPONENTS OF DAMAGE	-0.3	0.0	0.05	0.1	0.2	0.3	0.6	0.9	1.2	1.5	1.8	2.1	2.4	2.7	3.0	(M)
PATHS & PAVED AREAS	0	3	8	11	25	40	52	60	60	63	64	66	69	72	75	(£)
GARDENS/FENCES/SHEDS	0	2	46	89	358	761	956	1020	1349	1749	2078	2497	2923	3036	3098	(£)
EXT. MAIN BUILDING	0	84	89	142	264	379	586	770	1041	1323	1707	2079	2523	2881	3251	(£)
PLASTERWORK	0	0	124	189	466	669	866	1068	1131	1163	1096	1182	1182	1182	1182	(£)
FLOORS	0	40	131	274	428	507	582	605	633	663	663	663	664	664	664	(£)
JOINERY	0	0	0	33	99	317	452	562	908	909	909	910	910	910	910	(£)
INTERNAL DECORATIONS	0	3	23	113	162	176	176	177	184	184	184	185	187	192	192	(£)
PLUMBING & ELECTRICAL	0	0	0	0	27	42	52	71	71	71	92	92	82	98	107	(£)
DOMESTIC APPLIANCES	0	0	0	0	1	228	542	567	567	567	567	567	567	567	567	(£)
HEATING EQUIPMENT	0	0	0	0	0	0	0	0	0	0	0	0	0	0	0	(£)
AUDIO/VIDEO	0	0	1	1	1	2	3	14	14	14	14	14	14	14	14	(£)
FURNITURE	0	0	8	16	28	28	33	33	33	33	33	33	33	33	33	(£)
PERSONAL EFFECTS	0	0	0	0	0	0	0	0	0	0	0	0	0	0	0	(£)
FLOOR COVER/CURTAINS	0	0	630	630	631	631	651	672	680	680	680	680	680	680	680	(£)
GARDEN/DIY/LEISURE	0	0	32	64	95	132	178	202	210	222	227	227	235	235	235	(£)
DOMESTIC CLEAN-UP	0	0	72	74	156	162	169	176	186	198	221	235	235	235	235	(£)

FLOOD DURATION MORE THAN 12 HOURS

	-0.3	0.0	0.05	0.1	0.2	0.3	0.6	0.9	1.2	1.5	1.8	2.1	2.4	2.7	3.0	
TOTAL BUILDING FABRIC	0	134	423	853	1830	2890	3730	4340	5380	6130	6800	7680	8540	9040	9480	(£)
COL PERCENT	0.0	100.0	36.2	52.0	66.7	70.9	70.2	72.2	76.1	78.1	79.6	81.3	82.9	83.6	84.3	
ROW PERCENT	0.0	1.4	4.5	9.0	19.3	30.5	39.3	45.7	56.7	64.6	71.7	81.0	90.1	95.3	100.0	
TOTAL INVENTORY	0	0	745	787	914	1190	1580	1670	1690	1720	1750	1760	1770	1770	1770	(£)
COL PERCENT	0.0	0.0	63.8	48.0	33.3	29.1	29.8	27.8	23.9	21.9	20.4	18.7	17.1	16.4	15.7	
ROW PERCENT	0.0	0.0	42.2	44.6	51.8	67.1	89.3	94.3	95.8	97.1	98.8	99.6	100.0	100.0	100.0	
TOTAL DAMAGE	0	134	1170	1640	2750	4080	5300	6000	7070	7840	8540	9440	10310	10810	11250	(£)
ROW PERCENT	0.0	1.2	10.4	14.6	24.4	36.3	47.2	53.4	62.9	69.7	75.9	83.9	91.7	96.1	100.0	
U.C. LIMIT	0	155	1300	1830	3060	4560	5920	6700	7870	8710	9470	10460	11410	11960	12470	(£)
L.C. LIMIT	0	112	1040	1450	2440	3600	4680	5300	6270	6970	7610	8420	9210	9660	10030	(£)
TOTAL DAMAGE/SQ.METRE	0.0	5.6	46.1	64.5	109.9	164.5	214.4	243.3	287.2	320.4	350.4	389.1	427.1	448.5	467.7	(£)
U.C. LIMIT	0.0	6.5	50.7	70.7	120.5	182.0	237.1	269.2	316.0	351.2	382.9	424.5	465.1	488.5	510.4	(£)
L.C. LIMIT	0.0	4.7	41.5	58.3	99.3	147.0	191.7	217.4	258.4	289.6	317.9	353.7	389.1	408.5	425.0	(£)

APPENDIX 5.2 347

PRE 1918 BUNGALOW   LAND USE CODE 141   JANUARY 1990 PRICES   SALT WATER FLOOD DURATION MORE THAN 12 HOURS

DEPTH ABOVE UPPER SURFACE OF GROUND FLOOR

COMPONENTS OF DAMAGE	-0.3	0.0	0.05	0.1	0.2	0.3	0.6	0.9	1.2	1.5	1.8	2.1	2.4	2.7	3.0	(M)
PATHS & PAVED AREAS	0	1	13	28	42	67	76	92	106	123	139	159	158	161	165	(£)
GARDENS/FENCES/SHEDS	0	0	32	67	251	329	539	603	781	1886	2224	2691	3166	3241	3282	(£)
EXT. MAIN BUILDING	498	533	563	644	751	837	935	1094	1288	1661	2266	2564	3231	4048	5376	(£)
PLASTERWORK	57	146	209	267	427	589	925	1267	2038	2666	2814	2718	2759	2805	2801	(£)
FLOORS	0	89	828	1193	1744	1832	1959	1959	2634	2965	2901	2901	2901	2901	2901	(£)
JOINERY	0	0	73	179	580	949	1382	1853	1873	1906	1943	2522	2522	2522	2522	(£)
INTERNAL DECORATIONS	0	1	13	107	268	272	282	294	294	304	311	313	320	338	335	(£)
PLUMBING & ELECTRICAL	14	25	46	140	195	210	303	303	303	303	303	303	303	303	303	(£)
DOMESTIC APPLIANCES	0	0	3	95	236	514	624	875	875	885	885	885	885	885	885	(£)
HEATING EQUIPMENT	0	0	5	28	53	119	156	178	195	195	195	201	201	201	201	(£)
AUDIO/VIDEO	0	0	70	195	277	345	554	711	729	729	729	729	729	729	729	(£)
FURNITURE	0	0	68	245	1061	1426	2233	2263	2362	2367	2367	2367	2367	2367	2367	(£)
PERSONAL EFFECTS	0	0	42	285	436	1372	2011	2477	2554	2566	2566	2566	2566	2566	2566	(£)
FLOOR COVER/CURTAINS	0	0	753	769	786	786	832	854	875	875	875	875	875	875	875	(£)
GARDEN/DIY/LEISURE	0	0	64	116	166	209	259	289	297	310	317	317	325	325	325	(£)
DOMESTIC CLEAN-UP	0	0	139	143	300	311	322	335	356	378	423	447	447	447	447	(£)

FLOOD DURATION MORE THAN 12 HOURS

	-0.3	0.0	0.05	0.1	0.2	0.3	0.6	0.9	1.2	1.5	1.8	2.1	2.4	2.7	3.0	
TOTAL BUILDING FABRIC	570	796	1780	2630	4260	5090	6410	7470	9320	11820	12900	14170	15360	16320	17690	(£)
COL PERCENT	100.0	100.0	60.8	58.3	56.2	50.0	47.8	48.3	53.1	58.7	60.7	62.8	64.7	66.0	67.8	
ROW PERCENT	3.2	4.5	10.1	14.9	24.1	28.8	36.2	42.2	52.7	66.8	73.0	80.1	86.9	92.3	100.0	
TOTAL INVENTORY	0	0	1150	1880	3320	5080	7000	7990	8250	8310	8360	8390	8400	8400	8400	(£)
COL PERCENT	0.0	0.0	39.2	41.7	43.8	50.0	52.2	51.7	46.9	41.3	39.3	37.2	35.3	34.0	32.2	
ROW PERCENT	0.0	0.0	13.7	22.4	39.5	60.5	83.3	95.1	98.2	98.9	99.5	99.9	100.0	100.0	100.0	
TOTAL DAMAGE	570	796	2930	4510	7580	10170	13400	15450	17570	20130	21270	22570	23760	24720	26090	(£)
ROW PERCENT	2.2	3.1	11.2	17.3	29.1	39.0	51.4	59.2	67.3	77.2	81.5	86.5	91.1	94.8	100.0	
U.C. LIMIT	947	853	3260	4990	8330	11220	14770	16980	19180	21810	22970	24260	25460	26450	27870	(£)
L.C. LIMIT	192	738	2600	4030	6830	9120	12030	13920	15960	18450	19570	20880	22060	22990	24310	(£)
TOTAL DAMAGE/SQ.METRE	11.3	15.7	57.3	88.1	148.2	198.9	262.0	302.3	343.8	394.2	416.4	442.1	465.7	484.7	511.6	(£)
U.C. LIMIT	19.5	17.7	64.1	97.6	164.4	221.5	291.3	335.2	377.6	430.5	453.9	479.7	503.2	522.8	551.2	(£)
L.C. LIMIT	3.1	13.7	50.5	78.6	132.0	176.3	232.7	269.4	310.0	357.9	379.3	404.5	428.2	446.6	472.0	(£)

INTER WAR BUNGALOW   LAND USE CODE 142   JANUARY 1990 PRICES   SALT WATER FLOOD DURATION MORE THAN 12 HOURS

DEPTH ABOVE UPPER SURFACE OF GROUND FLOOR

COMPONENTS OF DAMAGE	-0.3	0.0	0.05	0.1	0.2	0.3	0.6	0.9	1.2	1.5	1.8	2.1	2.4	2.7	3.0	(M)
PATHS & PAVED AREAS	0	0	14	25	48	102	178	205	216	226	236	244	255	271	283	(£)
GARDENS/FENCES/SHEDS	0	6	64	143	392	633	1107	1219	2055	2622	3101	3803	4522	4763	4894	(£)
EXT. MAIN BUILDING	340	397	454	586	736	913	1094	1359	1519	2292	3200	4168	5500	7026	9602	(£)
PLASTERWORK	0	0	0	406	665	926	1373	1856	2302	2785	3303	3857	3847	3847	3847	(£)
FLOORS	0	152	1353	2007	2867	3007	3232	3232	4371	4945	4804	4804	4804	4804	4804	(£)
JOINERY	0	0	91	219	805	1397	2093	2471	2697	2862	2967	3157	3124	3124	3124	(£)
INTERNAL DECORATIONS	31	37	61	391	522	584	607	644	649	665	688	676	692	713	713	(£)
PLUMBING & ELECTRICAL	38	60	80	91	135	160	208	280	343	372	401	422	441	461	480	(£)
DOMESTIC APPLIANCES	0	0	2	81	188	392	470	650	650	654	654	654	654	654	654	(£)
HEATING EQUIPMENT	0	0	5	30	42	115	144	175	198	198	198	205	205	205	205	(£)
AUDIO/VIDEO	0	0	49	136	200	246	408	517	530	530	530	530	530	530	530	(£)
FURNITURE	0	0	82	317	1180	1513	2302	2388	2493	2496	2496	2496	2496	2496	2496	(£)
PERSONAL EFFECTS	0	0	64	375	573	1609	2192	2683	2757	2765	2765	2765	2765	2765	2765	(£)
FLOOR COVER/CURTAINS	0	0	1012	1035	1059	1059	1118	1146	1167	1167	1167	1167	1167	1167	1167	(£)
GARDEN/DIY/LEISURE	0	0	47	87	125	157	195	214	221	230	236	236	242	242	242	(£)
DOMESTIC CLEAN-UP	0	0	279	288	604	627	649	676	717	762	852	902	902	902	902	(£)

FLOOD DURATION MORE THAN 12 HOURS

	-0.3	0.0	0.05	0.1	0.2	0.3	0.6	0.9	1.2	1.5	1.8	2.1	2.4	2.7	3.0	
TOTAL BUILDING FABRIC	409	653	2120	3870	6170	7730	9890	11270	14160	16770	18700	21140	23190	25010	27750	(£)
COL PERCENT	100.0	100.0	57.9	62.2	60.8	57.5	56.9	57.1	61.8	65.6	67.8	70.2	72.1	73.6	75.6	
ROW PERCENT	1.5	2.4	7.6	14.0	22.2	27.8	35.7	40.6	51.1	60.4	67.4	76.2	83.6	90.1	100.0	
TOTAL INVENTORY	0	0	1540	2350	3970	5720	7480	8450	8740	8810	8900	8960	8960	8960	8960	(£)
COL PERCENT	0.0	0.0	42.1	37.8	39.2	42.5	43.1	42.9	38.2	34.4	32.2	29.8	27.9	26.4	24.4	
ROW PERCENT	0.0	0.0	17.2	26.3	44.3	63.8	83.5	94.3	97.5	98.2	99.3	99.9	100.0	100.0	100.0	
TOTAL DAMAGE	409	653	3660	6230	10150	13450	17380	19720	22890	25580	27610	30090	32150	33980	36720	(£)
ROW PERCENT	1.1	1.8	10.0	17.0	27.6	36.6	47.3	53.7	62.3	69.7	75.2	82.0	87.6	92.5	100.0	
U.C. LIMIT	680	700	4080	6890	11150	14830	19150	21680	24990	27720	29810	32340	34450	36360	39220	(£)
L.C. LIMIT	137	605	3240	5570	9150	12070	15610	17760	20790	23440	25410	27840	29850	31600	34220	(£)
TOTAL DAMAGE/SQ.METRE	4.5	7.1	39.8	67.6	110.1	145.9	188.6	214.1	248.5	277.6	299.6	326.7	349.1	368.9	398.7	(£)
U.C. LIMIT	7.8	8.0	44.5	74.9	122.1	162.5	209.7	237.4	273.0	303.2	326.4	354.5	377.2	397.9	429.5	(£)
L.C. LIMIT	1.2	6.2	35.1	60.3	98.1	129.3	167.5	190.8	224.0	252.0	272.7	298.9	321.0	339.9	367.9	(£)

348 THE ECONOMICS OF COASTAL MANAGEMENT

```
POST WAR BUNGALOW        LAND USE CODE 143      JANUARY 1990 PRICES   SALT WATER FLOOD DURATION MORE THAN 12 HOURS
                                     DEPTH ABOVE UPPER SURFACE OF GROUND FLOOR
COMPONENTS OF DAMAGE  -0.3   0.0   0.05   0.1    0.2    0.3    0.6    0.9    1.2    1.5    1.8    2.1    2.4    2.7    3.0  (M)
--------------------  ----  ----  -----  -----  -----  -----  -----  -----  -----  -----  -----  -----  -----  -----  -----
  PATHS & PAVED AREAS    0     6    13     19     53     86    117    139    147    158    169    178    195    211    225 (£)
  GARDENS/FENCES/SHEDS   0     6    71    150    570    996   1317   1411   2022   2737   3350   4132   4913   4913   4913 (£)
  EXT. MAIN BUILDING     0   379   386    444    630    755    908   1084   2338   3528   4828   6712   8581  10108  11509 (£)
  PLASTERWORK            0     0     0    336    555    771   1546   2324   3099   3473   3099   3364   3531   3531   3531 (£)
  FLOORS                 0     0   375   1156   1769   2307   2497   2657   3557   4004   3883   3883   3883   3883   3883 (£)
  JOINERY                0     0   204    524    899   1094   1381   2227   2886   2919   3275   3275   3275   3275   3275 (£)
  INTERNAL DECORATIONS   0    10    72    336    484    524    534    557    568    576    582    597    616    670    668 (£)
  PLUMBING & ELECTRICAL 38    70   105    167    289    314    559    596    606    615    632    642    647    657    661 (£)

  DOMESTIC APPLIANCES    0     0     2     85    203    429    518    721    721    728    728    728    728    728    728 (£)
  HEATING EQUIPMENT      0     0     5     29     46    125    159    192    217    217    217    224    224    224    224 (£)
  AUDIO/VIDEO            0     0    56    155    226    281    458    589    603    603    603    603    603    603    603 (£)
  FURNITURE              0     0    85    328   1268   1630   2525   2617   2745   2749   2749   2749   2749   2749   2749 (£)
  PERSONAL EFFECTS       0     0    71    406    607   1737   2375   2920   3003   3013   3013   3013   3013   3013   3013 (£)
  FLOOR COVER/CURTAINS   0     0  1153   1179   1206   1206   1274   1305   1331   1331   1331   1331   1331   1331   1331 (£)
  GARDEN/DIY/LEISURE     0     0    53     96    138    173    214    237    244    255    260    260    267    267    267 (£)
  DOMESTIC CLEAN-UP      0     0   243    251    526    545    565    589    624    663    742    785    785    785    785 (£)

                                                                          FLOOD DURATION MORE THAN 12 HOURS
TOTAL BUILDING FABRIC   38   471  1230   3140   5250   6850   8860  11000  15230  18010  19820  22790  25640  27250  28670 (£)
    COL PERCENT       100.0 100.0  42.4   55.3   55.4   52.8   52.3   54.5   61.6   65.3   67.3   70.1   72.5   73.7   74.7
    ROW PERCENT         0.1   1.6   4.3   10.9   18.3   23.9   30.9   38.4   53.1   62.8   69.1   79.5   89.5   95.1  100.0
TOTAL INVENTORY          0     0  1670   2530   4220   6130   8090   9170   9490   9560   9650   9700   9700   9700   9700 (£)
    COL PERCENT         0.0   0.0  57.6   44.7   44.6   47.2   47.7   45.5   38.4   34.7   32.7   29.9   27.5   26.3   25.3
    ROW PERCENT         0.0   0.0  17.2   26.1   43.5   63.2   83.4   94.5   97.8   98.5   99.4   99.9  100.0  100.0  100.0

TOTAL DAMAGE            38   471  2900   5670   9480  12980  16960  20170  24720  27580  29470  32480  35350  36960  38370 (£)
    ROW PERCENT         0.1   1.2   7.6   14.8   24.7   33.8   44.2   52.6   64.4   71.9   76.8   84.7   92.1   96.3  100.0
    U.C. LIMIT          63   504  3230   6270  10420  14310  18690  22170  26980  29890  31820  34910  37880  39550  40980 (£)
    L.C. LIMIT          12   437  2570   5070   8540  11650  15230  18170  22460  25270  27120  30050  32820  34370  35760 (£)
TOTAL DAMAGE/SQ.METRE  0.4   5.3  32.6   63.7  106.5  145.9  190.5  226.6  277.7  309.8  331.1  364.9  397.1  415.2  431.1 (£)
    U.C. LIMIT         0.7   6.0  36.5   70.5  118.1  162.5  211.8  251.3  305.0  338.3  360.8  395.9  429.1  447.9  464.4 (£)
    L.C. LIMIT         0.1   4.6  28.7   56.9   94.9  129.3  169.2  201.9  250.4  281.3  301.4  333.9  365.1  382.5  397.8 (£)


MODERN BUNGALOW          LAND USE CODE 144      JANUARY 1990 PRICES   SALT WATER FLOOD DURATION MORE THAN 12 HOURS
                                     DEPTH ABOVE UPPER SURFACE OF GROUND FLOOR
COMPONENTS OF DAMAGE  -0.3   0.0   0.05   0.1    0.2    0.3    0.6    0.9    1.2    1.5    1.8    2.1    2.4    2.7    3.0  (M)
--------------------  ----  ----  -----  -----  -----  -----  -----  -----  -----  -----  -----  -----  -----  -----  -----
  PATHS & PAVED AREAS    0    15    32     49    110    180    237    279    291    307    321    334    364    392    415 (£)
  GARDENS/FENCES/SHEDS   0     6    68    142    551    998   1326   1434   2059   2749   3392   4164   4966   5058   5080 (£)
  EXT. MAIN BUILDING     0   150   164    255    492    666    867   1028   2017   2909   3896   5325   6777   8030   9285 (£)
  PLASTERWORK            0     0     0    483    792   1102   2208   3313   4419   4951   4419   4572   5040   5040   5040 (£)
  FLOORS                 0   289   954   1544   2143   2473   2766   2792   2993   3097   3097   3097   3097   3097   3097 (£)
  JOINERY                0     0   144    552    865   1662   2134   2842   3888   3954   4026   4026   4026   4026   4026 (£)
  INTERNAL DECORATIONS   0    15    94    500    725    802    821    858    866    868    872    882    892    924    924 (£)
  PLUMBING & ELECTRICAL  0    40    87    187    316    379    650    717    735    745    825    835    853    872    917 (£)

  DOMESTIC APPLIANCES    0     0     2     86    205    435    526    732    732    739    739    739    739    739    739 (£)
  HEATING EQUIPMENT      0     0     5     29     46    125    160    194    219    219    219    227    227    227    227 (£)
  AUDIO/VIDEO            0     0    57    158    230    286    468    601    616    616    616    616    616    616    616 (£)
  FURNITURE              0     0    85    331   1289   1658   2575   2670   2802   2806   2806   2806   2806   2806   2806 (£)
  PERSONAL EFFECTS       0     0    72    411    614   1759   2408   2959   3047   3057   3057   3057   3057   3057   3057 (£)
  FLOOR COVER/CURTAINS   0     0  1620   1645   1673   1673   1742   1774   1800   1800   1800   1800   1800   1800   1800 (£)
  GARDEN/DIY/LEISURE     0     0    54     98    140    176    218    241    248    259    265    265    272    272    272 (£)
  DOMESTIC CLEAN-UP      0     0   301    311    651    676    700    729    773    822    919    972    972    972    972 (£)

                                                                          FLOOD DURATION MORE THAN 12 HOURS
TOTAL BUILDING FABRIC    0   516  1550   3720   6000   8270  11010  13270  17270  19580  20850  23240  26020  27440  28790 (£)
    COL PERCENT       100.0 100.0  41.3   54.7   55.3   54.9   55.6   57.3   62.8   65.5   66.7   68.9   71.3   72.3   73.3
    ROW PERCENT         0.0   1.8   5.4   12.9   20.8   28.7   38.3   46.1   60.0   68.0   72.4   80.7   90.4   95.3  100.0
TOTAL INVENTORY          0     0  2200   3070   4850   6790   8800   9900  10240  10320  10420  10490  10490  10490  10490 (£)
    COL PERCENT         0.0   0.0  58.7   45.3   44.7   45.1   44.4   42.7   37.2   34.5   33.3   31.1   28.7   27.7   26.7
    ROW PERCENT         0.0   0.0  21.0   29.3   46.3   64.7   83.9   94.4   97.6   98.4   99.4   99.9  100.0  100.0  100.0

TOTAL DAMAGE             0   516  3740   6790  10850  15060  19810  23170  27510  29900  31270  33720  36510  37930  39280 (£)
    ROW PERCENT         0.0   1.3   9.5   17.3   27.6   38.3   50.4   59.0   70.0   76.1   79.6   85.9   93.0   96.6  100.0
    U.C. LIMIT           0   553  4170   7510  11920  16610  21830  25470  30030  32400  33770  36240  39120  40590  41960 (£)
    L.C. LIMIT           0   478  3310   6070   9780  13510  17790  20870  24990  27400  28770  31200  33900  35270  36600 (£)
TOTAL DAMAGE/SQ.METRE  0.0   4.1  29.9   54.3   86.8  120.4  158.4  185.3  220.0  239.1  250.1  269.7  292.0  303.3  314.1 (£)
    U.C. LIMIT         0.0   4.6  33.5   60.1   96.3  134.1  176.1  205.5  241.7  261.1  272.5  292.6  315.5  327.2  338.4 (£)
    L.C. LIMIT         0.0   3.6  26.3   48.5   77.3  106.7  140.7  165.1  198.3  217.1  227.7  246.8  268.5  279.4  289.8 (£)
```

APPENDIX 5.2 349

POST 1975 BUNGALOW LAND USE CODE 145 JANUARY 1990 PRICES SALT WATER FLOOD DURATION MORE THAN 12 HOURS

DEPTH ABOVE UPPER SURFACE OF GROUND FLOOR

COMPONENTS OF DAMAGE	-0.3	0.0	0.05	0.1	0.2	0.3	0.6	0.9	1.2	1.5	1.8	2.1	2.4	2.7	3.0	(M)
PATHS & PAVED AREAS	0	5	10	15	34	55	72	83	85	88	90	93	99	104	107	(£)
GARDENS/FENCES/SHEDS	0	6	88	169	700	1243	1541	1631	2258	2987	3584	4299	5014	5014	5014	(£)
EXT. MAIN BUILDING	0	127	138	226	466	675	1035	1357	1848	2358	2909	3736	4557	5255	6003	(£)
PLASTERWORK	0	0	276	421	1305	2024	2731	3446	3717	3851	3566	3941	3941	3941	3941	(£)
FLOORS	0	205	526	932	1337	1605	1878	1941	2022	2107	2107	2107	2107	2107	2107	(£)
JOINERY	0	0	207	659	1062	1659	2010	2488	3193	3226	3262	3262	3262	3262	3262	(£)
INTERNAL DECORATIONS	0	12	85	383	557	612	630	673	671	673	674	680	690	712	721	(£)
PLUMBING & ELECTRICAL	38	40	72	163	290	339	617	652	677	704	754	779	806	838	867	(£)
DOMESTIC APPLIANCES	0	0	2	87	209	446	540	753	753	760	760	760	760	760	760	(£)
HEATING EQUIPMENT	0	0	5	28	46	127	163	197	223	223	223	231	231	231	231	(£)
AUDIO/VIDEO	0	0	59	163	237	296	483	622	637	637	637	637	637	637	637	(£)
FURNITURE	0	0	86	335	1322	1703	2657	2754	2893	2897	2897	2897	2897	2897	2897	(£)
PERSONAL EFFECTS	0	0	74	422	627	1804	2473	3040	3134	3144	3144	3144	3144	3144	3144	(£)
FLOOR COVER/CURTAINS	0	0	1075	1102	1131	1131	1202	1235	1263	1263	1263	1263	1263	1263	1263	(£)
GARDEN/DIY/LEISURE	0	0	55	101	145	182	225	250	257	268	274	274	281	281	281	(£)
DOMESTIC CLEAN-UP	0	0	216	223	468	485	503	524	555	590	660	698	698	698	698	(£)

FLOOD DURATION MORE THAN 12 HOURS

	-0.3	0.0	0.05	0.1	0.2	0.3	0.6	0.9	1.2	1.5	1.8	2.1	2.4	2.7	3.0	
TOTAL BUILDING FABRIC	38	395	1410	2970	5750	8220	10520	12270	14470	16000	16950	18900	20480	21240	22030	(£)
COL PERCENT	100.0	100.0	47.1	54.6	57.9	57.1	56.0	56.7	59.8	62.0	63.2	65.6	67.4	68.2	69.0	
ROW PERCENT	0.2	1.8	6.4	13.5	26.1	37.3	47.7	55.7	65.7	72.6	77.0	85.8	93.0	96.4	100.0	
TOTAL INVENTORY	0	0	1580	2470	4190	6180	8250	9380	9720	9790	9860	9910	9920	9920	9920	(£)
COL PERCENT	0.0	0.0	52.9	45.4	42.1	42.9	44.0	43.3	40.2	38.0	36.8	34.4	32.6	31.8	31.0	
ROW PERCENT	0.0	0.0	15.9	24.9	42.3	62.3	83.2	94.6	98.0	98.7	99.5	99.9	100.0	100.0	100.0	
TOTAL DAMAGE	38	395	2980	5440	9940	14390	18770	21650	24190	25780	26810	28810	30400	31150	31940	(£)
ROW PERCENT	0.1	1.2	9.3	17.0	31.1	45.1	58.8	67.8	75.7	80.7	83.9	90.2	95.2	97.5	100.0	
U.C. LIMIT	63	423	3320	6010	10920	15870	20690	23800	26410	27930	28950	30960	32570	33330	34120	(£)
L.C. LIMIT	12	366	2640	4870	8960	12910	16850	19500	21970	23630	24670	26660	28230	28970	29760	(£)
TOTAL DAMAGE/SQ.METRE	0.5	5.0	37.7	68.8	125.8	182.1	237.4	273.9	306.0	326.2	339.2	364.5	384.5	394.1	404.1	(£)
U.C. LIMIT	0.9	5.6	42.2	76.2	139.5	202.8	263.9	303.7	336.1	356.3	369.6	395.5	415.5	425.1	435.4	(£)
L.C. LIMIT	0.1	4.4	33.2	61.4	112.1	161.4	210.9	244.1	275.9	296.1	308.8	333.5	353.5	363.1	372.8	(£)

PRE 1918 FLAT LAND USE CODE 151 JANUARY 1990 PRICES SALT WATER FLOOD DURATION MORE THAN 12 HOURS

DEPTH ABOVE UPPER SURFACE OF GROUND FLOOR

COMPONENTS OF DAMAGE	-0.3	0.0	0.05	0.1	0.2	0.3	0.6	0.9	1.2	1.5	1.8	2.1	2.4	2.7	3.0	(M)
PATHS & PAVED AREAS	0	0	0	0	0	0	0	0	0	0	0	0	0	0	0	(£)
GARDENS/FENCES/SHEDS	0	0	0	0	0	0	0	0	0	0	0	0	0	0	0	(£)
EXT. MAIN BUILDING	262	288	311	361	428	490	559	663	794	1066	1431	1702	2157	2713	3616	(£)
PLASTERWORK	34	136	187	238	306	375	426	528	767	937	1078	1461	1742	1908	2112	(£)
FLOORS	0	128	496	683	978	1024	1114	1114	1651	1850	1794	1762	1762	1762	1762	(£)
JOINERY	0	0	94	132	305	419	547	810	846	879	916	1119	1119	1119	1119	(£)
INTERNAL DECORATIONS	0	5	14	253	330	359	362	366	368	374	384	387	387	396	396	(£)
PLUMBING & ELECTRICAL	38	48	74	98	98	98	98	98	98	184	251	464	464	464	464	(£)
DOMESTIC APPLIANCES	0	0	0	78	197	431	524	735	735	743	743	743	743	743	743	(£)
HEATING EQUIPMENT	0	0	5	5	9	31	38	52	64	64	64	66	66	66	66	(£)
AUDIO/VIDEO	0	0	29	40	53	71	100	153	153	153	153	153	153	153	153	(£)
FURNITURE	0	0	43	148	760	773	1366	1391	1468	1472	1472	1472	1472	1472	1472	(£)
PERSONAL EFFECTS	0	0	36	253	395	1086	1571	1928	1987	1997	1997	1997	1997	1997	1997	(£)
FLOOR COVER/CURTAINS	0	0	364	364	364	364	382	401	418	418	418	418	418	418	418	(£)
GARDEN/DIY/LEISURE	0	0	54	99	142	179	221	245	252	263	269	269	276	276	276	(£)
DOMESTIC CLEAN-UP	0	0	109	112	236	245	254	264	280	298	333	353	353	353	353	(£)

FLOOD DURATION MORE THAN 12 HOURS

	-0.3	0.0	0.05	0.1	0.2	0.3	0.6	0.9	1.2	1.5	1.8	2.1	2.4	2.7	3.0	
TOTAL BUILDING FABRIC	334	605	1180	1770	2450	2770	3110	3580	4530	5290	5860	6900	7630	8360	9470	(£)
COL PERCENT	100.0	100.0	64.6	61.6	53.1	46.5	41.1	40.9	45.8	49.4	51.8	55.7	58.2	60.4	63.3	
ROW PERCENT	3.5	6.4	12.4	18.7	25.8	29.2	32.8	37.8	47.8	55.9	61.8	72.8	80.6	88.3	100.0	
TOTAL INVENTORY	0	0	643	1100	2160	3180	4460	5170	5360	5410	5450	5470	5480	5480	5480	(£)
COL PERCENT	0.0	0.0	35.4	38.4	46.9	53.5	58.9	59.1	54.2	50.6	48.2	44.3	41.8	39.6	36.7	
ROW PERCENT	0.0	0.0	11.7	20.1	39.4	58.1	81.4	94.4	97.8	98.7	99.5	99.9	100.0	100.0	100.0	
TOTAL DAMAGE	334	605	1820	2870	4610	5950	7570	8750	9890	10700	11310	12370	13120	13850	14950	(£)
ROW PERCENT	2.2	4.0	12.2	19.2	30.8	39.8	50.6	58.5	66.1	71.6	75.6	82.7	87.7	92.6	100.0	
U.C. LIMIT	474	679	2100	3300	5330	6880	8740	10050	11280	12150	12780	13920	14700	15500	16740	(£)
L.C. LIMIT	193	530	1540	2440	3890	5020	6400	7450	8500	9250	9840	10820	11540	12200	13160	(£)
TOTAL DAMAGE/SQ.METRE	8.3	15.1	45.5	71.7	115.0	148.5	188.9	218.5	246.7	267.1	282.2	308.8	327.4	345.6	373.3	(£)
U.C. LIMIT	12.0	17.8	50.4	78.1	126.4	164.5	209.2	241.8	270.9	291.2	306.9	334.7	353.6	372.6	401.9	(£)
L.C. LIMIT	4.6	12.4	40.6	65.3	103.6	132.5	168.6	195.2	222.5	243.0	257.5	282.9	301.2	318.6	344.7	(£)

350 THE ECONOMICS OF COASTAL MANAGEMENT

```
INTER WAR FLAT            LAND USE CODE 152      JANUARY 1990 PRICES   SALT WATER FLOOD DURATION MORE THAN 12 HOURS

                                           DEPTH ABOVE UPPER SURFACE OF GROUND FLOOR
COMPONENTS OF DAMAGE   -0.3    0.0   0.05    0.1    0.2    0.3    0.6    0.9    1.2    1.5    1.8    2.1    2.4    2.7    3.0  (M)
---------------------------------------------------------------------------------------------------------------------------------
   PATHS & PAVED AREAS    0      0      0      0      0      0      0      0      0      0      0      0      0      0      0  (£)
   GARDENS/FENCES/SHEDS   0      0      0      0      0      0      0      0      0      0      0      0      0      0      0  (£)
   EXT. MAIN BUILDING   318    380    433    533    655    768    900   1064   1293   1759   2187   2683   3380   4176   5471  (£)
   PLASTERWORK            0      0      0    378    620    863   1279   1729   2147   2597   3083   3966   3943   3943   3943  (£)
   FLOORS                 0      0    911   1506   2216   2367   2547   2547   3466   3926   3756   3756   3756   3756   3756  (£)
   JOINERY                0      0     55    140    636   1143   1734   2113   2201   2365   2471   2504   2471   2471   2471  (£)
   INTERNAL DECORATIONS   0      4     24    401    525    552    565    582    584    592    605    597    607    618    618  (£)
   PLUMBING & ELECTRICAL 46     74     75     90    105    116    132    158    198    221    246    270    292    292    292  (£)

   DOMESTIC APPLIANCES    0      0      1     75    167    339    403    552    552    553    553    553    553    553    553  (£)
   HEATING EQUIPMENT      0      0      5     32     38     98    119    142    160    160    160    165    165    165    165  (£)
   AUDIO/VIDEO            0      0     40    109    159    192    324    400    410    410    410    410    410    410    410  (£)
   FURNITURE              0      0     69    281    986   1261   1799   1822   1887   1889   1889   1889   1889   1889   1889  (£)
   PERSONAL EFFECTS       0      0     54    330    520   1427   1928   2352   2403   2411   2411   2411   2411   2411   2411  (£)
   FLOOR COVER/CURTAINS   0      0    713    722    732    732    768    793    808    808    808    808    808    808    808  (£)
   GARDEN/DIY/LEISURE     0      0     40     74    107    134    165    180    186    194    198    198    203    203    203  (£)
   DOMESTIC CLEAN-UP      0      0    224    232    487    505    523    545    577    613    686    727    727    727    727  (£)

                                                                                           FLOOD DURATION MORE THAN 12 HOURS

TOTAL BUILDING FABRIC  364    458   1500   3050   4760   5810   7160   8190   9890  11460  12350  13780  14450  15260  16550  (£)
         COL PERCENT  100.0  100.0   56.6   62.1   59.8   55.3   54.3   54.7   58.6   61.9   63.4   65.8   66.8   68.0   69.8
         ROW PERCENT    2.2    2.8    9.1   18.4   28.8   35.1   43.2   49.5   59.8   69.2   74.6   83.2   87.3   92.2  100.0
TOTAL INVENTORY          0      0   1150   1860   3200   4690   6030   6790   6990   7040   7120   7160   7170   7170   7170  (£)
         COL PERCENT    0.0    0.0   43.4   37.9   40.2   44.7   45.7   45.3   41.4   38.1   36.6   34.2   33.2   32.0   30.2
         ROW PERCENT    0.0    0.0   16.0   26.0   44.6   65.5   84.2   94.7   97.5   98.2   99.3   99.9  100.0  100.0  100.0

TOTAL DAMAGE           364    458   2650   4910   7960  10500  13190  14980  16880  18500  19470  20940  21620  22430  23720  (£)
         ROW PERCENT    1.5    1.9   11.2   20.7   33.6   44.3   55.6   63.2   71.1   78.0   82.1   88.3   91.1   94.5  100.0
         U.C. LIMIT   516    514   3060   5650   9200  12140  15220  17200  19260  21000  22000  23560  24230  25100  26560  (£)
         L.C. LIMIT   211    401   2240   4170   6720   8860  11160  12760  14500  16000  16940  18320  19010  19760  20880  (£)
TOTAL DAMAGE/SQ.METRE   4.4    5.6   32.3   59.8   96.9  127.9  160.5  182.4  205.4  225.2  237.0  254.9  263.2  273.1  288.9  (£)
         U.C. LIMIT    6.3    6.6   35.8   65.1  106.5  141.6  177.7  201.8  225.5  245.6  257.8  276.3  284.3  294.4  311.0  (£)
         L.C. LIMIT    2.5    4.6   28.8   54.5   87.3  114.2  143.3  163.0  185.3  204.8  216.2  233.5  242.1  251.8  266.8  (£)


POST WAR FLAT             LAND USE CODE 153      JANUARY 1990 PRICES   SALT WATER FLOOD DURATION MORE THAN 12 HOURS

                                           DEPTH ABOVE UPPER SURFACE OF GROUND FLOOR
COMPONENTS OF DAMAGE   -0.3    0.0   0.05    0.1    0.2    0.3    0.6    0.9    1.2    1.5    1.8    2.1    2.4    2.7    3.0  (M)
---------------------------------------------------------------------------------------------------------------------------------
   PATHS & PAVED AREAS    0      1      3      5     12     20     26     32     31     32     33     33     35     35     36  (£)
   GARDENS/FENCES/SHEDS   0      0     33     71    153    198    328    366    488    768   1002   1302   1603   1649   1673  (£)
   EXT. MAIN BUILDING     0    298    306    349    450    561    747    903   1115   1371   1631   2038   2435   2771   3082  (£)
   PLASTERWORK            0      0      0    123    204    283    568    854   1139   1277   1139   1246   1312   1312   1312  (£)
   FLOORS                 0      0    146    407    601    820    891    961   1368   1572   1525   1525   1525   1525   1525  (£)
   JOINERY                0      0      0      0     24     53    121    302    487    487    852    852    852    852    852  (£)
   INTERNAL DECORATIONS   0      4      9    129    166    189    204    235    231    232    235    239    243    258    257  (£)
   PLUMBING & ELECTRICAL 38     50     78    148    238    253    471    471    471    471    471    471    471    471    471  (£)

   DOMESTIC APPLIANCES    0      0      2     74    162    328    366    517    517    518    518    518    518    518    518  (£)
   HEATING EQUIPMENT      0      0      0     26     30     73     89    100    109    109    109    112    112    112    112  (£)
   AUDIO/VIDEO            0      0     23     88    133    161    281    339    350    350    350    350    350    350    350  (£)
   FURNITURE              0      0     44    113    245    495    652    710    717    719    719    719    719    719    719  (£)
   PERSONAL EFFECTS       0      0      0      0     25     90    121    174    174    183    183    183    183    183    183  (£)
   FLOOR COVER/CURTAINS   0      0    271    291    312    312    340    345    349    349    349    349    349    349    349  (£)
   GARDEN/DIY/LEISURE     0      0     23     46     69     95    128    143    149    158    162    162    167    167    167  (£)
   DOMESTIC CLEAN-UP      0      0     96     99    207    215    223    232    246    261    292    309    309    309    309  (£)

                                                                                           FLOOD DURATION MORE THAN 12 HOURS

TOTAL BUILDING FABRIC   38    353    576   1230   1850   2380   3360   4130   5330   6210   6890   7710   8480   8880   9210  (£)
         COL PERCENT  100.0  100.0   55.6   62.5   60.9   57.3   60.4   61.7   67.1   70.1   72.0   74.0   75.8   76.6   77.3
         ROW PERCENT    0.4    3.8    6.3   13.4   20.1   25.8   36.5   44.8   57.9   67.4   74.8   83.7   92.1   96.4  100.0
TOTAL INVENTORY          0      0    460    740   1190   1770   2200   2570   2620   2650   2680   2700   2710   2710   2710  (£)
         COL PERCENT    0.0    0.0   44.4   37.5   39.1   42.7   39.6   38.3   32.9   29.9   28.0   26.0   24.2   23.4   22.7
         ROW PERCENT    0.0    0.0   17.0   27.3   43.8   65.4   81.3   94.7   96.5   97.8   99.1   99.8  100.0  100.0  100.0

TOTAL DAMAGE            38    353   1040   1980   3040   4150   5560   6690   7950   8860   9570  10410  11190  11590  11920  (£)
         ROW PERCENT    0.3    3.0    8.7   16.6   25.5   34.8   46.7   56.1   66.7   74.3   80.3   87.4   93.9   97.2  100.0
         U.C. LIMIT    53    396   1200   2280   3510   4800   6420   7680   9070  10060  10810  11710  12540  12970  13340  (£)
         L.C. LIMIT    22    309    877   1680   2570   3500   4700   5700   6830   7660   8330   9110   9840  10210  10500  (£)
TOTAL DAMAGE/SQ.METRE   1.1   10.1   29.7   56.4   86.7  118.5  158.7  191.0  226.8  252.9  273.2  297.2  319.3  330.7  340.3  (£)
         U.C. LIMIT    1.6   11.9   32.9   61.4   95.3  131.2  175.8  211.4  249.0  275.8  297.2  322.1  344.9  356.5  366.4  (£)
         L.C. LIMIT    0.6    8.3   26.5   51.4   78.1  105.8  141.6  170.6  204.6  230.0  249.2  272.3  293.7  304.9  314.2  (£)
```

APPENDIX 5.2   351

MODERN FLAT   LAND USE CODE 154   JANUARY 1990 PRICES   SALT WATER FLOOD DURATION MORE THAN 12 HOURS

```
                                    DEPTH ABOVE UPPER SURFACE OF GROUND FLOOR
COMPONENTS OF DAMAGE  -0.3   0.0   0.05   0.1    0.2    0.3    0.6    0.9    1.2    1.5    1.8    2.1    2.4    2.7    3.0   (M)
---------------------------------------------------------------------------------------------------------------------------------
PATHS & PAVED AREAS     0     9     17     24     50     80    106    124    122    124    125    126    129    131    133   (£)
GARDENS/FENCES/SHEDS    0     0      8     20     79    135    192    210    259    348    436    554    678    726    748   (£)
EXT. MAIN BUILDING      0    48     52     99    189    272    387    488    979   1428   1933   2657   3388   3991   4596   (£)
PLASTERWORK             0     0      0    272    446    620   1242   1864   2486   2782   2486   2576   2858   2858   2858   (£)
FLOORS                  0   189    542    834   1137   1331   1508   1523   1648   1713   1713   1713   1713   1713   1713   (£)
JOINERY                 0     0    116    316    498    855   1065   1400   1856   1889   1925   1925   1925   1925   1925   (£)
INTERNAL DECORATIONS    0     8     51    213    357    413    438    489    483    486    487    495    503    526    525   (£)
PLUMBING & ELECTRICAL  38    40     56    132    227    268    539    571    571    571    607    607    613    619    634   (£)

DOMESTIC APPLIANCES     0     0      2     78    176    361    433    595    595    598    598    598    598    598    598   (£)
HEATING EQUIPMENT       0     0      5     31     40     96    118    138    154    154    154    159    159    159    159   (£)
AUDIO/VIDEO             0     0     44    120    174    213    355    442    453    453    453    453    453    453    453   (£)
FURNITURE               0     0     59    215    797   1061   1560   1580   1631   1634   1634   1634   1634   1634   1634   (£)
PERSONAL EFFECTS        0     0     29    214    348   1018   1465   1789   1830   1838   1838   1838   1838   1838   1838   (£)
FLOOR COVER/CURTAINS    0     0    716    726    737    737    770    792    806    806    806    806    806    806    806   (£)
GARDEN/DIY/LEISURE      0     0     43     80    115    144    179    196    202    210    215    215    220    220    220   (£)
DOMESTIC CLEAN-UP       0     0    197    203    426    442    458    477    506    537    602    637    637    637    637   (£)

                                                                                        FLOOD DURATION MORE THAN 12 HOURS

TOTAL BUILDING FABRIC  38   295    844   1910   2990   3980   5480   6670   8410   9340   9720  10660  11810  12490  13140   (£)
     COL PERCENT    100.0 100.0   43.5   53.4   51.5   49.4   50.6   52.6   57.6   60.0   60.7   62.7   65.0   66.3   67.4
     ROW PERCENT      0.3   2.3    6.4   14.6   22.7   30.3   41.7   50.8   64.0   71.1   74.0   81.1   89.9   95.1  100.0
TOTAL INVENTORY         0     0   1100   1670   2820   4080   5340   6010   6180   6230   6300   6340   6350   6350   6350   (£)
     COL PERCENT      0.0   0.0   56.5   46.6   48.5   50.6   49.4   47.4   42.4   40.0   39.3   37.3   35.0   33.7   32.6
     ROW PERCENT      0.0   0.0   17.3   26.3   44.4   64.2   84.1   94.7   97.4   98.2   99.3   99.9  100.0  100.0  100.0

TOTAL DAMAGE           38   295   1940   3580   5810   8050  10820  12690  14590  15580  16020  17000  18160  18840  19480   (£)
     ROW PERCENT      0.2   1.5   10.0   18.4   29.8   41.3   55.5   65.1   74.9   80.0   82.2   87.3   93.2   96.7  100.0
     U.C. LIMIT       53   331   2240   4120   6710   9310  12490  14570  16650  17690  18100  19130  20350  21090  21810   (£)
     L.C. LIMIT       22   258   1640   3040   4910   6790   9150  10810  12530  13470  13940  14870  15970  16590  17150   (£)
TOTAL DAMAGE/SQ.METRE 0.5   4.1   26.9   49.7   80.5  111.7  150.1  176.0  202.3  216.1  222.2  235.8  251.9  261.4  270.3   (£)
     U.C. LIMIT      0.7   4.8   29.8   54.1   88.5  123.7  166.2  194.8  222.1  235.6  241.7  255.6  272.1  281.8  291.0   (£)
     L.C. LIMIT      0.3   3.4   24.0   45.3   72.5   99.7  134.0  157.2  182.5  196.6  202.7  216.0  231.7  241.0  249.6   (£)
```

POST 1975 FLAT   LAND USE CODE 155   JANUARY 1990 PRICES   SALT WATER FLOOD DURATION MORE THAN 12 HOURS

```
                                    DEPTH ABOVE UPPER SURFACE OF GROUND FLOOR
COMPONENTS OF DAMAGE  -0.3   0.0   0.05   0.1    0.2    0.3    0.6    0.9    1.2    1.5    1.8    2.1    2.4    2.7    3.0   (M)
---------------------------------------------------------------------------------------------------------------------------------
PATHS & PAVED AREAS     0     9     15     20     42     64     86    102     99     99     99     99     99     99     99   (£)
GARDENS/FENCES/SHEDS    0     0     11     15     49     95    122    133    177    237    279    319    363    405    427   (£)
EXT. MAIN BUILDING      0    46     49     80    143    204    312    406    612    814   1052   1364   1683   1945   2196   (£)
PLASTERWORK             0     0    162    248    722   1098   1467   1840   1976   2043   1904   2106   2106   2106   2106   (£)
FLOORS                  0   118    266    426    585    725    876    916    964   1017   1017   1017   1017   1017   1017   (£)
JOINERY                 0     0    108    268    437    766    961   1286   1738   1771   1807   1807   1807   1807   1807   (£)
INTERNAL DECORATIONS    0     6     42    168    277    322    326    334    334    334    335    339    344    358    358   (£)
PLUMBING & ELECTRICAL  38    40     56    132    227    297    540    572    572    572    608    608    614    620    634   (£)

DOMESTIC APPLIANCES     0     0      1     74    166    336    380    529    529    531    531    531    531    531    531   (£)
HEATING EQUIPMENT       0     0      0     26     31     85    105    124    138    138    138    143    143    143    143   (£)
AUDIO/VIDEO             0     0     34    102    153    183    312    389    399    399    399    399    399    399    399   (£)
FURNITURE               0     0     55    171    539    791   1155   1216   1246   1248   1248   1248   1248   1248   1248   (£)
PERSONAL EFFECTS        0     0     27    129    213    563    683    862    879    886    886    886    886    886    886   (£)
FLOOR COVER/CURTAINS    0     0    324    344    364    364    402    417    426    426    426    426    426    426    426   (£)
GARDEN/DIY/LEISURE      0     0     40     75    108    135    166    181    187    195    199    199    204    204    204   (£)
DOMESTIC CLEAN-UP       0     0     93     96    201    208    216    225    239    254    284    301    301    301    301   (£)

                                                                                        FLOOD DURATION MORE THAN 12 HOURS

TOTAL BUILDING FABRIC  38   219    710   1360   2480   3570   4690   5590   6470   6890   7100   7660   8040   8360   8650   (£)
     COL PERCENT    100.0 100.0   55.1   57.1   58.3   57.3   57.8   58.6   61.5   62.8   63.3   64.9   66.0   66.9   67.6
     ROW PERCENT      0.4   2.5    8.2   15.7   28.7   41.3   54.3   64.7   74.9   79.7   82.2   88.6   92.9   96.7  100.0
TOTAL INVENTORY         0     0    578   1020   1780   2670   3420   3950   4050   4080   4120   4140   4140   4140   4140   (£)
     COL PERCENT      0.0   0.0   44.9   42.9   41.7   42.7   42.2   41.4   38.5   37.2   36.7   35.1   34.0   33.1   32.4
     ROW PERCENT      0.0   0.0   14.0   24.7   42.9   64.4   82.6   95.3   97.7   98.5   99.4   99.9  100.0  100.0  100.0

TOTAL DAMAGE           38   219   1290   2380   4260   6240   8120   9540  10520  10970  11220  11800  12180  12500  12790   (£)
     ROW PERCENT      0.3   1.7   10.1   18.6   33.3   48.8   63.5   74.6   82.3   85.8   87.7   92.3   95.2   97.8  100.0
     U.C. LIMIT       53   245   1490   2740   4920   7220   9370  10960  12000  12450  12680  13280  13650  13990  14320   (£)
     L.C. LIMIT       22   192   1090   2020   3600   5260   6870   8120   9040   9490   9760  10320  10710  11010  11260   (£)
TOTAL DAMAGE/SQ.METRE 1.1   6.4   37.9   69.9  125.0  183.2  238.1  279.9  308.7  321.9  329.2  346.3  357.4  366.9  375.3   (£)
     U.C. LIMIT      1.6   7.5   42.0   76.1  137.4  202.9  263.7  309.7  339.0  351.0  358.1  375.3  386.0  395.6  404.0   (£)
     L.C. LIMIT      0.6   5.3   33.8   63.7  112.6  163.5  212.5  250.1  278.4  292.8  300.3  317.3  328.8  338.2  346.6   (£)
```

Appendix 5.3 Standard questionnaire for the assessment of site-specific flood damage potential: the business site interview schedule

Middlesex University
Flood Hazard Research Centre

BUSINESS SITE SURVEY INTERVIEW SCHEDULE:
FOR ESTIMATION OF DIRECT AND INDIRECT FLOOD LOSS POTENTIAL

CONFIDENTIAL

Interview number ....

Location ....

Date ....

Warning lead time (hrs) ....

Duration flooded (hrs) ....

Flood type (circle appropriate number)   Saltwater — 1
                                         Freshwater — 2

1. Name of company ....

    Address ....

    Name and position of respondent ....

    ....

2. Type of business including specialisations ....

    ....

3. Total area of buildings and any outside stores .... (m²)

4. a) What is the total present value of all plant, fixtures and equipment
       held on site? .... (£)

    b) What proportion of this is sited on ground floor? .... (%)

5. a) What is the value on average of all raw materials, work in progress,
       and finished goods (stock) on site? .... (£)

    b) What proportion of this is sited on ground floor? .... (%)

6. a) Is production (are sales) concentrated at any particular time of year?
       Yes — 1   No — 2   If yes, when? ....

    b) Are stock levels high at any particular time of year?
       Yes — 1   No — 2   If yes, when? ....

7. a) Has this site ever been flooded? Yes — 1   No — 2

    b) If yes, when was it last flooded ....
       and to what depth? .... (m)

8. a) Are there any basements or pits below the floor level here?
       Yes — 1   No — 2

    b) What is their depth below ground floor level? ....

9. At what depth of floodwater would damage or disruption begin here?
    .... (m)

354 THE ECONOMICS OF COASTAL MANAGEMENT

DIRECT DAMAGE

10. What would you estimate the cost of damage to be to **plant and equipment** (fixtures, fittings and equipment) at this site following flooding to the following depths? **Separate** a) costs of replacement of entire items, less scrap value, and b) costs of repairs and parts replacement.

Assume a flood warning lead time of .... hours and a flood duration of .... hours.

Depth	Description and costs
-0.03m (-0.13")	
0.15m (6")	
0.30m (12")	
0.60m (24")	
1.00m (39")	

54 | 0 | . | 1 | 5 |
60 | 0 | . | 3 | |
66 | 0 | . | 6 | |
72 | 1 | . | 0 | |

Dep
Der

11. What would you estimate the cost of damage to be to raw materials, work-in-progress and finished goods (stocks and materials) at this site following flooding to the following depths?

    Notes note the cost of damage to each category of stock. Materials may be recovered by reworking (loss is extra cost involved) or may be sold off cheaply (loss is fall in sales revenue).

    Assume a flood warning lead time of .... hours and a flood duration of .... hours.

Depth	Description and costs	Ds
-0.03m		3/1
0.15m		8
0.30m		17
0.60m		26
1.00m		36

12. Would there be any **additional costs** incurred in cleaning up after flooding? For example: overtime, hire of equipment, cost of contractors. Please estimate the cost for each flood depth.

Depth	Details of costs	C
-0.03m		46
0.15m		52
0.30m		58
0.60m		64
1.00m		70

INDIRECT LOSS

13. What particularly damageable item(s) of plant or materials (what factors) would determine how long it would be before you could return to normal/full production (business)?

    ....

    ....

14. For each depth of flooding, would there be any period of time when you could produce nothing (do no business)?
    Yes — 1   No — 2   If yes, how long would this be?

Depth	Length of time with no production (specify units of measurement e.g. hours)	Dn
-0.03m		1
0.15m		5
0.30m		9
0.60m		13
1.00m		17

15. For each depth of flooding, how long would it be from starting to produce something (business) after flooding to regaining normal/full production (business)?
    (Complete interview schedule Annex).

Depth	Additional information (if any)	Dp
-0.03m		21
0.15m		25
0.30m		29
0.60m		33
1.00m		37

16. For each depth of flooding we would like you to mark on the interview schedule Annex how you expect production (business) to build up during the recovery period of partial production (business) which you have just indicated.

Depth	Additional information (if any) e.g. estimated percentage production lost
-0.03m	
0.15m	
0.30m	
0.60m	
1.00m	

Pn

41
45
49
53
57

17. a) Do you think this firm could make up any of the production (business) you have indicated it would loose? i.e. between the time of flooding to the restoration of normal production (business) Yes — 1   No — 2
How would you do this?

Overtime — 1   Increased productivity — 2   Extra staff — 3
Sub-contracting within this region — 4   Other — 5

b) For each depth, what percentage of production (business) otherwise lost could you recover in these ways? (Reminder: estimate amount of lost production/business from interview schedule Annex and Question 14).

R

Depth	Percentage of lost production/business recovered
-0.03m	
0.15m	
0.30m	
0.60m	
1.00m	

61
65
69
73
77

Only asked if answer to Q17a is yes

18. Do you expect that making up production (business) in any of these ways would result in any additional cost?   Yes — 1   No — 2
How much above normal would this cost be? (For overtime note the number of hours and wage rate and calculate the additional costs later).

A

Depth	Details of costs
-0.03m	
0.15m	
0.30m	
0.60m	
1.00m	

5/1
7
13
19
25

19. If you could not supply your customers or take on new customers after a flood, what proportion would

    Produce from (run down) their stocks until normal deliveries? _____ (%)

    Buy from your competitors? _____ (%)

    Be held up and lose their normal production (operations)? _____ (%)

    Other? (give details) _____ (%)

20. What proportions of sales (business) lost to this plant (firm) would you expect to be taken up by each of the following groups of competitors?

    Competitors within this region _____ (%)

    Other competitors in the U.K. _____ (%)   Rg

    Foreign firms or plants _____ (%)   F

21. Would there be any other impact of flooding here on this firm, such as long run effects?   Yes — 1   No — 2   If yes, please give details.

_____

_____

_____

_____

22. How many people work on this site? (Full-time equivalents)   E

_____   T

23. What is the approximate annual turnover of this site?

    (£) _____

Interviewer's name _____

Additional notes

INTERVIEW SCHEDULE ANNEX

FOR USE WITH Q's 15 & 16 OF THE BUSINESS SITE SURVEY INTERVIEW SCHEDULE

Instructions

On the graphs below indicate how production (business) builds up following a flood. For each depth:

1. Indicate by drawing a vertical line e.g. A-B, the point in time when you expect to retain normal production (business). Time is counted from the date at which you first start to produce something (do some business).

2. Construct a line (e.g. C-A) from the bottom left corner of the graph to the top right hand corner of the box constructed by drawing in the vertical line A-B. This line C-A indicates the rate of production (business) build up after flooding.

3. Where applicable superimpose -0.03m curve on 0.15m graph.

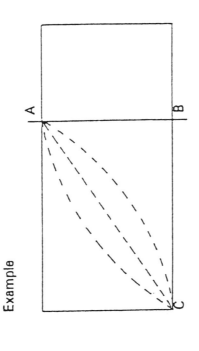

Example

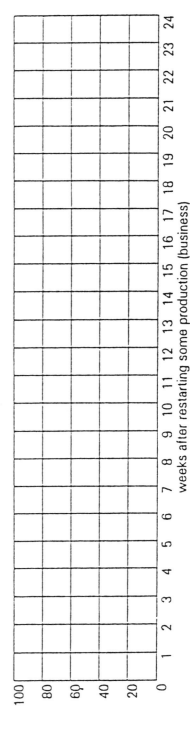

Depth 0.15m

Percentage of normal production (business)

weeks after restarting some production (business)

PLEASE TURN OVER FOR 0.30M, 0.60M AND 1.00M DEPTHS

360 THE ECONOMICS OF COASTAL MANAGEMENT

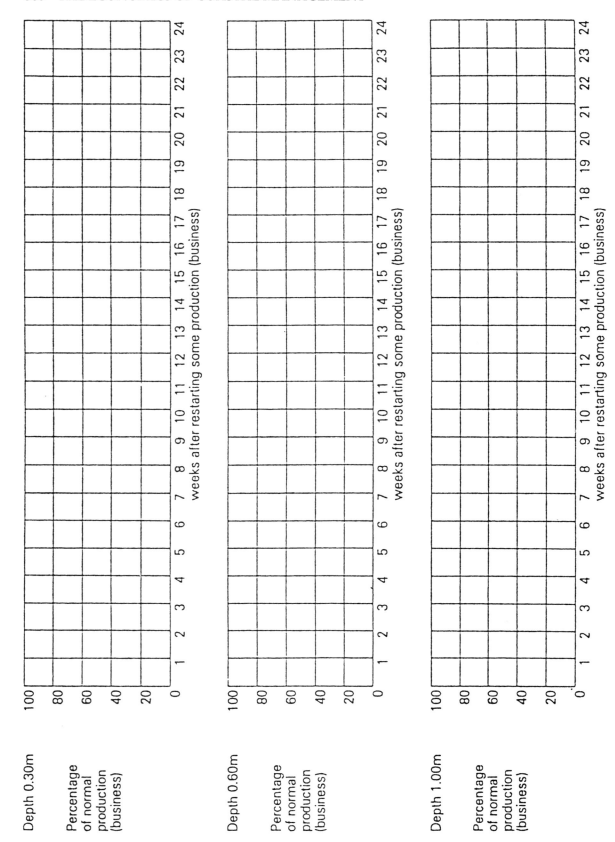

Appendix 5.4 Land use coding form for the recording of data on the flood affected benefit area

362 THE ECONOMICS OF COASTAL MANAGEMENT

Middlesex University Flood Hazard Research Centre — Coding Form: Residential land use only

NAME:
ADDRESS:
DATE:
Page ___ of ___

Ø 1 2 5 FIGURES   0 1 2 5 LETTERS
Scheme location and codes, etc.

Columns	Field
1–12	**Plan reference** — This is optional but helps to keep a large data set organised by having a reference to the number of the plan on which the properties are located
13–20	**Grid reference** — A 8-figure grid reference is preferable for maximum accuracy
21	Land Use Sector
22	House type code
23	House age code
24	Social class code
25,26	Not used for residential
27,28	A/B ⎫
29,30	C1 ⎬ Social Class Proportions OPTIONAL
31,32	C2 ⎪
33,34	D/E ⎭
35,36	Spare
37–41	**Height of properties in metres OD.** Floor height is coded
42–48	**Area of house (not grounds).** Code in square metres. This is only used for very large houses
49,50	Number of properties coded same.
51–72	**Address/es of properties.** This is very useful as aid to organisation. See Manual for format which is most convenient. Print out will list this if required
73,74	Reach number
75–77	
78,79,80	Comments numbers

APPENDIX 5.4 363

Coding of land use data for flood alleviation benefit analysis
Residential land uses:

All letters and numbers should be clearly written. Zeros should be crossed Ø to differentiate from the letter 'O'. It is wise to use a soft *pencil*, to facilitate alteration.

The following notes should assist the coding of properties. Other details are on pages 66 to 69 of *The Benefits of Flood Alleviation*.

Plan reference This is the reference number of the plan (ie large-scale map) on which the property is to be found. It is useful to have this information, to facilitate cross-checking after the data is coded. For example, the illustration in the Manual (page 67) shows a code of 5671NE5771NW but it might be equally applicable to code 'Map 1' or 'East Bristol.'

Grid reference Code with 8-figure grid reference. This is important to secure greatest accuracy, especially when using the rectangle method of sub-area definition.

Land Use Sector This form is *just for residential properties* so the number 1 has been completed right down the form, being the number of the residential sector.

House type Detached = 1; Semi = 2; Terraced = 3; Bungalow = 4; Flat = 5.

House age Pre-1918 = 1; 1918-1938 = 2; 1939-1965 = 3; 1965+ = 4. For examples see Manual pages 20-23.

Social Class A/B = 1; C1 = 2; C2 = 3; D/E = 4. For definitions see Manual page 27. This section can be used when the social class is known, whereas columns 27 to 34 are used when this is not known and data is taken from the national census (see page 68). For *Social Class proportions* see page 68. This feature is *optional*.

Height of properties This is the *floor* height. The depth/damage data recognises that in old properties with suspended floors damage commences below this height. Accuracy here is important: code in Metres OD. 26.32 would be the code for 26.32m.

Area of house This is generally *not* used and only is applicable where a very large house clearly does not fit any of the 21 main categories of age and type. In this case damage per square metre *can* be allocated (but this is not automatic—consult Middlesex Polytechnic before using this option).

Number of properties If a number of properties have the same characteristics (height, grid reference, house age and type, etc) they can be lumped together (up to 99). Naturally this gets less precise the larger the number since it is unlikely that they are flooded to the same depth or do in fact have the same characteristics. *Right justify*.

Address This is invaluable for checking. The address is printed out at various stages in the computer compilations and can be used to check at that stage whether errors have crept into other data codes (particularly the height codes).

Reach number It is simplest to identify which areas of the flood plain to be separated for benefit assessment purposes into sub-areas and then give each property a code based on those sub-areas (eg 1 to 10 for 10 sub-areas). For greater sophistication see page 69 of the Manual (and good luck!); the permutations are endless but invaluable when sub-areas need to be re-combined and otherwise altered without changing the large amounts of land use data. In this case the 'bank' column is used as well (see page 67). *Remember to right justify*.

Comments It is used to keep comments in some sort of order in a separate file (not computer file). Thus a code sequence starting 001 could refer to pages in that file on which details of correspondence or particular house or damage data is kept.

──────── at the top of the columns indicates that this information is *essential* for any assessment.

Middlesex University Flood Hazard Research Centre

Coding form: Land Use Sectors 2-9

NAME:
ADDRESS:

DATE:
Page of

Ø 1 2 5 FIGURES 0 1 2 5 LETTERS

Scheme location and codes, etc.

Columns	Field
1–12	**Plan reference** — This is optional but helps to keep a large data set organised by having a reference to the number of the plan on which the properties are located.
13–20	**Grid reference** — A 8-figure grid reference is preferable for maximum accuracy
21–24	**Land Use Sector** — Type land use code, Category code, Sub-category code
25–26	Site survey code
27–34	Spare
35–41	Height of properties in metres OD. Floor height is coded
42–48	Area of property (not grounds). Code in square metres. Can be estimated from large scale plans.
49–50	Number of properties coded same
51–72	Address/es of properties. This is very useful as aid to organisation. See Manual for format which is most convenient. Print out will list this if required.
73–74	Reach number
75–77	
78–80	Comments numbers

Coding of land use data for flood alleviation benefit analysis

Land Use Sectors 2-9 (all non-residential uses):

All letters and numbers should be clearly written. Zeros should be crossed Ø to differentiate from the letter 'O'. It is wise to use a soft *pencil*, to facilitate alterations.

The following notes should assist the coding of properties. Other details are on pages 66 to 69 of *The Benefits of Flood Alleviation*.

Plan reference As on reverse of residential coding form.

Grid reference As on reverse of residential coding form.

Land Use Sector
Type land use code Consult Appendix 1.1 of the Manual (pages 135 to 148)

Category code
Sub-category code Do not worry at this stage whether or not flood damage data is contained in the Manual appendices.

Site survey code Whenever a site potential damage survey is undertaken, the data for which will be incorporated into the assessment rather than using standard data, enter numbers here which are unique for each property. Numbering must begin at 11 and proceed through to 19 and recommence at 21, etc. Do not code any zeros in these columns.

Height of properties This is the floor height and is assumed to be the flood height at which damage begins. If properties have more than one floor height, code as two or more separate properties. Accuracy is important here. Code in metres OD and to at least one and preferably two decimal places.

Area of properties This is crucial since damage is allocated in sectors 2-9 per square meter. Code *without* decimal places (1000 = one thousand square metres). Estimates can often be made with sufficient accuracy from large-scale plans. *Right justify*.

Number of properties If a number of properties have the same land use code (all four digits) they can be lumped together (up to 99). Right justify the number.

Address As on reverse of residential coding form.

Reach number As on reverse of residential coding form.

Comments numbers As on reverse of residential coding form.

───── ───── at the top of the columns indicates that this information is *essential* for any assessment.

Appendix 5.5 Standard household questionnaire for flood affected households

… APPENDIX 5.5 367

Middlesex University
Flood Hazard Research Centre
STANDARD HOUSEHOLD QUESTIONNAIRE

CONFIDENTIAL

Questionnaire number..

Location..

Date..

Depth zone..

1. Address ....

2. House type

    Basement flat — 1
    Bungalow/ground
       floor flat — 2
    Other — 3

3. How many live in this house/flat? (Ascertain number in each category).

	Total	Number infirm/disabled
School/pre-school children		
Elderly		
Other adults		

4. Have you experienced flooding in this house/flat?   Yes — 1
                                                         No — 2

    If Yes go to Question 6
    If No end interview

5. When was the last time you were flooded here?

PLEASE REFER TO THIS FLOOD WHEN ANSWERING THE REMAINING QUESTIONS

6. Which rooms were flooded and to what depth?

	No.	Depth
Living rooms		
Bedrooms		
Other		
Garden		

    specify whether metres or feet

7. For how long was this house flooded? (hrs)

8. Did your house or contents suffer from flood damage?   Yes — 1
                                                           No — 2

    If no go to Question 15

9. Did you do all the repairs and redecoration yourself or did you pay workmen to do it?

    Self — 1
    Workmen — 2
    Mixture — 3

10. Did you recover the costs of repair, redecoration and replacement from insurance?

    Yes — 1
    No — 2
    Partly — 3

11. Did you receive any financial help from relief funds?

    Yes — 1
    No — 2

12. Have you now completed repairing and redecorating your house and replacing the contents damaged in the flood?

    Yes — 1
    No — 2

13. If no to Question 14
What proportion of the damage has not yet been made good? (%)

14. Did you lose anything that is impossible to replace?
(e.g. photos, letters etc.)
If so what? _____

    Yes — 1
    No — 2

15. People sometimes find that worry and stress is caused by the possibility of more flooding in the future. I am now going to read you some statements — for each please say whether you agree or disagree. (Note: using wording which matches event). Agree — 1 Disagree — 2

    1. We stay up all night when it rains heavily/there is a gale
    2. If we go away we arrange with neighbours how they can contact us in case of a flood
    3. When it rains, we check the level of the water in the river/stream/ on the sea front
    4. We are afraid to go out when it rains heavily/there is a gale
    5. We are afraid to go out when water levels are high
    6. When we go away on holiday or a visit, we move important things above possible floodwater levels
    7. We are too worried to sleep at night when it rains heavily/when tide levels are high
    8. When it rains heavily/when river/tide levels are high we move the car to a safe area
    9. We are afraid to go out when heavy rain/high river/high tide levels are forecast
    10. We would move to another house if we could
    11. We are worried every time it rains heavily/river/tide levels are high

16. a) Did you stay in the house during and after the flood or did you have to go and stay with friends, relatives or elsewhere?

    Stayed — 1
    Left — 0

    b) If you had to leave the house, for how long did you stay elsewhere? (days)

17. Where or with whom did you stay?

    Relatives — 1
    Friends — 2
    Council provided — 3
    Other — 4

18. How long did it take to get the house clean and normal after the flood? (weeks)

19. Do you feel that the flood caused you financial problems? (SHOW CARD 1)   Code 1 — 10

20. Would your home be more difficult to sell because of the flood problem?

         Yes — 1
         No — 2
         DK — 3
         NA — 4

21. a) Has your house lost value because of the flood problem?

         Yes — 1
         No — 2
         DK — 3
         NA — 4

     b) If yes, by roughly how much?

22. What is the approximate market value of your house?

23. Did you have to use up savings to pay for repairs and so on?

         Yes — 1
         No — 2

24. Has the risk of flooding meant that you have had to do any of the following?

         Yes — 1
         No — 2

     1. Keep a stock of candles, matches and/or emergency supplies in case the electricity or gas supply is interrupted.
     2. Lose use of the cellar because of the likelihood of flooding.
     3. Give up trying with the garden.
     4. Spend money trying to stop the floodwater getting into the house.

25. Do you think that the flood affected the health of any members of the household?

         Yes — 1
         No — 2
         DK — 3

26. What were these effects? (Bronchitis, rheumatism, gastric problems, depression, trouble sleeping etc.)

_____

_____

_____

_____

27. Did they see their doctor about these problems?   Yes — 1
         No — 2

28. How great a risk to health and safety do you think there is here from flooding? (SHOW CARD 2) Code 0 — 6

29. How worried are you about the possibility of flooding in the future?
(SHOW CARD 3) Code 0 – 6

30. How likely do you think it is that there will be flooding in the future?
(SHOW CARD 4) Code 0 – 6

WE WOULD LIKE YOU TO TRY TO SUMMARISE ALL THESE DIFFERENT AFFECTS OF THE FLOOD UPON YOU

31. Overall how serious were the effects of the flood upon your household?
(SHOW CARD 5)   Code 0 – 6

32. Please rate the following consequences of the flood upon your household on a scale of 1 – 10 for each affect.
(SHOW CARD 6)
Code 0 – 10
Not applicable   – 11
  1. Affect upon your health
  2. Having to leave home
  3. Damage to replaceable furniture and contents
  4. Worry about flooding in the future
  5. Loss of irreplaceable objects (photos etc.)
  6. All the problems and discomfort whilst trying to get the house back to normal afterwards
  7. Damage to the house itself
  8. Stress of the flood itself
  9. Other (specify) _____

33. Below is a list of comments made by people after stressful events. Please rate those comments with regard to the flood which have been true for you DURING THE PAST 7 DAYS on a scale of 1 – 4. (SHOW CARD 7)
Code 1 – 4
  1. I thought about it when I didn't mean to
  2. I avoided letting myself get upset when I thought about it or was reminded of it
  3. I tried to remove it from memory
  4. I had trouble falling asleep or staying asleep
  5. I had waves of strong feelings about it
  6. I had dreams about it
  7. I felt as it hadn't happened or that it wasn't real
  8. I tried not to talk about it
  9. Pictures about it popped into my mind
  10. Other things kept making me think about it
  11. I tried not to think about it
  12. Any reminder brought back feelings about it
  13. My feelings about it were rather numbed

34. Could you tell me what is the current, or most recent, occupation of the Head of the Household?

_____

35. Do you own your house?
  Own outright   – 1
  Own on a mortgage – 2
  Rent   – 3
  Other   – 4

36. Sex of respondent
  Male   – 1
  Female – 2

37. Respondent's age group	less than 30 — 1
	31 — 59 — 2
	60 or over — 3

Interviewer's name _____

Additional notes

CARD 1

DO YOU FEEL THAT THE FLOOD CAUSED YOU FINANCIAL PROBLEMS?

1	2	3	4	5	6	7	8	9	10

NONE WHATSOEVER EVEN IN SHORT TERM

SEVERE CONTINUING PROBLEM

CARD 2

HOW GREAT A RISK TO HEALTH AND SAFETY DO YOU THINK THERE IS HERE FROM FLOODING?

0	1	2	3	4	5	6

NONE

VERY GREAT

CARD 3

HOW WORRIED ARE YOU ABOUT THE POSSIBILITY OF FLOODING IN THE FUTURE?

0	1	2	3	4	5	6

NOTHING TO WORRY ABOUT

VERY WORRIED

CARD 4

HOW LIKELY DO YOU THINK IT IS THAT THERE WILL BE FLOODING IN THE FUTURE?

0	1	2	3	4	5	6

VERY UNLIKELY

VERY LIKELY

CARD 5

OVERALL HOW SERIOUS WERE THE EFFECTS OF THE FLOOD UPON YOUR HOUSEHOLD?

0	1	2	3	4	5	6

NONE   VERY SEVERE

CARD 6

PLEASE RATE THE FOLLOWING CONSEQUENCES OF THE FLOOD UPON YOUR HOUSEHOLD ON A SCALE OF 1 – 10 FOR EACH AFFECT.

1. Affect upon your health
2. Having to leave home
3. Damage to replaceable furniture and contents
4. Worry about flooding in the future
5. Loss of irreplaceable objects (photos etc.)
6. All the problems and discomfort whilst trying to get the house back to normal afterwards
7. Damage to the house itself
8. Stress of the flood itself
9. Other (specify)

0	1	2	3	4	5	6	7	8	9	10	11

NO AFFECT   MOST SERIOUS AFFECT

11 – NOT APPLICABLE

CARD 7

BELOW IS A LIST OF COMMENTS MADE BY PEOPLE AFTER STRESSFUL EVENTS. PLEASE RATE ON A SCALE OF 1 — 4 THOSE COMMENTS WITH REGARD TO THE FLOOD WHICH HAVE BEEN TRUE FOR YOU DURING THE PAST 7 DAYS.

	NOT AT ALL 1	RARELY 2	SOMETIMES 3	OFTEN 4
I thought about it when I didn't mean to				
I avoided letting myself get upset when I thought about it or was reminded of it				
I tried to remove it from memory				
I had trouble falling asleep or staying asleep				
I had waves of strong feelings about it				
I had dreams about it				
I felt as it hadn't happened or that it wasn't real				
I tried not to talk about it				
Pictures about it popped into my mind				
Other things kept making me think about it				
I tried not to think about it				
Any reminder brought back feelings about it				
My feelings about it were rather numbed				

Index

Note: Page numbers in **bold** and *italics* refer to figures and tables respectively.

advisors, environmental 135
agriculture 12, 33, 114-16, 115-16,**116**, *117*
  see also land, agricultural
Aldeburgh 32, 97
Ancient Monuments 123, 132, 183
appraisal approach 2
archaeological sites *24*, 123, 125
  Hengistbury Head 183, **188**, 188-9, 191
  value 131-2
    assessment procedure 136-7, 142
archaeology 144-5
Areas of Outstanding Natural Beauty (AONBs) 123
assessment
  archaeological 136-7, 142
  benefits **50**, 171-4, *175*
    recreational **59**, 79-86, *91-2*, 92, 167-8
  coastal erosion *42*
  ecological 138-41, **139**, 184-8
  environmental significance 119-20
  operational methods 57
  site 140-1
  see also Environmental Assessment (EA); flood alleviation; recreation
audio/video equipment 102

Bank of England Housing Model 29-30
Barnes, R.S.K. 140, 184
Barton Beds 189
Bateman, I. *et al* 33
beach(es) 68
  characteristics 71, *164*
  coast protection effects 33, 60
  drawings **72**, 85, **164**, **177-8**
  enjoyment values *75*, *165*
  erosion 60, 61-2, 71-4
  functionality 60
  Herne Bay **177-8**
  material 61
    longshore drift removal **158**, **159**, 159
  nourishment/replenishment 17, 63, 161, 183
  profile 61
  recreation 33, 61, 157
    surveys *68-9*, 69-74, 74, 87
      Clacton 77-8
      Hastings 162-6
  valuation 71, *165*
benefit-cost analysis (BCA) 2-3, 16, *173*
  aims and assumptions 6-8
  'extended' 8-10, *9*
  timing and phasing 11-12
benefits
  Above Design Standard Benefits 32
  agricultural 12, 33, 114-16, **116**, *117*
  assessment 167-8
    cliff-top recreation 82, 167-70
    coastal erosion 42-57, 151-4
    economic 64
    Fairlight Cove 171-4
    framework **50**
    Herne Bay 175-7
    see also feasibility studies
  average annual 32, 33
  as delayed loss 46
  negative 33
  sea defence project 32, 94-117
  see also flood alleviation; recreation
Benson, J.F. and Willis, K.G. 122
bird diversity 130
boat, access by 190-1
Boscombe sands 189
boundaries analysis 22-3
Bournmouth 183
breaches
  coastal 97, 180
    Double Dykes 183, 187, 189-91
breakwaters, offshore 64
brickwork 101
Brooke, J.S.
  et al 33
  and Turner, R.K. 32
Brundtland Commission Report 21
buildings
  fabric damage 101-2
  loss 43, 108, 177
  see also property
businesses 57, 176
  site interview 357-65
  valuation 45

Canvey Island 107
Carson, R.T. and Mitchell, R.C. 31, 79
case studies 6
  characteristics *157*
  cliff-top recreation 74-7
  see also Fairlight Cove; Hastings; Hengistbury Head; Herne Bay; Peacehaven
chalk cliffs 167
change valuation 25
Chatterton, J.B. and Penning-Rowsell, E.C. 112
chemical release 107
Chesil Beach 97
Chesil sea defence project 105
children, as coastal visitors 70, 80
Christchurch Harbour 110, **181**, 181-91
Clacton **72**, 77-8
Clausen, L. 106
Clawson, M. 26
  see also Travel Cost Method (TCM) [Clawson]
cliffs and cliff-tops 62
  falls 170
  habitats 125
  the Naze 74
  recreation 68-9
    benefits 82, 167-70
    surveys 74-7
  regrading 64
  see also erosion
coast protection projects 33-6, 43, 87, 157, 159
  benefits
    assessment models 151-4
    and flood alleviation 175-7
  evaluation **5**
  externalities **23**
  Fairlight Cove 171-4, *173*
  gains with 88-9
  Hastings **160**, 160-1
  Hengistbury Head 182, 183-4, 189, 191
  Peacehaven 167
  planning
    integrated approach 1-2
    regional perspective 1
  recreation effects 60-1, 62-4, 177-9, 181, 189
  and sea defence 10
  strategies 8-10
  see also sea defence projects
coastal change 84-5
coastlines, soft and hard 133
Coker, A. *et al* 190
combinatory value, site 141
commercial production 120
communities, significant 141
compensation 19, 122
computer
  data processing 146
  systems 154
computer programs
  AUTOROUTE 74
  BOCDAM 12, *148*, 148-9
  DDAS (Depth-Damage Data Assembly) 149
  ESTDAM 6, 11, 12, 104, *148*, **149**, 149-51
    Herne Bay 175, 177

376 INDEX

typical output **113**
micro-ESTDAM **149**, 149-51
spreadsheets 151-2, **152**, **153**, **172**, 177
VUEDAM 149
confidence intervals 93
conservation areas, marine *124*
consultation, environmental **134**, 134-6, 144
consumption 18, 20-1
Contingent Valuation Method (CVM) *25*, 30-1, 74, 120-1, 181
  and environmental valuation 144
  Hastings 157-66
  Hengistbury Head 189, 190
  Herne Bay 177-9
  and recreation enjoyment 12, 66-9
  results 77-9, 89
  surveys 83-6
COPRES *see* Middlesex University
corrosion 102
counselling, post-disaster 111
counters, infra-red 86, 179
Countryside Commission 137, 138
Countryside Council for Wales 137, 138
Countryside Recreation Survey (1986) 68
counts, manual 86
County Archaeological Officer 137
criteria
  biological/ecological 128
  monument evaluation 142
  site assessment 140-1

Dalvi, M.Q. 107
damage
  debris 103
  event 112, **113**, 114
  health 107-8
  savings 112
  site-specific surveys 12, 110-11
  storm 102-3
  structural 32
  see also flood damage
damageable property surveys 11
data
  archaeological 132
  depth/drainage 6, 98
  field 154
  flood damages 104-8
  processing 12-13
  project specific 108-12
  recreational usage 65, 80-1
  requirements, and availability 55-7
  seasonal 154
  survey population 89-90
  updating 13
  visit number 86
  visitor types 81-2
deaths 33, 105-7, **106**
debris 32, 103
decision making 1, 15, 131
  and environmental values 134-8
decorations 101-2
depth/damage data 6, 98
designations 119, 122-3, *124*, 126
desk top review 79-80
Dicks, M.J. 49-50
disaster research 111
disbenefits 33
discount rate 19-21, 38, 199
discounting 16, 19-21, *40*, **153**, 154
diversity index 129-30
domestic appliances 102
Double Dykes 183, 187, 189-91
downdrift sites 125
drainage
  agricultural 114-15, **115**
  data 6, 98
drawings, survey use **72**, 85, **164**, **177-8**
Dunwich 71, 74
Dwarf Spike-rush (*Eleocharis parvula*) 187

ecological

  impacts on Hengistbury Head 184-8
  role concept 122
  sites 24, 36, 55, 125
    designations *124*
    matrix method assessment 135, 138-41, **139**, 144, 184-8
economic
  analysis 14-15, *17*, 17
    application stages 22-5
    bases of 17-22
  benefits, national/local 64
  impacts 175, 180, 181
  loss 190
educational use 121, 141
EEC Habitats Directive 129, 141
electrical fittings 102
emergency services 103-4
energy flow 131
English Heritage (formerly Historic Buildings and Monuments Commission) 137, 142
English Nature (formerly Nature Conservancy Council) 126, 128, 136, 137, 189
enjoyment values 74, 87, 121, 162, 163
  beach *75*, *165*
  changes 81
  cliff-top *75*, 75, *169*, 169
  Herne Bay seafront *178*, 178, *179*
  recreational 66-8, *75*, 75, 92, 93
    CVM surveys *68*
environment
  components **121**, 135, 191
    value 119-22
  effects consultation **134**, 134-6, **144**
  gains 120
  priorities 123-6
  significance evaluation 118-45
  significance site 7
  valuation 120, 143-4, *144*
Environmental Assessment (EA) [previously Environmental Impact Assessment (EIA)] 2-3
Environmental Impact Assessment (EIA) 16, 126, 135
equity, inter-generational 21
erosion 22-3, 120, 182-3
  alleviation 10
  beach 60, 61-2, 71-4
  cliff 12, 24, 55, 62, 74, 75
    Fairlight Cove **170**, **171**, 171
    Peacehaven 167, **168**, 168, 170
  control 11
  effects 33, 60
    on recreation 61-2
  and flooding 10, 33
  Hengistbury Head 189
  losses 10, 55, 180-1
    assessment 87-8
    and benefits evaluation 42-57
  protection benefits, Herne Bay *176*, 177-9, *180*, 180
  seafront *73*, 159
erosion contours 44-6, **45**, 155, 172-3, 176
erosion rate 43-4, 44-6, 80, 82-3, 174
evacuation 107
evaluation 5, 16, 142
  consensual 136
  definition 119
  ecological site criteria 128
  environmental significance 118-45
  erosion losses and benefits 42-57
  goods *25*, 25-32
event damage 112, **113**, 114
explanatory text, survey use 85
Exploratory Data Analysis techniques 86-7
exposures 142
  cliff 183, 189
  geological 133, 145, 189
  'type' 142
extension of life factors (ELFs) 42, 57, 199-200
  discounted 201-9

Fairlight Coastal Preservation Society 156, 173-4
Fairlight Cove 49, 156, *157*, **158**, **170**, **171**, 171-4
farm interviews 12

feasibility studies 11, 147
　environmental effects 135
　recreational benefit assessment **59**, 79, 83-6, *91-2*, 92
Filey 71
financial analysis *17*, 17
FLAIR (Flood Loss Assessment Information Report) report 13, 104
flood alleviation 10
　benefits 94-117
　　agricultural *117*
　　analysis, land coding form 361-5
　　area definition 109-10
　　assessment 6, *117*, 148-51
　　average annual 112-14, **113**
　　calculation 94-8, 112-14
　　computation 112
　　non-monetary *117*
　　and coast protection benefits 175-7
　　land use classification 279-309
　　see also computer software, ESTDAM
flood damage 11, 95
　annual average 95, **97**, 114
　data 104-8, 309-51
　direct and indirect 105, *117*
　estimation by ESTDAM 150
　and losses 98-104
　normal 175
　potential assessment questionnaire 352-60
　savings 112
　sea water 32-3, 98-102, *99*, *100*
　site-specific surveys 12, 104-5, 110-11
flood losses 10, 23, 43, 176, 180-1
　annual average 97-8
　contents, estimation 26
　and damages 98-104
　Herne Bay 177
　indirect 95, *117*
flood(ing)
　and erosion 10
　event losses assessment 37-8
　frequency 94-5, 109-10
　impacts and severity model **107**
　levels 176
　non-monetary impacts on households 105-8
　post-project residual 109
　predictions 154
　proofing 112
　protection benefits *176*, 177-9
　protection projects *147*
　recurrent 10
　river 32
　sea 32-4, 187
floor
　coverings 102
　height levels 110, 116, 154, 177
　standard damage data 104
flooring and joists 101
fossils 143
Freeman, A.M. 30
furniture 102

garden plants *101*, 101
Geological Conservation Review (GCR) 137
geological sites 125
　Hengistbury Head 189
　value 132-3, 137, 142-3
geology 145
geomorphological sites 125
geomorphology 138
global warming 97
golf courses 55
goods *40*
　private 18
　public 18, 71
grant aid 126
Grant, E. *et al* 190
grassland production 116
Green, C.H. *et al* 110, 111
'green' image 126
growth rates 21
groynes 71, **159**, 159, **160**
　conventional 63

fishtail 63-4
　reconstruction 161, **163**
　timber and rock 183
Guidelines for Selection of Biological SSSIs 141
Habitat Evaluation Procedure (HEP) 131
habitats 125, 127, 138-41, 182-3, 187
　coastal 127, 187
　diversity 140, 184
　Hengistbury Head 184-8, **185-6**
　marine 123, 127, 138-40
　rare 141
　terrestrial 127, **139**, 140
　types 127-8
　see also matrices
Harrison, A.J.M. and Stabler, M.J. 29
Hastings 71-3, 156, *157*, 157-66, **158**
health damage 107-8
heating equipment 102
Hedonic Price Method (HPM) **25**, 29-30, 48
Hengistbury Beds 189
Hengistbury Head 97, 108, 110, 135, 156, *157*, **158**, **181**, 181-92
　habitat matrices 141
　spreadsheet use 151
Heritage Coasts 123
heritage sites 7, *24*, 24
Herne Bay 64, 71, 108, 156, *157*, **158**, **175**, 175-81
　flood alleviation benefits assessment 6
　infrastructure 51
　Pier Pavilion 176
　seafront user and residents survey 73-4
　spreadsheet use 151
Hess, T.M. *et al* 115, **116**
Hicks-Kaldor Compensation principle 19
Hill, M.I. *et al* 47, 109-10
house prices 11, 29-30, 48-9, 49-50, 174
households
　flooded
　　non-monetary impacts 105-8
　　questionnaire 366-79
　　surveys 111
　surveys 36
Hydraulics Research Ltd. 183
hydrographic analysis 97

income multiplier model 180
infrastructure
　and at-risk properties 51-2
　benefit assessment procedure **50**, 50-3
　declining 180
　and distant erosion-prone properties 52
　and erosion 45-6
　and erosion-free areas 52-3
　integral to coast 51
　losses 42, 44, 50-1, 57
　methods 53
　removal 173, 174
'integrity' sites 133, 142
interest groups, minority 62
inundation loss 43, 177
invertebrates 127
investment 15, 20-1, 146, 154
　agricultural appraisal **115**, 115
Iron Age 189
irrationals 88, 179, 181

Jones-Lee, M. 107

Kent 105
Kosmin, R. 49

lagoons 140, 184, 187
land
　agricultural 42
　　conservation value 32
　　erosion rates 45
　　sea defence projects benefits 114-16
　　valuing 44, 45, 53-5, 57
　expected use of 44
　valuation 53-5
　values 26, 46

land use 10
  classification, flood alleviation projects 279-350
    coding 110
    coding form 361-65
    survey 12, 97-8
landscape 133-4, 138, 145
  values 137-8, 143
Least Cost Alternative Method 25, 31-2, 33, 39
Leeds-Harrison, P.B. *et al* 115, **116**
life, loss of 33, 105-7, **106**
Local Nature Reserve 184
Long Groyne 183
longshore drift **158**, *159*, 159
loss probability relationship 95-7, 98, 112, **113**, **150**
losses
  cliff-top 55
  environmental 120
  erosion 10, 42-57, 180-1
    assessment 87-8
  event assessment 37-8
  infrastructure 42, 50-1, 57
    valuing 44
  see also flood losses

Mackinder, I.H. 151
Margules, C. and Usher, M.B. 128
marine sites 123
Markandya,A. and Pearce, D.W. 21
market
  distortion 18
  perfectly competitive 18
  prices 25, 26, 47, 103, 199-200
Martello tower 33
matrices
  ecological summary 184, **185-6**
  environmental 135
  habitat 138-41, **139**, 141, 144
meadows 122
measurement, in economic analysis 23-5
Middlesex University
  Flood Hazard Research Centre
    COPRES project (Coastal Protection Research Project) 2, 6, 148, 149, 156, 177
    spreadsheet 151-2, **152**, *153*, *172*, 177
Ministry of Agriculture
  Fisheries and Food (MAFF) 115, 116, 171
    agricultural land grades 54
Mitchell, R.C. and Carson, R.T. 31, 79
monument evaluation method 142
Morecambe **72**
Morris, J. *et al* 115, **116**
mortar 101

National Curriculum 143
National Scheme for Geological Site Documentation (NSGSD) 138
National Vegetation Classification (NVC) 126-7, 140
natural diversity 131
Nature Conservancy Council (now English Nature) 189
Nature Conservation Review 128, 141
nature reserves 25, 36, *124*, 183
Naze, the 74
Netherlands 131
'Newton-on-cliff', hypothetical example 152-4
'no access' scenario 190
non-responses 87, 90, 93
numeraire, money as 16, 18

open spaces, valuing 44, 55, 57
opportunity cost 18, 20, 21
  travel 27
options
  'do nothing' 16
  'do something' 16
Orfordness 97
outliers (extreme responses) 87, 93

palaeontological sites 143
Parker, D.J. 38, 109-10, 112
  et al 47, 110, 111, 190
paving 101
Peacehaven 74-6, 156, *157*, **158**, **166**, 166-71

Pearce, D.W. and Markandya,A. 21
pecuniary externalities 19
Penning-Rowsell, E.C.
  and Chatterton, J.B. 112
  et al 47, 109-10, 190
personal effects 102
physico-chemical specialisation 141
plasterwork 101
plumbing 102
Popper, K.R. 27
populations (wildlife) 141
Potential Pareto Improvement 19
pre-feasibility studies 10-11
  and BOCDAM programs 12
  environmental effects 134
  flood alleviation on agricultural land 116
  Hengistbury Head 191
  recreation benefit **59**, 65, 79-83, *91-2*
precision 37-8
prices 18
  business property 57
  erosion-free property 47-8
  house 11, 29-30, 48-9, 49-50, 174
  market 25, 26, 47, 103, 199-200
  residential property *56*, 56-7
  shadow 25, 26
problem identification 134
programme allocation 15
project
  appraisal 15, 147
    methods 7, 15-17, *40*
    stages **22**, 22-5, *40*
  area definition 83
  characteristics 80
  design 125, 126, 147
  failure 39
  planning 8, 146-7
  prioritisation 15
  time profile **22**
promenades 51, 157, 159, 161, 163
  and erosion 62
  grass cliff-top 167
property 43, 101-2
  at risk 51-2, 103, 176
    Fairlight Cove **170**, **171**, 173
  erosion contours 45
  erosion-free prices 47-8
  expected life 44
  flood damage 98
  flood-prone 183
  lifetime 47
  losses 23
    discounted extension of life factors 201-9
  prices *see* house prices; prices
  salt water damage 98-102, *99*, *100*, 102
  social value 47
  values 44, 46-7, 47, 103, *152*, 176, 199-200
    business 57
    life extension 46-7
    residential *56*, 56-7
  writing off 108, 177

questionnaires 110-11, 121
  Contingent Valuation Method 85-6
  flood affected household 366-379
  landscape value 138
  recreational 189-90, 210-38
  residents 168-9, 253-65, 266-75
  site user 229-52, 266-75
  site-specific flood damage potential 352-60
  visit enjoyment 162

Ramsar sites 123, *124*
rare species 129, 141
Ratcliffe, D.A. 140, 141
rationality, axiom of 88
recreation 1, 58-93
  benefits
    annual 82, 89-90
    assessment 64-8, *91*
      Hastings coast protection project 161-2, 165-6

Hengistbury Head 191-2
procedure **59**
recommended approach 79-86, 86-92
see also Contingent Valuation Method (CVM)
of beach protection 33
of cliff protection 167-70
and coast protection projects 60-1, 62-4, 177-9, 181
and pre-feasibility studies **59**, 65, 79-83, *91-2*
total project calculation 90-2
see also Travel Cost Method (TCM)
coast protection effects on 62-4, 189
enjoyment value 66-8, *68*, 75, 75, 92, 93
erosion effects 61-2
facilities 51
non-developed sites 66
surveys **67**
Hengistbury Head 189-90
Herne Bay 177-9
questionnaire 210-38
seafront users 162-6
willingness to pay *69*
use 80, 159, 189-90, 278
levels 65-6
users 65, **67**
values 6, 92
see also beach(es); cliffs and cliff-tops
Recreation Site Survey Manual 86
'Red Manual' 105
Regionally Important Geological and Geomorphological Sites (RIGS) 137
regression analysis 89
research 111, 122, 126
programme 68
see also Middlesex Polytechnic
residential property 98-102, *99*, *100*, 309-51
prices *56*, 56-7
residents
as beach visitors **67**, 70-4
questionnaire 253-65, 266-75
surveys **67**, 83-4
Herne Bay 178-9
Peacehaven 167-9
seafront user 73-4, 178-9
resorts 70-1, 81
resources 16, 126
distribution 19
responses
extreme 87, 93
irrational 88, 179, 181
risk
aversion 21
of death 105-7
and house prices 48-9
and uncertainty 37-9
river flooding 32, 97
roads 21, 173
'robustness analysis' 38
rock armour protection 63
rock strata erosion 142
Royal Commission on Historical Monuments 137
rural landscapes 138
Ryan, A.M. *et al* 115, **116**

safety margins 45, 46
salt marshes 187
salt water 32, 114
flood damage 98-102, *99*, *100*
residential properties flooding data 309-51
sample size 84, 93
sand-dune systems *127*, 127
Scarborough 71
scientific significance 24
geological sites 74
Scotland 123
sea defence
benefits 33, 94-117
breaches 97
and coast protection 10
economic analysis application 32
event losses 32-3
inadequate 97

non-existent 97
policies planning 1-2
problem definition 109
strategies 8-10
sea defence projects 38
benefits
agricultural 114-16
assessment **95**, **96**
evaluation **5**
recreation effects 62-4
sea walls 64, 183
sea water *see* salt water
seafront 65
breaching 180
losses/gains with erosion/protection *73*
physical changes by erosion 159
users survey 73-4, 162-6, 178-9, 229-52, 266-75
visitors 71
seawalls 159, 160-1, **161**, **162**
sedimentation 125, 127, 133
sensitivity analysis 38, 44-5, 155
service utilities 11-12, 45-6, 52, 173, 176-7
sewage 102, *108*, 108
sewerage system 176-7
shadow prices *25*, *26*
'Shadow Project' concept 31-2
Shannon-Weaver index 130, 188
shingle diminishment 171
Silsoe College 12, **115**, 115, **116**
site
characteristics 134-5
classification *34-5*, 126
scoring systems 128-9
user surveys 83, 84, 89, 170
questionnaire 229-52, 266-75
sites
theoretical conditions **61**
undeveloped 82-3
Sites of Special Scientific Interest (SSSIs) 33-6, 123, *124*, 183, 184
assessment 140
earth science 138
geological 189
Smithson, M. 37
specialist users 65, 80, 82, 83, 121
species
agricultural and medical value 130-1
diversity 123, 129-30, 188
maritime 130
population estimates 131
rarity 129, 140
richness 130
spray damage 32, 102-3
spreadsheets 6, 11, 12-13, 151-2, **152**, **153**
COPRES **172**, 177
use at Herne Bay 175
Spurn Head 71, 74
Stabler, M.J. and Harrison, A.J.M. 29
standard cost data, emergency services 104
state, change of 23-4, 33
statuatory bodies 136
storm damage 102-3
stress 107-8, **108**
study area definition 134
Suleman, M. 190
surveys
conduct of 83
resources 12-13
site-specific 11, 104-5
see also individual subjects; questionnaires
sustainability 16, 21, 39, *41*
Swalecliffe 151

Teesdale 122
Test Discount Rate (TDR) 20-1, 200
Thompson, P.M. *et al* 47, 109-10, 110, 111, 190
tides 103
timber 101
time 21, 126
timing
project construction 39
survey 84

tourism 64, 180
traffic disruption 110
Travel Cost Method (TCM) [Clawson] *25*, 26-9, 66, 74, 121, 189, 190
  for geological sites 145
Turner, R.K.
  and Brooke, J. 32
  et al 33

uncertainty 8, *40-1*
  managing 38-9
  and risk 37-9, **38**
United kingdom 32
United States 29, 30, 66, 106, 111
  ecological site evaluation 130-1
  Endangered Species Act 131
  Fish and Wildlife Service 130
urban flood protection projects *147*, 149-51
use
  informal 80
  values 18-19
Usher, M.B. 130
  and Margules, C. 128
utilities 11-12, 45-6, 52, 173, 176-7

valuation
  beach 71, *165*
  definition 119
  environmental 143-4
  methods 7, *40*
  non built-up 53-5
  recreational 92
  techniques 8
value *40*
  concept 7
  economic 17-19
  environmental components 119-22, *144*
  existence 19
  intrinsic 19, 120, 122

land 26
non-use 6, 7, 19, 23, 36, 39, 122
option 18-19
quasi-option 18-19
of a statistical life 107
use 6, 7, 18-19
  see also enjoyment value; property
vegetation
  diversity 130
  succession 125
velocity damage 103
victims, flood 111
visitors
  attitudes and behaviours *76*, *77*, *78*, 92
  beach user **67**, 69-74
  Herne Bay 179
  the Naze 74
  types 66, 69-70, *76-8*, 92, 162
visits
  count record sheet 276-7
  number of 80, 86, 93, 179
  site 26-9

Wales 123
warnings, flood 107, 111-12, *112*
Washington, H.G. 130
water systems 131
waterlogging 132
waves 103
Whitstable 97, 108
wildlife 130
Wildlife and Countryside Act (19810 136
willingness to pay 121-2
  boat trips 190-1
  for coast protection *36*, 36, *165*, 165
  Peacehaven 169, 170-1
  recreation *69*, 93
    cliff-top 74, 75
Willis, K.G. and Benson, J.F. 122
writing off 108, 177